Theoretical Physics: Advanced Concepts

Edited by
Aydan Bailey

Willford Press
www.willfordpress.com

Published by Willford Press,
118-35 Queens Blvd., Suite 400,
Forest Hills, NY 11375, USA

ISBN: 978-1-68285-597-3

Cataloging-in-Publication Data

Theoretical physics : advanced concepts / edited by Aydan Bailey.
p. cm.
Includes bibliographical references and index.
ISBN 978-1-68285-597-3
1. Physics. I. Bailey, Aydan.
QC21.3 .T44 2019
530--dc23

For information on all Willford Press publications
visit our website at www.willfordpress.com

Willford Press

Printed in China.

Contents

Preface

Theoretical physics is the study of simple and complex natural phenomena occurring in the universe using abstraction and mathematical models. There are myriad areas of research in theoretical physics that are actively being pursued in the world today like string theory, chaos theory, astrophysics, cosmology and biophysics. The pillars of modern theoretical physics are relativity theory and quantum mechanics. These help in understanding astronomical, cosmological phenomena as well as phenomena at the particle level. This book traces the progress of this field and highlights some of its key concepts and applications. Researchers and students in this field will find it a valuable text to stay updated with the latest developments across all areas of theoretical physics. It unfolds the innovative aspects of this field, which will be crucial for the progress of this discipline in the future.

After months of intensive research and writing, this book is the end result of all who devoted their time and efforts in the initiation and progress of this book. It will surely be a source of reference in enhancing the required knowledge of the new developments in the area. During the course of developing this book, certain measures such as accuracy, authenticity and research focused analytical studies were given preference in order to produce a comprehensive book in the area of study.

This book would not have been possible without the efforts of the authors and the publisher. I extend my sincere thanks to them. Secondly, I express my gratitude to my family and well-wishers. And most importantly, I thank my students for constantly expressing their willingness and curiosity in enhancing their knowledge in the field, which encourages me to take up further research projects for the advancement of the area.

Editor

The effect of varying temperature and O_2 flow rate in ex situ annealed tin-doped indium for fabrication of commercial grade indium tin oxide

M. R. Kiyani · Y. S. Jalili

Abstract In this work, tin-doped indium was deposited on a glass substrate via the electron beam evaporation technique. Then, the as-grown thin films were baked in the presence of oxygen at different O_2 flow rates and temperatures inside a furnace to obtain transparent conducting oxide thin film structures. The electrical and optical properties of the layers were investigated, the thickness of all samples was kept at 500 nm and the rate of deposition was set at 0.1 nm min^{-1}. The best optical and electrical properties were obtained at O_2 flow rate of 1.5 Nl min^{-1} and temperatures of 500 °C where above 90 % optical transparency and $\leq 4 \times 10^{-3}$ohm cm^{-1} electrical resistivity were achieved.

Keywords ITO · Transparent conductors · Annealing · Transparency · Resistivity · Sputtering

Introduction

Transparent conducting oxides (TCO) are used mainly in optoelectronic applications such as flat panel displays, thin film transistors, electroluminescent devices, solar cells, gas sensors and light emitting diodes [1–5]. The value of electrical conductivity depends on various factors, one of which is the concentration of the dopant incorporated. However, the increase in conductivity usually lowers the transparency level. Hence, it is necessary to find a balance between high optical transparency and high electrical conductivity.

M. R. Kiyani · Y. S. Jalili
Department of Physics, Science and Research Branch, IAU. P. O. Box: 14665-678, Tehran, Iran

Y. S. Jalili (✉)
SheykhBahaee Experimental Research Center, Science & Research Branch, IAU P. O. Box: 14665-678, Tehran, Iran
e-mail: seyedjalili@srbiau.ac.ir

Indium tin oxide (ITO) is the most popular among the TCO materials, used particularly for transparent electrodes in LCD, PLED and OLED displays, solar cells and also as antistatic coatings. ITO is usually grown by RF-sputtering reactively in the presence of oxygen. The strong dependence of ITO electro-optical properties on oxygen pressure and stoichiometry ratio during the growth process [4] adds extra difficulties to this technique during the growth process. The basic properties of ITO oxides are primarily low electrical resistivity (of the order of 10^{-3} ohm cm^{-1} at room temperature) and high optical transparency (about 90 % in the visible range or the optical spectrum) and high infrared reflectance [2].

In the current study, indium-tin thin films are grown by electron beam evaporation technique, with O_2 being a reactive gas and hence advantageous to be excluded at the growth stage, and implemented under O_2-free conditions. The film samples were then baked inside a furnace under exposure to O_2 flow to obtain ITO thin films. This process was carried out under varying furnace temperatures and O_2 flow rates, and their effects on the electro-optical properties of the ITO thin film samples were investigated.

Experimental

In (Aldrich 99.99 %) and Sn (Aldrich 99.99 %) powder were purchased from Sigma, and were mixed to obtain indium-tin with a ratio of 9:1. After having been fully grinded and mixed in a mortar, it was then cast into pellets of 1 cm in diameter using a pellet press under force of 2,500 N. To enhance cohesion in the pellets, a drop of acetone was added to the powder before being pressed.

Prior to deposition, the lamellae substrates were first rinsed with deionized water and then they were sonicated

Table 1 Bake setting according to oxygen flow rates, baking temperature and time duration of as-grown tin-doped indium thin films

Sample no.	Temperature (°C)	Oxygen flow ($l\ min^{-1}$)	Time (min)	Oxygen pressure (bar)
1	500	0.5	30	1.5
2	500	1	30	1.5
3	500	1.5	30	1.5
4	400	0.5	30	1.5
5	400	1	30	1.5
6	400	1.5	30	1.5
7	300	0.5	30	1.5
8	300	1	30	1.5
9	300	1.5	30	1.5

in propanol for 10 min. Finally, they were washed with acetone and dried using pure N_2 compressed gas. The resulting pellets and substrates were placed under electron beam evaporation gun (EDS-160). The distance between the indium-tin pellets target and substrate was set at 10 cm. Pressure was set at 1.5×10^{-5} tor and the atmosphere was rendered oxygen-free. Voltage was kept constant at 6,000 V and the current was gradually increased to 16 A, where evaporation was initiated. The deposition was carried at a rate of 0.1 nm min^{-1}. According to the thickness monitor, once the thickness of the as-grown layers reached 500 nm, deposition was terminated.

The as-grown tin-doped indium samples were then baked at varying temperatures and times inside a furnace under exposure to oxygen with varying flow rates to obtain ITO thin films. The baking settings for each substrate sample are given in Table 1; the rate of heating/cooling was kept at 4 °C min^{-1} throughout the baking process.

All the analyses were carried out in 1 day to prevent any aging effects. SEM analysis was carried out with the aid of Hitachi model S-4160 and AFM analysis was implemented using a Park Scientific Instrument model auto probe CP, the optical transmittance spectra of films in the UV–Vis–NIR region were recorded by a spectrophotometer (Cary 500 Scan).

Results and discussion

The crystallinity and crystal orientation of ITO thin films were measured using X-ray diffraction (STADI MP) with a

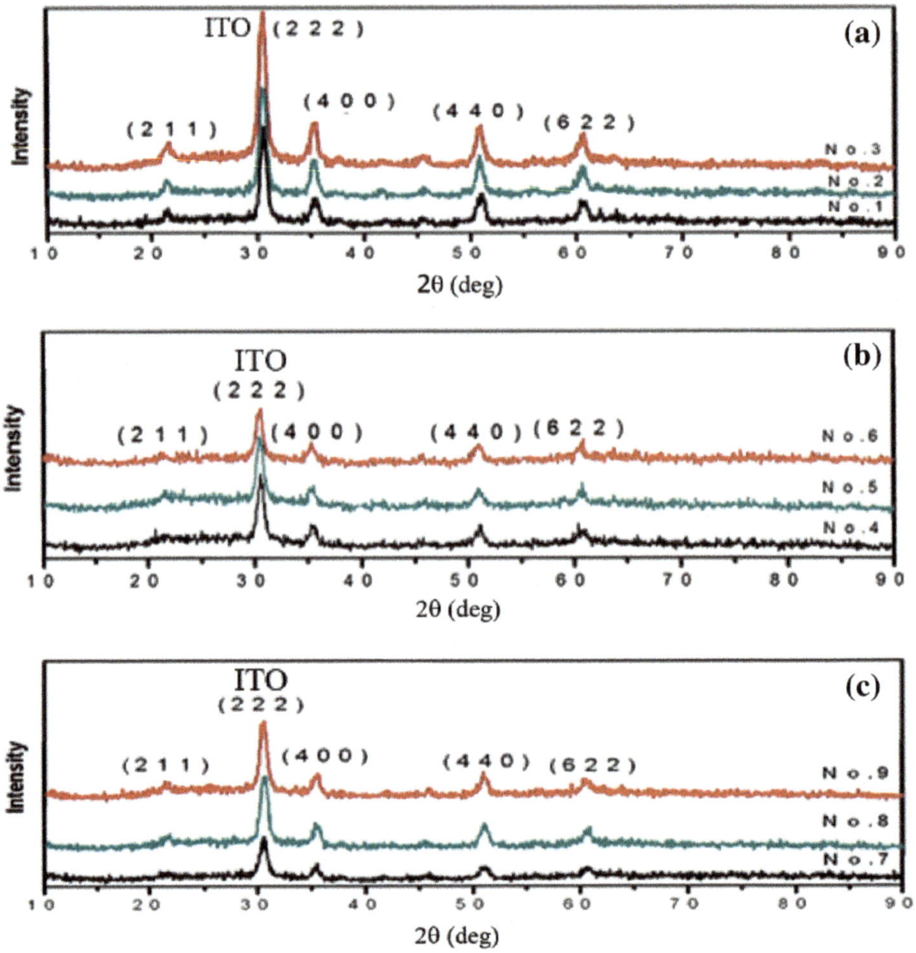

Fig. 1 X-ray diffraction pattern of ITO thin films by ex situ baking of as-grown tin-doped indium thin films onto glass substrate at **a** at 500°, **b** at 400° and **c** at 300°. The corresponding sample number is indicated on each spectra, see Table 1 for details

$CuKa_1$ source ($\lambda = 1.54$). XRD spectra are shown in Fig. 1.

After baking under exposure to oxygen, five distinct peaks (211), (222), (400), (440) and (622) appeared for all of the ITO thin film samples in comparison to the as-grown tin-doped indium thin films which demonstrated amorphous structures as expected. This indicates crystallization within the thin film samples during the baking process in the presence of oxygen. The peaks at $2\theta = 30^\circ$, 35°, 51° and 60° indicate the existence of ITO. For some layers, raising the flow rate of oxygen results in the appearance of indium oxide peak at $2\theta = 21^\circ$. The fact that all spectra demonstrate a predominant peak at (222) plane indicates that the prepared films had a tendency to crystallize along a favorable orientation.

The UV–Vis–NIR spectra measurement is illustrated in Fig. 2. From the spectra, for the baked sample under O_2 at 400 °C (samples no. 4–6), visible range transparency of more than 90 % was achieved.

The results indicate that the transmittance of the films was significantly influenced by increasing the furnace temperature and raising the flow rate of oxygen. In general, the transmittance is highly dependent on the scattering effect. It is assumed that the increase in temperature accelerated the occurrence degree of crystallization, bringing the system to the fibrous microstructure more rapidly. Accordingly, the microstructure with the fibrous pattern had a higher value of transmittance due to the reduction of light scattering sites in the thin film structure.

Using the relation below, the band gap of the thin film ITO layers can be obtained and the results are illustrated in Fig. 3.

$$\alpha = \frac{A}{h\nu}\left(h\nu - E_g\right)^n \tag{1}$$

In this equation, α is the absorption coefficient, A is a constant, $h\nu$ is the incident photon energy, E_g the band gap and n is the number dependent on the type of excitation. We have set $n = 0.5$ [10, 11].

In Fig. 3, the band gap of the samples (numbers 1–9) is plotted and compared. As illustrated, with the increase in the flow rate of oxygen at different temperatures, the band gap decreases. This is because as the flow rate of oxygen increases, ITO thin films are better and more efficiently crystallized and as the thin film structures become more crystalline, it is more prone to have a relatively smaller band gap. Also it is worth noting that, the layer baked at 500 °C and flow rate of 1.5 Nl min^{-1} demonstrates higher crystallinity (see Fig. 1) and hence will be expected to have a relatively smaller band gap and increased conductivity.

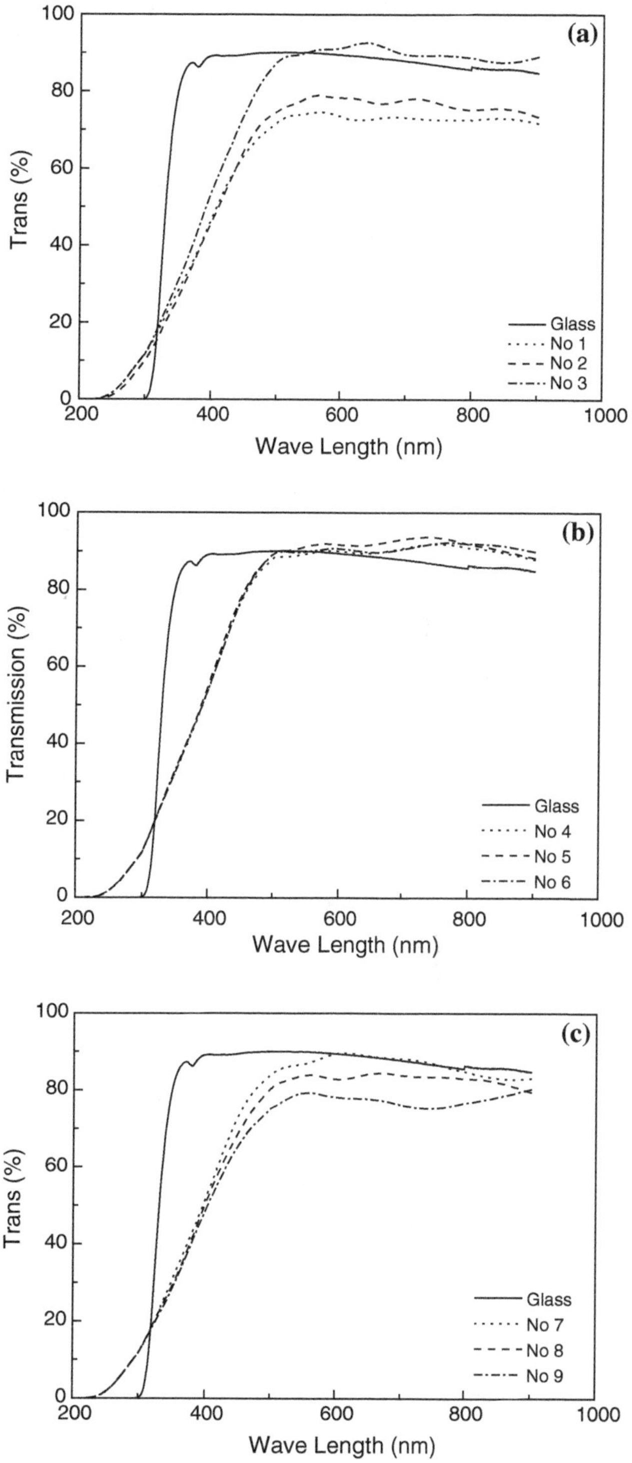

Fig. 2 Transmission spectra of the ITO thin film samples baked under exposure of O_2 of varying flow rates at temperatures **a** 500°, **b** 400° and **c** 300°. Note that Glass spectra are also included for reference

Conductivity measurement results indicate that resistivity of commercial grade magnitude of the order of 10^{-3} ohm cm^{-1} was achieved. Considering the omission of oxygen during the indium-tin growth process and its subsequent introduction into the ex situ thermal treatment stage, the 90 % increase in transparency (Fig. 4) and the

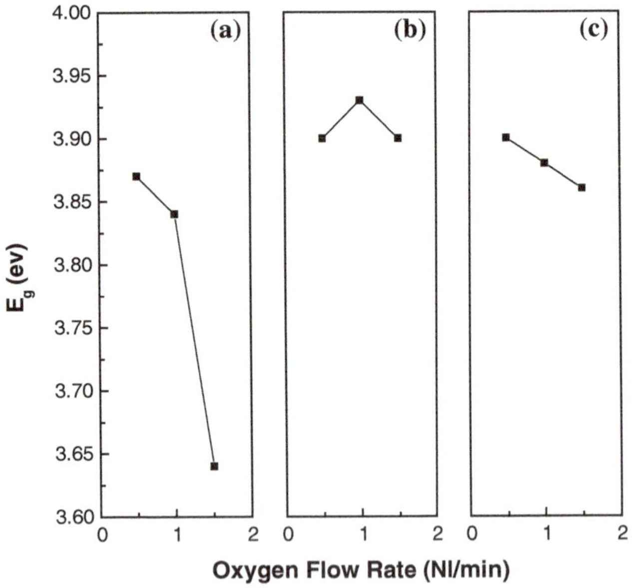

Fig. 3 The band gap of ITO samples baked at temperatures **a** 300 °C, **b** 400 °C and **c** 500 °C using oxygen flow rates of 0.5, 1.0 and 1.5 Nl min^{-1}, respectively

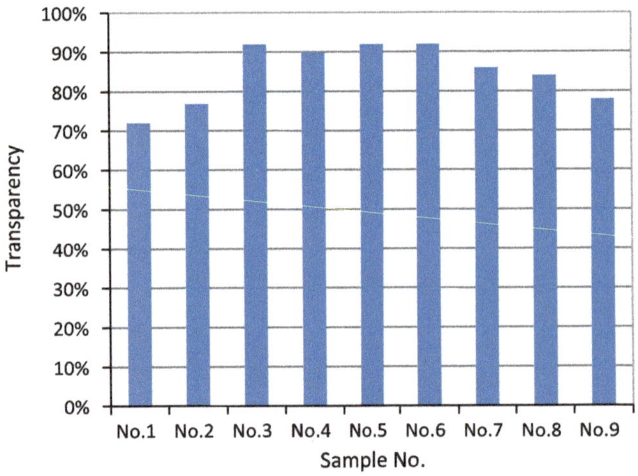

Fig. 4 Peak transmission of ITO thin films at 550 nm according to sample number, see Table 1 for details

decrease in resistance (Fig. 5) indicate that the foretold sample (no. 3) has a resistance of approximately commercial grade magnitude $\leq 4 \times 10^{-3}$ ohm cm^{-1}. This was the best result obtained in this investigation.

In general, at 500 °C, by increasing flow rate of oxygen, the resistivity of layers decreases systematically. It is clear that crystallization with higher flow rate of oxygen causes a reduction in sheet resistivity.

At 400 °C, sheet resistivity of layers is approximately constant and higher than that at other temperatures. The transmittance of these layers is over 90 %. This indicates that the band gap increases and thereby sheet resistance also increases. Therefore, if we want to have layers with

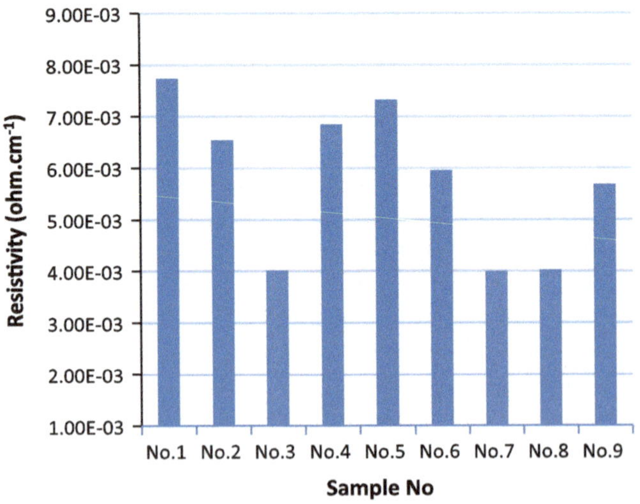

Fig. 5 Comparison of resistivity of ITO thin films according to sample number, see Table 1 for details

low resistivity, this temperature is not suitable for our purpose.

At 300 °C, the initial resistivity of the ITO films is found to be approximately constant with oxygen flow rates of up to 1 Nl min^{-1}. Under this condition, layers exhibited the lowest sheet resistance. The resistivity of ITO reached a minimum at O_2 flow rate of 1 Nl min^{-1}, from there on, increasing the flow rate of oxygen resulted in a gradual increase in resistivity which is due to the formation of non-conducting O-rich clusters inside the film. Excess oxygen forms non-conducting O-rich clusters in the films resulting in crystal disorder, which act as trap sites and consequently increases the resistivity of the ITO thin films.

Overall, because of a relatively lower baking temperature at 300 °C, as the O_2 flow rate increases, non-conducting O-rich clusters in the film also increase; hence an increase in resistivity was observed [7, 8, 9].

The high conductivity of the ITO films has been attributed to both substitutional tin and oxygen vacancies, created either during film growth or post-deposition annealing [6, 10, 12]. Therefore, at this temperature, the baking conditions seem not to be comparatively suitable for ITO formation from the as-grown tin-doped indium thin films.

The SEM images demonstrate that the unbaked as-grown tin-doped indium sample is void of crystallites. From Fig. 6b, d which correspond to sample with high resistance, crystallites are observed but the more obvious matter is the existence of trap sites between separate crystallites which bring about the observed increase in the electrical resistivity, this is also indicated by the AFM results.

As Fig. 6c corresponds to a sample with low resistivity and high transparency, it is well crystallized, as expected,

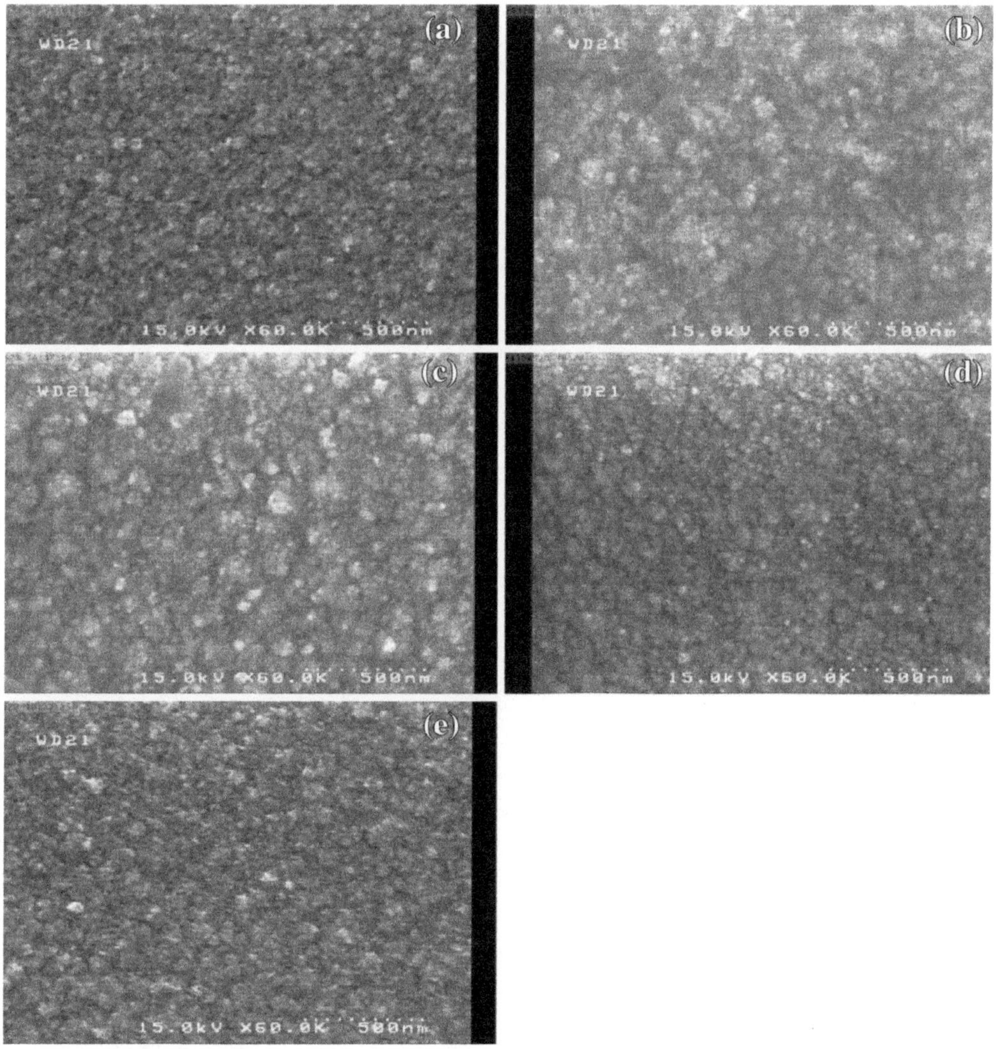

Fig. 6 **a** Unbaked (as-grown) indium tin-doped thin film SEM image, **b** ITO thin film (sample no. 1) with low transparency and high resistance, **c** ITO thin film (sample no. 3) with high transparency and low resistivity, **d** ITO thin film (sample no. 5) with high transparency and high resistivity and **e** ITO thin film (sample no. 7) with low transparency and low resistivity

and one can easily observe the crystallites and the boundaries in between. Therefore, proper crystallization results in high transparency and low resistivity. As Fig. 6e illustrates, crystallites with low order result in low transparency and high resistivity.

AFM analysis was also performed for the three samples discussed above. The results are illustrated below in Fig. 7.

As indicated by the scanned images, the unbaked sample seems to be free of crystallites indicating a relatively much higher degree of roughness. In comparing the AFM image of sample no. 3 to that of sample no. 5, the particle size in sample no. 3 seems relatively larger than that observed in sample no. 5 and the degree of roughness observed is less than that of sample 5. Moreover, from the SEM image of sample no 5, trap sites are more visible indicating a relative increase in the resistivity of the sample. Finally, it is worth noting that comparison of the roughnesses of samples 3 and 5 observed in AFM results may be an indication of the relative lower resistivity of sample no. 3, considering the experimental observations given above.

The optoelectronic and structural properties of the ITO thin films are highly dependent on the deposition and baking conditions of the layers, such that any minuscule changes in these conditions result in notable changes in structural and conclusively in optoelectronic properties [5].

Regarding the lower resistivity of sample no. 3, if we consider the comparison of the indentations observed within the two samples 3 and 5 AFM images; sample no. 3 illustrates a lower degree of indentation to that observed in sample no. 5 this seems to ease the conduction process of the particles by omitting the necessity of tunneling.

Conclusion

Tin-doped indium oxide thin films were deposited using electron beam evaporation technique with an oxygen-free

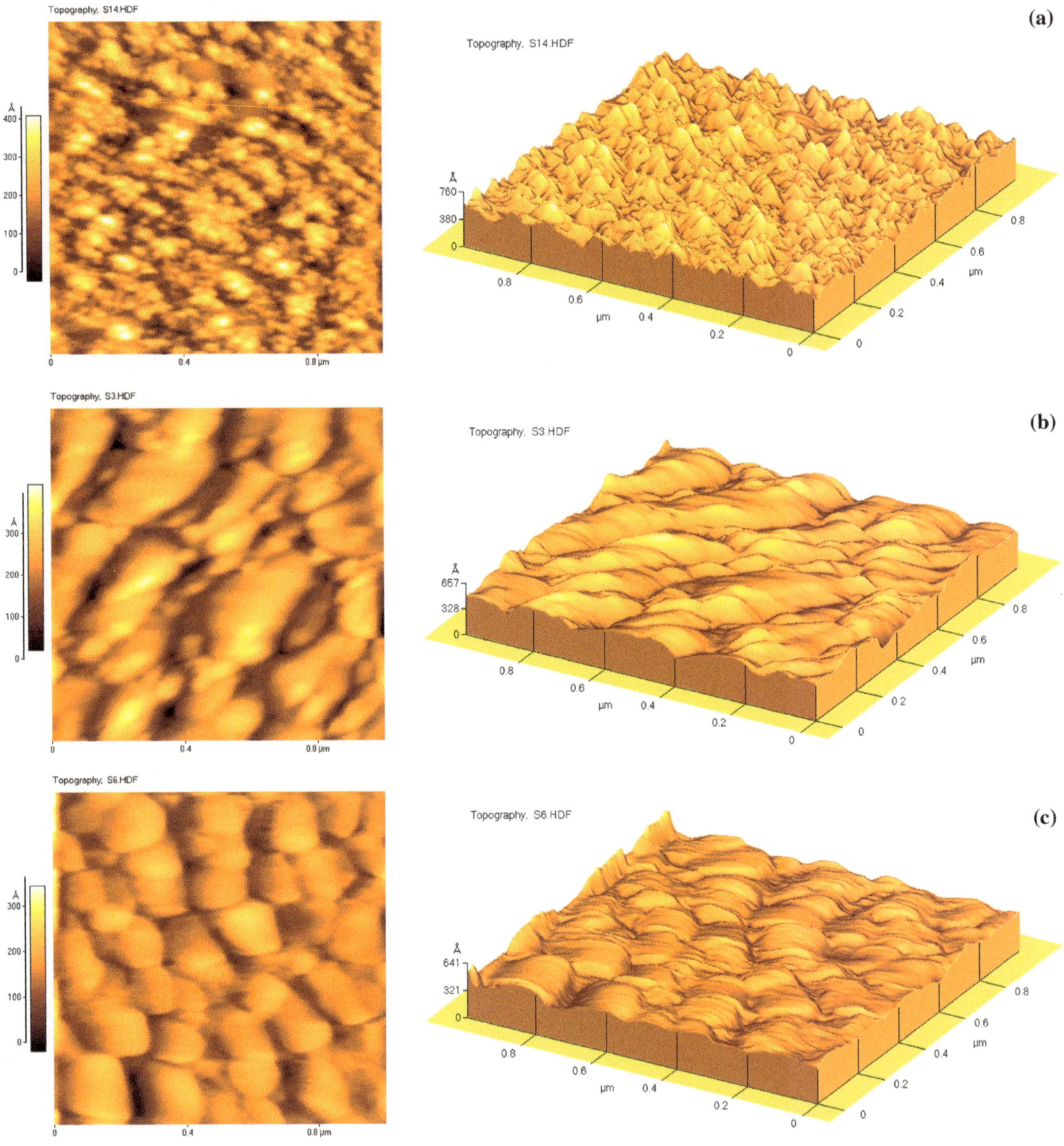

Fig. 7 **a** The as-grown (unbaked) tin-doped indium thin film AFM image, **b** ITO thin film baked at 500 °C and oxygen flow rate of 1.5 Nl min^{-1} and **c** ITO thin film baked at 400 °C and oxygen flow rate of 1.0 Nl min^{-1}

environment. The resulting layers were baked in the presence of oxygen with different flow rates for 30 min at 300, 400 and 500 °C. XRD, UV–Vis spectrophotometry and four-point probe analysis show varying results. The experimental results indicate that the thin film baked at 500 °C with the oxygen flow rate of 1.5 Nl min^{-1} is the layer with the optimal characteristics. After baking in an oxygen-containing environment, this layer exhibits low sheet resistance $\leq 4 \times 10^{-3}$ ohm cm^{-1} and a transmittance of >90 % within the visible range of the spectrum.

References

1. Hamberg, I., Granqvist, C.G.: Evaporated Sn-doped In2O3 films: basic optical properties and applications to energy-efficient windows. J. Appl. Phys. **60**, R123 (1986)
2. You, Y.Z., kim, Y.S., Choi, D.H., jang, H.S.: Electrical and optical study of ITO films on glass and polymer substrates prepared by DC magnetron sputtering type negative metal ion beam deposition. Mater. Chem. Phys. **107**, 444–448 (2008)
3. Ohta, H., Hosono, H.: Deposition of ITO using R.F magnetron sputtering. Thin Solid Films **413**, 181–185 (2002)
4. Kim, H., Gilmore, C.M., Pique, A., Horwitz, J.S., Mattoussi, H., Murata, H., Kafafi, Z.H., Chrisey, D.B.: J. Appl. Phys. **86**, 6451–6461 (1999)
5. Venkatachalam, S., Nanjo, H., Kawasaki, K., Hayashi, H., Ebina T., Mangalaraj, D.: Optoelectronic Properties of ZnSe, ITO, TiO_2 and ZnO thin films, optoelectronics—materials and techniques. In: Predeep, P. (ed.) InTech. doi:10.5772/18418. ISBN: 978-953-307-276-0. http://www.intechopen.com/books/optoelectronics-materials-and-techniques/optoelectronic-properties-of-znse-ito-tio2-and-zno-thin-films
6. Cho, B.R., Park, I.Y.: Effect of post-annealing on the formation of ITO films. Korean Phys. Soc. **53**, 19–22 (2008)
7. Pokaipisit, A., Horprathum, M., Limsuwan, P.: Influence of annealing temperature on the properties of ITO prepared by electron beam evaporation, Kasetsart. J. Nat. Sci. **42**, 362–366 (2008)
8. Trejo-Cruz, C., Mendoza-Galvan, A., Lopez-Beltran, A.M.: Effect of air annealing on the optical, electrical, and structural properties of indium tin oxide film. Thin Solid Films **517**, 4615–4620 (2009)
9. Zhou, Y., Kelly, P.J., Postill, A., Abu-zeid, O., Alnajjar, A.A.: The characteristics of aluminum-doped zinc oxide films prepared by pulsed magnetron sputtering from powder targets. Thin Solid Films **447**, 33–39 (2009)
10. Aiempanakit, K., Rakkwamsuk, P., Grattana, D.: Characterization of indium tin oxide films after annealing in vacuum, Kasetsart. J. Nat. Sci. **42**, 351 (2008)
11. Ho, C.Y., Tu, T.Y., Wang, C.C., Kang, Y: "Investigation of post-annealing indium tin oxide for future electro-optical device application". In: Proceedings of the 10th WSEAS international conference on Telecommunications and informatics and microelectronics, nanoelectronics, optoelectronics, and World Scientific and Engineering Academy and Society (WSEAS) international conference on signal processing, pp. 154–158 (2011)
12. Fan, J.C.C., Goodenough, J.B.: X-ray photoemission spectroscopy studies of Sn-doped indium-oxide films. J. Appl. Phys. **48**, 3524–3532 (1977)

2

Thermodynamics of charged rotating solutions in Brans–Dicke gravity with Born–Infeld field

J. Pakravan[1] · **M. V. Takook**[1]

Abstract We derive new exact charged rotating solutions of $(n+1)$-dimensional Brans–Dicke theory in the presence of Born–Infeld field and investigated their properties. Because of the coupling between scalar field and curvature, the field equations cannot to be solved directly. Using a new conformal transformation, which transforms the Einstein-dilaton–Born–Infeld gravity Lagrangian to the Brans–Dicke–Born–Infeld gravity one, the field equations are solved. We also compute temperature, charge, mass, electric potential, and entropy; entropy, however, does not obey the area law. These quantities are invariant under conformal transformation and satisfy the first law of thermodynamics.

Keywords Brans–Dicke · Black holes · Born–Infeld · Thermodynamics

Introduction

Brans–Dicke theory [1] is one of the most important alternative theories of gravity to modify Einstein's theory to incorporate Mach's principle into the theory of gravitation. In this theory, gravity is described through a metric tensor $g_{\mu\nu}$ and a scalar field Φ, which replace Newton's gravitational constant. This theory involves a dimensionless parameter ω which represents the strength of coupling between scalar field and curvature. Recently, this theory has obtained more attention as it arises as the low energy limit of many theories of quantum gravity such as the supersymmetric string theory or the Kaluza–Klein theory [2]. Besides, recent discoveries show that the universe is accelerating [3–5], so scalar-tensor theories, including Brans–Dicke theory, can be used to explain some features of dark energy (the cause of the acceleration). Cylindrically symmetric solutions are one of the most studied solutions of various gravity theories. For instance, the cylindrical symmetry is used in the study of gravitational waves, cosmological models, and gravitational collapse of non-spherical matter distributions. The first black hole solutions of Brans–Dicke theory in four dimensions were obtained by Brans [6]. Four-dimensional static cylindrical vacuum solutions of Brans–Dicke theory were obtained in [7, 8]. Static solutions of Brans–Dicke–Maxwell theory were presented in [9]. Recently, higher dimensional cylindrically symmetric solutions were investigated by many authors [10–12]. Charged rotating solutions in $(n+1)$ -dimensions for an arbitrary value of ω were presented in [19], but charged rotating solutions for an arbitrary value of ω in the presence of Born–Infeld field have not been constructed. In this paper, we will obtain the $(n+1)$-dimensional charged rotating solutions in Brans–Dicke–Born–Infeld theory and investigate their properties.

The organization of this paper is as follows. In Sect. 2, we introduce the action of Brans–Dicke theory and dilaton gravity in the presence of Born–Infeld field and obtain field equations and conformal transformations between them. In Sect. 3, a new charged rotating solution in $(n+1)$-dimensions with Liouville-type potential is constructed. In Sect. 4, temperature, charge, electric potential, and entropy are obtained. In Sect. 5, by calculating the Euclidean action method, we obtain the conserved quantities and study the

✉ J. Pakravan
jalal_pakravan@yahoo.com

M. V. Takook
takook@razi.ac.ir

[1] Department of Physics, Razi University, Kermanshah, Iran

first law of thermodynamics. We finish our paper with some concluding remarks.

The action and field equations

The gravitation action in $(n+1)$-dimensions for the Brans–Dicke theory with scalar field Φ coupled to a Born–Infeld field can be written as

$$I_{BD} = -\frac{1}{16\pi}\int_{\mathcal{M}} d^{n+1}x\sqrt{-g}\Big(\Phi R - \frac{\omega}{\Phi}(\nabla\Phi)^2 - V(\Phi) + L(F)\Big), \tag{1}$$

where R is the Ricci scalar, ω is the coupling constant, Φ denotes the BD scalar field, and $V(\Phi)$ is a potential for the scalar field Φ. The last term in Eq. (1) is the Born–Infeld term and is given by the following:

$$L(F) = 2\gamma\left(1 - \sqrt{1 + \frac{F_{\mu\nu}F^{\mu\nu}}{\gamma}}\right). \tag{2}$$

Here, $F_{\mu\nu} = \partial_{[\mu}A_{\nu]}$ is the electromagnetic field tensor and A_μ is the electromagnetic vector potential. γ is called the Born–Infeld parameter with dimension of mass. Notice that when $\gamma \to \infty$, $L(F)$ reduces to Maxwell electrodynamics

$$L(F) = -F_{\mu\nu}F^{\mu\nu}. \tag{3}$$

In the limit $\gamma \to 0$, $L(F) \to 0$. It is convenient to set

$$L(F) = 2\gamma L(Y), \tag{4}$$

where

$$L(Y) = 1 - \sqrt{1 + \frac{F_{\mu\nu}F^{\mu\nu}}{\gamma}}. \tag{5}$$

Varying the action (1) with respect to the gravitational field $g_{\mu\nu}$, the scalar field Φ and the electromagnetic field A_μ give the following field equations:

$$\begin{aligned} G_{\mu\nu} =& \frac{\omega}{\Phi^2}\left(\nabla_\mu\Phi\nabla_\nu\Phi - \frac{1}{2}g_{\mu\nu}(\nabla\Phi)^2\right) \\ &- \frac{g_{\mu\nu}}{2\Phi}\left[V(\Phi) - 2\nabla^2\Phi + L(Y)\right] \\ &+ \frac{1}{\Phi}\left[\nabla_\mu\nabla_\nu\Phi - 4\partial_Y L(Y) F_{\mu\lambda}F_\nu^\lambda\right], \end{aligned} \tag{6}$$

$$\begin{aligned} \nabla^2\Phi =& -\frac{4\partial_Y L(Y)}{[(n-1)\omega + n]}F^2 + \frac{(n+1)}{[(n-1)\omega+n]}\left[L(Y) - \frac{V(\Phi)}{2}\right] \\ &+ \frac{(n-1)}{2[(n-1)\omega + n)]}\left[\Phi\frac{dV(\Phi)}{d\Phi}\right], \end{aligned} \tag{7}$$

$$\nabla_\mu(\partial_Y L(Y) F^{\mu\nu}) = 0, \tag{8}$$

where $G_{\mu\nu}$ and ∇_μ are, respectively, the Einstein tensor and covariant differentiation corresponding to the metric $g_{\mu\nu}$. Because the right-hand side of Eq. (6) appears as second derivatives of the scalar field Φ, so we cannot solve it directly and use a suitable conformal transformation such as:

$$\begin{aligned} g_{\mu\nu} &= \Omega^2\tilde{g}_{\mu\nu}, \\ \Phi &= \exp\left(\frac{4\alpha\tilde{\Phi}}{n-3}\right), \\ F_{\mu\nu} &= \tilde{F}_{\mu\nu}, \end{aligned} \tag{9}$$

where $\Omega = \Phi^{-\frac{1}{n-1}}$. Using this conformal transformation, the action (1) becomes

$$\begin{aligned} \tilde{I} = -\frac{1}{16\pi}\int_{\mathcal{M}} d^{n+1}x\sqrt{-\tilde{g}}\Bigg(\tilde{R} - \frac{16\alpha^2}{(n-3)^2} \\ \left(\omega + \frac{n}{n-1}\right)(\tilde{\nabla}\tilde{\Phi})^2 - \tilde{V}(\tilde{\Phi}) + \tilde{L}(\tilde{F},\tilde{\Phi})\Bigg), \end{aligned} \tag{10}$$

this action equals the action of Einstein–dilaton gravity coupled to a Born–Infeld field, which has been studied in [13]:

$$\begin{aligned} \tilde{I}_{ED} = -\frac{1}{16\pi}\int_{\mathcal{M}} d^{n+1}x\sqrt{-\tilde{g}}\Bigg(\tilde{R} - \frac{4}{n-1}(\tilde{\nabla}\tilde{\Phi})^2 \\ - \tilde{V}(\tilde{\Phi}) + \tilde{L}(\tilde{F},\tilde{\Phi})\Bigg), \end{aligned} \tag{11}$$

provided that

$$\begin{aligned} \omega =& \frac{(n-3)^2}{4(n-1)\alpha^2} - \frac{n}{n-1}, \\ \gamma =& \exp\left(\frac{8\alpha\tilde{\Phi}}{n-3}\right)\tilde{\gamma}. \end{aligned} \tag{12}$$

In action (11), $\tilde{R}$ and $\tilde{\nabla}$ are, respectively, the Ricci scalar and covariant differentiation corresponding to the metric $\tilde{g}_{\mu\nu}$, and α is a constant which determines the strength of coupling between the scalar and electromagnetic field. The transformed Born–Infeld field $\tilde{L}(\tilde{F},\tilde{\Phi})$ and $\tilde{V}(\tilde{\Phi})$ are given by

$$\tilde{L}(\tilde{F},\tilde{\Phi}) = 2\tilde{\gamma}e^{4\alpha\tilde{\Phi}/(n-1)}\left(1 - \sqrt{1 + \frac{e^{-8\alpha\tilde{\Phi}/(n-1)}\tilde{F}_{\mu\nu}\tilde{F}^{\mu\nu}}{\tilde{\gamma}}}\right), \tag{13}$$

$$\tilde{V}(\tilde{\Phi}) = V(\Phi)\Phi^{-(n+1)/(n-1)} = \exp\left(\frac{4(n+1)\alpha\tilde{\Phi}}{(n-1)(n-3)}\right), \tag{14}$$

for convenience, we set

$$\tilde{L}(\tilde{F},\tilde{\Phi}) = 2\tilde{\gamma}e^{4\alpha\tilde{\Phi}/(n-1)}(1 - \sqrt{1+\tilde{Y}}), \tag{15}$$

where

$$\tilde{Y} = \frac{e^{-8\alpha\tilde{\Phi}/(n-1)}\tilde{F}_{\mu\nu}\tilde{F}^{\mu\nu}}{\tilde{\gamma}}. \tag{16}$$

Varying the action (11) with respect to $\tilde{g}_{\mu\nu}$, $\tilde{\Phi}$ and $\tilde{F}_{\mu\nu}$, the equations of motion are obtained as

$$\begin{aligned}\tilde{R}_{\mu\nu} =& \frac{4}{n-1}\left(\tilde{\nabla}_\mu\tilde{\Phi}\tilde{\nabla}_\nu\tilde{\Phi} + \frac{1}{4}\tilde{V}\tilde{g}_{\mu\nu}\right)\\ &- 4e^{-4\alpha\tilde{\Phi}/(n-1)}\tilde{\partial}_{\tilde{Y}}\tilde{L}(\tilde{Y})\tilde{F}_{\mu\eta}\tilde{F}^{\eta}_{\nu}\\ &+\frac{2\tilde{\gamma}}{n-1}e^{4\alpha\tilde{\Phi}/(n-1)}\left[2\tilde{Y}\tilde{\partial}_{\tilde{Y}}\tilde{L}(\tilde{Y}) - \tilde{L}(\tilde{Y})\right]\tilde{g}_{\mu\nu},\end{aligned} \tag{17}$$

$$\tilde{\nabla}^2\tilde{\Phi} = \frac{n-1}{8}\frac{\partial\tilde{V}}{\partial\tilde{\Phi}} + \tilde{\gamma}\alpha e^{4\alpha\tilde{\Phi}/(n-1)}\left[2\tilde{Y}\tilde{\partial}_{\tilde{Y}}\tilde{L}(\tilde{Y}) - \tilde{L}(\tilde{Y})\right], \tag{18}$$

$$\tilde{\nabla}_\mu\left[e^{-4\alpha\tilde{\Phi}/(n-1)}\tilde{\partial}_{\tilde{Y}}\tilde{L}(\tilde{Y})\tilde{F}^{\mu\nu}\right] = 0. \tag{19}$$

In the present work, we wish to find charge rotating solutions of Eqs. (6)–(8) with potential $V(\Phi)$. Therefore, we use conformal transformation (9) and solutions of Eqs. (17)–(19).

Charge rotating solutions in $(n+1)$-dimensions

The solutions of the field Eqs. (17) and (18) have been obtained by many authors. Here, we want to obtain the charge rotating solutions in $(n+1)$-dimensional of Brans–Dicke theory with Born–Infeld field. The $(n+1)$-dimensional charge rotating solution to the field equations (17)–(19) has been obtained by [13] for a Liouville-type potential:

$$\tilde{V}(\tilde{\Phi}) = 2\Lambda\exp\left(\frac{4\alpha\tilde{\Phi}}{n-1}\right). \tag{20}$$

By applying the conformal transformation (9), the potential $\tilde{V}(\tilde{\Phi})$ becomes $V(\Phi) = 2\Lambda\Phi^2$. The metric for this solution was written as [13]:

$$\begin{aligned}d\tilde{s}^2 =& -F(r)\left(\Xi dt - \sum_{i=1}^{k}a_i d\varphi_i\right)^2\\ &+\frac{r^2}{l^4}R^2(r)\sum_{i=1}^{k}\left(a_i dt - \Xi l^2 d\varphi_i\right)^2\\ &-\frac{r^2}{l^2}R^2(r)\sum_{i=1}^{k}\left(a_i d\varphi_j - a_j d\varphi_i\right)^2 + \frac{dr^2}{F(r)} + \frac{r^2}{l^2}R^2(r)dX^2,\end{aligned} \tag{21}$$

where a_is are k rotation parameters and

$$\Xi^2 = 1 + \sum_{i=1}^{k}\frac{a_i^2}{l^2}.$$

In metric (21), $F(r)$ and $R(r)$ are functions of r which should be determined. The modified Maxwell equation (19) for the metric (21) can be integrated, where all of the components are zero except

$$\begin{aligned}\tilde{F}_{tr} =& \frac{q\Xi e^{4\alpha\tilde{\Phi}/(n-1)}}{(rR)^{n-1}\sqrt{1+\frac{2q^2}{\tilde{\gamma}(rR)^{2n-2}}}},\\ \tilde{F}_{\varphi r} =& -\frac{a_i}{\Xi}\tilde{F}_{tr}.\end{aligned} \tag{22}$$

It can be showed that $F(r)$, $R(r)$, and $\tilde{\Phi}(r)$ have solutions of the form [13]:

$$\begin{aligned}F(r) =& \frac{2\Lambda(\alpha^2+1)^2c^{2\beta}}{(n-1)(\alpha^2-n)}r^{2(1-\beta)} - \frac{m}{r^{(n-1)(1-\beta)-1}}\\ &-\frac{2\tilde{\gamma}(\alpha^2+1)^2c^{2\beta}r^{2(1-\beta)}}{(n-1)(\alpha^2-n)}\\ &\times\left(1 - {}_2F_1\left(-\frac{1}{2}, \left[\frac{\alpha^2-1}{2(n-1)} - \frac{1}{2}\right], \left[\frac{\alpha^2-1}{2(n-1)} + \frac{1}{2}\right]; \frac{-2q^2}{\tilde{\gamma}c^{2\beta(n-1)}r^{2(n-1)(1-\beta)}}\right)\right),\end{aligned} \tag{23}$$

$$R(r) = \exp\left(\frac{2\alpha\tilde{\Phi}}{n-1}\right) = \left(\frac{c}{r}\right)^\beta, \tag{24}$$

$$\tilde{\Phi}(r) = \frac{(n-1)\alpha}{2(1+\alpha^2)}\ln\left(\frac{c}{r}\right), \tag{25}$$

where c and m are integration constants, and $\beta = \alpha^2/(\alpha^2+1)$. One may note that in the limit $\gamma \to \infty$, these solutions reduce to the solutions presented in [14]. In the absence of dilaton field ($\alpha = \beta = 0$), these solutions reduce to the charged rotating black brane solutions of Einstein–Born–Infeld theory [15]. Using the conformal transformation (9), the $(n+1)$-dimensional charged rotating solutions of BD theory in the presence of Born–Infeld field can be obtained as follows:

$$\begin{aligned}ds^2 =& -A(r)\left(\Xi dt - \sum_{i=1}^{k}a_i d\varphi_i\right)^2\\ &+\frac{r^2}{l^4}H^2(r)\sum_{i=1}^{k}\left(a_i dt - \Xi l^2 d\varphi_i\right)^2\\ &-\frac{r^2}{l^2}H^2(r)\sum_{i=1}^{k}\left(a_i d\varphi_j - a_j d\varphi_i\right)^2 + \frac{dr^2}{B(r)} + \frac{r^2}{l^2}H^2(r)dX^2,\end{aligned} \tag{26}$$

where $A(r)$, $B(r)$, $H(r)$, and $\Phi(r)$ are

$$
\begin{aligned}
A(r) =& \frac{2\Lambda(\alpha^2+1)^2 c^{2\beta\left(\frac{n-5}{n-3}\right)}}{(n-1)(\alpha^2-n)} r^{2\left(1-\frac{\beta(n-5)}{n-3}\right)} \\
& - \frac{mc^{\left(\frac{-4\beta}{n-3}\right)}}{r^{(n-2)}} r^{\beta\left((n-1)+\frac{4}{n-3}\right)} + \frac{2\gamma(\alpha^2+1)^2 c^{-4\beta\left(\frac{n}{n-3}\right)}}{(n-1)(\alpha^2-n) r^{-2\left(1+\beta\frac{n\pm3}{n-3}\right)}} \\
& \times \left(1 - {}_2F_1\left(-\frac{1}{2}, \left[\frac{\alpha^2-1}{2(n-1)} - \frac{1}{2}\right],\right.\right. \\
& \left.\left.\left[\frac{\alpha^2-1}{2(n-1)} + \frac{1}{2}\right]; \frac{-2q^2 r^{2(n-1)\left(\beta-\frac{n-1}{n-3}\right)}}{\gamma c^{\frac{2\beta(n-1)(n-5)}{(n-3)}}}\right)\right),
\end{aligned} \tag{27}
$$

$$
\begin{aligned}
B(r) =& \frac{2\Lambda(\alpha^2+1)^2 c^{2\beta\left(\frac{n-1}{n-3}\right)}}{(n-1)(\alpha^2-n)} r^{2\left(1-\frac{\beta(n-1)}{n-3}\right)} - \frac{mc^{\left(\frac{2\beta}{n-3}\right)}}{r^{(n-2)}} r^{\beta\left((n-1)-\frac{4}{n-3}\right)} \\
& + \frac{2\gamma(\alpha^2+1)^2 c^{-4\beta\left(\frac{2n-1}{n-3}\right)}}{(n-1)(\alpha^2-n) r^{-2\left((1-\beta)-2\beta\frac{n}{n-3}\right)}} \\
& \times \left(1 - {}_2F_1\left(-\frac{1}{2}, \left[\frac{\alpha^2-1}{2(n-1)} - \frac{1}{2}\right],\right.\right. \\
& \left.\left.\left[\frac{\alpha^2-1}{2(n-1)} + \frac{1}{2}\right]; \frac{-2q^2 c^{2(n-1)\left(\beta-\frac{n-1}{n-3}\right)}}{\gamma r^{\frac{2\beta(n-1)(n-5)}{(n-3)}}}\right)\right),
\end{aligned} \tag{28}
$$

$$
H(r) = \left(\frac{c}{r}\right)^{\frac{(n-5)\beta}{n-3}}, \tag{29}
$$

$$
\Phi(r) = \left(\frac{c}{r}\right)^{\frac{2(n-1)\beta}{n-3}}. \tag{30}
$$

In the above solutions, c and m are integration constants. We obtain for the electromagnetic fields

$$
\begin{aligned}
F_{tr} &= \frac{q\Xi c^{(3-n)\beta}}{r^{2+(n-3)(1-\beta)}} \frac{1}{\sqrt{1+\frac{2q^2 c^{2\beta(n-3)}}{\gamma r^{(2n-6)(1-\beta)+4}}}}, \\
F_{\varphi r} &= -\frac{a_i}{\Xi} F_{tr}.
\end{aligned} \tag{31}
$$

Let us notice that as $r \to \infty$, the electromagnetic fields (31) become zero and, in the large γ limit, reduce to the Brans–Dicke–Maxwell theory [19]:

$$
\begin{aligned}
F_{tr} &= \frac{q\Xi c^{(3-n)\beta}}{r^{2+(n-3)(1-\beta)}}, \\
F_{\varphi r} &= -\frac{a_i}{\Xi} F_{tr}.
\end{aligned} \tag{32}
$$

Using the fact that ${}_2F_1(a,b,c;z)$ converges for $|z| < 1$, so we can study the behavior of the $A(r)$ in the limiting case where $r \to \infty$. The Function $A(r)$ for large r is given by

$$
\begin{aligned}
A(r) =& \frac{2\Lambda(\alpha^2+1)^2 c^{2\beta\left(\frac{n-5}{n-3}\right)}}{(n-1)(\alpha^2-n)} r^{2\left(1-\frac{\beta(n-5)}{n-3}\right)} - \frac{mc^{\left(\frac{-4\beta}{n-3}\right)}}{r^{(n-2)}} r^{\beta\left((n-1)+\frac{4}{n-3}\right)} \\
& + \frac{2q^2(\alpha^2+1)^2 c^{-2\beta\left((n-2)+\frac{2}{n-3}\right)} r^{2\left((1-\beta)+\frac{2\beta}{n-3}\right)}}{(n-1)(\alpha^2+n-2) r^{2\left((n-2)(1-\beta)-\frac{2\beta}{n-3}\right)}} \\
& - \frac{q^4(\alpha^2+1)^2 c^{-4\beta\left(\frac{n}{n-3}\right)}}{\gamma(n-1)^2(\alpha^2+3n-4) r^{2\left((2n-3)(1-\beta)-2\beta\frac{n}{n-3}\right)}}.
\end{aligned} \tag{33}
$$

Note that in the limit $\gamma \to \infty$ and $\alpha = 0$, it has the form of the asymptotically AdS black holes. The last term in the above equation is due to the Born–Infeld field in the large γ limit. We can see from Eq. (27) that the solution is well-defined except for $\alpha = \sqrt{n}$. Therefore, we investigate two cases $\alpha > \sqrt{n}$ and $\alpha < \sqrt{n}$ separately. When $\alpha > \sqrt{n}$, as $r \to \infty$, the second term is dominant term, and therefore, the spacetime has a horizon for positive values of the mass parameter. In the second case, where $\alpha < \sqrt{n}$, as $r \to \infty$, the first term is dominant term, and therefore, there exists a horizon provided that $\Lambda > 0$. If $\Lambda < 0$, it is possible to have horizon depending on the different values of the parameters m, q, and α. Because exponential terms appear in (26), it is not straightforward task to find the location of horizons for an arbitrary value of α. However, we can obtain some information by studying the temperature of the horizons.

Thermodynamic quantities

Temperature

By taking $t \to i\tau$ and $a_i \to ia_i$, we define the Euclidean section of the metric (26). The regularity of the metric at r_+ requires that we must identify $\tau \to \tau + \beta_+$ and $\phi_i \sim \phi_i + \beta_+\Omega_i$, where β_+ and Ω_i are the inverse temperature and the angular velocity of the outer horizon [16]

$$
\Omega_i = \frac{a_i}{\Xi l^2}. \tag{34}
$$

The Killing horizon is a null hypersurface whose null generators are tangent to a Killing field. It is easy to see that the Killing vector:

$$
\chi = \frac{\partial}{\partial_t} + \sum_{i=1}^{k} \Omega_i \frac{\partial}{\partial_{\phi_i}}, \tag{35}
$$

is the null generator of the event horizon. The temperature on the outer horizon r_+, is defined through the use of definition of surface gravity κ,

$$
T_+ = \frac{\kappa}{2\pi}, \tag{36}
$$

where the surface gravity $\kappa = \frac{2\pi}{\beta_+}$ is given by

$$
\kappa = \sqrt{-\frac{1}{2}(\nabla_\mu \chi_\nu)(\nabla^\mu \chi^\nu)} = \frac{1}{2\Xi}\left(\frac{\mathrm{d}F(r)}{\mathrm{d}r}\right)_{r_+}. \tag{37}
$$

Then, it is a matter of calculation to show that

$$\begin{aligned}\frac{1}{\beta_+} =T_+ = &\frac{(n-\alpha^2)m}{4\pi\Xi(\alpha^2+1)}r_+^{(n-1)(\beta-1)} \\ &-\frac{q^2(\alpha^2+1)c^{2(2-n)\beta}}{\pi\Xi(\alpha^2+n-2)}r_+^{2(2-n)(1-\beta)-1} \\ &\times {}_2F_1\left(\frac{1}{2},\left[\frac{\alpha^2-1}{2(n-1)}+\frac{1}{2}\right],\left[\frac{\alpha^2-1}{2(n-1)}+\frac{3}{2}\right];\right. \\ &\left.-\frac{2q^2r_+^{2(n-1)(\beta-1)}}{\gamma c^{2(n-1)(\beta-1-\frac{2\beta}{n-3})}}\right),\end{aligned} \tag{38}$$

which shows that the temperature of the solution is invariant under the conformal transformation (9). This result concludes from this point that the conformal parameter is regular at the horizon. There is also an extreme value for the mass parameter in which the temperature of the black hole is zero. Using the fact that $F(r_+)=0$, it is easy to show that

$$\begin{aligned}m_{\text{ext}} = &\frac{4q^2(\alpha^2+1)^2c^{2(2-n)\beta}}{(n-\alpha^2)(\alpha^2+n-2)}r_+^{(3-n)(1-\beta)-1} \\ &\times {}_2F_1\left(\frac{1}{2},\left[\frac{\alpha^2-1}{2(n-1)}+\frac{1}{2}\right],\left[\frac{\alpha^2-1}{2(n-1)}+\frac{3}{2}\right];\right. \\ &\left.-\frac{2q^2r_+^{2(n-1)(\beta-1)}}{\gamma c^{2(n-1)(\beta-1-\frac{2\beta}{n-3})}}\right).\end{aligned} \tag{39}$$

Depending on the value of m, there are three cases to consider separately. In the first case, where $m > m_{\text{ext}}$, the metric of (26) has two inner and outer horizons (r_- and r_+). In the case of $m = m_{\text{ext}}$, we have an extreme black brane and a naked singularity if $m<m_{\text{ext}}$. It is notable to mention that in the absence of scalar field, where ($\alpha = 0$) and $\gamma \to \infty$, m_{ext} reduces to the equation obtained in [17, 18], while in the limiting case, where $\gamma \to \infty$, m_{ext} reduces to the result in [19].

Charge and electric potential

Here, we want to calculate the electric charge and potential of the solutions. We should consider the projections of the electromagnetic field tensor on special hypersurfaces to determine the electric field. The normal to such hypersurfaces is

$$u^0 = \frac{1}{N}, \quad u^r = 0, \quad u^i = -\frac{V^i}{N}, \tag{40}$$

where N and V^i are the lapse and shift function and the electric field is $E^\mu = g^{\mu\alpha}F_{\alpha\lambda}u^\lambda$. The electric charge per unit volume, Q, can be obtained by calculating the flux of the electric field at infinity, obtaining

$$Q = \frac{\Xi q}{4\pi l^{n-2}}. \tag{41}$$

By comparing the charge (41) with the charge of black brane solutions of Einstein–Born–Infeld theory obtained in [13], we notice that the charge Q is invariant under the conformal transformation (9). The electric potential, U, measured at infinity with respect to the horizon is defined by the following [20, 21]:

$$U = A_\mu\chi^\mu\big|_{r\to\infty} - A_\mu\chi^\mu\big|_{r=r_+}. \tag{42}$$

Here, χ is the null generators of the horizon. The vector potential A_μ corresponding to electromagnetic field (31) can be obtained as

$$\begin{aligned}A_\mu = &\frac{qc^{(3-n)\beta}}{\Delta r^\Delta}\,{}_2F_1\left(\frac{1}{2},\left[\frac{\alpha^2-1}{2(n-1)}+\frac{1}{2}\right],\left[\frac{\alpha^2-1}{2(n-1)}+\frac{3}{2}\right];\right. \\ &\left.-\frac{2q^2}{\gamma c^{2\beta\frac{(n-1)(n-5)}{n-3}}r^{2(n-1)(n-3-\beta(n-5))}}\right)\times\left(\Xi\delta^t_\mu - a_i\delta^i_\mu\right),\end{aligned} \tag{43}$$

where $\Delta = (1-\beta)(n-3)+1$. Therefore, using (42) and (43), the electric potential can be obtained as

$$\begin{aligned}U = &\frac{qc^{(3-n)\beta}}{\Xi\Delta r_+^{\Delta}}\,{}_2F_1\left(\frac{1}{2},\left[\frac{\alpha^2-1}{2(n-1)}+\frac{1}{2}\right],\left[\frac{\alpha^2-1}{2(n-1)}+\frac{3}{2}\right];\right. \\ &\left.-\frac{2q^2}{\gamma c^{2\beta\frac{(n-1)(n-5)}{n-3}}r_+^{2(n-1)(n-3-\beta(n-5))}}\right).\end{aligned} \tag{44}$$

Entropy

For a charged rotating black hole, the first law of thermodynamics takes the form

$$TdS = \mathrm{d}M - \sum_{i=1}^{k}\Omega_i\mathrm{d}J_i + U\mathrm{d}Q, \tag{45}$$

where T and S are the horizon temperature and entropy, M, J, U, and Q are the mass, angular momentum, electric potential, and charge measured at infinity, and Ω_H is the angular velocity of the horizon. The first law of thermodynamics connects the quantities M, J, U, and Q measured at infinity with the local quantities S, T, A, and Ω_i on the horizon. As studied in [22], in alternative theories of gravity, the first law of thermodynamics (45) still holds true, but the area law for the entropy $S = \frac{A}{4G}$ is no longer valid in these theories which has been known since 1980s [23–27]. Some numerical studies showed that the area law during the collapse of dust to black holes in alternative theories of gravity is violated [28–30]. Black hole entropy in Brans–Dicke theory was studied by Kang [31]. Kang noticed that the problem is in the expression of the black hole entropy, which is not one quarter of the area. The relation for the entropy is

$$S = \frac{1}{4}\int_{\Sigma} d^{n-1}x\sqrt{g^{(n-1)}}\Phi = \frac{A\Phi}{4}, \tag{46}$$

where Φ is the Brans–Dicke scalar and $g^{(n-1)}$ is the determinant of the metric $g_{\mu\nu}$ to the horizon surface Σ. This expression can be obtained by the replacement of the Newton constant G with the inverse of Brans–Dicke scalar Φ. In the Einstein frame, the gravitational coupling is a constant, but in Brans–Dicke frame matter couples to the scalar field. Massive test particles which follow time-like geodesics of the metric $g_{\mu\nu}$ do not follow geodesics of the rescaled metric $\tilde{g}_{\mu\nu}$. Null geodesics remain unchanged under conformal transformation, and also null vectors and all forms of conformally invariant matter. Therefore, a black hole event horizon which is a null surface remains unchanged. The area of an event horizon is not a null surface. Therefore, the change in the entropy formula can be obtained as the change in the area due to the conformal transformation of the metric $g_{\mu\nu}$. In fact, $g_{\mu\nu} = \Omega^2\tilde{g}_{\mu\nu}$, and the relation between areas in two frames is

$$\tilde{A} = \int_{\Sigma} d^{n-1}x\sqrt{\tilde{g}^{(n-1)}} = \int_{\Sigma} d^{n-1}x\Omega^{-(n-1)}\sqrt{g^{(n-1)}} = \Phi A, \tag{47}$$

where $\Omega = \Phi^{\frac{-1}{n-1}}$ according to (9). Therefore, the entropy in the Brans–Dicke and Einstein theory becomes equal

$$S = \frac{A\Phi}{4} = \frac{\tilde{A}}{4} = \tilde{S}. \tag{48}$$

Denoting the volume of the hypersurface boundary at constant t and r by V_{n-1}, it is easy to show that the entropy per unit volume is

$$S = \frac{\Xi c^{(n-1)\beta}}{4l^{n-2}} r_+^{(n-1)(1-\beta)}. \tag{49}$$

It is notable to mention that the equality between black hole entropies in the Brans–Dicke and Einstein theory is not confined to scalar-tensor gravity but is valid in all theories with action $\int d^{n+1}x\sqrt{g}f(R_{\mu\nu}, g_{\mu\nu}, \varphi, \nabla_\mu\varphi)$ [32].

Action and conserved quantities

In this section, we obtain the action and the thermodynamic quantities of our solutions. In general, the action (1) does not have a well-defined variational principle as well as is divergent when evaluated on the solution. By variation of the action (1), one encounters a total derivative which gives rise to a surface integral involving the normal derivative of $\delta g_{\mu\nu}$. These normal derivative terms are removed by the variation of the surface term

$$I_b = -\frac{1}{8\pi}\int_{\partial\mathcal{M}} d^n x\sqrt{-h}\Theta\Phi. \tag{50}$$

Here, h_{ab} is the determinant of the boundary metric and Θ is the trace of extrinsic curvature Θ^{ab} of the boundary. In general, the action $I_{BD} + I_b$ is divergent when evaluated on the solutions. For asymptotically (A)dS solutions of Einstein gravity, one can remove these divergences through the use of counterterm method. In this method, we add a finite number of surface integral to leave the action finite [33]. In this paper, our solutions have zero curvature boundary, and therefore, all the counterterms containing the curvature invariants of the boundary are zero. Thus, there exists one counterterm as

$$I_c = -\frac{1}{8\pi}\int_{\partial\mathcal{M}} d^n x\sqrt{-h}\left(\frac{(n-1)\Phi V(\Phi)}{\alpha^2 - n}\right)^{\frac{1}{2}}, \tag{51}$$

it is notable that the counterterm has the same form as in the case of asymptotically AdS solutions with zero curvature boundary, where $\alpha \to 0$ ($\Phi = 1$). The total finite action can be written as

$$I_{\text{tot}} = I_{BD} + I_b + I_c. \tag{52}$$

Using Eqs. (1), (50), and (51), we obtain the Euclidean action per unit volume V_{n-1} as

$$\begin{aligned}\frac{I_{\text{tot}}}{\beta_+} = &-\frac{n(\alpha^4-1)mc^{(n+1)\beta}}{16\pi l^n(\alpha^2-n)} r_+^{(n+1)(1-\beta)-1} \\ &- \frac{2\gamma(\alpha^4-1)c^{\beta\left(\frac{n^2-6n+1}{n-3}\right)} r_+^{n-4\beta\left(\frac{n^2-6n+1}{n-3}\right)}}{16\pi l^{n-2}(n-1)(\alpha^2-n)} \\ &\times\left(1 - {}_2F_1\left(-\frac{1}{2}, \left[\frac{\alpha^2-1}{2(n-1)} - \frac{1}{2}\right], \left[\frac{\alpha^2-1}{2(n-1)} + \frac{1}{2}\right]; \right.\right. \\ &\left.\left.\frac{-2q^2}{\gamma c^{2\beta\frac{(n-1)(n-5)}{n-3}} r^{2(n-1)(1-\beta\frac{n-5}{n-3})}}\right)\right).\end{aligned} \tag{53}$$

According to Refs. [34–36], we can calculate the entropy, the mass and angular momentum through the relation

$$I_{\text{tot}} = \beta_+ M - S - \beta_+(\Omega_i J_i - QU), \tag{54}$$

by comparing Eqs. (53) and (54), we can easily find that

$$S = \frac{\Xi c^{(n-1)\beta}}{4l^{n-2}} r_+^{(n-1)(1-\beta)}, \tag{55}$$

$$M = \frac{c^{(n-1)\beta}}{16\pi l^{n-2}}\left(\frac{(n-\alpha^2)\Xi^2 + \alpha^2 - 1}{1+\alpha^2}\right)m, \tag{56}$$

$$J_i = \frac{c^{(n-1)\beta}}{16\pi l^{n-2}}\left(\frac{n-\alpha^2}{1+\alpha^2}\right)\Xi m a_i. \tag{57}$$

Comparing the thermodynamic quantities calculated in this section with those obtained in the previous sections, we find that they are invariant under the conformal transformation. When $a_i = 0$, the angular momentum vanishes, and so, a_i is the ith rotational parameter of the spacetime. Black hole entropy typically satisfies the so-called area law of the entropy [37, 38], but it does not follow the area law in Brans–Dicke theory [39–45]. Nevertheless, the entropy remains invariant under conformal transformations. It is notable to mention that the thermodynamic quantities calculated above satisfy the first law of thermodynamics:

$$\mathrm{d}M = T\mathrm{d}S + \sum_{i=1}^{k} \Omega_i \mathrm{d}J_i + U\mathrm{d}Q. \tag{58}$$

Closing remarks

We presented the $(n+1)$-dimensional BD-BI action coupled to a scalar field Φ and obtained the field equations by varying this action with respect to the gravitational field $g_{\mu\nu}$, the dilaton field Φ, and the gauge field A_μ. In the special case of the linear electrodynamics where we have $\mathcal{L}(Y) = -\frac{1}{2}Y$, the system of equations reduced to the equations of Brans–Dicke–Maxwell theory [19]. Because of the coupling between the scalar field and curvature, solving the field equations is complicated. Therefore, to solve field equations, we present new conformal transformations. In this paper, using these conformal transformations, we obtained the charged rotating solutions of $(n+1)$-dimensional $(n \geq 4)$ BD-BI equations in the presence of a potential and studied their properties. These solutions are neither asymptotically flat nor (A)dS. In the particular case $\gamma \to \infty$, these solutions reduce to the $(n+1)$-dimensional charged rotating dilaton black brane in Brans–Dicke theory with quadratic scalar field potential [19]. These solutions are ill-defined for $\alpha = \sqrt{n}$ (corresponding to $\omega = \frac{-3(n+3)}{4n}$), so we investigated two cases $\alpha > \sqrt{n}$ and $\alpha < \sqrt{n}$ separately. We also computed the entropy, temperature, charge, mass, and electric potential, and found that these quantities are invariant under conformal transformations and satisfy the first law of thermodynamics. We show that the entropy does not obey the area law, but remain invariant under the conformal transformation. In addition, we found that when we have $(\alpha = 0)$, the scalar field becomes a constant $(\Phi = 1)$ and the BD theory degenerates into the Einstein theory of gravitation.

References

1. Brans, C.H., Dicke, R.H.: Mach's principle and a relativistic theory of gravitation. Phys. Rev. **124**, 925 (1961)
2. Adhav, K.S., Dawande, M.V., Borikar, S.M.: Kaluza–Klein interacting cosmic fluid cosmological model. J. Theor. Appl. Phys. **6**, 33 (2012)
3. Perlmutter, S., et al.: Measurements of Ω and Λ from 42 high-redshift supernovae. Astrophys. J. **517**, 565–586 (1999)
4. Riess, A.G., et al.: Observational evidence from supernovae for an accelerating universe and a cosmological constant. Astron. J. **116**, 1009–1083 (1998)
5. Riess, A.G., et al.: The farthest known supernova: support for an accelerating universe and a glimpse of the epoch of deceleration. Astrophys. J. **560**, 49–71 (2001)
6. Brans, C.H.: Mach's principle and a relativistic theory of gravitation II. Phys. Rev. **125**, 2194 (1962)
7. Hindmarsh, M.B., Kibble, T.W.B.: Cosmic strings. Rep. Prog. Phys. **58**, 477 (1995)
8. Dahia, F., Romero, C.: Line sources in Brans–Dicke theory of gravity. Phys. Rev. D **60**, 104019 (1999)
9. Cai, R.G., Myung, Y.S.: Black holes in the Brans–Dicke–Maxwell theory. Phys. Rev. D **56**, 3466 (1997)
10. Bronnikov, K.A., Meierovich, B.E.: Global strings in extra dimensions: the full map of solutions, matter trapping, and the hierarchy problem. J. Exp. Theor. Phys. **106**, 247 (2008)
11. Baykal, A., Delice, O., Ciftci, D.K.: Cylindrically symmetric vacuum solutions in higher dimensional Brans–Dicke theory. J. Math. Phys. **51**, 072505 (2010)
12. Ciftci, D.K., Delice, O.: Brans–Dicke–Maxwell solutions for higher dimensional static cylindrical symmetric spacetime. J. Math. Phys. **56**, 072502 (2015)
13. Dehghani, M.H., Hendi, S.H., Sheykhi, A., Rastegar Sedehi, H.: Thermodynamics of rotating black branes in Einstein–Born–Infeld–dilaton gravity. J. Cosmol. Astropart. Phys. **0702**, 020 (2007)
14. Sheykhi, A., Dehghani, M.H., Riazi, N., Pakravan, J.: Thermodynamics of rotating solutions in $(n+1)$-dimensional Einstein–Maxwell–dilaton gravity. Phys. Rev. D **74**, 084016 (2006)
15. Dehghani, M.H., Rastegar Sedehi, H.R.: Thermodynamics of rotating black branes in $(n+1)$-dimensional Einstein–Born–Infeld gravity. Phys. Rev. D **74**, 124018 (2006)
16. Hawking, S.W., Hunter, C.J., Taylor-Robinson, M.M.: Rotation and the AdS-CFT correspondence. Phys. Rev. D **59**, 064005 (1999)
17. Dehghani, M.H.: Thermodynamics of rotating charged black strings and (A) dS/CFT correspondence. Phys. Rev. D **66**, 044006 (2002)
18. Dehghani, M.H., Khodam-Mohammadi, A.: Thermodynamics of a d-dimensional charged rotating black brane and the AdS/CFT correspondence. Phys. Rev. D **67**, 084006 (2003)
19. Dehghani, M.H., Pakravan, J., Hendi, S.H.: Thermodynamics of charged rotating black branes in Brans–Dicke theory with quadratic scalar field potential. Phys. Rev. D **74**, 104014 (2006)
20. Cvetic, M., Gubser, S.S.: Phases of R-charged black holes, spinning branes and strongly coupled gauge theories. J. High Energy Phys. **04**, 024 (1999)
21. Caldarelli, M.M., Cognola, G., Klemm, D.: Thermodynamics of Kerr–Newman-AdS black holes and conformal field theories. Class. Quantum Grav. **17**, 399 (2000)
22. Jacobson, T., Kang, G., Meyers, R.C.: On black hole entropy. Phys. Rev. D **49**, 6587–6598 (1994)
23. Raychaudhuri, A.K., Bagchi, B.: Temperature dependent gravitational constant and black hole physics. Phys. Lett. B **124**, 168–170 (1983)
24. Callan, C., Meyers, R.C., Perry, M.: Black holes in string theory. Nucl. Phys. B **311**, 673–698 (1988)
25. Meyers, R.C., Simon, J.Z.: Black-hole thermodynamics in Lovelock gravity. Phys. Rev. D **38**, 2434–2444 (1988)
26. Lu, M., Wise, M.: Black holes with a generalized gravitational action. Phys. Rev. D **47**, 3095–3098 (1993)

27. Visser, M.: Dirty black holes: entropy as a surface term. Phys. Rev. D **48**, 5697–5705 (1993)
28. Scheel, M.A., Shapiro, S.L., Teukolsky, S.A.: Collapse to black holes in Brans–Dicke theory: I. Horizon boundary conditions for dynamical spacetimes. Phys. Rev. D **51**, 4208–4235 (1995)
29. Kerimo, J., Kalligas, D.: Gravitational collapse of collisionless matter in scalar-tensor theories: scalar waves and black hole formation. Phys. Rev. D **58**, 104002 (1998)
30. Kerimo, J., Kalligas, D.: Dynamical black holes in scalar-tensor theories. Phys. Rev. D **62**, 104005 (2000)
31. Kang, G.: Black hole area in Brans–Dicke theory. Phys. Rev. D **54**, 7483–7489 (1996)
32. Koga, J., Maeda, K.: Equivalence of black hole thermodynamics between a generalized theory of gravity and the Einstein theory. Phys. Rev. D **58**, 064020 (1998)
33. Awad, A.M.: Higher-dimensional charged rotating solutions in (A) dS spacetimes. Class. Quantum Grav. **20**, 2827 (2003)
34. Brown, J.D., Martinez, E.A., York, J.W.: Complex Kerr–Newman geometry and black-hole thermodynamics. Phys. Rev. Lett. **66**, 2281 (1991)
35. Braden, H.W., Brown, J.D., Whiting, B.F., York, J.W.: Charged black hole in a grand canonical ensemble. Phys. Rev. D **42**, 3376 (1990)
36. York, J.W.: Black-hole thermodynamics and the Euclidean Einstein action. Phys. Rev. D **33**, 2092 (1986)
37. Beckenstein, J.D.: Black holes and entropy. Phys. Rev. D **7**, 2333 (1973)
38. Gibbons, G.W., Hawking, S.W.: Cosmological event horizons, thermodynamics, and particle creation. Phys. Rev. D **15**, 2738 (1977)
39. Visser, M.: Dirty black holes: entropy as a surface term. Phys. Rev. D **48**, 5697 (1993)
40. Englert, F., Houart, L., Windey, P.: The black hole entropy can be smaller than A4. Phys. Lett. B **372**, 29 (1996)
41. Kang, G.: Black hole area in Brans–Dicke theory. Phys. Rev. D **54**, 7483 (1996)
42. Martinez, C., Zanelli, J.: Conformally dressed black hole in 2+ 1 dimensions. Phys. Rev. D **54**, 3830 (1996)
43. Henneaux, M., Martinez, C., Troncoso, R., Zanelli, J.: Black holes and asymptotics of 2 + 1 gravity coupled to a scalar field. Phys. Rev. D **65**, 104007 (2002)
44. Ashtekar, A., Corichi, A., Sudarsky, D.: Non-minimally coupled scalar fields and isolated horizons. Class. Quantum Grav. **20**, 3413 (2003)
45. Martinez, C., Troncoso, R., Staforelli, J.P.: Topological black holes dressed with a conformally coupled scalar field and electric charge. Phys. Rev. D **74**, 044028 (2006)

3

A parametric study on the PD pulses activity within micro-cavities

Alireza A. Ganjovi[1]

Abstract A two-dimensional kinetic model has been used to parametric investigation of the spark-type partial discharge pulses inside the micro-cavities. The model is based on particle-in-cell methods with Monte Carlo Collision techniques for modeling of collisions. Secondary processes like photo-emission and cathode-emission are considered. The micro-cavity may be sandwiched between two metallic conductors or two dielectrics. The discharge within the micro-cavity is studied in conjunction with the external circuit. The model is used to successfully simulate the evolution of the discharge and yield useful information about the build-up of space charge within the micro-cavity and the consequent modification of the applied electric field. The phase-space scatter plots for electrons, positive, and negative ions are obtained in order to understand the manner in which discharge progresses over time. The rise-time and the magnitude of the discharge current pulse are obtained and are seen to be affected by micro-cavity dimensions, gas pressure within the micro-cavity, and the permittivity of surrounding dielectrics. The results have been compared with existing experimental, theoretical, and computational results, wherever possible. An attempt has been made to understand the nature of the variations in terms of the physical processes involved.

Keywords Partial discharges · Submillimeter cavities · PIC-MCC simulation · Spark-type discharges

✉ Alireza A. Ganjovi
alirezaganjovi@yahoo.com; ganjovi@kgut.ac.ir

[1] Photonics Research Institute, Institute of Science, High Technology & Environmental Sciences, Graduate University of Advanced Technology, Kerman, Iran

Introduction

Some of the principal factors that can influence partial discharges (PD's) and other such discharges inside micro-cavities are the dimension of micro-cavity, the pressure of the gaseous environment and gas composition, the magnitude and frequency of the applied voltage, the electrode arrangement and geometry, and the properties, condition, and age of the insulation. The different character of the discharges affects in turn the response of conventional PD pulse detectors, which is dependent to the pulse shape and its rise-time [1–3].

The rise-time of PD pulses is one of the most important subjects. The rise-time of the discharge pulses determines the type of the PD detection circuit that must be employed to the detection of the discharge pulses and record their pulse shape attributes with specified precision and accuracy [4]. While it is evident that the rise-time of the detected PD pulses is depend upon the bandwidth of the discharge detection equipment as well as location of PD source, the rise-time is also very much dependent to the discharge mechanism as such and the availability of free electrons to initiate that [4]. In contrast to the long rise-times characterizing the discharge pulse by glow discharges, spark-type discharges exhibit typically short rise-times and large magnitude pulses [2, 4]. Bartnikas et al. [2] studied the effect of over-voltage on PD rise-time. They reported that with greater over-voltages, the PD pulses assume shorter rise-times and pulse widths as well as increasing magnitudes, thereby inferring larger charge transfers.

It has been demonstrated earlier that the discharge mechanisms in a micro-cavity are essentially of the Townsend type and the form of the ensuing discharge current pulse at the time of a spark-type breakdown is controlled by the electric field created by the ion space charge in the micro-cavity [1].

A few studies deal with cavities in the submillimeter range; these include modeling of avalanches inside cylindrical micro-cavities by Burgener et al. [5] and experimental studies by Gu et al. [6]. As already reported in the literature, in large disk-shaped voids, transition takes place from streamer mechanism to Townsend mechanism due to a conducting layer formed on the dielectric walls by discharge by-products [7–11]. However, an important point to note is that streamer discharges do not occur in submillimeter cavities, in the absence of the critical length required for formation of a streamer head [1].

Kinetic PIC-MCC simulation methods are known to be better suited for problems in which the particle distribution may deviate significantly from Maxwellian, such as when trapping or particle stochastic heating occurs [12–14]. Additionally, the PIC-MCC method guarantees high accuracy for both transient and stationary problems [15], and yields more detailed information about the evolution of the discharge. Radjenovic et al. [16, 17] used a one-dimensional PIC-MCC model to study Paschen's law for gas breakdown at small electrode separations.

A two-dimensional kinetic model based on PIC-MCC technique to study the behavior of spark-type PDs inside the micro-cavities has been developed [18]. The same model was used to investigate the filamentary distribution of spark-type PD's inside a micro-cavity [19]. In another work, a radiation transport (RT) model integrated with the PIC-MCC model was developed to study the behavior of excited (radiative) states of air within discharge in the narrow channel [20]. The radiative state atoms were described by a fluid model combined with the Holstein–Biberman equation.

Using this model, in this paper, the influences of different parameters of micro-cavity on the different features of a spark-type PD pulse such as discharge current, current rise-time, etc., with two different electrode configurations, i.e., metallic–metallic (MM) and dielectric–dielectric (DD) are examined [18]. The model has been employed for performing an extensive parametric study relevant to effect of quantities such that micro-cavity dimensions, gas pressure inside the micro-cavity, and dielectric permittivity on the PD development within MM and DD micro-cavity. The intent is to discuss the results obtained systemically using the model to establish the effect of above quantities on the nature and shape of PD pulses in micro-cavities undergoing a spark-type discharge in air.

Computational techniques

In the PIC scheme, particles are defined in a continuum position and velocity space. Field values are defined at discrete locations in space. Particle and field values are advanced sequentially in time, starting from initial conditions. The particle equations of motion are solved at every time-step, using field values interpolated from the discrete grid to particle locations. The force on every particle is computed by interpolation of the field values from the grid position to the given particle positions. The position and velocities of each particle is next updated based on the solution of the classical equation of motion [21]. Next, particle boundary conditions are applied. For modeling collisions, the Monte Carlo collision (MCC) scheme is applied [22]. Source terms for the field equations are accumulated from the particle locations to the grid locations. The field values are then advanced by one time-step, and the time loop starts again (Fig. 1).

In the PIC-MCC scheme, each super-particle, representative of one or a much larger number of real particles, is designated in continuum space by its position and velocity. Initially, a Maxwellian distribution is assumed for electrons and ions. Further, the particles are uniformly distributed in the gap space. Particle boundary conditions such as absorption and emission are used to account for the relation between the discharge current in the gap and the current in the external circuit. When an electron or ion passes from the discharge into an end wall, it adds to the wall charge and is deleted from the list of active particles [13]. Additionally, secondary emissions occur when a charged particle impacts a surface (metallic or dielectric) causing ejection of electrons from the surface.

In the current work, the gas within the cavity is assumed to be air at atmospheric pressure. Therefore, photo-emission at the cathode is likely to be the dominant secondary

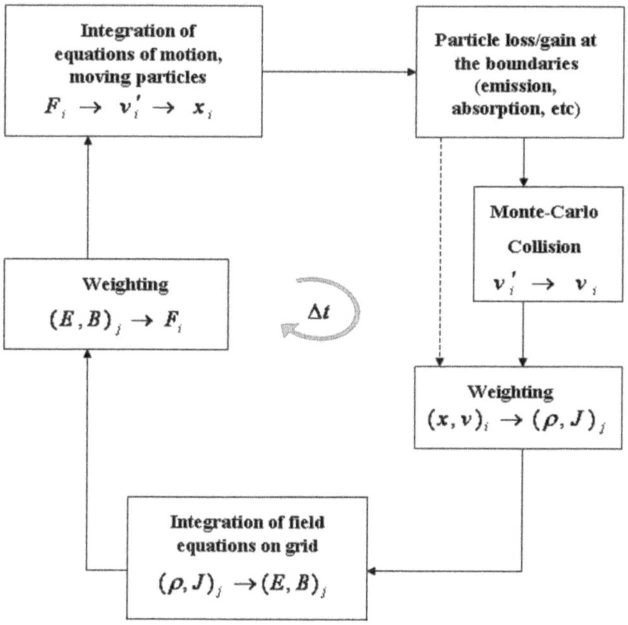

Fig. 1 Schematic representation of the computation sequence for particle in cell-Monte Carlo (PIC-MCC) method

emission process, rather than electron emission from cathode or anode due to ion or electron bombardment. However, all of the above processes have been accounted for in the model. After emission of a new electron, its energy is also assigned from an assumed Maxwell–Boltzmann distribution. The interaction of charged particles with neutral atoms and molecules, and other collisional processes are included using a Monte Carlo Collision technique [12]. Three main charged particle species, viz. electrons, positive ions, and negative ions, have been considered. The major electron-neutral molecule impact reactions included are (1) elastic, (2) excitation, (3) ionization (including all important ionization reactions), and (4) electron attachment. Related cross-sectional values are extracted from experimental data available in the literature [23]. Integral cross-section values for electron interaction in dry air have been determined from its molecular constituents by assuming a target composition of 78.09 % N_2 and 21.19 % O_2 [24]. Only elastic and charge transfer collisions between ions and neutrals are considered. Because ion–molecule cross-sections are not readily available, a constant cross-section (10^{-20}) has been assumed based on a hard-sphere collision model [25].

A significant contribution of this work is that the discharge within the cavity is simulated in conjunction with an external circuit comprising of a voltage source, and a blocking capacitor that stops the flow of any average current. The coupling between the charge flow within the micro-cavity and the external current is obtained via the parallel-plane electrodes, which may be metallic or dielectric, depending on the location of the discharge site.

The potentials and fields are obtained using Poisson's equation:

$$\nabla \cdot \varepsilon_0 \nabla \Phi\,(x,\,t) \;=\; \rho\,(x,\,t). \tag{1}$$

The potential (Φ) can be separated as follows [12, 26, 27]:

$$\Phi = \Phi_{\mathrm{P}} + \sum_{\text{Boundaries}} \Phi_{\mathrm{L}}, \tag{2}$$

where Φ_{P} and Φ_{L} represent the Poisson and Laplacian parts of electrical potential (Φ).

Therefore, the field may be described by the combination of the following equations:

$$\nabla \cdot \varepsilon_0 \nabla \Phi_{\mathrm{P}} = -\rho, \tag{3}$$

$$\nabla \cdot \varepsilon_0 \nabla \Phi_{\mathrm{Li}} = 0. \tag{4}$$

The boundary condition for Eq. (3) is $\Phi = 0$ on all boundaries as the Poisson field is solely due to charge in the medium. For each boundary with a Dirichlet condition, Eq. (4) is solved for $\Phi_{\mathrm{i}} = 0$ on the equi-potential surface, and $\Phi = 0$ elsewhere which gives us Φ_{Li} (the potential due to the ith Laplacian field). Φ_{L} is obtained by the superposition of all the Laplacian fields. Neumann boundary conditions are included through $\Phi_{\mathrm{P}} = 0$. This method neglects charges induced by a driven electrode on other boundaries which are connected to an external circuit. It is also possible to solve the field Poisson equation with boundaries and circuits [12, 21, 28].

In this work, identical cavities, as might occur at discharge sites, with two different electrode configurations have thus been considered. These are the two generic cases described below. The model is two-dimensional with Neumann boundary conditions on the side-walls. Dirichlet boundary conditions (potential drop specified) are assumed on the electrodes.

Metallic–metallic configuration

The first case considers a cavity between MM electrodes. For this configuration, the potential at the anode can be obtained as follows [29]:

$$\varphi_0^t = \left(\frac{\varepsilon_0}{\Delta x} N + C/A\right)^{-1} \times \left(\sigma_{\mathrm{T}}^{t-1} + \frac{1}{A}\left(CV(t) - Q_{\mathrm{c}}^{t-1} + Q_{\mathrm{Conv}}^{t}\right) + \frac{\Delta x}{2} \sum_{j=0}^{j=N-1} \rho_{0j} + \frac{\varepsilon_0}{\Delta x} \sum_{j=0}^{j=N-1} \varphi_{1j}^t \right), \tag{5}$$

where σ_T, A, C, $V(t)$, Q_{c}, Q_{r}, Q_{conv}, ρ, φ_{ij}, and N_j are, respectively, the total surface charge density on the electrodes at a given time 't', electrode cross-sectional area, capacitance of the blocking capacitor, driving voltage, charge in the external circuit, charge convection due to discharge, space charge density within the discharge, the electric potential at each grid point, and the number of grid divisions along the y-axis.

Dielectric–dielectric configuration

The second case we implement is that of DD where the micro-cavity is bounded in both sides by dielectrics. This necessitates the incorporation of a second dielectric slab between the metallic cathode and the void wall at the cathode end (Fig. 2a). The equivalent circuit is shown in Fig. 2b, where C_{g}, C_{d1}, and C_{d2} represent the equivalent capacitances of the micro-cavity and the dielectric layers, respectively. Assuming $C_{\mathrm{d1}} = C_{\mathrm{d2}} = C_{\mathrm{d}}$, using KCL, it is possible to obtain the following relations:

$$C_{\mathrm{d}} \frac{\mathrm{d}U_{\mathrm{d}}}{\mathrm{d}t} = I_{\mathrm{ext}}, \tag{6}$$

$$A \frac{\mathrm{d}\sigma_{\mathrm{s}}}{\mathrm{d}t} = I_{\mathrm{ext}} + AJ_{\mathrm{conv}}, \tag{7}$$

$$V_{\mathrm{c}} = V(t) - 2U_{\mathrm{d}} - \varphi_0, \tag{8}$$

where U_{d}, I_{ext}, σ_{s}, J_{Conv}, and V_{c} are, respectively, the potential across each dielectric slab, total external current,

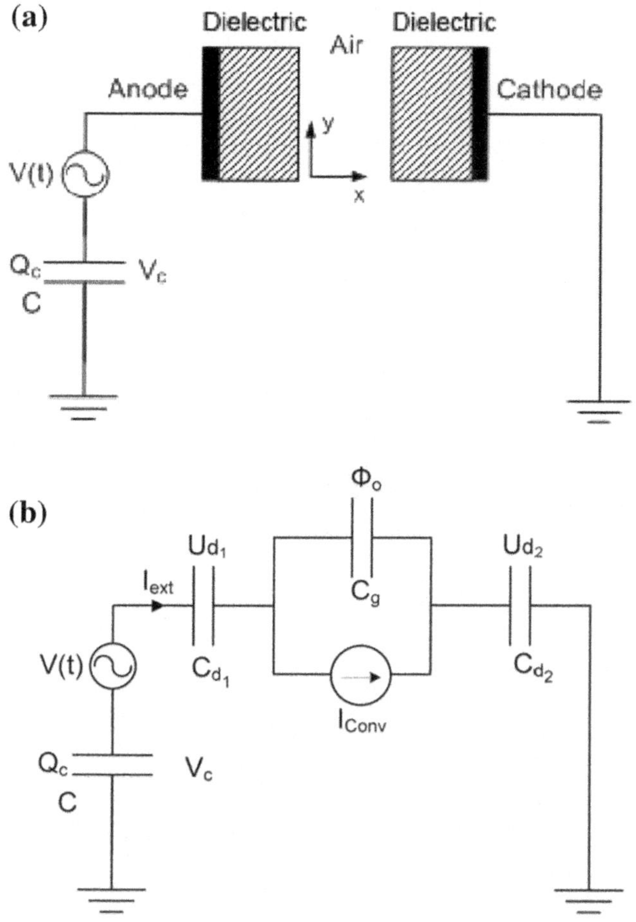

Fig. 2 **a** Schematic representation of a dielectric–dielectric micro-cavity; the external circuit consists of a voltage source and a blocking capacitor which stops the flow of any average current, **b** the corresponding equivalent circuit

charge density on the dielectric surface, discharge current density ($I_{\mathrm{conv}} = I_{\mathrm{d}} = AJ_{\mathrm{Conv}}$), and voltage across blocking capacitor. By combining the above equations, Eq. (5) may be rewritten for DD case as follows

$$\varphi_0^t = \left(\sigma_{s_T}^{t-1} + \frac{1}{A}\left(\alpha CV(t) - \alpha Q_c^{t-1} - 2\alpha CU_d^{t-1} + Q_{\mathrm{Conv}}^t\right) + \frac{\Delta x}{2}\sum_{j=0}^{j=N}\rho_{0j} + \frac{\varepsilon}{\Delta x}\sum_{j=0}^{j=N}\varphi_{1j}^t\right) \cdot \left(\frac{\varepsilon}{\Delta x}N + \alpha C/A\right)^{-1}, \quad (9)$$

where $\alpha = (1 + 2C/C_{\mathrm{d}})^{-1}$. The derivation of Eq. (9) is presented in "Appendix" section.

Simulation results

This section is devoted to the presentation of the results obtained by numerical modeling. A grid-size of $\Delta x = \Delta y = 2 \times 10^{-6}$ m, and a time-step of $\Delta t = 10^{-14}$ s are used; typical simulation times of 24 h are required with a dual processor SUN workstation. As a first step, in this section, we trace the evolution of the distribution of charged particles within the submillimeter void.

The scatter plots in phase space for velocities in x direction i.e., (V_x, x, y) for electrons and positive ions in Figs. 3 and 4 at particular instants of time. As shown here, drift velocity of electrons formed during the discharge is much higher than ions (10^3 times) and quickly traverse the micro-cavity. All charged particles are distributed about the center-line $y = 50$ μm. Most of the charged particles are created very close to the anode. In the early stages, positive ions are distributed evenly over the inter-electrode distance, except for a small region near the cathode. It is also uniformly distributed along y-direction. The number of ions near the anode increases by ionization with passage of time. The positive ions slowly drift toward the cathode, till at about 13 ns, the entire gap distance is almost evenly populated. From this time onward, ions gradually leave the micro-cavity, thus depleting the ions resident in the gap. This is similar to the evolution of the electrons within the gap, as seen in Fig. 3. Electrons move faster and travel only the distance between the ionization region and the anode. As is clear from Fig. 4, the ion velocities increase as they reach near the cathode, accelerated by the electric field.

In following, the effects of micro-cavity dimensions, gas pressure, and dielectric permittivity on the PD pulse features such as discharge current magnitude and its rise-time have been investigated. Obviously, not all PD pulses are characterized by very a rapid rise-time; with a high probability of occurrence of a free electron within the gas of occluded cavities, only small or moderate over-voltages may be required to precipitate a breakdown event. Unless otherwise mentioned, all simulations reported in this work are performed for operating conditions specified in Table 1.

Micro-cavity length

For a relatively short anode–cathode gap distance, the effect of the micro-cavity dimensions on the discharge characteristics, from experimental as well as theoretical findings, is evident [7, 30–41]. In this paper, the length of the micro-cavity is varied from 50 to 600 μm, while the applied electric field across micro-cavity is held constant at $E = 10^6$ V/m. As the length of the micro-cavity increases, for a given pressure, the residence time of ionizing particles inside the micro-cavity is higher. Each electron suffers a larger number of ionizing collisions as it travels from the cathode to the anode; as a result, the production of charged particles inside the micro-cavity is much higher at higher lengths. This is evident as peak value of charged particles

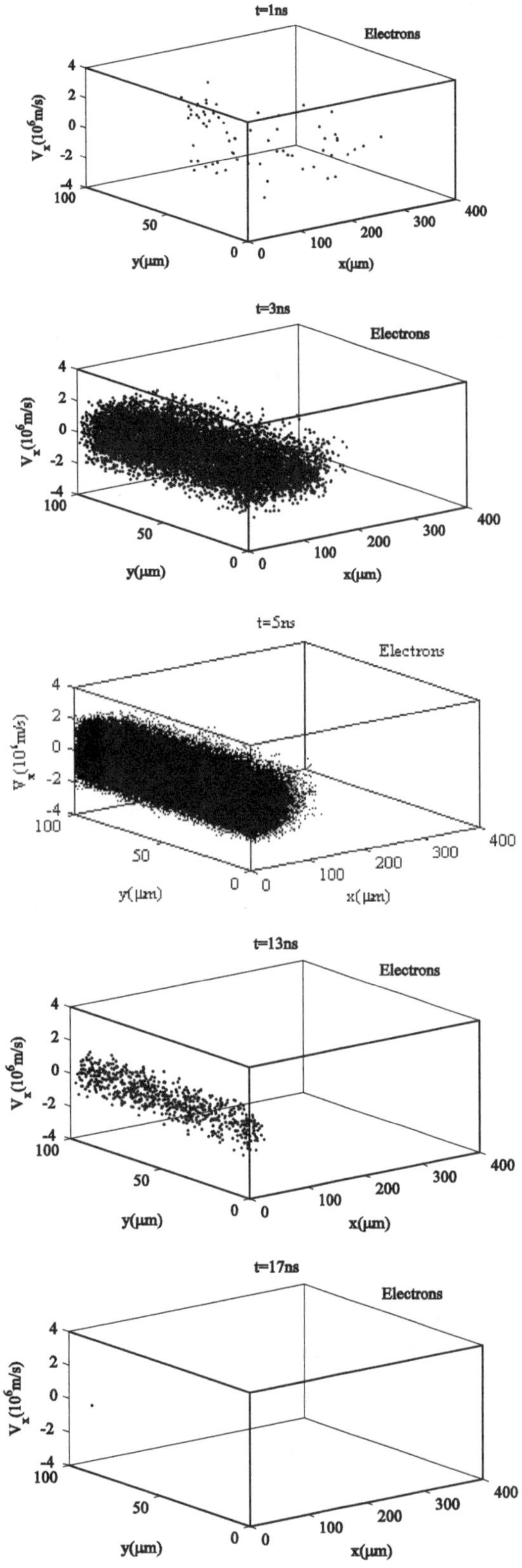

Fig. 3 Scatter plot of electrons in the phase-space (V_x, x, y) inside the micro-cavity at particular instants of time; $L_X = 400$ μm, $L_y = 100$ μm, $L_z = 100$ μm, applied electric field across the micro-cavity: $E = 1.292 \times 106$ V/m, gas pressure: $P_g = 760$ Torr

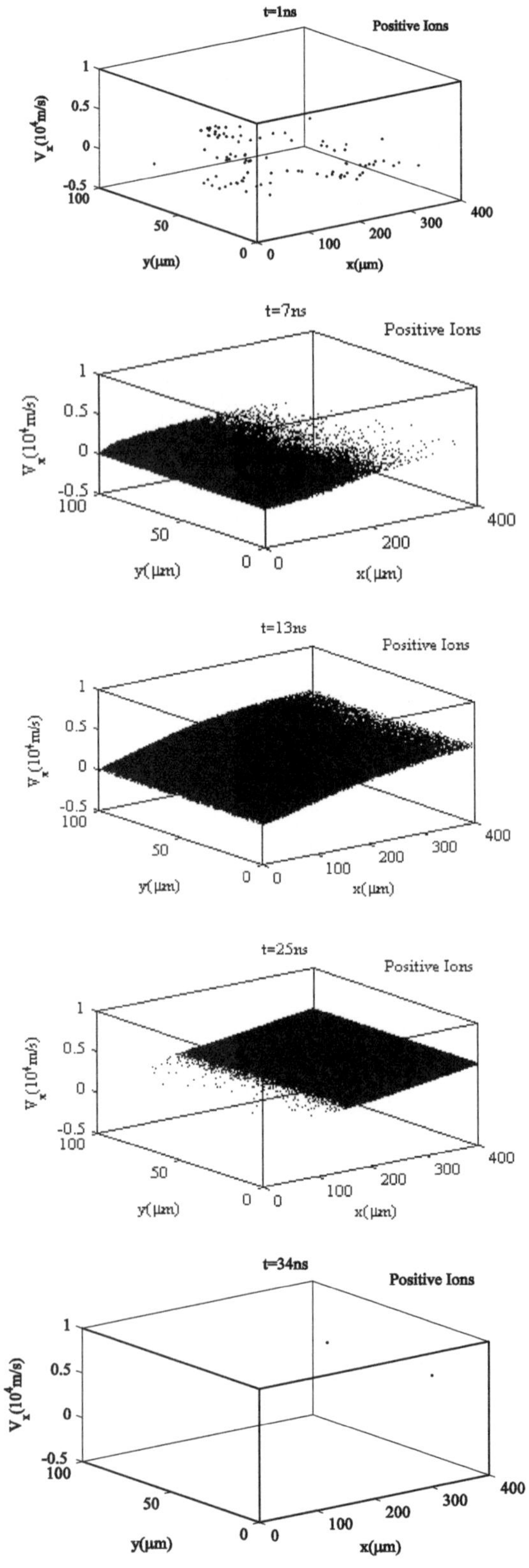

Fig. 4 Scatter plot of positive ions in the phase-space (V_y, x, y) inside the micro-cavity at particular instants of time; $L_X = 400$ μm, $L_y = 100$ μm, $L_z = 100$ μm, applied electric field across the micro-cavity: $E = 1.292 \times 106$ V/m, gas pressure: $P_g = 760$ Torr

Table 1 Simulation data

Property	Symbol	Value
Micro-cavity length	L_X	200 μm
Micro-cavity width	L_y	100 μm
Micro-cavity depth	L_z	100 μm
Dielectric thickness	L_d	200 μm
Background gas pressure	P_g	101,325 Pa
Background gas temperature	T_g	0.026 eV
External capacitor	C	40 pF
Initial number of charged particles	n_f	10
Photo-emission coefficient	λ_{ph}	5×10^{-5} [30]
Ion-induced emission coefficient	λ_i	1.5×10^{-3} [30]
Dielectric constant of air	ε_0	8.85×10^{-12} F/m

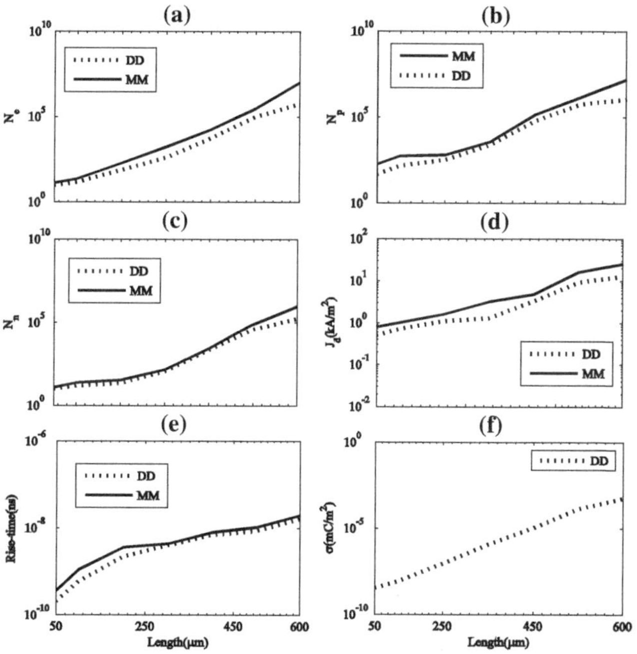

Fig. 5 Variation with micro-cavity length of **a** number of electrons, **b** number positive ions, **c** number of negative ions, **d** discharge current, **e** current rise-time, for both MM and DD configurations, and **f** the magnitude of density of charge accumulated on the dielectrics

in Fig. 5a–c. Results obtained by Serra et al. similarly show that the increase in length of micro-cavity results in increase in the number of electrons generated inside the micro-cavity [41]. Aging and life models by Montanari and coworkers [34, 35] have similarly shown that the number of electrons generated inside the cavity increases with cavity length. The electrons generated inside the cavity are the main carriers of PD current [40, 41]. Increase in the number of charged particles is reflected in an increase in the magnitude of discharge current shown in Fig. 5d. This finding is in agreement with experimental results reported by Devins [33], which shows that the discharge current is sensitive to length of micro-cavity, increases with micro-cavity length. Bartnikas [32] varied gap length from 0.1 mm to 1 mm and reports that a reduction in the gap length results in the disappearance of the pulse discharge (spark-type) and the onset of a diffused glow (pseudo-glow and glow discharge).

Figure 5e shows the variation of PD pulse current rise-time as a function of micro-cavity length. The rise-time increases as length increases. This is understandable as the ionization process in longer micro-cavities require a longer time to reach peak values. These finding corroborates the experimental results by Morshuis [36]. He studied the pulse shape of PD in disk-shaped cavities in polyethylene and other polymers. He varied the length of the cavities between 30 μm and 3.7 mm. He reports that for cavities with a diameter/length ratio $\gg$1, PD rise-time is lower [7, 40]. Ramachandra et al. measured fast discharge pulses with rise-times below 1 ns [42] for void depths (lengths) of 80, 120, and 200 μm. His results show that discharge current rise-times are faster for smaller void depths. Experimental results by Wetzer et al. for PD waveforms show a longer rise-time and pulse duration for micro-cavities (submillimeter) with greater lengths [37].

The variation of charge accumulation at the dielectric anode and cathode at the end of discharge pulse is shown as a function of micro-cavity length in Fig. 5f. The amount of accumulated charge density is calculated as the difference between the charges at the beginning and end of discharge, ($\Delta\sigma_s = \sigma_{si} - \sigma_{sf}$). As larger numbers of charged particles are deployed in the discharge with increased length, the amount of charge accumulated on the dielectric surfaces also increases. This value is always negative for the anode, as the absorbed negative charges lower the applied potential on the anode. For the MM case, the accumulated charge is soon dissipated into the external circuit, and there is no charge accumulation at discharge extinction.

Figure 6 depicts the number of charged particles for both MM and DD configurations and for two different micro-cavity lengths. This is in agreement with simulation results obtained by Novak et al. [43] using a fluid model, for a 0.5-mm gap and DD configuration. Negative ions are depleted fast. As is seen here, the number of negative ions is much lower than that of electrons or positive ions. For the DD configuration, a smaller number of charged particles are involved. Interestingly, for the smaller micro-cavity length, the rise-time for the DD is smaller, and so is the fall-time.

The time evolution of the anode potential for different micro-cavity lengths, φ_0, is presented in Fig. 7a, b. A similar decrease in anode potential with time was observed by Bartnikas et al. for 0.5 mm gap and DD configuration. It may be noted that the effect of external circuit is ignored by Bartnikas et al. [1]. The values remain constant with time for MM configuration as expected, because the

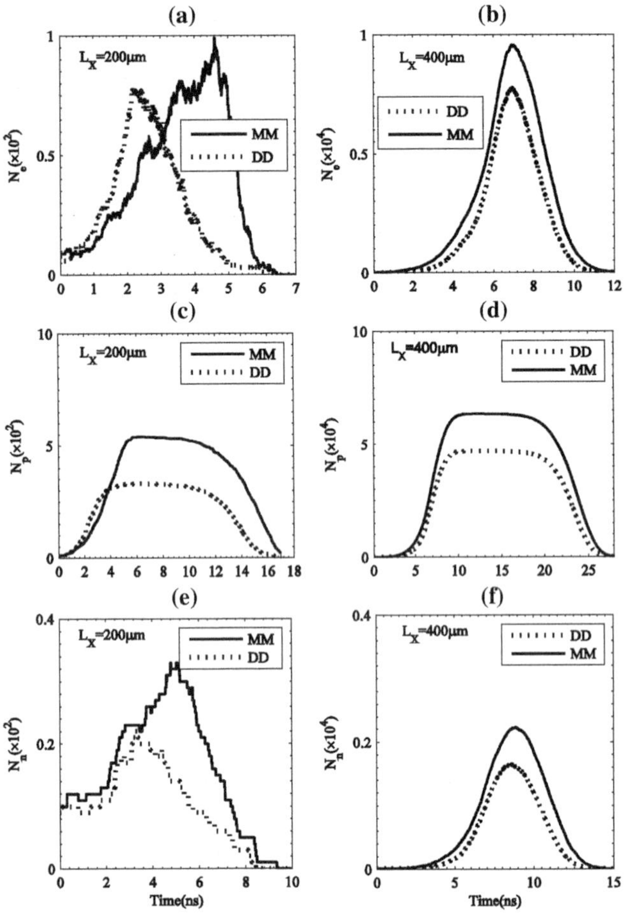

Fig. 6 Time history of (**a** and **b**) the number of electrons, (**c** and **d**) the number of positive ion, (**e** and **f**) the number of negative ions for two different micro-cavity lengths: $L_X = 200$ μm and $L_X = 400$ μm

conductors are maintained at a constant applied potential by the external circuit. However, for DD configuration, due to the role of dielectric on the electric field next to dielectric anode in the discharge medium and also initially presence of positive-charged particles in this region, the dielectric anode electrode potential first increases. Then, because of accumulation of negative-charged particles (i.e., electrons and negative ions) on anode dielectric, the anode potential decreases, settling at considerably decreased values at the end of the discharge-period. The anode dielectric potential, for DD configuration, at the end of the discharge (φ_{0f}) for micro-cavity length of 200 μm is reduced to 99.9 % of its initial value (φ_{0i}) and for length of 400 μm, it has reached to 97 % of its initial value. Thus, while the applied electric field across micro-cavity is kept constant at different lengths, the drop ($\Delta\varphi_0 = \varphi_{0i} - \varphi_{0f}$) at higher micro-cavity lengths is higher.

The change in potential is brought about by charge accumulation at the dielectric electrodes, and as it is clear in Fig. 7c, d, the charge accumulated on the dielectrics for

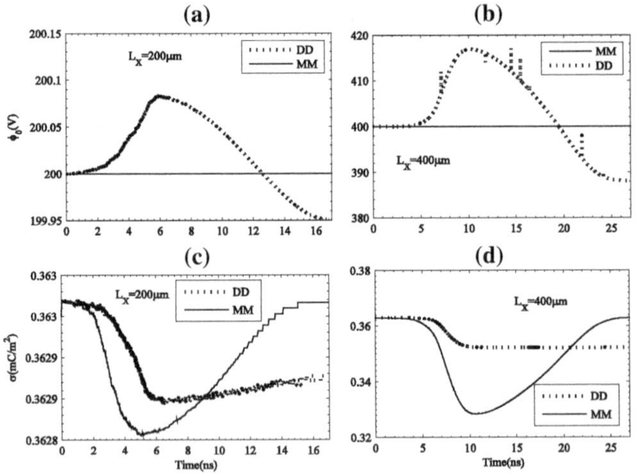

Fig. 7 Time history of the electric potential on the anode dielectric for **a** length of $L_X = 400$ μm, **b** length of $L_X = 400$ μm, and the charge density on the anode dielectric for **c** length of $L_X = 400$ μm, **d** length of $L_X = 400$ μm

200-μm micro-cavity length (10^{-4} mC/m^2) is much <400 μm micro-cavity length (0.01 mC/m^2). Nikonov presented the variation of the field at the anode with time [44], and this shows a trend similar to change in σ. It may be noted that $\sigma = \varepsilon_0 E_{\text{anode}}$, where E_{anode} is the field at the anode, and ε_0 the permittivity of free space. The variation of E_{anode} in Nikonov's work may therefore be taken to be a corroboration of the general trend in variation of surface charge density accumulated on the anode surface.

It is interesting to study the effect of length on the PD pulse by looking at the evolution of the electric field inside the micro-cavity. Figure 8 shows the variation in the axial field (component in the direction of the field) along the micro-cavity length, at different moments of time. As may be seen, the field within the micro-cavity is originally uniform for both lengths i.e., $L_X = 200$ μm and $L_X = 400$ μm. Due to less strength of ionization at lower micro-cavity lengths, the effect of space charge on the applied electric field in case of $L_X = 200$ μm is much less than $L_X = 400$ μm (Fig. 8a, b). Wu et al. [45] showed that as length of micro-cavity increases, the effect of the field due to charged particles inside the micro-cavity becomes more evident. Generally, for the MM case, the field at the dielectric anode is reduced considerably within a very short time. Positive charges take longer to travel the gap distance, and a positive space charge region forms ahead of the anode where they accumulate. This results in a negative field ahead of the anode. The field in the rest of the micro-cavity is low and positive, being directed from the positive space charge region toward the cathode. As positive ions gradually travel toward the cathode, the field in the near vicinity of the anode is relieved and gradually reaches initial values. Simultaneously the field near the cathode is enhanced in the direction of the applied electric field, due

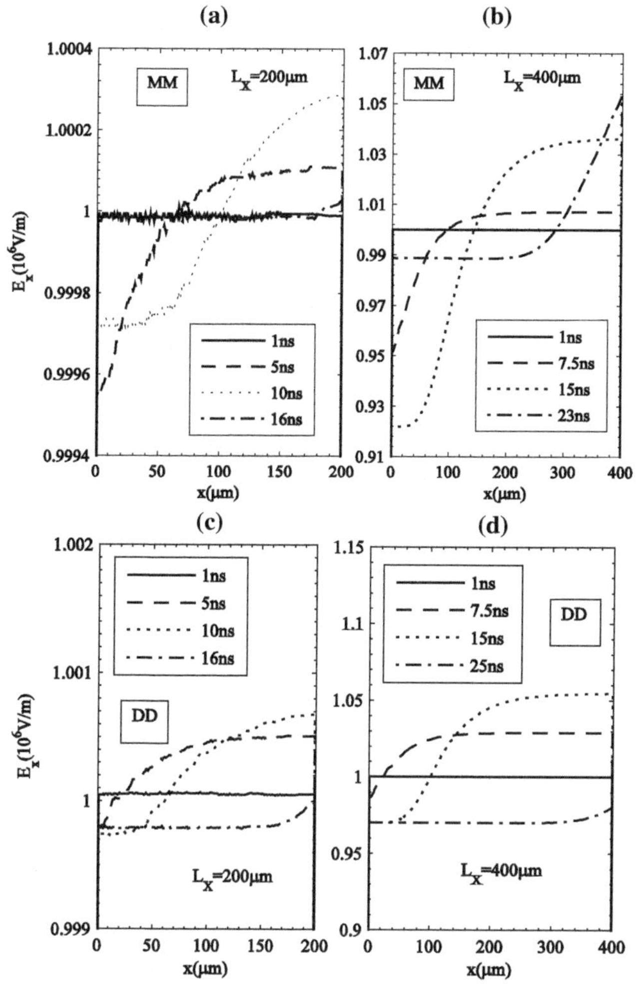

Fig. 8 Variation of E_x along micro-cavity length at different times for different configurations at two different values of micro-cavity length

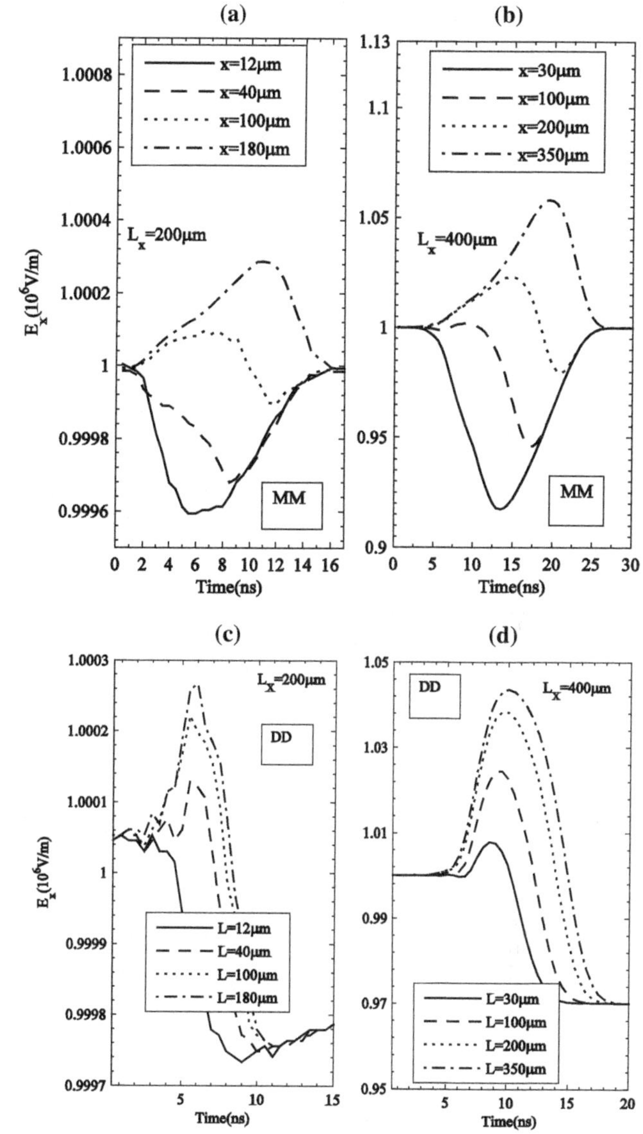

Fig. 9 Time history of E_x as a function of time at different points along x-axis for different configurations for two different values of pressure

to the positive charge cloud nearing the negatively charged cathode. Ultimately after 25 ns (for length of $L_X = 400$ - μm), with the gap being cleared of all charges, original field values are regained throughout the micro-cavity.

For the DD case, shown in Fig. 8c, d, the evolution of the field values is similar to the previous case till about 6 ns. However, the field at the anode never recovers the original values owing to accumulation of negative charges at the anode, lowering the potential thereof. Here too, the field within the micro-cavity becomes uniform; the ultimate field value is, however, lower than in the MM case. Again due to higher effect of space charges at higher micro-cavity lengths, change in the applied electric field is much higher at higher micro-cavity lengths. This is because of the space charge field (produced due to negative charge accumulation at the dielectric anode) opposes the applied field.

Figure 9 presents the electric field (E_x) as a function of time at different axial distances from the anode (along the length of the micro-cavity) for two lengths of micro-cavity i.e., $L_X = 200$ μm and $L_X = 400$ μm. In the MM case, the field variation is similar in both cases; the magnitudes though are different, the field level near the anode falls initially, followed by a rise which is due to lowering of field values owing to an initial positive charge accumulation in front of the anode electrode. While the electric field near the cathode is almost close to its initial value, the electric field near the anode is at a minimum (at $t = 6$ ns) when most of the negative charges have left the micro-cavity. After this moment, the breakdown mechanism becomes much more space charge controlled, and the electric field near the cathode will be under influence of the positive charges arriving in its vicinity and consequently increases. At distances near the cathode, there is a considerable enhancement in the electric field between the positive ions and cathode. Similar observations were made by Serdyuk et al. in modeling of electric discharges in a 3-mm air-gap between parallel plate solid dielectric barriers [46]. They showed that the interaction of the avalanche with the solid dielectric led to charging of the barriers.

Optical studies by Morshuis et al. revealed a large number of separate discharging sites dependent on the void length [7, 40]. They have reported that the current pulses of these separate discharges were integrated by the relatively slow detector, resulting in extremely high discharge values. However, in this work, it has been assumed that only one discharging site on dielectric anode (DD configuration).

Thus, it could be concluded that greater PD pulse current rise-time and charge accumulation on anode and cathode dielectric surfaces occurs at higher micro-cavity lengths.

Micro-cavity width

It has been shown that on increasing the micro-cavity width, there is a greater likelihood of movement of charge carriers in the direction perpendicular to applied electric field (y-direction) [18, 19]. This results in increased dissipation of energy of the PD pulse. Ionization and production of new charged particles therefore, reduces, as was seen in the peak values of charged species.

Figure 10 shows the PD pulses for MM and DD configurations for two different micro-cavity widths. For a width of $L_y = 300$ μm, the PD pulse current peak occurs at 350 and 300 μA for MM and DD configurations, respectively. For a width of $L_y = 600$ μm, the values for PD pulse current occur at 273 and 256 μA for MM and DD configurations which are lower than the corresponding values for the case of $L_y = 300$ μm. Thus in the wider micro-cavity, reduced peak values are observed. The current is lower for a DD configuration. The variation in micro-cavity width does not affect the resultant PD pulse current shape. Gu et al. varied the micro-cavity diameter in the range of 50–600 μm [6] and found that the ionization coefficient increases rapidly as micro-cavity diameter decreases. This is in agreement with the presented results in this paper.

Figure 11a, b shows the time evolution of the anode potential, φ_0, for different micro-cavity widths. The anode dielectric potential, for DD configuration, at the end of the discharge for micro-cavity width of 300 μm is reduced to 92.2 % of its initial value (φ_{0i}) and for width of 600 μm, it has reached to 95.2 % of it is initial value. Thus, a lower potential drop ($\Delta\varphi_0 = \varphi_{0i} - \varphi_{0f}$) occurs at greater widths, and the probability of re-strike is therefore higher. Consequently, in Fig. 11c, d, the charge 0.0694 mC/m^2 accumulated in the 300-μm wide micro-cavity is higher than that in the 600-μm wide micro-cavity (0.0432 mC/m^2). These findings are in agreement with experimental results reported by Auckland et al. [47–49]. He studied the influence of diameter on the quantity of charge accumulating on the side-walls of a micro-cavity in specimens with

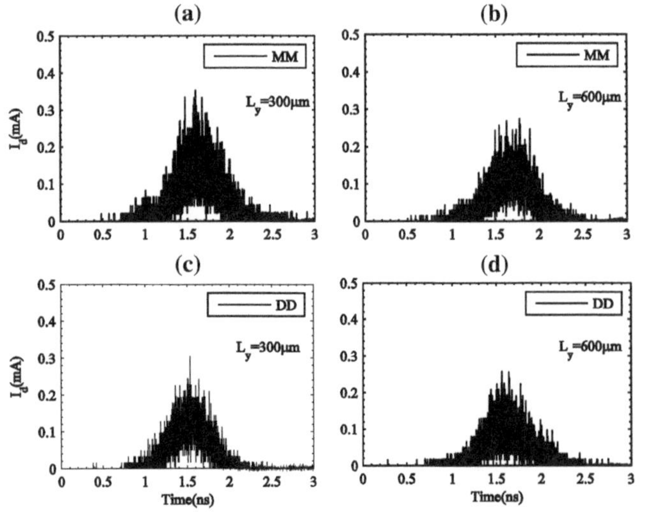

Fig. 10 Time history of PD pulse current for **a** MM configuration for width of $L_y = 300$ μm, **b** MM configuration for width of $L_y = 600$ μm, **c** DD configuration for width of $L_y = 300$ μm, and **d** DD configuration for width of $L_y = 600$ μm

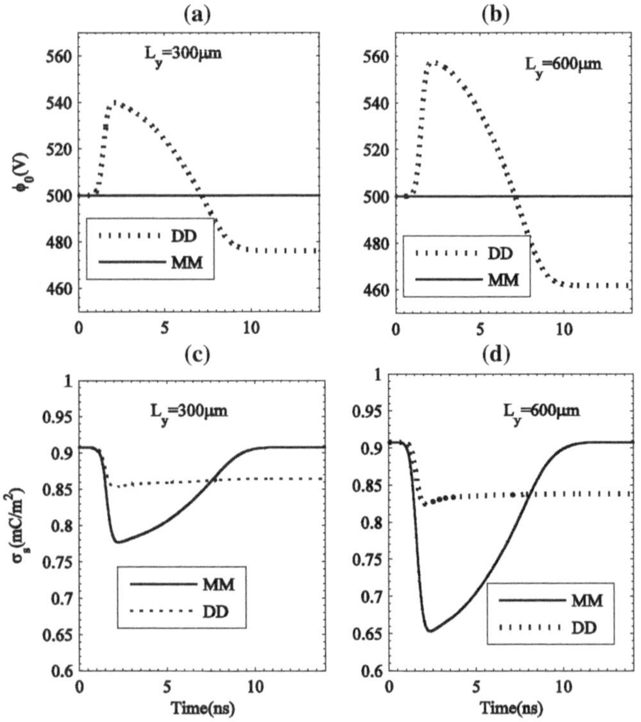

Fig. 11 Time history of the electric potential on the anode dielectric for **a** width of $L_y = 300$ μm, **b** width of $L_y = 600$ μm, and the charge density on the anode dielectric for **c** width of $L_y = 300$ μm, and **d** width of $L_y = 600$ μm

diameters in the range 50–126 μm and length of 10 mm. While the model of discharge propagation in their experiments is different (streamer as opposed to pulses), he observed a reduction in the amount of accumulated charges with increasing in diameter of micro-cavity.

These results corroborate the experimental results obtained by Auckland et al. [48], indicating that the electrical strength of gas confined in narrow micro-cavities increases with the length of the micro-cavity and decreases with micro-cavity diameter.

Morshuis et al. [7] reports that tests on voids with small diameters (1–4 mm) show discharge mechanisms similar to larger widths. These findings also show that the effect of width on discharge is negligible.

Gas pressure inside the micro-cavity

The effect of gas pressure on the nature of PD current pulse is studied in this section. The gas pressure varies from 76 to 1260 Torr, while other parameters are kept constant (Table 1).

The discharge process occurring in cavities, depending upon gas pressure, may assume either a spark, glow or pseudo-glow character. As Morshuis [36] pointed out, in small cavities, the gas plays an important role in the ionization process. The effective ionization coefficient is a

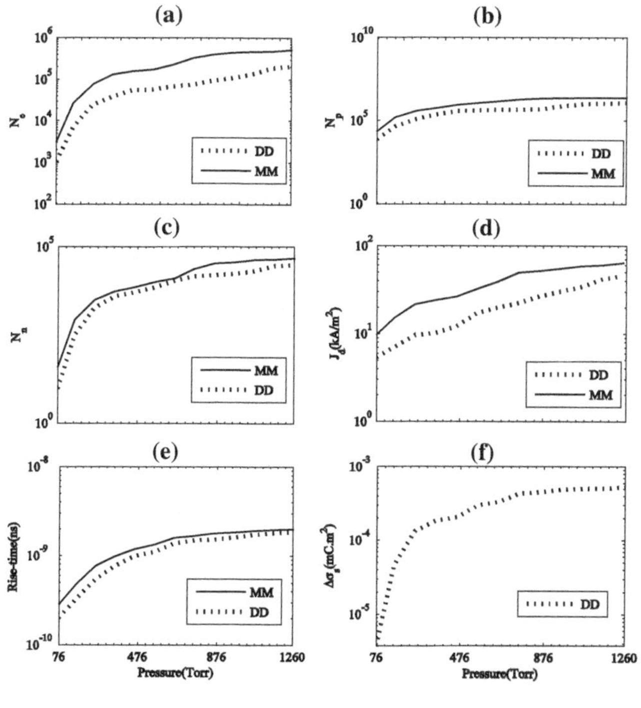

Fig. 12 Variation with gas pressure inside the micro-cavity **a** number of electrons, **b** number of positive ions, **c** number of negative ions, **d** PD current, **e** discharge current rise-time for both MM and DD configurations, and **f** charge accumulated on the dielectrics

property of the gas and is therefore affected by the changing gas chemistry and by a change of pressure.

It is seen in Fig. 12a–c that the maximum number of charged particles residing inside the micro-cavity increases with gas pressure. Therefore in Fig. 12d, the magnitude of PD pulse increases. The rise-time of the PD pulse versus gas pressure is presented in Fig. 12e. The effect of gas pressure on the rise-time is significantly high. At lower gas pressures, e.g., 76 Torr, the resultant PD pulse current amplitudes are too low but the rise-time is too rapid and it might be possible to detect them by conventional PD detectors. Gaxiola et al. has reported that with increasing pressure, the time-to-breakdown rapidly increases over two orders of magnitude [30].

However, the experimental findings by Kasten et al. for a cavity with length of 20 mm show that the time to peak of discharge current decreases as pressure increases [50]. His findings relate to cavities in the millimeter range, where streamers may be taken to control discharge propagation and transfer the results correspond to a different physical mechanism.

The variation of charge accumulated on the anode and cathode dielectrics ($\Delta\sigma_s = \sigma_{si} - \sigma_{sf}$) versus gas pressure inside the micro-cavity is depicted in Fig. 12f. Owing to higher ionization at higher pressures, larger number of charges (mostly electrons) will reach the dielectric at higher pressures.

As is clear in Fig. 13, for a gas pressure of 260 Torr, the maximum values of PD pulse current is around 416 μA for MM configuration and 256 μA for DD configuration, respectively, while for a gas pressure of 860 Torr, the corresponding values are 520 and 288 μA, respectively. It may be noted that lower pressure values lead not only to reduced peak values of the PD pulse current, but also cause shorter delays in the formation as well as reduction in the width of PD pulse.

In summary, at higher gas pressures, greater production of charged particles within micro-cavity over time and consequently increase in discharge current as well as charge accumulated on the anode and cathode dielectrics occurs.

Dielectric permittivity

In homogenous insulating materials, the electric field is distributed uniformly within the entire volume. In nonhomogenous insulating materials, due to manufacturing defects, aging effects, etc., the distribution of the electric field is no longer uniform.

The electric field distributes across two layers of dielectric connected in series in the ratio of their permittivities and thickness (Fig. 2a). In this section, the effects of dielectric permittivity on the behavior of PD pulse are

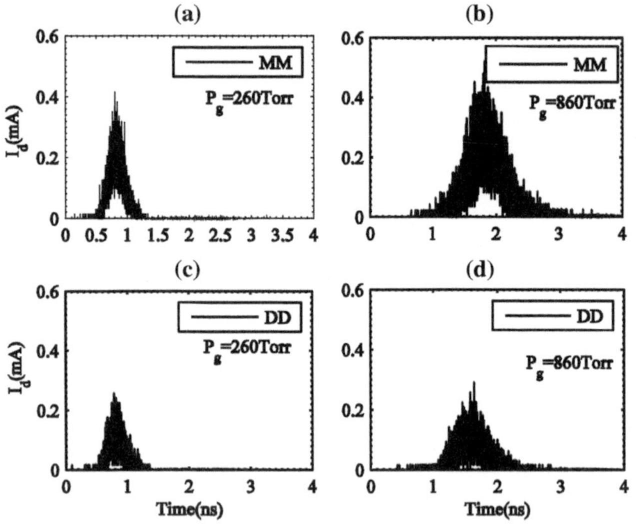

Fig. 13 Time history of PD pulse current for **a** MM configuration for gas pressure of $P_g = 260$ Torr, **b** MM configuration for gas pressure of $P_g = 860$ Torr, **c** DD configuration for gas pressure of $P_g = 260$ - Torr, and **d** DD configuration for gas pressure of $P_g = 860$ Torr

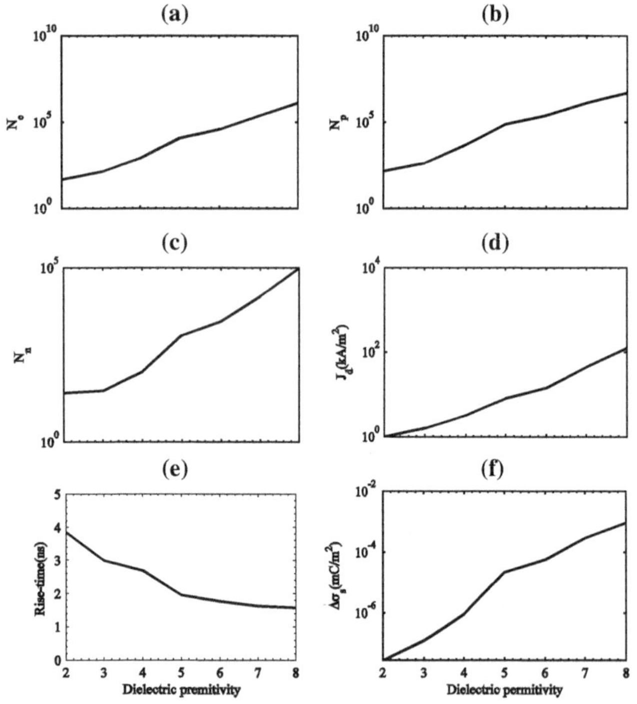

Fig. 14 Variation with dielectric permittivity of **a** peak value of number of electrons, **b** peak value of number of positive ions, **c** peak value of number of negative ions, **d** PD current density, **e** PD pulse rise-time, and **f** charge accumulated on the dielectrics

studied. The applied electric potential between the metallic electrodes has been kept constant (e.g., $E_a = 20$ kV/mm) for the DD configuration, i.e., the electric potential across the dielectric-gap combination. The permittivity of surrounding material is varied, while keeping the dimensions and geometry constant. The applied electric field inside the micro-cavity (E_a) increases and ionization inside the micro-cavity becomes stronger. Owing to this, as is clear from Fig. 14a–c, the maximum number of charged particles produced inside the micro-cavity increases. The magnitude of the PD pulse as a function of dielectric permittivity is presented in Fig. 14d. Figure 14e shows that the rise-time of the PD pulse falls with ε_r, probably due to increased space charge effects at the dielectric cathode, and decreases with dielectric permittivity as effective electric field along micro-cavity increases.

The variation of charge accumulated on the dielectric anode and cathode ($\Delta\sigma_s = \sigma_{si} - \sigma_{sf}$) with dielectric permittivity is depicted in Fig. 14f. Owing to increase in the strength of ionization with permittivity of dielectric, the amount of charge accumulated on the dielectrics surface increases.

A point to be noted is that the effect of dielectric permittivity on charge accumulation on the dielectric anode and cathode, PD pulse magnitude, and its rise-time is significant. All simulations reported in the previous sections have been performed with a dielectric permittivity of $\varepsilon_r = 5$.

Discussions

An important contribution of this work is its ability to investigate of the effects of different parameters on PDs inside the micro-cavities for two different dielectric configurations (MM and DD). The calculated PD pulse forms are very much affected by values of micro-cavity length, gas pressure, and dielectric permittivity, while the effect of variations in micro-cavity width is relatively less. These results show that for a kept fix each one of the above parameters, the detected PD pulse form, amplitude and rise-time can exhibit considerable variation due to possible changes in each every above-mentioned parameter. Since the response of a conventional PD detector is a function of the rise-time and pulse amplitude of the PD pulse, variations in the detected pulse would result.

The accuracy of peak pulse measurement would thus be affected in that PD pulse detector would respond more favorably to pulses having rapid rise-times. Consequently, it is quite conceivable that a lower amplitude discharge pulse with a more rapid rise-time may produce a larger indication on that detector than a higher amplitude pulse with a slower rise-time. It would be difficult to resolve this measurement inaccuracy in practice.

By increasing some parameters such as micro-cavity length, gas pressure the space charge density accumulated at the cathode increases and leads to less rapid breakdowns while by increase micro-cavity width and dielectric permittivity charge accumulation on the cathode leads to more rapid breakdowns and results in pulses with correspondingly reduced rise-time and augmented magnitudes.

Conclusions

A two-dimensional kinetic model based on PIC-MCC has been developed to a parametric analysis of nature of PD pulse within micro-cavities. Indeed, this parametric analysis focused especially on micro-cavity dimensions, gas pressure inside the micro-cavity, and dielectric permittivity. It proves the ability of the model in reproducing what is known about the dependence of the nature of PD pulse on the various physical parameters. All the parameters that characterize the proposed model have physical meaning and they can be measured resorting analytical to analytical techniques.

So, determining the dependency of the nature of highly localized intense spark-type PDs to different parameters governing them gives an idea of the physical processes involved in their operations within the micro-cavities. It is expected that the employed model will be a useful tool in designing the solid insulating materials.

In summary, while the peak magnitude of PD pulse current increases with micro-cavity length, gas pressure, and dielectric permittivity, it decreases with increase in width. The rise-time of the PD pulse increases with micro-cavity length and pressure, but it decreases with micro-cavity width and dielectric permittivity.

Acknowledgments I would like to acknowledge the Institute of Science, High Technology & Environmental Sciences for financial support within the Project 1.140-90/2/24.

Appendix

The total charge density σ_T on the anode metallic electrode may be expressed by:

$$\sigma_T^t = \sum_{j=0}^{N-1} \sigma_{0j}^t, \tag{A1}$$

where σ_{0j}^t is the surface charge density on anode electrode at time 't'. The simplified form of Eq. (26) of [29] for a 'simple metallic-metallic' configuration is:

$$\sigma_T^t = \frac{\varepsilon_0}{\Delta x} N \varphi_0^t - \frac{\varepsilon_0}{\Delta x} \sum_{j=0}^{N-1} \varphi_{1j}^t - \frac{\Delta x}{2} \sum_{j=0}^{N-1} \rho_{0j}, \tag{A2}$$

where φ_0^t, φ_{1j}^t, and N are the potential on the anode at 't', potential at x_1 (= $1 \times \Delta x$) and number of divisions along y-axis. Particle 'p' contributes to Q_{ij} through the space charge density ρ_{ij} (bilinear weighting: [29]). By combining equations (A1) and (A2), the Eq. (27) of [29] is obtained and is simplified for a MM configuration as follows:

$$\varphi_0^t = (\sigma_T^{t-1} + \frac{1}{A}(CV(t) - Q_c^{t-1} + Q_{Conv}^t) + \frac{\Delta x}{2} \sum_{j=0}^{j=N-1} \rho_{0j} + \frac{\varepsilon_0}{\Delta x} \sum_{j=0}^{j=N-1} \varphi_{1j}^t)(\frac{\varepsilon_0}{\Delta x} N + C/A)^{-1}. \tag{A3}$$

Note that this is Eq. (5) in the paper and is used for the MM case.

The equivalent circuit for DD configuration is shown in Fig. 3b. C_g and C_d represent the equivalent capacitances of the gap and the dielectric layers, respectively, and J_{Conv} is the discharge current density. The time variation of σ_T, the total charge density on the driven dielectric electrode, may be obtained from the Kirchhoff's Current Law:

$$A \frac{d\sigma_T}{dt} = I_{ext} + AJ_{conv}, \tag{A4}$$

where I_{ext} and σ_T are the total external current and charge on the dielectric surfaces.

The time variation of voltage drop (U_d) across the dielectric slab is obtained as:

$$C_d \frac{dU_d}{dt} = I_{ext}. \tag{A5}$$

Figure 3b shows a simple external circuit consisting of a voltage source in series with a capacitor coupled to the left electrode. The voltage drop V_c across the capacitor in the external circuit is obtained from voltage balance as,

$$V_c = V(t) - 2U_d - \varphi_0 \tag{A6}$$

$V(t)$ is the applied voltage source. The discrete finite differenced form of Eq. (A4) can be expressed as:

$$A(\sigma_T^t - \sigma_T^{t-1}) = Q_c^t - Q_c^{t-1} + Q_{conv}^t, \tag{A7}$$

where $\int I dt = Q_c = CV_c$ is the charge on the capacitor and Q_{conv}^t is the charge deposited on the electrode from the discharge in the cavity during the time interval (t − 1, t). Combining Eqs. (A6) and (A7), we obtain:

$$\sigma_T^t = \sigma_T^{t-1} + \frac{1}{A}(CV(t) - C\varphi_0^t - 2CU_d^t - Q_c^{t-1} + Q_{conv}^t). \tag{A8}$$

Similarly, the discrete finite differenced form of Eq. (A5) can be expressed as:

$$U_{\mathrm{d}}^{t}-U_{\mathrm{d}}^{t-1}=\frac{1}{C_{\mathrm{d}}}\int I\mathrm{d}t=\frac{Q_{\mathrm{c}}^{t}-Q_{\mathrm{c}}^{t-1}}{C_{\mathrm{d}}}. \tag{A9}$$

If we combine Eqs. (A6) and (A9), then we have:

$$U_{\mathrm{d}}^{t}=\frac{U_{\mathrm{d}}^{t-1}+\frac{1}{C_{\mathrm{d}}}\left(CV(t)-C\varphi_{0}^{t}-Q_{\mathrm{c}}^{t-1}\right)}{1+C/C_{\mathrm{d}}}. \tag{A10}$$

By inserting the value of U_{d}^{t} to the Eq. (A8), the value of σ_{T}^{t} can be written as follow:

$$\sigma_{\mathrm{T}}^{t}=\sigma_{\mathrm{T}}^{t-1}+\frac{1}{A}\left(\alpha CV(t)-\alpha C\varphi_{0}^{t}-2\alpha CU_{\mathrm{d}}^{t-1}-\alpha Q_{\mathrm{c}}^{t-1}+Q_{\mathrm{conv}}^{t}\right) \tag{A11}$$

Equations (A1), (A2) for σ_{T}^{t} and (A11) can be combined and solved for φ_{0}^{t} on dielectric anode to produce:

$$\varphi_{0}^{t}=\left(\sigma_{\mathrm{ST}}^{t-1}+\frac{1}{A}\left(\alpha CV(t)-\alpha Q_{\mathrm{c}}^{t-1}-2\alpha CU_{\mathrm{d}}^{t-1}+Q_{\mathrm{Conv}}^{t}\right)+\frac{\Delta x}{2}\sum_{j=0}^{j=N}\rho_{0j}+\frac{\varepsilon}{\Delta x}\sum_{j=0}^{j=N}\varphi_{1j}^{t}\right)\cdot\left(\frac{\varepsilon}{\Delta x}N+\alpha C/A\right)^{-1}. \tag{A12}$$

References

1. Bartnikas, R., Novak, J.P.: On the character of different forms of partial discharge and their related terminologies. IEEE Trans. Electr. Insul. **28**(6), 956–968 (1993)
2. Bartnikas, R., Novak, J.P.: Effect of over-voltage on the rise-time and amplitude of PD pulses. IEEE Trans. Dielectr. Electr. Insul. **2**(4), 557–566 (1995)
3. Bartnikas, R.: Partial discharges: their mechanism, detection and measurement. IEEE Trans. Dielectr. Electr. Insul. **9**(5), 763–808 (2002)
4. Bartnikas, R.: A comment governing the rise-times of partial discharge pulses. IEEE Trans. Dielectr. Electr. Insul. **12**(2), 196–202 (2005)
5. Burgener, H. P., Teich, T. H. Frohljch, K. (2002), Simulations of partial discharges of small micro-cracks parallel to the electrical field in polymeric materials, *Annual report CEIDP*, pp. 146–150
6. Gu, W.Y., Laurent, C., Mayoux, C.: Characteristics of discharges inside simulated tree micro-cavities under impulse voltage. J. Phys. D **19**, 2197–2207 (1986)
7. Morshuis, P.H.F., Kreuger, F.H.: Transition from streamer to Townsend mechanisms in dielectric voids. J. Phys. D **23**, 1562–1568 (1990)
8. Hudon, C., Bartnikas, R., Wertheimer, M.R.: Effect of physico-chemical degradation of epoxy resin on partial discharge behavior. IEEE Trans. Dielectr. Electr. Insul. **2**(6), 1083–1094 (1995)
9. Hudon, C., Bartnikas, R., Wertheimer, M.R.: Spark-to-glow discharge transition due to increased surface conductivity on epoxy resin specimens. IEEE Trans. Dielectr. Electr. Insul. **28**(1), 1–8 (1993)
10. Hudon, C., Bartnikas, R., Wertheimer, M. R.: Surface conductivity of epoxy specimens subjected to partial discharges. *IEEE International Symposium on Electrical Insulation, Toronto, Canada*, pp. 3–6 (1990)
11. Hudon, C., Bartnikas, R., Wertheimer, M. R.: Chemical and physical degradation effects on epoxy surfaces exposed to partial discharges. *4th International Conference on Properties and Applications of Dielectric Materials, Brisbane Australia*, pp. 811–814 (1994)
12. Birdsall, C.K., Langdon, A.B.: Plasma Physics via Computer Simulation. McGraw-Hill, New York (1985)
13. Lawson, W.: Particle simulation of bounded 1D plasma systems. J. Comput. Phys. **80**, 253–276 (1989)
14. Lawson, W.: Artificial cooling due to quiet injection in bounded plasma particle simulations. J. Comput. Phys. **77**, 330–339 (1988)
15. Hockney, R.W., Eastwood, J.W.: Computer Simulation Using Particles. Taylor & Francis Group, New York (1988)
16. Radmilovic-Radjenovic, M., Radjenovic, B.: A particle-in-cell simulation of the breakdown mechanism in micro-discharges with an improved secondary emission model. Contrib. Plasma Phys. **47**(3), 165–172 (2007)
17. Radmilovic-Radjenovic, M., Lee, J.K., Iza, F., Park, G.Y.: Particle-in-cell simulation of gas breakdown in micro-gaps. J. Phys. D **38**(6), 950–954 (2005)
18. Ganjovi, A., Gupta, N., Raju, G.R.G.: A kinetic model of a PD pulse within voids of sub-millimeter dimensions. IEEE Trans. Dielectr. Electr. Insul. **16**(6), 1743–1754 (2009)
19. Ganjovi, A., Rastpour, G.: Modeling of distribution of PD pulses within micro-cavities. Int. Rev. Model. Simul. **4**(2), 219–227 (2011)
20. Ganjovi, A.: Estimation of photo-degradation of dielectrics surrounding the narrow channel due to PD activity. J. Theor. Appl. Phys. (2014). doi:10.1007/s40094-014-0144-6
21. Birdsall, C.K.: Particle-in-cell charged-particle simulations, plus Monte Carlo collisions with neutral atoms, PIC–MCC. IEEE Trans. Plasma Sci. **19**, 65–85 (1991)
22. Vahedi, V., Surendra, M.: Monte-Carlo collision model for particle-in-cell method: application to argon and oxygen discharges. Comput. Phys. Commun. **87**, 179–198 (1995)
23. Raju, G.R.G.: Gaseous Electronics: Theory and Practice. Taylor and Francis, Boca Raton (2005)
24. Roldan, A., Perez, J.M., Williart, A., Garcia, B.: Energy deposition model for low-energy electrons (10–10 000 eV) in air. J. Appl. Phys. **95**(10), 5865–5870 (2004)
25. Zhang, W., Fisher, T.S., Garimella, S.V.: Simulation of ion generation and breakdown in atmospheric air. J. Appl. Phys. **96**(11), 6066–6072 (2004)
26. Mardahl, P.J., Cartwright, K.L.: "Reducing Numerical Heating in 1-D PIC Simulations", Annual Progress Report for 1993, Plasma Theory and Simulation Group. University of California Technical Memorandum, Electronics Research Laboratory, Berkeley (1993)
27. Mardahl, P.J., Verboncoeur, J.P.: Charge conservation in electromagnetic PIC codes; spectral comparison of Boris/DADI and Langdon-Marder methods. Comput. Phys. Commun. **106**, 219–229 (1997)
28. Verboncoeur, J.P.: Symmetric spline weighting for charge and current density in particle simulation. J. Comput. Phys. **174**, 421–427 (2001)
29. Vahedi, V., DiPeso, G.: Simultaneous potential and circuit solution for two-dimensional bounded plasma simulation codes. J. Comput. Phys. **131**, 149–163 (1997)
30. Gaxiola, E. H. R., Wetzer, J. M., Partial discharges in micrometer voids PTFE. *IEEE International Symposium on Electrical Insulation, Quebec, Canada*, pp. 16–19 (1996)
31. Novak, J. P., Bartnikas, R., Quantitative model of a short-gap breakdown in air. *XVIIth International symposium on discharge and electrical insulation in vacuum, Berkeley, U.S.*, pp. 60–64 (1996)

32. Bartnikas, R.: Some observations on the character of corona discharges in short gap spaces. IEEE Trans. Electr. Insul. **2**(2), 63–75 (1971)
33. Devins, J.C.: The physics of partial discharges in solid dielectrics. IEEE Trans. Electr. Insul. **EI-19**(5), 475–495 (1984)
34. Mazzanti, G., Montanari, G.C., Civenni, F.: Model of inception and growth of damage from microvoids in polyethylene-based materials for HVDC cables part 1: theoretical approach. IEEE Trans. Dielectr. Electr. Insul. **14**(5), 1242–1254 (2007)
35. Mazzanti, G., Montanari, G.C., Civenni, F.: Model of inception and growth of damage from microvoids in polyethylene-based materials for HVDC cables part 2: parametric investigation and data fitting. IEEE Trans. Dielectr. Electr. Insul. **14**(5), 1255–1263 (2007)
36. Morshuis, P.: Assessment of dielectric degradation by ultrawide-band PD detection. IEEE Trans. Dielectr. Electr. Insul. **2**(5), 744–760 (1995)
37. Wetzer, J.M., van der Laan, P.C.T.: Pre-breakdown currents: basic interpretation and time-resolved. IEEE Trans. Electr. Insul. **24**(2), 297–308 (2989)
38. Morshuis, P.: Partial discharge mechanisms in voids related to dielectric degradation. IEE Proc. Sci. Meas. Technol. **142**(1), 62–68 (1995)
39. Novak, J.P., Bartnikas, R.: Effect of dielectric surfaces on the nature of partial discharges. IEEE Trans. Dielectr. Electr. Insul. **7**(1), 146–151 (2000)
40. Morshuis, P.: Partial discharge mechanisms in voids related to dielectric degradation. IEE Proc. Sci. Meas. Technol. **142**(1), 62–68 (1995)
41. Serra, S., Montanari, G.C., Mazzanti, G.: Theory of inception mechanism and growth of defect-induced damage in polyethylene cable insulation. J. Appl. Phys. **98**, 034102-1–03410203410215 (2005)
42. Ramachandra, B., Nema, R.S.: Characterization of partial discharge pulses in artificial voids in polypropylene films used in capacitors. *IEEE International Symposium on Electrical Insulation, Montreal, Canada*, pp. 517–520 (1996)
43. Novak, J.P., Bartnikas, R.: Ionization and excitation behavior in a micro-cavity. IEEE Trans. Dielectr. Electr. Insul. **2**(5), 724–728 (1995)
44. Nikonov, V.: Master Thesis, Influence of Electrode Surface Charge Accumulation upon Partial Discharge Behavior. University of Montreal, Canada (1999)
45. Wu, K., Suzuoki, Y., Mizutani, T., Xie, H.: Model for partial discharges associated with treeing breakdown: I. PDs in tree channels. J. Phys. D **33**, 1197 (2000)
46. Serdyuk, Y.V., Gubanski, S.M.: Computer modeling of interaction of gas discharge plasma with solid dielectric barriers. IEEE Trans. Dielectr. Electr. Insul. **12**(4), 725 (2005)
47. Auckland, D. W., Borishade, A. B., Cooper, R.: The breakdown characteristics of air-filled tubules in solid insulation. *IEEE Conference Publication, Dielectric Materials, Measurements and Applications, Cambridge*, Vol. 129, p. 15 (1975)
48. Auckland, D.W., Borishade, A.B., Cooper, R.: Photographic investigation of breakdown of composite insulation. Proc. IEE **124**(12), 1263 (1977)
49. Auckland, D.W., Kabir, S.M.F., Varlow, B.R.: Charge deposition in gas filled channels with insulating walls. IEE Proc. A **140**(6), 509–516 (1993)
50. Kasten, D.G., Liu, X., Sebo, S.A., Grosjean, D.F., Schweickart, D.L.: Partial discharge measurements in air and argon at low pressures with and without a dielectric barrier. IEEE Trans. Dielectr. Electr. Insul. **12**(2), 362–373 (2005)

4

Physical and mathematical justification of the numerical Brillouin zone integration of the Boltzmann rate equation by Gaussian smearing

Christian Illg[1] · Michael Haag[1] · Nicolas Teeny[2] · Jens Wirth[3] · Manfred Fähnle[1]

Abstract Scatterings of electrons at quasiparticles or photons are very important for many topics in solid-state physics, e.g., spintronics, magnonics or photonics, and therefore a correct numerical treatment of these scatterings is very important. For a quantum-mechanical description of these scatterings, Fermi's golden rule is used to calculate the transition rate from an initial state to a final state in a first-order time-dependent perturbation theory. One can calculate the total transition rate from all initial states to all final states with Boltzmann rate equations involving Brillouin zone integrations. The numerical treatment of these integrations on a finite grid is often done via a replacement of the Dirac delta distribution by a Gaussian. The Dirac delta distribution appears in Fermi's golden rule where it describes the energy conservation among the interacting particles. Since the Dirac delta distribution is a not a function it is not clear from a mathematical point of view that this procedure is justified. We show with physical and mathematical arguments that this numerical procedure is in general correct, and we comment on critical points.

Keywords Electron scattering · Boltzmann rate equations · Brillouin zone integration · Treatment of Diracs delta distribution

✉ Manfred Fähnle
faehnle@is.mpg.de

Christian Illg
illg@is.mpg.de

Michael Haag
mhaag@is.mpg.de

Nicolas Teeny
nicolas.teeny@mpi-hd.mpg.de

Jens Wirth
Jens.Wirth@mathematik.uni-stuttgart.de

1 Max Planck Institute for Intelligent Systems, Heisenbergstr. 3, 70569 Stuttgart, Germany

2 Max Planck Institute for Nuclear Physics, Saupfercheckweg 1, 69117 Heidelberg, Germany

3 Institut für Analysis, Dynamik und Modellierung, Universität Stuttgart, Pfaffenwaldring 57, 70569 Stuttgart, Germany

Introduction

In solid-state physics, scatterings of electrons at periodic perturbations (quasiparticles or photons) are very important for many research fields and we give three examples in the following:

1. In all-optical switching experiments [1] a thin ferrimagnetic film, e.g., GdFeCo, is irradiated by a femtosecond laser pulse which can be linearly or circularly polarized and thereafter a demagnetization with subsequent switching of the magnetization can be observed under certain preconditions. The fundamental mechanisms are strongly debated at the moment, however, electron–photon scatterings, electron–phonon scatterings and electron–magnon scatterings certainly play a big role for the demagnetization of the ferrimagnetic film.
2. In ultrafast demagnetization experiments [2] a thin ferromagnetic film, e.g., Ni or Fe, is irradiated by a femtosecond laser pulse which is normally linearly polarized and thereafter an ultrafast demagnetization (on the timescale of about 100 fs) without switching of the magnetization can be observed. The magnetization recovers on a timescale of several picoseconds. Despite many years of research the fundamental

mechanisms are still unclear but scatterings of electrons at phonons [3, 4] or at magnons [5] or at electrons [6] have been discussed intensively.

3. Spin-polarized currents are important for devices in spintronics [7], e.g., spin-transistors or spin-diodes. The lifetime of the spin-polarized electrons is crucial for the spintronics devices. The lifetimes are determined by scatterings of electrons at quasiparticles and at interfaces or defects.

A correct numerical calculation of the various scattering processes is important for the understanding of these effects in solid-state physics. In quantum mechanics, Fermi's golden rule gives the transition rate $W^{\lambda}_{j\mathbf{k},j'\mathbf{k}'}$ from an initial electronic state $\Psi_{j\mathbf{k}}$ in a solid with energy $\varepsilon_{j\mathbf{k}}$ to a final electronic state $\Psi_{j'\mathbf{k}'}$ with energy $\varepsilon_{j'\mathbf{k}'}$ (j,j': band indices; $\mathbf{k}$, $\mathbf{k}'$: wave vectors) due to a periodic perturbation arising from a (quasi)particle [8]

$$W^{\lambda}_{j\mathbf{k},j'\mathbf{k}'} = \frac{2\pi}{\hbar}\left|M^{\lambda}_{j\mathbf{k},j'\mathbf{k}'}\right|^2 \cdot \delta\left(\varepsilon_{j'\mathbf{k}'} - (\varepsilon_{j\mathbf{k}} \pm \hbar\omega_{\mathbf{q}\lambda})\right). \quad (1)$$

$\pm\hbar\omega_{\mathbf{q}\lambda}$ is the energy of the involved (quasi)particle ($\mathbf{q}$: wave vector, λ: polarization) which may be, e.g., photons, phonons, magnons, plasmons etc., with frequency $\omega_{\mathbf{q}\lambda}$ for absorption (plus sign) or emission (minus sign), and $M^{\lambda}_{j\mathbf{k},j'\mathbf{k}'}$ is the scattering matrix element

$$M^{\lambda}_{j\mathbf{k},j'\mathbf{k}'} = \langle F'\Psi_{j'\mathbf{k}'}|W_{\mathbf{q}\lambda}|F\Psi_{j\mathbf{k}}\rangle, \quad (2)$$

where $|F\rangle$ and $|F'\rangle$ are the initial and final (quasi)particle states and $W_{\mathbf{q}\lambda}$ is the scattering operator. Thereby, momentum conservation $\mathbf{k} \pm \mathbf{q} = \mathbf{k}' + \mathbf{G}$ is demanded ($\mathbf{G}$: reciprocal lattice vector). Fermi's golden rule is the first-order approximation of the time-dependent quantum-mechanical perturbation theory. It implies that the scattering processes are Markovian which means that a scattering process does not depend on preceding scattering processes. Fermi's golden rule is only valid in a time window where the perturbation time on the one hand must be short enough because of the first-order approximation and on the other hand must be long enough to replace the $\sin(x)/x$-function appearing in the derivation of Fermi's golden rule by the Dirac delta distribution. The validity of Fermi's golden rule for a magnetization dynamics on the 100 fs timescale is critically discussed in Ref. [4].

Normally, one is not interested in a specific transition rate $W^{\lambda}_{j\mathbf{k},j'\mathbf{k}'}$ from an initial state $\Psi_{j\mathbf{k}}$ to a final state $\Psi_{j'\mathbf{k}'}$ but in the total transition rate W_{total} from all initial states to all final states. Thereby $\mathbf{k}$ and $\mathbf{k}'$ are related via $\mathbf{k} \pm \mathbf{q} = \mathbf{k}' + \mathbf{G}$ if the scattering is at a quasiparticle with wave vector $\mathbf{q}$. This is calculated with Boltzmann rate equations [4, 9]

$$W_{\text{total}} = W_{\text{in}} - W_{\text{out}} \quad (3)$$

where

$$W_{\text{in}} = \frac{1}{\Omega^2_{\text{BZ}}} \sum_{j,j',\lambda} \int_{\text{BZ}} \mathrm{d}^3k \int_{\text{BZ}} \mathrm{d}^3k'\; n_{j'\mathbf{k}'}\left[1 - n_{j\mathbf{k}}\right] W^{\lambda}_{j'\mathbf{k}',j\mathbf{k}} \quad (4)$$

$$W_{\text{out}} = \frac{1}{\Omega^2_{\text{BZ}}} \sum_{j,j',\lambda} \int_{\text{BZ}} \mathrm{d}^3k \int_{\text{BZ}} \mathrm{d}^3k'\; n_{j\mathbf{k}}\left[1 - n_{j'\mathbf{k}'}\right] W^{\lambda}_{j\mathbf{k},j'\mathbf{k}'}. \quad (5)$$

Ω_{BZ} is the Brillouin zone (BZ) volume and n is the distribution function for the electrons.

Often one is also interested in the rate of change of the distribution function $n_{j\mathbf{k}}$ due to scattering which is also calculated with Boltzmann rate equations [10]

$$\frac{\mathrm{d}n_{j\mathbf{k}}}{\mathrm{d}t} = \frac{1}{\Omega_{\text{BZ}}} \sum_{j',\lambda} \int_{\text{BZ}} \mathrm{d}^3k' \left\{ n_{j'\mathbf{k}'}\left[1 - n_{j\mathbf{k}}\right] W^{\lambda}_{j'\mathbf{k}',j\mathbf{k}} - n_{j\mathbf{k}}\left[1 - n_{j'\mathbf{k}'}\right] W^{\lambda}_{j\mathbf{k},j'\mathbf{k}'} \right\}. \quad (6)$$

So we have to calculate Brillouin zone integrals of the form

$$\int_{\text{BZ}} \mathrm{d}^3k\; g(\mathbf{k})\; \delta(\varepsilon(\mathbf{k})). \quad (7)$$

Because the quantities $\varepsilon_{j\mathbf{k}}$, $\varepsilon_{j'\mathbf{k}'}$, $W^{\lambda}_{j'\mathbf{k}',j\mathbf{k}}$, $W^{\lambda}_{j\mathbf{k},j'\mathbf{k}'}$ can be calculated numerically only for a finite number of $\mathbf{k}$-points, finite $\mathbf{k}$-point grids have to be used for the numerical calculation of the total transition rate W_{total} or of the rate of change of the distribution function $\mathrm{d}n_{j\mathbf{k}}/\mathrm{d}t$. Thereby, energy conservation $\varepsilon_{j'\mathbf{k}'} = \varepsilon_{j\mathbf{k}} \pm \hbar\omega_{\mathbf{q}\lambda}$ and momentum conservation $\mathbf{k} \pm \mathbf{q} = \mathbf{k}' + \mathbf{G}$ have to be fulfilled; however, energy conservation in combination with momentum conservation is in general never fulfilled for a finite $\mathbf{k}$-point grid. Therefore, the Dirac delta distribution has to be replaced by a "smeared" delta function to obtain a result which approximates the integral (which is done, e.g., in Refs. [3, 4, 11–13] and in very many other papers). To do this, often the following equation is used

$$\int_{\text{BZ}} \mathrm{d}^3k\; g(\mathbf{k})\; \delta(\varepsilon(\mathbf{k})) \approx \int_{\text{BZ}} \mathrm{d}^3k\; g(\mathbf{k})\; \frac{1}{\sqrt{\pi}\sigma}\exp\left(-\frac{\varepsilon^2(\mathbf{k})}{\sigma^2}\right) \quad (8)$$

and the smearing parameter σ has to be chosen appropriately, see Sect. 3. This means that the contribution of a certain grid point to the total transition rate W_{total} or to the rate of change of the distribution function $\mathrm{d}n_{j\mathbf{k}}/\mathrm{d}t$ is small if the energy conservation is fulfilled very badly, and vice versa the contribution is large if the energy conservation is fulfilled very well. However, from a mathematical point of view it is not obvious that Eq. (8) holds since the Dirac delta distribution is not a function and the smearing is with respect to the energy ε but the integration is with respect to the wave vector **k**. The problem is explained in more detail in Sect. 2.

Mathematical proofs of Eq. (8) under certain preconditions can be found in Ref. [14], Theorem 7.2.1, and in Ref. [15], Theorem 6.1.5; however, the proofs are for general distributions and are very abstract. We want to show in this article that Eq. (8) is correct using also physical arguments.

In Sect. 2 we explain in detail the problem which arises when in a Brillouin zone integration Dirac's delta distribution is approximated by a Gaussian. This is done in many papers without giving any justification. We, therefore, think that the outline of this problem is a novelty per se. Then we give a justification of the Gaussian smearing method by mathematical and physical arguments. Each of these arguments has been used in other contexts in previous papers. The novelty of our paper is that we use these arguments to justify the Gaussian smearing method for the Brillouin zone integration. First, we consider a coordinate transformation from the wave vector variables $\mathbf{k} = (k_x, k_y, k_z)$ to the variables $\epsilon, \vartheta, \varphi$ where ϵ is the energy and ϑ, φ are variables for the surface of constant energy. This transformation involves the Jacobian $J(k_x, k_y, k_z)$. The inverse function theorem [14] says that if this Jacobian is non-zero at a **k**-point, this transformation is invertible. Then the integration in **k**-space including $\delta(\epsilon)$ can be represented by an integration over $\epsilon, \vartheta, \varphi$ of a function which involves the Jacobian $\tilde{J}(\epsilon, \vartheta, \varphi) = [J(k_x, k_y, k_z)]^{-1}$ and, which now can without any problem be replaced by a Gaussian. The problem is that there are special **k**-points where $\nabla_{\mathbf{k}}\epsilon(\mathbf{k}) = 0$. For these special points the Jacobian $J(k_x, k_y, k_z)$ is zero, and the reverse transformation involving $\tilde{J}(\epsilon, \vartheta, \varphi)$ is not defined in a rigorous mathematical interpretation. According to a general theorem of M. Morse the dispersion relation $\epsilon(\mathbf{k})$ exhibits such special points because it is periodic in all components. There are special points which can be identified easily, e.g., the Γ-point and points on the Brillouin zone boundary. These points can be avoided by shifting the grid of **k**-points considered in the Brillouin zone integration accordingly [16]. Other special points cannot be easily found, and they might be in the shifted k-point grid. Van Hove has shown [17] that for three dimensions the appearance of these special points does not appreciably modify the result of a numerical integration. We motivate these steps by physical reasoning.

In Sect. 3 we give practical hints for the appropriate choice of the smearing parameter σ. Finally, our results are summarized in Sect. 4.

Numerical integration of the Dirac delta distribution

It is very well known that in integrals involving the Dirac delta distribution, the distribution can be replaced by a Gaussian for the limes $\sigma \to 0$. It reads

$$\int_{-\infty}^{+\infty} \mathrm{d}\varepsilon\; g(\varepsilon)\; \delta(\varepsilon) = \lim_{\sigma\to 0} \int_{-\infty}^{+\infty} \mathrm{d}\varepsilon\; g(\varepsilon)\; \frac{1}{\sqrt{\pi}\sigma} \exp\left(-\frac{\varepsilon^2}{\sigma^2}\right) \tag{9}$$

where $g(\varepsilon)$ is a continuously differentiable function which depends on the energy ε. The Dirac delta distribution is approximated by a Gaussian

$$\int_{-\infty}^{+\infty} \mathrm{d}\varepsilon\; g(\varepsilon)\; \delta(\varepsilon) \approx \int_{-\infty}^{+\infty} \mathrm{d}\varepsilon\; g(\varepsilon)\; \frac{1}{\sqrt{\pi}\sigma} \exp\left(-\frac{\varepsilon^2}{\sigma^2}\right) \tag{10}$$

for a numerical calculation of the integral and σ has to be chosen appropriately, see Sect. 3.

However, the integrals in Eqs. (3)–(6) are not over the energy ε but over the wave vector **k**. For the sake of simplicity we discuss the following integral

$$\int_{\mathrm{BZ}} \mathrm{d}^3k\; g(\mathbf{k})\; \delta(\varepsilon(\mathbf{k})) \tag{11}$$

where $g(\mathbf{k})$ is a continuously differentiable function of the wave vector **k**, and the generalization to the expression $\delta(\varepsilon_{j'\mathbf{k}'} - (\varepsilon_{j\mathbf{k}} \pm \hbar\omega_{\mathbf{q}\lambda}))$ used in Eqs. (3)–(6) is straightforward. In explicit numerical calculations it is always assumed that also the relation

$$\int_{\mathrm{BZ}} \mathrm{d}^3k\; g(\mathbf{k})\; \delta(\varepsilon(\mathbf{k})) = \lim_{\sigma\to 0} \int_{\mathrm{BZ}} \mathrm{d}^3k\; g(\mathbf{k})\; \frac{1}{\sqrt{\pi}\sigma} \exp\left(-\frac{\varepsilon^2(\mathbf{k})}{\sigma^2}\right) \tag{12}$$

holds without giving any justification, reference or comment and that this may be approximated by

$$\int_{\mathrm{BZ}} \mathrm{d}^3k\; g(\mathbf{k})\; \delta(\varepsilon(\mathbf{k})) \approx \int_{\mathrm{BZ}} \mathrm{d}^3k\; g(\mathbf{k})\; \frac{1}{\sqrt{\pi}\sigma} \exp\left(-\frac{\varepsilon^2(\mathbf{k})}{\sigma^2}\right). \tag{13}$$

However, the Dirac delta distribution is defined by Eq. (9) and not by Eq. (12). We show in the following how the use of Eq. (13) can be justified.

We consider a coordinate transformation from the wave vector variables $\mathbf{k} = (k_x, k_y, k_z)$ to the variables $\varepsilon, \vartheta, \varphi$ (ε: energy; ϑ, φ: variables for the surface of constant energy)

$$\begin{aligned} \varepsilon &= \varepsilon(k_x, k_y, k_z) \\ \vartheta &= \vartheta(k_x, k_y, k_z) \\ \varphi &= \varphi(k_x, k_y, k_z) \end{aligned} \tag{14}$$

where the energy dispersion relation $\varepsilon(k_x, k_y, k_z)$ is known and the surfaces of constant energy can be parametrized with two variables ϑ and φ. The inverse function theorem says [18] that every continuously differentiable, vector-valued function which maps values from an open set of $\mathbb{R}^n$ to other values of an open set of $\mathbb{R}^n$ [so-called coordinate transformation, e.g., Eq. (14)] and whose Jacobian determinant

$$J(k_x, k_y, k_z) = \det\begin{pmatrix} \frac{\partial \varepsilon}{\partial k_x} & \frac{\partial \varepsilon}{\partial k_y} & \frac{\partial \varepsilon}{\partial k_z} \\ \frac{\partial \vartheta}{\partial k_x} & \frac{\partial \vartheta}{\partial k_y} & \frac{\partial \vartheta}{\partial k_z} \\ \frac{\partial \varphi}{\partial k_x} & \frac{\partial \varphi}{\partial k_y} & \frac{\partial \varphi}{\partial k_z} \end{pmatrix} \tag{15}$$

is non-zero at a point is invertible in the neighborhood of this point, i.e., the reverse transformation of Eq. (14)

$$\begin{aligned} k_x &= k_x(\varepsilon, \vartheta, \varphi) \\ k_y &= k_y(\varepsilon, \vartheta, \varphi) \\ k_z &= k_z(\varepsilon, \vartheta, \varphi) \end{aligned} \tag{16}$$

exists and can in principle be given in the neighborhood of every point (k_x, k_y, k_z) if the above-mentioned conditions are fulfilled.

If Eq. (14) is invertible in the neighborhood of every point $\mathbf{k} = (k_x, k_y, k_z)$—whereby only points $\mathbf{k}$ with $\varepsilon(\mathbf{k}) = 0$ are relevant because of the Dirac delta distribution in Eq. (11)—it is possible to make for this neighborhood a local coordinate transformation (using Eq. (16)) for the function $g(\mathbf{k}) = \tilde{g}(\varepsilon, \vartheta, \varphi)$ appearing in Eq. (11). Then, the integral over the wave vector $\mathbf{k}$ can be replaced by the integral over the variables $\varepsilon, \vartheta, \varphi$

$$\begin{aligned} &\int_{\mathrm{BZ}} \mathrm{d}^3k\; g(\mathbf{k})\; \delta(\varepsilon(\mathbf{k})) \\ &= \int \mathrm{d}\varepsilon \int \mathrm{d}\vartheta \int \mathrm{d}\varphi\; |\tilde{J}(\varepsilon, \vartheta, \varphi)| \cdot g\big(k_x(\varepsilon, \vartheta, \varphi), \\ &\qquad\qquad k_y(\varepsilon, \vartheta, \varphi), k_z(\varepsilon, \vartheta, \varphi)\big) \cdot \delta(\varepsilon) \\ &= \int \mathrm{d}\varepsilon \int \mathrm{d}\vartheta \int \mathrm{d}\varphi\; |\tilde{J}(\varepsilon, \vartheta, \varphi)| \cdot \tilde{g}(\varepsilon, \vartheta, \varphi) \cdot \delta(\varepsilon) \end{aligned} \tag{17}$$

where $\tilde{J}(\varepsilon, \vartheta, \varphi)$ is the Jacobian determinant of the reverse transformation (16)

$$\tilde{J}(\varepsilon, \vartheta, \varphi) = \det\begin{pmatrix} \frac{\partial k_x}{\partial \varepsilon} & \frac{\partial k_x}{\partial \vartheta} & \frac{\partial k_x}{\partial \varphi} \\ \frac{\partial k_y}{\partial \varepsilon} & \frac{\partial k_y}{\partial \vartheta} & \frac{\partial k_y}{\partial \varphi} \\ \frac{\partial k_z}{\partial \varepsilon} & \frac{\partial k_z}{\partial \vartheta} & \frac{\partial k_z}{\partial \varphi} \end{pmatrix}. \tag{18}$$

Note that $\tilde{J}(\varepsilon, \vartheta, \varphi) = J^{-1}(k_x, k_y, k_z)$ with $(\varepsilon, \vartheta, \varphi)$ expressed by Eq. (14). It is now definitely allowed to approximate the Dirac delta distribution by a Gaussian in analogy to Eqs. (9) and (10)

$$\begin{aligned} &\int_{\mathrm{BZ}} \mathrm{d}^3k\; g(\mathbf{k})\; \delta(\varepsilon(\mathbf{k})) \\ &\approx \int \mathrm{d}\varepsilon \int \mathrm{d}\vartheta \int \mathrm{d}\varphi\; |\tilde{J}(\varepsilon, \vartheta, \varphi)| \cdot \tilde{g}(\varepsilon, \vartheta, \varphi) \cdot \frac{1}{\sqrt{\pi}\sigma} \exp\left(-\frac{\varepsilon^2}{\sigma^2}\right) \\ &= \int \mathrm{d}^3k\; g(\mathbf{k}) \frac{1}{\sqrt{\pi}\sigma} \exp\left(-\frac{\varepsilon^2(\mathbf{k})}{\sigma^2}\right) \end{aligned} \tag{19}$$

where in the last step the integration variables are changed back again to an integration over the wave vector using Eq. (14). This is exactly what we wanted to show in Eq. (13).

One must keep in mind that the Jacobian determinant $J(k_x, k_y, k_z)$ given by Eq. (15) is zero for special points $\mathbf{k} = (k_x, k_y, k_z)$, where $\nabla_{\mathbf{k}}\varepsilon(\mathbf{k}) = 0$. This is the case for the Γ-point and usually for points on the Brillouin zone boundary [19], and even the transformation (14) could be not continuously differentiable for special points. Then, the reverse transformation (16) used in Eq. (19) is not defined anymore in a rigorous mathematical interpretation. However, these problems arise because of two idealizations, the long-time idealization and the infinite-solid idealization, and the following remarks have to be considered:

1. In a physical interpretation the Dirac delta distribution appearing in Fermi's golden rule, Eq. (1), is only a long-time idealization which should be replaced by a $\sin(x)/x$-function for realistic physical calculations. However, this would yield time-dependent rates which is usually not desired.
2. For a numerical calculation an infinite periodicity of the lattice is assumed (infinite-solid idealization). **k**-points of this numerical calculation only sometimes coincide exactly with a point where the reverse transformation (16) is not defined and the **k**-point grid can always be shifted so that there is no point where the reverse transformation is not defined. For an arbitrary $g(\mathbf{k})$ it is not clear that one gets a correct result when omitting these **k**-points. In a real solid in the ground state only a finite number of energy levels are occupied. These energy levels do not correspond to states with defined wave vectors **k**. In the numerical treatment of these finite systems the energy levels are approximated by the energies of a lattice with infinite periodicity at a number of discrete points on a **k**-point grid. For sufficiently large systems the result must be independent of the detailed choice of the **k**-point grid. If we know the critical points, we can shift the **k**-point grid in such a way that it does not include the above-

defined critical points (see also Ref. [16]). One could argue that a **k**-point very close to a critical **k**-point could yield an extremely large contribution because the Jacobian determinant $\widetilde{J}(\varepsilon, \vartheta, \varphi)$ (18) appearing in Eq. (17) may be very large for this point. This means that the second line of Eq. (19) would not be a good approximation for a choice of the **k**-point grid which contains points very close to critical points, but the near equality of the first line with the third line of Eq. (19) still holds because for the transition between the first and the third line the integration variables have been changed back again and this corresponds to the multiplication with $\widetilde{J}^{-1}(\varepsilon, \vartheta, \varphi) = J(k_x, k_y, k_z)$, so altogether, a possibly large value of $\widetilde{J}(\varepsilon, \vartheta, \varphi)$ does not matter. However, it is extremely complicated to identify all critical points and, therefore, it is not clear whether a chosen **k**-point grid contains critical points or not. To avoid these cumbersome investigations one can also do the following: one performs calculations for denser and denser grids and/or for shifted and rotated grids and compares the results. If the results are very similar, this means that the grids either do not contain critical points or that the critical points do not make a big contribution so that they do not falsify the results, in agreement with Ref. [17].

Practical hints for the appropriate choice of the smearing parameter

The appropriate choice of the smearing parameter σ appearing in Eq. 10 is crucial for the correct numerical calculation of the Boltzmann rate equation. The appropriate choice of σ depends on two quantities:

1. First, the smearing parameter σ depends on the energy scale of the involved (quasi)particle which may be, e.g., a photon, phonon, magnon, plasmon (see Sect. 1). The energy scale for a phonon is in the order of some mRy (about 40 meV) and for a magnon the energy scale is a factor 10 larger than for a phonon. The energy scale for a plasmon is much larger, in the order of 700 mRy (about 10 eV). An appropriate choice of the smearing parameter is in the same order as the energy scale of the involved (quasi)particle.
2. Second, the smearing parameter σ depends on the grid spacing. A typical ansatz is $\sigma = p/N_1$ where N_1 is the number of **k**-points in one direction and the total number of **k**-points is N_1^3. For a fixed proportionality constant p it is guaranteed that the smearing parameter is smaller, the larger the N_1.

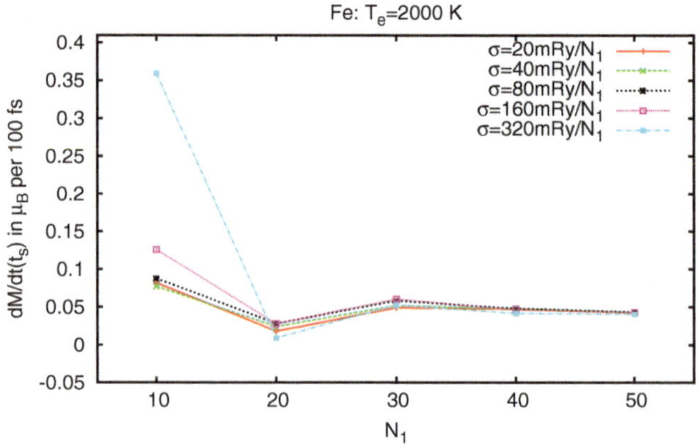

Fig. 1 Rate of the magnetic moment change per atom $\mathrm{d}M/\mathrm{d}t(t_s)$ vs. number of k-points in one direction N_1 for different smearing parameters σ. For iron and an excitation temperature of $T_e = 2000$ K

In the following we discuss the choice of σ for the case of electron–phonon scatterings. In Ref. [3] the smearing parameter is fixed to 15 meV (about 1 mRy) for the numerical calculation of the electron–phonon Boltzmann rate equation. In Ref. [4] we tested our numerical results of the electron–phonon Boltzmann rate equation for many different grids and smearing parameters. In this publication we considered the case of ultrafast demagnetization, see Sect. 1. Among other quantities we calculated the rate of the magnetic moment change per atom $\mathrm{d}M/\mathrm{d}t(t_s)$ for a time t_s (see Eq. (14) of Ref. [4]). t_s is the time after the laser pulse irradiation where the electron system has thermalized, i.e., the electron distribution can be described by a Fermi–Dirac distribution with the electron temperature T_e. Figure 1 shows the rate of the magnetic moment change per atom $\mathrm{d}M/\mathrm{d}t(t_s)$ for Fe and an electron temperature of $T_e = 2000$ K. We calculated $\mathrm{d}M/\mathrm{d}t(t_s)$ for different grids (number of k-points in one direction N_1) and for different smearing parameters σ. One can see that the results for $\mathrm{d}M/\mathrm{d}t(t_s)$ depend hardly on the chosen grid and on the chosen smearing parameter except for $N_1 = 10$. Therefore, the above-discussed critical points do not falsify our results and the smearing parameter is in the right order of magnitude (about several mRy). Of course it is trivial that increasing the number of **k**-points increases the convergence. By Fig. 1 we just want to show that the results depend only very slightly on the specific choice of σ if the number of **k**-points is above a certain value.

Conclusions

Scattering processes of electrons at periodic perturbations are very important in solid-state physics and also a correct numerical treatment is crucial for the quantitative analysis of scattering processes in many research activities, e.g., spintronics. In quantum mechanics, Fermi's golden rule is used which contains the Dirac delta distribution. The Dirac delta distribution is usually replaced by a Gaussian to integrate numerically the Boltzmann rate equations on a finite grid of **k**-points. It is not obvious from the very beginning that this numerical treatment is correct since the Dirac delta distribution is not a function and the smearing variable differs from the integration variable. We have shown in the present article that this procedure is in general correct. There are special **k**-points for which it is in principle not justified to replace the Dirac delta distribution by a Gaussian; however, we have given mathematical and physical arguments why this procedure is nevertheless a good approximation for the integration of the Boltzmann rate equation and should not falsify the results, at least for three dimensions. It is not clear whether the same holds also for $d = 2$ or even for $d = 1$. In conclusion, the naive replacement of the Dirac delta distribution by a Gaussian gives in general correct results for the Boltzmann rate equation but this has to be checked for denser and/or for shifted and rotated grids to avoid wrong contributions from the above-described special **k**-points where the replacement is critical.

References

1. Kirilyuk, A., Kimel, A.V., Rasing, T.: Ultrafast optical manipulation of magnetic order. Rev. Mod. Phys. **82**, 2731 (2010). doi:10.1103/RevModPhys.82.2731
2. Koopmans, B., Malinowski, G., Longa, F.D., Steiauf, D., Fähnle, M., Roth, T., Cinchetti, M., Aeschlimann, M.: Nat. Mater. **9**, 259 (2010). doi:10.1038/nmat2593
3. Essert, S., Schneider, H.C.: Electron–phonon scattering dynamics in ferromagnetic metals and their influence on ultrafast demagnetization processes. Phys. Rev. B **84**, 224405 (2011). doi:10.1103/PhysRevB.84.224405
4. Illg, C., Haag, M., Fähnle, M.: Ultrafast demagnetization after laser irradiation in transition metals: Ab initio calculations of the spin-flip electron–phonon scattering with reduced exchange splitting. Phys. Rev. B **88**, 214404 (2013). doi:10.1103/PhysRevB.88.214404
5. Haag, M., Illg, C., Fähnle, M.: Role of electron-magnon scatterings in ultrafast demagnetization. Phys. Rev. B **90**, 014417 (2014). doi:10.1103/PhysRevB.90.014417
6. Mueller, B., Baral, A., Vollmar, S., Cinchetti, M., Aeschlimann, M., Schneider, H., Rethfeld, B.: Feedback effect during ultrafast demagnetization dynamics in ferromagnets. Phys. Rev. Lett. **111**, 167204 (2013). doi:10.1103/PhysRevLett.111.167204
7. Žuti ć, I., Fabian, J., Das Sarma, S.: Spintronics: fundamentals and applications. Rev. Mod. Phys. **76**, 323–410 (2004). doi:10.1103/RevModPhys.76.323
8. Fähnle, M., Illg, C.: Electron theory of fast and ultrafast dissipative magnetization dynamics. J. Phys.: Condens. Matter **23**(49), 493201 (2011). doi:10.1088/0953-8984/23/49/493201
9. Yafet, Y.: g factors and spin-lattice relaxation of conduction electrons. In: Seitz, F., Turnbull, D. (eds.) Solid State Physics, vol. 14. Academic, New York (1963)
10. Ashcroft, N.W., Mermin, N.D.: Solid State Physics, 1st edn. Brooks Cole, Oxford (1976)
11. AlShaikhi, A., Srivastava, G.P.: Theoretical investigations of phonon intrinsic mean-free path in zinc-blende and wurtzite AlN. Phys. Rev. B **76**, 195205 (2007). doi:10.1103/PhysRevB.76.195205
12. Franchini, A., Santoro, G., Bortolani, V., Maradudin, A.A., Wallis, R.F.: Theory of the anharmonic linewidths of surface phonons in aluminum. Phys. Rev. B **45**, 11982–11988 (1992). doi:10.1103/PhysRevB.45.11982
13. Paulatto, L., Mauri, F., Lazzeri, M.: Anharmonic properties from a generalized third-order abinitio approach: theory and applications to graphite and graphene. Phys. Rev. B **87**, 214303 (2013). doi:10.1103/PhysRevB.87.214303
14. Friedlander, G., Joshi, M.: Introduction to the Theory of Distributions, 2nd edn. Cambridge University Press, Cambridge (1998)
15. Hörmander, L.: The Analysis of Linear Partial Differential Operators I, 2nd edn. Springer, Berlin (1998)
16. Martin, R.M.: Electronic Structure: Basic Theory and Practical Methods, vol. 1. Cambridge University Press, Cambridge (2008)
17. Van Hove, L.: The occurrence of singularities in the elastic frequency distribution of a crystal. Phys. Rev. **89**, 1189–1193 (1953). doi:10.1103/PhysRev.89.1189
18. Spivak, M.: Calculus on Manifolds: a Modern Approach to Classical Theorems of Advanced Calculus, 5th edn. Westview Press, New York (1971)
19. Haug, A.: Theoretische Festkörperphysik, vol. 1, 1st edn. Franz Deuticke, Wien (1970)

5

Solutions of D-dimensional Schrodinger equation for Woods–Saxon potential with spin–orbit, coulomb and centrifugal terms through a new hybrid numerical fitting Nikiforov–Uvarov method

A. Niknam[1] · A. A. Rajabi[1] · M. Solaimani[2]

Abstract Solution of the radial Schrodinger equation for the Woods–Saxon potential together with spin–orbit interaction, coulomb and centrifugal terms by using usual Nikiforov–Uvarov (NU) method is not possible. Here, we have presented a new NU procedure with which we are able to solve this Schrodinger equation and any other one-dimensional ones with any shape of the potential profile. For this purpose, we have combined the NU method with numerical fitting schema. The energy eigenvalues and corresponding eigenfunctions for various values of n, l, and j quantum numbers have been obtained. Good agreement with experimental values is also achieved. We have calculated the $1/2^+$ state energy with more accuracy (our absolute error = 0.023 MeV and Hagen et al. absolute error = 0.0918 MeV), while Hagen et al. have calculated the $5/2^+$ state energy with higher accuracy (our absolute error = 0.71 MeV and Hagen et al. absolute error = 0.0337 MeV). Our wave functions are in agreement with Kim et al.'s work, too.

Keywords Nikiforov–Uvarov (NU) method · Eigenvalues and Eigenfunctions · Woods–Saxon potential · Spin–orbit interaction · Coulomb potential · Centrifugal term · Numerical fitting

✉ A. Niknam
niknamabbas@gmail.com

1 Physics Faculty, Shahrood University, P.O. Box 3619995161-316, Shahrood, Iran

2 Department of Physics, Faculty of Science, Qom University of Technology, Qom, Iran

Introduction

In the study of the breakup of ^{17}F into proton + ^{16}O, some potential model for ^{17}F has been used previously such as Woods–Saxon potential with spin–orbit and coulomb potentials [1] and M3Y interaction model [2]. Solution of the Schrodinger equation including the above potentials has been done by the numerical methods in the above-mentioned works. This is because analytical solution of these equations is not possible. However, some theoretical groups have tried to solve these Schrodinger equations analytically. For example, Pahlavani et al. have solved the Schrodinger equation including Woods–Saxon potential with spin–orbit and centrifugal terms by Nikiforov–Uvarov method [3]. They did not include the coulomb term to their calculations. Adding the coulomb potential (here for $r < R_C$; R_C = spherical nucleus radius) and solution of the Schrodinger equation by Nikiforov–Uvarov method is the main goal of the present work. In our previous works we have used the NU method to solve the Schrodinger equation with different potentials such as angle-dependent potential [4], Energy-dependent potential [5]; Dirac equation with NU method such as Hartmann potential [6]; Duffin–Kemmer–Petiau (DKP) equation with NU method such as Woods–Saxon potential [7], Hulthen vector potential [8] and Klein–Gordon equation with NU method such as energy-dependent potential [9]. Here, we have extended the Ref. [3] by adding a coulomb term to its potential. Adding this potential enhances the difficulty of the problem very much. Thus, we had to change the transformations and other formulas of the Ref. [3] to be able to solve the Schrodinger equation. Finally, energy eigenvalues and corresponding eigenfunctions obtained.

Parametric NU method

This powerful mathematical tool could be used to solve the second-order differential equations. Considering the following differential equation [10–12]

$$\psi_n''(s)+\frac{\tilde{\tau}(s)}{\sigma(s)}\psi_n'(s)+\frac{\tilde{\sigma}(s)}{\sigma(s)^2}\psi_n(s)=0, \quad (1)$$

where $\sigma(z)$ and $\tilde{\sigma}(z)$ are polynomials of second order at most, and $\tilde{\tau}(z)$ is a first-order polynomial. To make the application of the NU method simpler and the checking of the validity of solution unnecessary, we present a shortcut for the method. We begin the method by writing the general form of the Schrodinger-like Eq. (1) as

$$\psi_n''(s)+\left(\frac{c_1-c_2s}{s(1-c_3s)}\right)\psi_n'(s)+\left.\frac{-p_2s^2+p_1s-p_0}{s^2(1-c_3s)^2}\right)\psi_n(s)=0, \quad (2)$$

where the wave functions $\psi_n(s)$ satisfies

$$\psi_n(s)=\phi(s)y_n(s). \quad (3)$$

By comparing Eq. (3) with its counterpart Eq. (2), one can obtain

$$\begin{aligned}\tilde{\tau}(s)&=c_1-c_2s, \quad \sigma(s)=s(1-c_3s),\\ \tilde{\sigma}(s)&=-p_2s^2+p_1s-p_0.\end{aligned} \quad (4)$$

According to the NU method [10], one can obtain the bound state energy equation as [11, 12]

$$\begin{aligned}&c_2n-(2n+1)c_5+(2n+1)(\sqrt{c_9}+c_3\sqrt{c_8})\\ &+n(n-1)c_3+c_7+2c_3c_8+2\sqrt{c_8c_9}=0,\end{aligned} \quad (5)$$

where

$$\begin{aligned}c_4&=\frac{1}{2}(1-c_1),\\ c_5&=\frac{1}{2}(c_2-2c_3),\\ c_6&=c_5^2+p_2,\\ c_7&=2c_4c_5-p_1\\ c_8&=c_4^2+p_0,\\ c_9&=c_3(c_7+c_3c_8)+c_6.\end{aligned} \quad (6)$$

In addition, we also find that:

$$\rho(s)=s^{c_{10}}(1-c_3s)^{c_{11}} \quad (7)$$

$$\varphi(s)=s^{c_{12}}(1-c_3s)^{c_{13}} \quad (8)$$

$$c_{12}>0, c_{13}>0$$

$$y_n(s)=P_n^{(c_{10},c_{11})}(1-2c_3s), c_{10}>-1, c_{11}>-1 \quad (9)$$

are necessary in calculating the wave functions

$$\psi_{nl}(s)=N_{nl}s^{c_{12}}(1-c_3s)^{c_{13}}P_n^{(c_{10},c_{11})}(1-2c_3s), \quad (10)$$

where $P_n^{(\mu,\nu)}(x), \mu>-1, \nu>-1, x\in[-1,1]$ are Jacobi polynomials. All undefined constant parameters are as follows [13]:

$$\begin{aligned}c_{10}&=c_1+2c_4+2\sqrt{c_8}\\ c_{11}&=1-c_1-2c_4+\frac{2}{c_2}\sqrt{c_9}\\ c_{12}&=c_4+\sqrt{c_8}\\ c_{13}&=-c_4+\frac{1}{c_3}(\sqrt{c_9}+c_5),\end{aligned} \quad (11)$$

where

$$c_3\neq 0,\ c_{13}>0,\ c_{12}>0,\ s\in[1,1/c_{13}] \text{ and } \quad c_{13}\neq 0. \quad (12)$$

Solutions of Schrodinger equation

We treat ^{17}F as the combination of a proton and an inert ^{16}O core with spin 0 [1]. Schrodinger equation for the N-particles which interact with each other can be written as [17, 18]:

$$\begin{aligned}&\frac{\hbar^2}{2\mu}\left[\frac{d^2}{dr^2}+\frac{D-1}{r}\frac{d}{dr}-\frac{l(l+D-2)}{r^2}\right]R(r)+V(r)R(r)\\ &=\text{E}R(r),\end{aligned} \quad (13)$$

where $D=3N-3$. Here, D, l, and N denote, respectively, the space dimension, total angular momentum and the number of particles, μ is one of particle masses, and x is the hyper-radius. If we write this equation for ^{17}F as the combination of a proton and an inert ^{16}O core with spin 0 (we have two-particle systems. Thus, $N=2$ and then we find $D=3$), we find the radial part of the Schrodinger equation as,

$$\frac{d^2R(r)}{dr^2}+\frac{2}{r}\frac{dR(r)}{dr}+\frac{2\mu}{\hbar^2}\left[E-V(r)-\frac{\hbar^2 l(l+1)}{2\mu r^2}\right]R(r)=0. \quad (14)$$

The potential profile contain the following terms.

1. Woods–Saxon:

$$V_{WS}=\frac{-V_0}{1+e^{\frac{(r-R_0)}{a}}} \quad (15)$$

We have used $V_0=42.71$ MeV for ^{17}F atom [3]. There, V_0 is defined as $40.5+0.13A$ MeV, where A is the mass number of nuclei (here 17 for ^{17}F). R_0 is radius of the spherical nucleus and $a=0.65$ fm is surface diffuseness.

Woods–Saxon potential has vast applications in description of both spherical and deformed nuclei in nuclear and particle physics. Analytic solution of Schrödinger equation with Woods–Saxon potential is very useful which provide for us invaluable theoretical results. DKP equation is also solved analytically by means of NU method a vector deformed Woods–Saxon potential [18]. However, scientist have solved the Woods–Saxon potential for S states ($l = 0$), exactly and also by some approximation they have also found the solutions for any l states. [19–22]

2. Spin–orbit term as,

$$V_{LS}(r) = V_{LS}^{(0)}\left(\frac{r_0}{\hbar}\right)^2 \frac{1}{r}\left[\frac{d}{dr}\frac{1}{1+e^{\frac{(r-R_0)}{a}}}\right]\left(\vec{L}.\vec{S}\right) \tag{16}$$

where we have used $V_{LS}^{(0)} = 0.44V_0$.

3. Coulomb term,

$$V_C(r) = \begin{cases} \dfrac{e^2}{R_C}\left[3-\left(\dfrac{r}{R_C}\right)^2\right] & r \le R_C \\ \dfrac{8e^2}{r} & r \le R_C \end{cases} \tag{17}$$

Here we have solved the Schrodinger equation for $r < R_C$. By using of the change of variable as $\psi(r) = rR(r)$ we have,

$$\frac{d\psi}{dr} = R(r) + r\frac{dR(r)}{dr}. \tag{18}$$

Then the Schrodinger equation becomes,

$$\frac{d^2\psi(r)}{dr^2} + \frac{2\mu E}{\hbar^2}\psi(r) - \frac{2\mu V(r)}{\hbar^2}\psi(r) - \frac{l(l+1)}{r^2}\psi(r) = 0. \tag{19}$$

Now, complete Schrodinger equation reads,

$$\begin{aligned} &\frac{d^2\psi(r)}{dr^2} + \frac{2\mu E}{\hbar^2}\psi(r) \\ &- \frac{2\mu}{\hbar^2}\left[\frac{-V_0}{1+e^{\frac{(r-R_0)}{a}}} + V_{LS}^{(0)}r_0^2\frac{1}{r}\ \frac{d}{dr}\frac{1}{1+e^{\frac{(r-R_0)}{a}}}\right) \\ &\times\left(j(j+1) - l(l+1) - \frac{3}{4}\right) + \frac{3e^2}{\pi\varepsilon_0 R_C} - \frac{e^2r^2}{\pi\varepsilon_0 R_C^3}\Bigg]\psi(r) \\ &- \frac{l(l+1)}{r^2}\psi(r) = 0. \end{aligned} \tag{20}$$

We need another change of variable as $s = e^{-\delta r}$, which leads,

$$\frac{d\psi(r)}{dr} = \frac{d\psi(r)}{ds} \times \frac{ds}{dr} = -\delta s\frac{d\psi(r)}{ds} \tag{21}$$

$$\frac{d^2\psi(r)}{dr^2} = \delta^2 s^2\frac{d^2R(s)}{ds^2} + \delta^2 s\frac{dR(s)}{ds}. \tag{22}$$

We define $\delta = \frac{1}{a}$, $q = e^{\delta R_0}$, $V_0' = V_0 q$ and $V_{LS}'^0 = V_{LS}^0 q$. We have used the following approximation too,

$$\frac{1}{r^2} \approx \delta^2\left(\frac{e^{-\delta r}}{1-e^{-\delta r}}\right)^2. \tag{23}$$

To find the best value of the δ we have changed the δ and plotted both side of the above equation several times and found the best δ to be 0.4 fm^{-1}. Figure 1 shows two sides of the above equation to obtain δ. By using the later change of variable the Schrodinger equation reads,

$$\begin{aligned} &\delta^2 s^2\frac{d^2R(s)}{ds^2} + \delta^2 s\frac{dR(s)}{ds} + \left(\frac{2\mu E}{\hbar^2} + \frac{2\mu V_0'}{\hbar^2}\left(\frac{s}{1+qs}\right)\right. \\ &+ \frac{\mu r_0^2 V_{LS}'^0\delta^2}{\hbar^2}\left(j(j+1) - l(l+1) - \frac{3}{4}\right)\frac{s^2}{(1-s)(1+qs)^2} \\ &\left. - \frac{6\mu e^2}{\pi\varepsilon_0\hbar^2 R_C} + \frac{2\mu e^2}{\pi\varepsilon_0\hbar^2 R_C^3\delta^2}\frac{(1-s)^2}{s^2} - l(l+1)\delta^2\frac{s^2}{(1-s)^2}\right) \\ &\times R(s) = 0. \end{aligned} \tag{24}$$

After some simplifications we have,

$$\begin{aligned} &\frac{d^2R(s)}{ds^2} + \frac{1}{s}\frac{dR(s)}{ds} + \left(\frac{\frac{2\mu E}{\hbar^2\delta^2}(1-s)^2 + \frac{2\mu V_0'}{\hbar^2\delta^2}\frac{s(1-s)^2}{1+qs}}{s^2(1-s)^2}\right. \\ &+ \frac{\frac{\mu r_0^2 V_{LS}'^0}{\hbar^2}\left(j(j+1) - l(l+1) - \frac{3}{4}\right)\frac{s^2(1-s)}{(1+qs)^2} - \frac{6\mu e^2}{\pi\varepsilon_0\hbar^2 R_C\delta^2}(1-s)^2}{s^2(1-s)^2} \\ &\left. + \frac{\frac{2\mu e^2}{\pi\varepsilon_0\hbar^2 R_C^3\delta^4}\frac{(1-s)^4}{s^2} - l(l+1)s^2}{s^2(1-s)^2}\right)R(s) = 0. \end{aligned} \tag{25}$$

Here, we present a new procedure with which we can convert the Schrodinger equation with any shape of the potential profile into the NU type equation. To use the NU method we convert the numerator of the potential term to a second-order polynomial. For this purpose we have used the following approximations which we have found through numerical fitting,

$$\begin{aligned} \frac{s(1-s)^2}{1+qs} &= 0.006595579s^2 - 0.0132710857s \\ &\quad + 0.0067413175. \end{aligned} \tag{26}$$

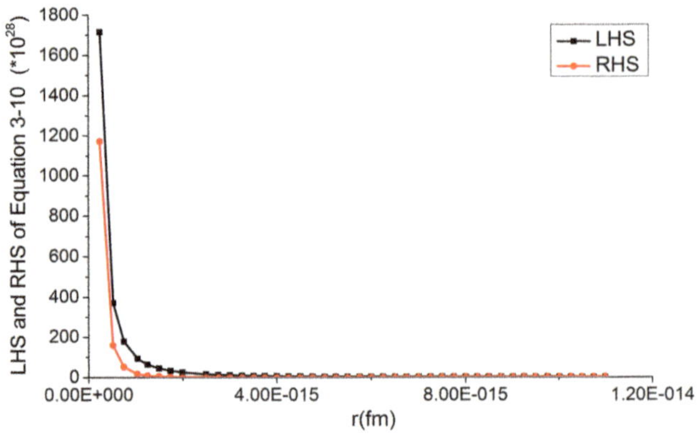

Fig. 1 Left hand side (LHS) and right hand side (RHS) of the Eq. (23) to obtain δ

Figure 2a shows the two sides of the equation above. As it is clear, this approximation is extremely good. We have also used from

$$\frac{s^2(1-s)}{(1+qs)^2} = -3.3579988 \times 10^{-9}s^2 - 0.432019240084 \times 10^{-4}s + 0.4341808999 \times 10^{-4}. \tag{27}$$

Figure 2b shows the two sides of the equation above. As it is clear, this approximation is extremely good. We have also used from

$$\frac{(1-s)^4}{s^2} = 0.981s^2 - 3.537s + 3.4. \tag{28}$$

Figure 2c shows the two sides of the equation above. As it is clear, this approximation is extremely good. To convert the last Schrodinger equation to the N-U type Schrodinger equation,

$$\psi_n''(s) + \frac{\tilde{\tau}(s)}{\sigma(s)}\psi_n'(s) + \frac{\tilde{\sigma}(s)}{\sigma(s)^2}\psi_n(s) = 0 \tag{29}$$

$$\psi_n''(s) + \left(\frac{1-s}{s(1-s)}\right)\psi_n'(s) + \left(\frac{-\gamma s^2 + \beta s - \varepsilon^2}{s^2(1-s)^2}\right)\psi_n(s) = 0. \tag{30}$$

We define three parameters ε, β, γ as follows:

$$\begin{aligned}-\gamma = {} & \frac{2\mu E}{\hbar^2\delta^2} + \frac{2\mu V_0'}{\hbar^2\delta^2}(0.006595579) \\ & + \frac{\mu V_{LS}'^0 r_0^2 \left(j(j+1) - l(l+1) - \frac{3}{4}\right)}{\hbar^2} \\ & \times (-3.3579988 \times 10^{-9}) \\ & - \frac{6\mu e^2}{\pi\varepsilon_0\hbar^2 R_C\delta^2} + \frac{2\mu e^2}{\pi\varepsilon_0\hbar^2 R_C^3\delta^4}(0.981) - l(l+1)\end{aligned} \tag{31}$$

$$\begin{aligned}\beta = {} & \frac{4\mu E}{\hbar^2\delta^2} + \frac{2\mu V_0'}{\hbar^2\delta^2}(0.0132710857) \\ & + \frac{\mu V_{LS}'^0 r_0^2 \left(j(j+1) - l(l+1) - \frac{3}{4}\right)}{\hbar^2} \\ & \times (0.432019240084 \times 10^{-4}) \\ & + \frac{12\mu e^2}{\pi\varepsilon_0\hbar^2 R_C\delta^2} - \frac{2\mu e^2}{\pi\varepsilon_0\hbar^2 R_C^3\delta^4}(3.537)\end{aligned} \tag{32}$$

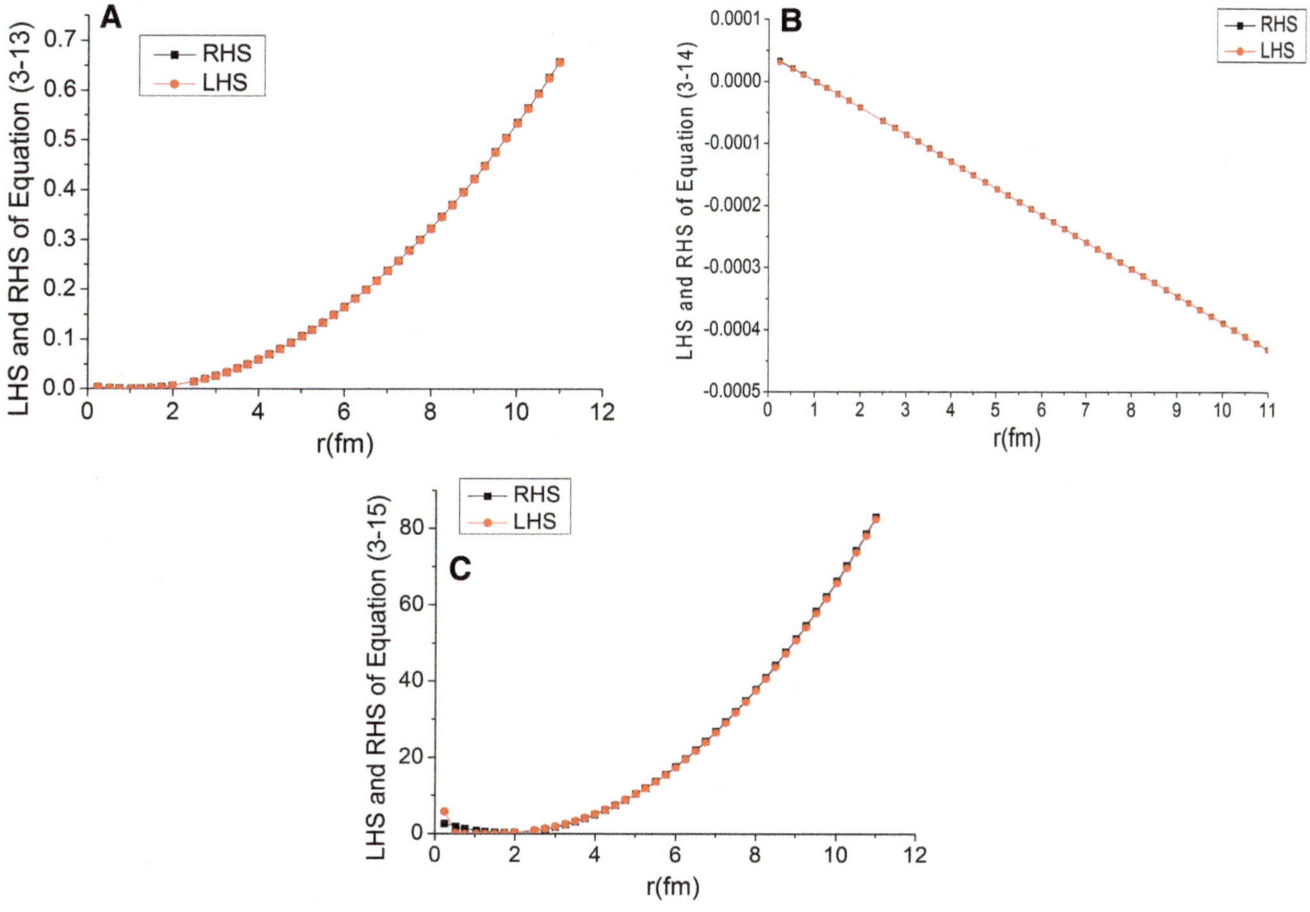

Fig. 2 **a** Left hand side (LHS, *filled circle* line) and right hand side (RHS, *filled square* line) of the Eq. (26) to show the accuracy of the approximation (27). **b** Same as **a** but for Eq. (27). **b** Same **a** but for Eq. (28)

$$-\varepsilon^2 = \frac{2\mu E}{\hbar^2\delta^2} + \frac{2\mu V_0'}{\hbar^2\delta^2}(0.0067413175) + \frac{\mu V_{LS}'^0 r_0^2 (j(j+1) - l(l+1) - \frac{3}{4})}{\hbar^2} \times (0.4341808999 \times 10^{-4}) - \frac{6\mu e^2}{\pi\varepsilon_0\hbar^2 R_C\delta^2} + \frac{6.8\mu e^2}{\pi\varepsilon_0\hbar^2 R_C^3\delta^4} \quad (33)$$

Now we write β, γ as function of ε,

$$\gamma = \varepsilon^2 + \frac{\mu V_0'}{\hbar^2\delta^2}(0.000291477) + \frac{\mu V_{LS}'^0 r_0^2 (j(j+1) - l(l+1) - \frac{3}{4})}{\hbar^2} (0.434214479888 \times 10^{-4}) + \frac{4.838\mu e^2}{\pi\varepsilon_0\hbar^2 R_C^3\delta^4} + l(l+1) \rightarrow \gamma = \varepsilon^2 + A \quad (34)$$

$$\beta = 2\varepsilon^2 + \frac{0.0004230986\mu V_0'}{\hbar^2\delta^2} + \frac{\mu V_{LS}'^0 r_0^2 (j(j+1) - l(l+1) - \frac{3}{4})}{\hbar^2} \times (0.436342559716 \times 10^{-4}) + \frac{6.526\mu e^2}{\pi\varepsilon_0\hbar^2 R_C^3\delta^4} \rightarrow \beta = 2\varepsilon^2 + B \quad (35)$$

Based on what we have described above in the N-U section, we can find the following second-order polynomial:

$$n - (2n+1)\left(-\frac{1}{2}\right) + (2n+1) \left(\sqrt{\gamma - \beta + \varepsilon^2 + \frac{1}{4}} + \varepsilon\right) + n(n+1) - \beta + 2\varepsilon^2 + 2\varepsilon\sqrt{\gamma - \beta + \varepsilon^2 + \frac{1}{4}} = 0. \quad (36)$$

After some simplifications we have,

$$\left[4\left(A - B + \frac{1}{4}\right) - (2n+1)^2\right]\varepsilon^2 + 2(2n+1)\left[2A - B - n^2 - n\right]\varepsilon + \left[(2n+1)^2\left(A - B + \frac{1}{4}\right) - \left(n^2 + n + \frac{1}{2} - B\right)^2\right] = 0 \quad (37)$$

$$h_2\varepsilon^2 + h_1\varepsilon + h_0 = 0. \quad (38)$$

By solving this equation and obtaining the ε we will wined the energies E as,

$$E_{nlj} = \frac{-\hbar^2\delta^2}{2\mu}\left(\varepsilon^2 + \frac{2\mu V_0'}{\hbar^2\delta^2}(0.0067413175) + \frac{\mu V_{LS}'^0 r_0^2 (j(j+1) - l(l+1) - \frac{3}{4})}{\hbar^2}(0.4341808999 \times 10^{-4}) - \frac{6\mu e^2}{\pi\varepsilon_0\hbar^2 R_C\delta^2} + \frac{6.8\mu e^2}{\pi\varepsilon_0\hbar^2 R_C^3\delta^4}\right) \quad (39)$$

Table 1 Comparison of the energies obtained in this work with ab initio computation [14] and experimental values [16]

^{17}F			
E_{SO}	$\frac{1}{2}^+$	$\frac{5}{2}^+$	
5.000	−0.105	−0.600	Exp. [16]
3.7	−0.082	0.11	GHF [14]
	0.023	0.71	GHF [14] absolute error
–	−0.1968	−0.937	Our work
	0.0918	0.337	Our absolute error

All energies stated in the MeV

Ground state energy of the ^{17}F can be obtained by using of $j = \frac{5}{2}$ $n = 0$, $l = 2$ and the first excited states can be calculated by using of the $j = \frac{1}{2} n = 1, l = 0$. Table 1 shows the results obtained in this work and Refs. [14, 16]. As it is clear from the Table 1, good accuracy in our work has been achieved. However, we have calculated the $\frac{1}{2}^+$ state energy with more accuracy (our absolute error = 0.023 MeV, Hagen et al. absolute error = 0.0918 MeV) while Hagen et al. have calculated the $\frac{5}{2}^+$ state energy with higher accuracy (our absolute error = 0.71 MeV, Hagen et al. absolute error = 0.0337 MeV). Now, the wave functions can be obtained through,

$$\psi(s) = y(s)\phi(s) \quad (40)$$

$$\rho(s) = s^{1+2\varepsilon}(1-s)^{2\sqrt{\gamma-\beta+\varepsilon^2+\frac{1}{4}}} \quad (41)$$

$$\phi(s) = s^{\varepsilon}(1-s)^{\left(\sqrt{\gamma-\beta+\varepsilon^2+\frac{1}{4}}-\frac{1}{2}\right)} \quad (42)$$

$$y_n(s) = P_n^{\left(1+2\varepsilon, 2\sqrt{\gamma-\beta+\varepsilon^2+\frac{1}{4}}\right)}(1-2s) \quad (43)$$

$$\psi_{nl}(s) = N_{nl}s^{\varepsilon}(1-s)^{\sqrt{\gamma-\beta+\varepsilon^2+\frac{1}{4}}+\frac{1}{2}} \times P_n^{\left(1+2\varepsilon, 2\sqrt{\gamma-\beta+\varepsilon^2+\frac{1}{4}}\right)}(1-2s), \quad (44)$$

where $P_n^{(\mu,\nu)}(x), \mu > -1, \nu > -1, x \in [-1, 1]$ are Jacobi polynomials. $5/2^+$ state and $1/2^+$ state wave functions are presented in the Fig. 3a, b, respectively. As we can see in these figures, good agreement exists. We have used the parameters of the [3], since they lead to the energies closer to the experimental values which we have presented in the Table 1.

Conclusion

In this study, the non-relativistic radial Schrodinger equation solved for Wood–Saxon potential together with coulomb potential (for $r < R_C$; R_C = spherical nucleus radius), spin–orbit interaction and centrifugal term through a new hybrid numerical fitting Nikiforov–Uvarov method. For this

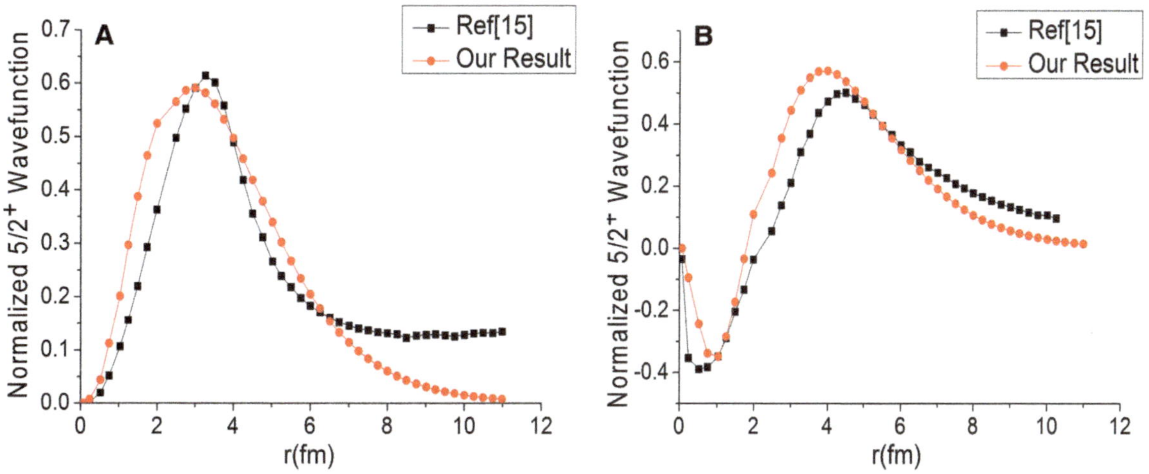

Fig. 3 **a** Normalized $5/2^+$ state wave function which we have calculated (*filled circle* line) compared with [15] (*filled square* line) as a function of the radius r (fm). **b** Normalized $1/2^+$ state wave function which we have calculated (*filled circle* line) compared with Ref. [15] (*filled square* line) as a function of the radius r (fm)

purpose, by using approximate expansion of $1/r^2$ different than that used by Pahlavani et al., Schrodinger equation has been transformed to the analytically solvable differential equation of NU method. Good agreement with experimental values is obtained. We have calculated the $1/2^+$ state energy with more accuracy (our relative error $= 0.023$ MeV and Hagen et al. relative error $= 0.0918$ MeV), while Hagen et al. have calculated the $5/2^+$ state energy with higher accuracy (our relative error $= 0.71$ MeV and Hagen et al. relative error $= 0.0337$ MeV). Our obtained wave functions are in agreement with Kim et al.'s too.

References

1. Esbensen, H., Bertsch, G.F.: Higher-order effects in the two-body breakup of 17F. Nucl. Phys. A **706**, 383 (2002)
2. Bertulani, C.A.: RADCAP: A potential model tool for direct capture reactions. Comp. Phys. Commun. **156**, 123 (2003)
3. Pahlavani, M.R., Alavi, S.A.: Solutions of woods—saxon potential with spin-orbit and centrifugal terms through Nikiforov—Uvarov method. Commun. Theor. Phys. **58**, 739 (2012)
4. Rajabi, A.A., Hamzavi, M.: A new Coulomb ring-shaped potential via generalized parametric Nikiforov-Uvarov method. J. Theor. Appl. Phys. **7**, 17 (2013)
5. Hassanabadi, H., Zarrinkamar, S., Rajabi, A.A.: Exact solutions of D-dimensional Schrödinger equation for an energy-dependent potential by NU method. Commun. Theor. Phys. **55**, 541 (2011)
6. Hamzavi, M., Hassanabadi, H., Rajabi, A.A.: Exact solutions of Dirac equation with Hartmann potential by Nikoforov-Uvarov method. Int. J. Mod. Phys. E **19**, 2189 (2010)
7. Oudi, R., Hassanabadi, S., Rajabi, A.A., Hasanabadi, H.: Approximate bound state solutions of DKP equation for any J state in the presence of woods—saxon potential. Commun. Theor. Phys. **57**, 15 (2012)
8. Zarrinkamar, S., Rajabi, A.A., Yazarloo, B.H., Hassanabadi, H.: An approximate solution of the DKP equation under the Hulthén vector potential. Chin. Phys. C **37**, 023101 (2013)
9. Hassanabadi, H., Zarrinkamar, S., Hamzavi, H., Rajabi, A.A.: Exact solutions of D-dimensional Klein–Gordon equation with an energy-dependent potential by using of Nikiforov–Uvarov method. Arab. J. Sci. Eng. **37**, 209 (2012)
10. Nikiforov, A.F., Uvarov, V.B.: Special functions of mathematical physics. Birkhausr, Berlin (1988)
11. Ikhdair, S.M.: Bound states of the Klein-Gordon equation for vector and scalar general Hulthen-type potentials in D-dimension. Int. J. Mod. Phys. C **20**, 25 (2009)
12. Tezcan, C., Sever, R.: A general approach for the exact solution of the Schrödinger equation. Int. J. Theor. Phys. **48**, 337 (2009)
13. Wang, Z.B., Zhang, M.C.: Spinless particles in the field of unequal scalar—vector Yukawa potentials. Acta Phys. Sin. **56**, 3688 (2007). **(in Chinese)**
14. Hagen, G., Papenbrock, T., Hjorth-Jensen, M.: Ab Initio Computation of the 17F Proton Halo State and Resonances in A=17 Nuclei. Phys. Rev. Lett. **104**, 182501 (2010)
15. Kim, K.-H.: Astrophysical S factors of the 16O(p, γ) 17F reaction at energies applicable in stellar cores. J. Korean Phys. Soc. **43**, 691 (2003)
16. Tilley, D.R., Weller, H.R., Cheves, C.M.: Energy levels of light nuclei A = 16–17. Nucl. Phys. A **564**, 1 (1993)
17. Rajabi, A.A.: Exact analytical solution of the Schrödinger equation for an N-identical body-force system. Few Body Syst. **37**, 197 (2005)
18. Hamzavi, M., Ikhdair, S.M.: Any J-state solution of the DKP equation for a vector deformed Woods-Saxon potential. Few-Body Syst. (2012). arXiv:1205.0938v2
19. Feizi, H., Rajabi, A.A., Shojaei, M.R.: Supersymmetric solution of the Schrödinger equation for Woods–Saxon potential by using the Pekeris approximation. Acta. Phys. Pol. B **42**, 2143 (2011)
20. Ikhdair, S.M., Sever, R.: Exact polynomial solution of non-symmetric and non-Hermitian modified Woods–Saxon potential by the Nikiforov–Uvarov method. Int. J. Theor. Phys. **46**, 1643 (2007)
21. Berkdemir, C., Berkdemir, A., Sever, R.: Polynomial solutions of the Schrödinger equation for the generalized Woods-Saxon potential. Phys. Rev. C **72**, 027001 (2005)
22. Berkdemir, C., Berkdemir, A., Sever, R.: Polynomial solutions of the Schrödinger equation for the generalized Woods-Saxon potential. Phys. Rev. C 74, 039902 (2006)

6

A short review about some exotic systems containing electrons, muons, and tauons

Mohsen Emami-Razavi

Abstract Some remarks about exotic systems consisting of various masses of fermions and antifermions have been presented. In paricular, few body systems containing electrons, muons and tauons, and the fundamental interest of the study of their bound states systems have been discussed.

Keywords Exotic Systems · Muons · Tauons · True Muonium · Bound states

Introduction

The purpose of the present paper is to present some remarks about the importance of the properties of exotic systems for arbitrary fermions and antifermions. We discuss also the bound states of the systems of two or more than two fermion–antifermion that include "muons" and "tauons" (systems like $\mu^+ \tau^-$, $e^- \tau^- \mu^+$, $e^- \mu^+ e^- \mu^+$). The investigation of the stability of some exotic systems, bound states energies and the properties of three-, four-, and five-body systems that contain muons and/or tauons is of fundamental interest in Quantum Electrodynamics (QED).

In a paper [1], it has been shown that the production and study of true muonium is possible at modern electron-positron colliders. The true muonium ($\mu^+ \mu^-$), true tauonium ($\tau^+ \tau^-$), and "mu-tauonium" ($\mu^\pm \tau^\mp$) bound states are not only the heaviest, but also the most compact QED systems. Hence, from fundamental point of view the study of the bound states systems that contain muons and tauons is of interest. The rapid weak decay of the τ makes the observation of the systems such as ($\tau^+ \tau^-$) or ($\mu^\pm \tau^\mp$) difficult.

A muon is a particle which has similar properties as the electron, except that it is about 207 times heavier than the electron ($m_\mu \simeq 207\ m_e$). Muon has a lifetime of around 2.2×10^{-6} second. Some effects, which play only a minor role for the electron in usual atoms, become important for the μ^- (or the μ^+) when it circles around the nucleus forming a muonic atom [2, 3]. These effects are linked to the large mass of the muon, which implies that the Bohr radius of the μ^- is smaller than of the electron by a factor of $\frac{1}{127}$. Similarly, the tauon (τ^-) is about 3,477 times heavier than the electron and since its interaction is very similar to that of the electron, a tauon can be thought of as a much heavier version of the electron. Even though a tauon has similar properties as the electron, its short lifetime (about 2.9×10^{-13} s) makes it much more difficult than a muon to study and thereby to obtain experimental results.

The simplest few-body problem for a system of fermions and antifermions of equal masses with electromagnetic interactions is that of positronium (Ps: $e^- e^+$). Deutsch [4] was the first person who observed Ps in 1951. Since the discovery of Ps, there have been major advances in understanding of the Ps system and in the use of Ps to explore the basic structure of QED [5, 6]. Theoretical studies of the Ps system are now well advanced. For example, accurate calculations of the positronium hyperfine interval contributions at the level of $O(\alpha^6)$ ground-state hyperfine splitting in positronium have been studied by Adkins et al. [5, 6]. An account of the history of QED has been written by Schweber [7] (see also Dyson [8]).

M. Emami-Razavi (✉)
Plasma Physics Research Center, Science and Research Branch, Islamic Azad University, P.O. Box 14665-678, Tehran, Iran
e-mail: mrazavi@srbiau.ac.ir

Positronium plays a unique role in the continuing development of the techniques of bound-state QED and high-precision tests of the standard model as a Coulombic bound system that is almost free of strong and weak interaction contaminations and that exhibits large recoil and annihilation (virtual and real) effects. In very recent works, Adkins et al. [9, 10] calculated the positronium hyperfine splitting and energy levels at order $O(\alpha^7)$. They obtained a new and more precise result for the light-by-light scattering correction to the real decay of parapositronium into two photons. Adkins et al. have also calculated the three-loop correction to the positronium hyperfine splitting due to light-by-light scattering in the exchange of two photons between the electron and positron.

The positronium negative ion (Ps^-), composed of three equal mass fermions (e^+ e^- e^-), is the simplest three-body system bound only by electromagnetic interactions. The existence of a bound Ps^- system was predicted by Wheeler [11]. Theoretical studies of the Ps^- are also well advanced (see, for example, Drake and Grigorescu [12] or Frolov [13]). The Ps^- system was first observed by Mills [14], and later on by Fleischer et al. [15].

With respect to the relativistic and QED corrections for the positronium negative ion system, we can quote the following example. In Ref. [12], the leading relativistic and QED corrections to the ground-state energy of the three-body system (e^+ e^- e^-) have been calculated numerically using a Hylleraas correlated basis set. The accuracy of the non-relativistic variational ground state in [12] is discussed with respect to the convergence of the energy with increasing size of the basis set, and also with respect to the variance of the Hamiltonian. The corrections to this energy include the lowest order Breit interaction, the vacuum polarization potential, one and two photon exchange contributions, the annihilation interaction and spin–spin contact terms.

Fundamental fermions antifermions systems with electromagnetic interactions are of interest because they are "pure" QED systems, with point-like constituents and no nuclear force or size effects. Experiments on such "exotic" atoms, though difficult, are being undertaken not only for positronium and the three-body Ps^- (e^+ e^- e^-) system, but also for the four-body "positronium molecule" (Ps_2: e^+ e^- e^+ e^-). The positronium molecule was observed in 2007 by Cassidy and Mills [16]. The existence of a bound-state Ps_2 was first predicted by Hylleraas and Ore [17]. There are many papers on this topic in the literature (see, for example, Emami-Razavi [18] for a review of earlier work). Moreover, the dipole excitation of the positronium molecule (the energy interval between ground and the P wave exited state of Ps_2) and thermal instability of Ps_2 have been recently studied, respectively, in [19] and [20]. For the positronium molecule system Bubin et al. [21] solved the non-relativistic problem variationally and used their solutions to calculate the relativistic corrections in first-order perturbation theory. Their result for the ground-state binding energy of Ps_2 is 0.01595425 Hartree (i.e., 0.4341373 eV). This includes the $O(\alpha^2)$ relativistic corrections to the non-relativistic ground-state energy of Ps_2 system.

The simplest system for fermions and antifermions of different flavor is muonium, e^- μ^+. This system (and muonium-like systems) have been investigated in the literature (see, for example, [22] and refs therein). The term "muonium" for the e^- μ^+ bound state and its first theoretical discussion appeared in Ref. [23], and the state was discovered soon thereafter [24].

For a three-body system of fermions and antifermions of various flavors, we can mention, for example, the system consisting of two identical particles and a different antiparticle (e.g., muonium negative ion: Mu^- or e^-, e^-, μ^+). The first observation of the negative muonium ion produced by electron capture in a beam-foil experiment has been done by Kuang et al. [25]. This system has been studied, for example, by Frolov [26], and by Barham and Darewych [27]. The Mu^- has only one bound (ground) state $^1S\,(L = 0)$ state [28]. The energies and other bound-state properties for the ground $^1S\,(L = 0)$ state in the Mu^- ion are known to very high accuracy [29].

The four-body system (m^{Z^+}, Ps^-), m is the mass of the particle to be specified with respect to Ps^- and Z is the number of the charge has been studied in Ref. [30]. The properties of some exotic five-particle systems have been studied in [31]. A proof of stability of four-body system, hydrogen, hydrogen-like molecules ($M^+M^+m^-m^-$), and some asymmetric molecules of the type (m_1^-, m_2^+, m_3^-, m_4^+) has been discussed by Richard [32] (see also [33]).

The prediction of the stability of Coulombic few-body systems requires sophisticated calculations [34]. As pointed out, for example in Ref. [31], the difficulty can largely be attributed to the fact that the correlations between like and opposite charges are quite different due to the attractive and repulsive interaction. Another factor which plays a crucial role in the binding mechanism for fermions and antifermions is the Pauli principle. The main forces to determine the stability domains are the Pauli principle and the mass ratios [31]. The Pauli principle severely restricts the available configuration space for fermionic systems.

As far as the two-body system interacting via Coulombic force is concerned, it is possible to find the binding energy of the systems analytically and there are many papers available in the literature. However, when one has a system of three- or four-body system, it is not possible to obtain analytical solution for the binding energy of the systems. Therefore, one needs to perform some numerical calculations. Those computations of stochastic variational method (SVM) or a variant of this method have been presented in [35, 37]. In these work, the diatomic basis sets were used to calculate energies or other properties of the exotic systems under study [37].

The presentation of this work is as following. The two-, three-, four- and five-body systems are discussed from section two to section five. Concluding remarks are included in the last section.

Two-body systems

Consider a system of N particles with masses (m_1, m_2, ..., m_N, and charges (q_1, q_2, ..., q_N). If we take the particles to be charged point masses with Coulombic interactions, then the non-relativistic Hamiltonian has the familiar form

$$H = -\sum_{i=1}^{N} \frac{\hbar^2}{2m_i} \nabla_i^2 + \sum_{i<j}^{N} \frac{q_i q_j}{|\mathbf{r_i} - \mathbf{r_j}|}. \tag{1}$$

The simplest few body exotic systems corresponds to the two-body case and analytical solutions can be obtained for the binding energy of a two-body system. However for a three-body system (or four-body system) numerical calculations are needed to obtain numerical results [36, 37].

The simplest two-body systems that contain muons are the true muonium $\mu^+ \mu^-$ and muonium $e^- \mu^+$ (discovered by Hughes [24]). Due to the close confinement in the bound state muonium can be used as an ideal probe of electroweak interaction, including particularly QED, and to search for additional yet unknown interactions acting on leptons. The term "muonium" for the $e^- \mu^+$ bound state and its first theoretical discussion appeared in Ref. [23], and the state was discovered soon thereafter. Since then, this system (and muonium-like systems) have been investigated in the literature by many authors, particularly for the non-relativistic case. For the relativistic and QED corrections of the muonium-like systems, there are some papers available in the literature (see, for example, [22] and refs therein).

Terekidi and Darewych [22] have considered a reformulation of QED in which covariant Green functions are used to solve for the electromagnetic field in terms of the fermion fields. The resulting modified Hamiltonian contains the photon propagator directly. The authors [22] obtained solutions of the two-body equations for muonium-like system to the order of $O(\alpha^4)$. Their results compare well with the observed muonium spectrum, as well as that for hydrogen and muonic hydrogen. Anomalous magnetic moment effects are also discussed in [22].

Unlike in the case of positronium, the true muonium ($\mu^+ \mu^-$) constituents themselves are unstable. However, the μ has an exceptionally long lifetime by particle physics standards (2.2×10^{-6} second), meaning that ($\mu^+ \mu^-$) annihilates long before its constituents weakly decay. Hence, the $\mu^+ \mu^-$ is unique as the heaviest metastable laboratory possible test for precision QED tests. It has a lifetime of 0.602 Ps in the $^1S_0$ state (decaying to $\gamma\gamma$) and 1.81 ps in the $^3S_1$ state (decaying to $e^+ e^-$) [38–39]. One should note those systems are the most compact two-body QED systems and therefore from fundamental point of view their studies are of interest.

In principle, the creation of true tauonium ($\tau^+ \tau^-$) and "mu-tauonium" ($\mu^\pm \tau^\mp$) are also possible but the relatively short lifetime of tauon makes it difficult to observe such systems with current experimental setups. The corresponding $^1S_0$ and $^3S_1$ lifetimes of ($\tau^+ \tau^-$) are 35.8 and 107 fs [1], respectively, to be compared with the free τ lifetime 291 fs (or half this for a system of two τ's). The ($\tau^+ \tau^-$) annihilation decay and the weak decay of the constituent τ's actually compete, making ($\tau^+ \tau^-$) not a pure QED system like ($e^+ e^-$) [1].

One should note that due to very short lifetime of tauon, the systems such as ($\mu^+ \tau^-$) have a very short lifetime and they are not true bound states; in other words, they constitute quasi-bound states (or resonances) and it is highly unlikely that they will be observed in near future experiments.

Three-body systems

The three-body bound states with electromagnetic interaction have been studied since many years ago by several authors (see, for example, Bhatia and Drachman [40]). A review of some particular three unit charge systems along with the domain of stability of those systems has been presented, for example in Ref. [33].

The positronium negative ion (Ps^-), composed of three equal mass fermions ($e^+ e^- e^-$), is the simplest three-body system with Coulombic interaction. The other three-body system of interest is muonium negative ions ($\mu^+ e^- e^-$). Note that for muonium negative ion (Mu^- : $\mu^+ e^- e^-$), the muonium conversion (or $\mu^+ e^- \longrightarrow \mu^- e^+$ conversion) in the Mu^- ion can produce a number of secondary atomic processes and it is the most interesting process which can be observed with μ^+ muons in atomic structures [26]. In fact, $\mu^+ e^- \longrightarrow \mu^- e^+$ conversion transforms the incident three-body system ($\mu^+ e^- e^-$) into a system ($\mu^- e^+ e^-$) which has no bound state [28]. The theoretical investigation of those systems are now well advanced (relativistic corrections of H^-, positronium and muonium negative ions have been studied, for example, by Barham and Darewych, see [27] and references therein).

In [27], the authors discussed relativistic three-fermion wave equations in QED. The equations are used to obtain relativistic $O(\alpha^2)$ corrections to the non-relativistic ground-state energy levels of the positronium and muonium negative ions, as well as H^-, using approximate variational three-body wavefunctions. The results of [27] have been compared with other calculations, where available.

In a paper by Frolov [41] using the approach of explicitly correlated exponential basis functions has studied the properties of some atomic and molecular three-body ions which contain positively charged muons (μ^+). In particular, the ground states in the muonium ion (Mu^- : $\mu^+ e^- e^-$) and muon-hydrogen molecular ($p^+ \mu^+ e^-$), ($d^+ \mu^+ e^-$), and ($t^+ \mu^+ e^-$) ions (d, t stand for deuteron and tritium) have been examined in details and the energies of these systems have been determined to high accuracy.

We remind again that due to short lifetimes of μ and τ, these systems are quasi-bound state and very difficult to detect by experimental equipments of nowadays. However, from theoretical point of view it would be of interest to see that some of these systems cannot even be bound and some of them can be stable enough to constitute a bound state.

Four-body systems

The solution of the four-body system is a very difficult problem. However, despite the complexity of the calculations, enormous progress has been made (see for, example [33], for a review) in this field which has been rapidly developing ever since of the birth of non-relativistic quantum mechanics (Schrödinger equation). Moreover, one should note that the QED and relativistic corrections of the four-body systems of different flavors still remain a very challenging problem. However, in a recent paper [42], the relativistic wave equations of systems consisting of fermions and antifermions of various masses have been presented. To our knowledge, there are no major work or study for the "relativistic" or QED calculations of four-body system of different masses.

The theoretical investigation of the "non-relativistic" four-body systems interacting through Coulombic potential has been discussed in several works and some studies have been done regarding the existence of bound states of those systems and also the properties of the domain of stability in the space masses or inverse masses [33] (those results are supplemented by numerical investigations using accurate variational methods). The stability domain of the system made up of four particles of different masses, two having the same positive charge and two having the opposite negative charge, interacting only through Coulombic forces, can be represented in terms of points in the interior, or on the surface of a regular tetrahedron [32].

Recent success in the production of trapped antihdyrogen atoms [43, 44] has renewed interest in the interaction of hydrogen–antihydrogen system. $H - \overline{H}$ is known to decay into protonium ($p\overline{p}$, proton or antiproton is considered fundamental particle here) and positronium ($e^+ e^-$). In Ref. [45] it was pointed out that from unstable $H - \overline{H}$ system, it also follows that the systems $p \mu^- e^+ e^-$, $\mu^+ \mu^- e^+ e^-$, $d \overline{p} e^+ e^-$, and $t \overline{p} e^+ e^-$ (d, t stand for deuteron and tritium) are unstable.

One of the system of interest consisting of fundamental particles is muonium molecule (Mu_2, $e^- \mu^+ e^- \mu^+$: system consisting of two electrons and two antimuons). The only exotic QED four-body system consisting of fundamental particles that has been observed is the positronium molecule [16]. Mu_2 system is not observed as yet. Some other exotic four-body systems are such as $e^- \mu^+ e^- \mu^+$, $e^- \mu^+ e^- p$, $e^- \mu^+ e^- \tau^+$, and $e^- \mu^+ \tau^- p$.

Bubin et al. [21] reported that they have obtained a very accurate variational wave function for non-relativistic binding energy of the positronium molecule (Ps_2), which they used to calculate the relativistic corrections. However, to our knowledge there is no study of relativistic or QED corrections for the four-body exotic systems for various masses of fermions and antifermions and it remains an open problem.

One should also note the following points about muonium molecule (Mu_2, $e^- \mu^+ e^- \mu^+$) in case of its existence. The muonium molecule system is H_2 (hydrogen molecule) like. There is no virtual annihilation in the Mu_2 and in this case it is very different from ($\mu^+ \mu^- e^+ e^-$) system. In the latter system, there are two different virtual annihilation interactions, namely, ($\mu^+ \mu^-$) and ($e^+ e^-$). The system ($\mu^+ \mu^- e^+ e^-$) is hydrogen–antihydrogen like and it is not bound according to Ref. [45].

Moreover, as it was pointed out for the three-body case, the muonium conversion (or $\mu^+ e^- \longrightarrow \mu^- e^+$ conversion) in the muonium molecule can transform the four-body system ($e^- \mu^+ e^- \mu^+$) into the system ($\mu^+ \mu^- e^+ e^-$), which has no bound state [45]. We remind that for the three-body case we mentioned earlier that the system ($\mu^- e^+ e^-$) has no bound state [28], on the contrary to muonium negative ion ($\mu^+ e^- e^-$), which has a bound state [26], and it was observed by Kuang et al. [25]. In case of a future observation of muonium molecule ($e^- \mu^+ e^- \mu^+$), the comparison of recent observed positronium molecule ($e^- e^+ e^- e^+$) [16] with muonium molecule may provide a better understanding of theory of QED.

Five-body systems

The study of exotic five-particle systems and the prediction of their stability requires very sophisticated calculations. Mezei et al. [31] did investigate the stability of a number of five-body systems using SVM. The small loosely bound systems require very accurate calculations. The properties of the most intriguing systems consisting of two electrons and two positrons (e.g., e^{+}PsH or $Li^{+}Ps_2$) have been investigated in great detail [31] for the non-relativistic Schrödinger equation.

The difficulty of the five-body calculations can largely be attributed to the fact that the correlations between like and opposite charges are quite different due to the attractive and repulsive interaction. Moreover, the Pauli principle plays a crucial role in the binding mechanism of the fermions. For systems with identical particles, the antisymmetry requirement seriously restricts phase space accessible to the particles by not allowing the energetically most favorable configurations. Furthermore, the total charges of a bound five particle system must be ± 1, that is, there is no bound system with $(+, -, -, -, -)$ charges [31].

The simplest five-body particles is a system consisting of three positively charged particles and two negatively charged identical fermions with spin 1/2 $(m^{+} m^{+} m^{+} m^{-} m^{-})$. The antisymmetry requirement restricts the configuration space and no bound state exists; and in particular, the system of three electrons and two positrons, is not bound [31]. However, the system $(m^{+} m^{+} m^{+} m^{-} m^{-})$ is bound if the three positively charged are bosons or if one of them is distinguishable, since the Pauli principle does not restrict the allowed states. An example is $(e^{+} e^{+} e^{+} e^{-} x^{-})$ system, where x^{-} is a fictitious particle which has the same mass of the electron (or it can have different mass) but it is distinguishable from both electron and the positron [31]. For example, a system made of positronium molecule and a proton $(e^{+} e^{+} e^{-} e^{-} p)$ is stable. The properties of some other exotic five-particle systems $(m_1^{+} m_2^{+} m_3^{+} m_4^{-} m_5^{-})$ where two or three of the particles may have the same mass have been discussed in [31] (see also [33] for a review).

On a practical level, it is very difficult to observe the five-body exotic systems that contains muon (or tauon) and electrons (for example: $e^{+} e^{+} e^{-} e^{-} \mu^{+}$). So far, to our knowledge the only exotic system that contains fundamental particles and antiparticles with electromagnetic interaction and has been observed is the positronium molecule $(e^{+} e^{+} e^{-} e^{-})$ [16]. However, from theoretical point of view, one should consider that the study of the exotic five-body systems and their application may be of interest [31] even though their observations will not be in the near future.

Concluding remarks

In this paper, we were particularly interested in presenting the study of the two-, three-, four-, and five-body exotic systems that contains electrons, muons, and tauons (with their corresponding antiparticles). The non-relativistic energies and bound states of those systems have been investigated by several authors. The relativistic or QED corrections of the energies of more than two-body exotic systems have not been considered by many authors and therefore remain a challenging problem.

For the two-body relativistic or QED corrections, such as for the positronium system, some advanced studies have recently been performed [9, 10] to calculate the positronium hyperfine splitting and energy levels at order $O(\alpha^7)$. Similarly, the relativistic or QED corrections for some particular three- and four-body systems have been investigated in various papers (see for examples, [12] and [21]).

The observation of positronium molecule by Cassidy and Mills [16] raises interest in other exotic systems. To our knowledge, there are not many papers in the literature that study the bound states of systems that contain a few muons and/or tauons combined with other particles such as e^{-} or e^{+}. Hence, at least from theoretical point of view, the study of such exotic systems would be of fundamental importance.

In a recent paper [46], the production and discovery of "True Muonium" (a bound state of a muon and antimuon) in fixed-target experiments have been discussed. According to the authors in Ref. [46], discovery and measurement prospects appear very favorable for the true muonium system.

To conclude, we can mention the following point. Experiments on "exotic" atoms or molecules, though difficult, will be undertaken in future even though it may not be near future. With the current experimental setup available at this time, it is expected the true muonium $(\mu^{+} \mu^{-})$ system can be observed in near future (see, for example, [1] or [46]). Therefore, further test of QED systems and their corresponding theories can be achieved using new observed systems.

Acknowledgments The author thanks Professors Jim Mitroy and Jurij W. Darewych for useful comments.

References

1. Brodsky, S.J., Lebed, R.F.: Phys. Rev. Lett. **102**, 213401 (2009)
2. Greiner, W.: Relativ. Quantum. Mech. Ch 9, p193, Springer-Verlag (1989)
3. Greiner W., Reinhardt, J.: Quantum. Electrodyn. Ch 5, p 200, Springer-Verlag (1989)
4. Deutsch, M.: Phys. Rev. **82**, 455 (1951)

5. Adkins, G.S., Fell, R.N., Mitrikov, P.M.: Phys. Rev. Lett. **79**, 3383 (1997)
6. Adkins, G.S., Fell, R.N., MitrikovPhys, P.M.: Rev. A **65**, 042103 (2002)
7. Schweber, S.S.: QED and the Men who Made It: 1) Freeman Dyson, 2) Richard Feynman, 3) Julian Schwinger, and 4) Sin-Itiro Tomonaga. Princeton University Press, Princeton (1994)
8. Freeman, D., Adv. Quantum. Mechan. World. Sci. (2007)
9. Adkins, G.S., Parsons, C., Salinger, M.D., Wang, R., Fell, R.N.: Phys. Rev. A **90**, 042502 (2014)
10. Adkins, G.S., Fell, R.N.: Phys. Rev. A **89**, 052518 (2014)
11. Wheeler, J.A.: Ann. NY Acad. Sci. **48**, 219 (1946)
12. Drake, G.W.F., Grigorescu, M.: J. Phys. B: At. Mol. Opt. Phys. **38**, 3377–3393 (2005)
13. Frolov, A.M.: Phys. Lett. A 342, 430 (2005). Phys. Rev. A **60**, 2834 (1999)
14. Mills Jr, A.P.: Phys. Rev. Lett. 46, 717 (1981). Phys. Rev. Lett. **50**, 671 (1983)
15. Fleischer, F., et al.: Phys. Rev. Lett. **96**, 063401 (2006)
16. Cassidy D.B., Mills A.P., Nature (London) 449
17. Hylleraas, E.A., Ore, A.: Phys. Rev. **71**, 493 (1947)
18. Emami-Razavi, M.: Phys. Rev. A **77**, 042104 (2008)
19. Puchalski, M., Czarnecki, A.: Phys. Rev. Lett. **101**, 183001 (2008)
20. Kylänpää, I., Rantala, T.T, Phys. Rev. A 80, 024504 (2009).
21. Bubin, S., Stanke, M., Kedziera, D., Adamowicz, L.: Phys. Rev. A **75**, 062504 (2007)
22. Terekidi, A.G., Darewych, J.W.: J. Math. Phys **46**, 032302 (2005)
23. Friedman, J.I., Telegdi, V.L.: Phys. Rev. **105**, 1681 (1957)
24. Hughes, V.W., McColm, D.W., Ziock, K., Prepost, R.: Phys. Rev. Lett. **5**, 63 (1960)
25. Kuang, Y., et al.: Phys. Rev. A **35**, 3172 (1987)
26. Frolov, A.M.: J. Phys. B: At. Mol. Opt. Phys. **37**, 2191 (2004)
27. Barham, M., Darewych, J.W.: J. Phys. B Mol. Opt. Phys. **41**, 185001 (2008)
28. Frolov, A.M.: J. Phys. B: At. Mol. Opt. Phys. **25**, 3059 (1992)
29. Frolov, A.M.: Phys. Rev. A 69, 022501 (2004). Phys. Rev. A **58**, 4479 (1998)
30. Mitroy, J., Novikov, S.A.: Phys. Rev. A **70**, 032511 (2004)
31. Mezei, Z.S.J., Mitroy, J., Lovas, R.G., Varga, K. Phys. Rev. A **64**, 032501 (2001)
32. Richard, J.M. Phys. Rev. A 49, 3573 (1994).
33. Armour, E.A.G., Richard, J.M., Varga, K.: Physics Reports **413**, 1–90 (2005)
34. Mitroy, J., Bromley, M.W., Ryzhikh, G.G.: J. Phys. B **35**, R81–R116 (2002)
35. Suzuki, Y., Varga, K.: Stochastic Variational Approach to Quantum Mechanical Few-Body Problems. Springer-Verlag, Berlin (1998)
36. Kinghorn, D.B., Adamowicz, L.: J. Chem. Phys. **110**, 7166 (1999)
37. Bubin, S., Cafiero, M., Adamowicz, L.: J. Chem. Phys. **131**, 377 (2003)
38. Bilen'kii, S., van Hieu, N., Nemenov, L., Tkebuchava, F.: Sov J. Nucl. Phys. **10**, 469 (1969)
39. Hughes, V.W., Maglic, B.: Bull. Am. Phys. Soc. **16**, 65 (1971)
40. A. K Bhatia and R. J. Drachman, Phys Rev A 35, 4051 (1987).
41. Frolov, A.M.: Phys Rev A **69**, 022505 (2004)
42. Emami-Razavi, M., Bergeron, N., Darewych, J.W.: Int. J. of Mod. Phys. E **21**(11), 1250091 (2012)
43. Gabrielse, G., et al.: Phys. Rev. Lett. **89**, 213401 (2002)
44. Amoretti, M., et al.: Nature (London) **419**, 456 (2002)
45. Gridnev, D.K., Greiner, C.: Phys. Rev. Lett. **94**, 223402 (2005)
46. Banburski, Andrzej, Schuster, Philip: Phys. Rev. D **86**, 093007 (2012)

7

Relation of the nonlinear Heisenberg algebras in two dimensions with linear ones

Won Sang Chung[1]

Abstract In this paper, we discuss the relation of the nonlinear Heisenberg algebras in two dimensions with linear ones following the Nowicki and Tkachuk's approach for one-dimensional case. For one-dimensional harmonic oscillator, we obtain the solution explicitly. For the nonlinear Heisenberg algebras in two dimensions, we introduce two generators to transform this algebra into the linear one. For the linear version of the nonlinear Heisenberg algebras in two dimensions, we obtain the eigenfunction for the position and angular momentum operator and solve the harmonic oscillator problem in two dimensions.

Keywords Nonlinear Heisenberg algebras · Algebra · Two dimensions

Introduction

The first form of the Heisenberg algebra giving the minimal length uncertainty was first introduced by Kempf, Mangano, and Mann [1] in the following form:

$$[X, P] = i\hbar(1 + \beta P^2), \tag{1}$$

which suggests the existence of the fundamental minimal length

$$(\Delta X)_0 = \hbar\sqrt{\beta}. \tag{2}$$

In this direction, much development has been accomplished in order to study the effect of minimal length on the quantum physical systems as well as on the classical ones, but only a few problems are shown to be solved exactly. They are one-dimensional harmonic oscillator with minimal length uncertainty in position [1, 2] and also with minimal length uncertainty in position and momentum [3, 4], D-dimensional isotropic harmonic oscillator [5, 6], three-dimensional Dirac oscillator [7], (1 + 1)-dimensional Dirac oscillator within Lorentz-covariant deformed algebra [8], one-dimensional Coulomb problem [9], and the singular inverse square potential with a minimal length [10, 11]. Three-dimensional Coulomb problem with deformed Heisenberg algebra was solved within the perturbation theory [12–15].

In this paper, we discuss the relation of the nonlinear Heisenberg algebras in two dimensions with linear ones following the Nowicki and Tkachuk's approach [16] for one-dimensional case. For one-dimensional harmonic oscillator, we obtain the solution explicitly. For the nonlinear Heisenberg algebras in two dimensions, we introduce two generators to transform this algebra into the linear one. For the linear version of the nonlinear Heisenberg algebras in two dimensions, we obtain the eigenfunction for the position and angular momentum operator and solve the harmonic oscillator problem in two dimensions.

One-dimensional deformed nonlinear Heisenberg algebra

Recently, Nowicki and Tkachuk [16] considered a one-dimensional deformed nonlinear Heisenberg algebra with function of deformation $f(P)$, namely

$$[X, P] = if(P), \tag{3}$$

where $f(P)$ is an positive function obeying

✉ Won Sang Chung
mimip4444@hanmail.net

[1] Department of Physics and Research Institute of Natural Science, College of Natural Science, Gyeongsang National University, Jinju 660-701, Korea

$$f(-P) = f(P). \tag{4}$$

It means that the space has the same properties in two opposite directions. The momentum representation reads

$$P = p, \quad X = if(p)\partial_p \tag{5}$$

and acts on the square integrable functions $\phi(p) \in \mathcal{L}^2(-a, a; f), (a \leq \infty)$ where the norm of ϕ is given by

$$||\phi||^2 = \int_{-a}^{a} \frac{\mathrm{dp}}{f(p)} |\phi(p)|^2. \tag{6}$$

More general momentum representation is given in "Appendix".

Nowicki and Tkachuk extended the algebra (3) into the three generator algebra by one additional generator $F = f(p)$ and obtain a concrete form of f as follows:

$$f(p) = \sqrt{1 + \beta p^2}. \tag{7}$$

Now let us consider the harmonic oscillator with the hamiltonian

$$H = \frac{1}{2\mu} P^2 + \frac{1}{2} \mu w^2 X^2. \tag{8}$$

The Schrödinger equation reads

$$\left[\frac{p^2}{2\mu} + \frac{1}{2}\mu w^2 (i\sqrt{1 + \beta p^2}\partial_p)^2\right]\psi = E\psi. \tag{9}$$

Let us change the variable like $\sqrt{\beta}p = \sinh \xi$, which means that ξ goes to zero when β approaches zero. Then, Eq. (9) becomes

$$\psi'' + \left(\epsilon - \frac{\sinh^2 \xi}{(\beta\mu w)^2}\right)\psi = 0, \tag{10}$$

where

$$\epsilon = \frac{2E}{\beta \mu w^2}. \tag{11}$$

When we consider the small value of ξ, we have

$$\psi'' + (\epsilon - \sigma\xi^2)\psi = 0, \tag{12}$$

where

$$\sigma = \frac{1}{(\beta \mu w)^2}. \tag{13}$$

Replacing $\xi^2 = z$, we get

$$4z\psi'' + 2\psi' + (\epsilon - \sigma z)\psi = 0. \tag{14}$$

If we set

$$\psi(z) = e^{-\frac{z}{2\mu w}} y(z),$$

we get

$$zy'' + \left(\frac{1}{2} - \frac{1}{\mu w} z\right) y' + \frac{1}{4}\left(\epsilon - \frac{1}{\mu w}\right) y = 0. \tag{15}$$

This equation can be solved using the Frobenius method. If we adopt

$$y(z) = \sum_{n=0}^{\infty} a_n z^{n+\lambda} \tag{16}$$

and insert it into Eq. (15), we have

$$a_0 \lambda(\lambda - 1/2) = 0 \tag{17}$$

$$(n + 1 + \lambda)(n + 1/2 + \lambda)a_{n+1} = \left[-\frac{1}{4}\left(\epsilon - \frac{1}{\mu w}\right) + \frac{1}{\mu w}(n + \lambda)\right] a_n, \quad (n = 0, 1, \ldots). \tag{18}$$

From the characteristic Eq. (17), we have two values of λ:

$\lambda = 0, 1/2.$

For $\lambda = 0$, we have

$$y_1(z) = {}_1F_1\left(\frac{1}{4}(1 - \mu w \epsilon); \frac{1}{2}; \frac{1}{\mu w}\right) \tag{19}$$

and for $\lambda = 1/2$, we have

$$y_2(z) = {}_1F_1\left(\frac{1}{4}(3 - \mu w \epsilon); \frac{3}{2}; \frac{1}{\mu w}\right), \tag{20}$$

where Kummer's function is defined as

$${}_1F_1(a; b; z) = \sum_{n=0}^{\infty} \frac{(a)_n z^n}{n!(b)_n} \tag{21}$$

and

$$(a)_0 = 1, \quad (a)_n = a(a + 1)(a + 2)\cdots(a + n - 1), \tag{22}$$

is the rising factorial.

Linearization of a two-dimensional deformed nonlinear Heisenberg algebra

Let us consider a two-dimensional deformed nonlinear Heisenberg algebra with deformation function $f(P)$:

$$[X_1, P_1] = if(P), \quad [X_2, P_2] = if(P), \quad [P_1, P_2] = 0, \tag{23}$$

where $f(P) = f(P_1, P_2)$ is an positive function obeying

$$f(-P_1, P_2) = f(P_1, P_2), \quad f(P_1, -P_2) = f(P_1, P_2). \tag{24}$$

It means that the space has the same properties in two opposite X_1- and X_2-directions. From the commutation relations (23), we have the momentum representation of the operators as follows:

$$P_1 = p_1, \quad P_2 = p_2$$

$$X_1 = if(p)\partial_{p_1}, \quad X_2 = if(p)\partial_{p_2}. \tag{25}$$

This fixes the remaining commutation relations, so the full algebra is then given by

$$[X_1, P_1] = if(P), \quad [X_2, P_2] = if(P), \quad [P_1, P_2] = 0$$

$$[X_1, X_2] = -f(\partial_{P_1} f \partial_{P_2} - \partial_{P_2} f \partial_{P_1}), \quad [X_1, P_2] = [X_2, P_1] = 0. \tag{26}$$

Indeed one can easily check that the relation (26) obeys the Jacobi identity.

Now we assume that the operators X_1, X_2, P_1 and P_2 act on the square integrable functions $\phi(p_1, p_2) \in \mathcal{L}^2(-a, a; f), (a \leq \infty)$ where the norm of ϕ is given by

$$||\phi||^2 = \int_{-a}^{a} \int_{-a}^{a} \frac{\mathrm{d}p_1 \mathrm{d}p_2}{f(p)} |\phi(p)|^2. \tag{27}$$

For the self-adjointness of X_1 and X_2, we have

$$\phi(-a, p_2) = \pm\phi(a, p_2)$$

and

$$\phi(p_1, -a) = \pm\phi(p_1, a).$$

Now we extend this algebra by two additional operators

$$F = f(P)$$

$$G = if(\partial_{P_1} f \partial_{P_2} - \partial_{P_2} f \partial_{P_1}). \tag{28}$$

Thus, the extended algebra $\mathcal{E}$ is generated by the six generators. Using representation (25), one can easily find

$$[X_1, F] = if\partial_{P_1} f, \quad [X_2, F] = if\partial_{P_2} f, \quad [P_1, F] = [P_2, F] = 0. \tag{29}$$

We require that both $\{X_1, P_1, F\}$ and $\{X_2, P_2, F\}$ form a subalgebras of $\mathcal{E}$. Then, one can put

$$f\partial_{P_1} f = \alpha + \beta P_1 + \gamma F \tag{30}$$

$$f\partial_{P_2} f = \alpha' + \beta' P_2 + \gamma' F, \tag{31}$$

where $\alpha, \beta, \gamma, \alpha', \beta', \gamma'$ are real parameters. Note that the linear combination in the right-hand side of (30) (or 31) does not contain X_1 (or X_2) because $f\partial_{P_1} f$ (or $f\partial_{P_2} f$) is a function of P_1, P_2 only. Using Eq. (24) and changing P_1, P_2 into $-P_1$, $-P_2$, respectively, one find

$$f\partial_{P_1} f = -\alpha + \beta P_1 - \gamma F \tag{32}$$

$$f\partial_{P_2} f = -\alpha' + \beta' P_2 - \gamma' F, \tag{33}$$

Comparing Eqs. (30) (31) or with Eqs. (32) (or 33), one can see

$$f\partial_{P_1} f = \beta P_1, \quad f\partial_{P_2} f = \beta' P_2. \tag{34}$$

From now on we will restrict our concern to the case of $\beta = \beta' = v^2 > 0$ for simplicity. From Eq. (34), we have

$$f(P) = \sqrt{1 + \beta P_1^2 + \beta P_2^2}. \tag{35}$$

Thus, the algebra $\mathcal{E}$ reads

$$\begin{aligned}
&[X_1, X_2] = iG, \quad [X_1, P_1] = iF, \quad [X_1, P_2] = 0, \\
&\quad [X_1, F] = i\beta P_1, \quad [X_1, G] = i\beta X_2, \\
&[X_2, P_1] = 0, \quad [X_2, P_2] = iF, \quad [X_2, F] = i\beta P_2, \quad [X_2, G] \\
&\quad = -i\beta X_1, \quad [P_1, P_2] = 0, \\
&[P_1, F] = 0, \quad [P_1, G] = i\beta P_2, \quad [P_2, F] = 0, \\
&\quad [P_2, G] = -i\beta P_1, \quad [F, G] = 0.
\end{aligned} \tag{36}$$

This is Lie algebra. One can find the Casimir operator (invariant) for this algebra

$$K = P_1^2 + P_2^2 - \frac{1}{\beta} F^2 \tag{37}$$

commuting with all elements of the algebra. Now let us define the following operators:

$$A_1 = vP_1, \quad A_2 = F, \quad A_3 = \frac{1}{v} X_1$$

$$A_4 = \frac{1}{v} X_2, \quad A_5 = G, \quad A_6 = vP_2. \tag{38}$$

Then, the algebra $\mathcal{E}$ can be written as

$$\begin{aligned}
&[A_1, A_2] = 0, \quad [A_3, A_1] = iA_2, \quad [A_3, A_2] = iA_1 \\
&[A_4, A_2] = iA_6, \quad [A_4, A_6] = iA_2, \quad [A_6, A_2] = 0 \\
&[A_5, A_1] = -iv^2 A_6, \quad [A_5, A_6] = iv^2 A_1, \quad [A_1, A_6] = 0 \\
&[A_3, A_4] = i\frac{1}{v^2} A_5, \quad [A_4, A_5] = -iv^2 A_3, \quad [A_5, A_3] = -iv^2 A_4 \\
&[A_1, A_4] = 0, \quad [A_2, A_5] = 0, \quad [A_3, A_6] = 0.
\end{aligned} \tag{39}$$

The algebra $\mathcal{E}$ possess some subalgebras:

1. subalgebra $\mathcal{E}_1$ generated by A_1, A_2, A_3
2. subalgebra $\mathcal{E}_2$ generated by A_2, A_4, A_6
3. subalgebra $\mathcal{E}_3$ generated by A_1, A_5, A_6
4. subalgebra $\mathcal{E}_4$ generated by A_3, A_4, A_5.

It is convenient to use two pairs of commuting hermitian operators $P_\pm$ and $Q_\pm$ defined as follows:

$$P_+ = A_1 + A_2 = vP_1 + F$$
$$P_- = A_2 - A_1 = F - vP_1$$
$$Q_+ = A_2 + A_6 = vP_2 + F$$
$$Q_- = A_2 - A_6 = F - vP_2. \quad (40)$$

Indeed one can easily check that

$$[P_\pm, Q_\pm] = 0. \quad (41)$$

In this case, we have the algebra $\mathcal{A}$ generated by $P_\pm, Q_\pm, A_3, A_4, A_5$. Algebra $\mathcal{E}$ has six generators, while $\mathcal{A}$ has seven ones. It seems to be nonsense because two algebras should be isomorphic. To cure this problem, let us consider the inverse relations of Eq. (40)

$$A_1 = \frac{1}{2}(P_+ - P_-), \quad A_2 = \frac{1}{2}(P_+ + P_-),$$
$$A_2 = \frac{1}{2}(Q_+ + Q_-), \quad A_6 = \frac{1}{2}(Q_+ - Q_-). \quad (42)$$

We know that A_2 can be expressed in terms of both $P_\pm$ and $Q_\pm$, which gives a constraint

$$P_+ + P_- = Q_+ + Q_-. \quad (43)$$

This constraint decreases the number of generators of the algebra $\mathcal{A}$, so two algebras are isomorphic.

Besides Eq. (41), the remaining commutation relations of the algebra $\mathcal{A}$ are

$$[A_3, P_\pm] = \pm iP_\pm, \quad [P_+, P_-] = 0$$
$$[A_4, Q_\pm] = \pm iQ_\pm, \quad [Q_+, Q_-] = 0$$
$$[A_3, Q_\pm] = \frac{i}{2}(P_+ - P_-), \quad [A_4, P_\pm] = \frac{i}{2}(Q_+ - Q_-),$$
$$[A_5, Q_\pm] = \frac{i}{2}v^2(P_+ - P_-), \quad [A_5, P_\pm] = \mp iv^2 A_6$$
$$[A_3, A_4] = \frac{i}{v^2}A_5, \quad [A_4, A_5] = iv^2 A_3, \quad [A_5, A_3] = -iv^2 A_4, \quad (44)$$

The Casimir operator is then given by

$$K = \frac{1}{v^2}(A_2^2 - P_+P_- - Q_+Q_-). \quad (45)$$

This algebra has two subalgebras (namely *ISO*(1, 1)) generated by $A_3, P_\pm$ and $A_4, Q_\pm$.

Two sets of the ladder operators can be expressed in terms of the momentum operators as follows:

$$P_\pm = \sqrt{1 + v^2(P_1^2 + P_2^2)} \pm vP_1,$$
$$Q_\pm = \sqrt{1 + v^2(P_1^2 + P_2^2)} \pm vP_2. \quad (46)$$

If we set

$$vP_1 = \sinh\xi \cos\eta, \quad vP_2 = \sinh\xi \sin\eta, \quad (47)$$

we can express $P_\pm$ and $Q_\pm$ as

$$P_\pm = \cosh\xi \pm \sinh\xi \cos\eta$$
$$Q_\pm = \cosh\xi \pm \sinh\xi \sin v\eta. \quad (48)$$

The ξ and η can be expressed in terms of the momentum operators as

$$\eta = \tan^{-1}\left(\frac{P_2}{P_1}\right), \quad \xi = \sinh^{-1}\sqrt{(vP_1)^2 + (vP_2)^2}. \quad (49)$$

Then, A_3, A_4, A_5 can be written as

$$A_3 = i[\cos\eta\partial_\xi - \coth\xi \sin\eta\partial_\eta]$$
$$A_4 = i[\sin\eta\partial_\xi + \coth\xi \cos\eta\partial_\eta]$$
$$A_5 = \frac{i}{v}\partial_\eta. \quad (50)$$

Eigenvectors of the position operator and angular momentum operator

In this section, we discuss the eigenvalue equation for the position operator and angular momentum operator. The eigenvalues for the position operators read

$$X_1\phi(\xi,\eta) = l_1 w(\xi,\eta)\phi(\xi,\eta)$$
$$X_2\phi(\xi,\eta) = l_2 w(\xi,\eta)\phi(\xi,\eta), \quad (51)$$

where $w(\xi,\eta)$ is a weight function. Inserting Eq. (50) into Eq. (51) yields

$$iv[\cos\eta\partial_\xi - \coth\xi \sin\eta\partial_\eta]\phi(\xi,\eta) = l_1 w(\xi,\eta)\phi(\xi,\eta)$$
$$iv[\sin\eta\partial_\xi + \coth\xi \cos\eta\partial_\eta]\phi(\xi,\eta) = l_2 w(\xi,\eta)\phi(\xi,\eta). \quad (52)$$

From Eq. (52), we have

$$\partial_\xi\phi = \frac{w}{iv}(l_1\cos\eta + l_2\sin\eta)\phi$$
$$\coth\xi\partial_\eta\phi = \frac{w}{iv}(l_2\cos\eta - l_1\sin\eta)\phi. \quad (53)$$

If we set

$$\phi(\xi,\eta) = X(\xi)Y(\eta), \quad w(\xi,\eta) = w_1(\xi)w_2(\eta),$$

we have

$$X' = \frac{1}{iv}w_1(\xi)w_2(\eta)(l_1\cos\eta + l_2\sin\eta)X$$
$$Y' = \frac{1}{iv}w_1(\xi)w_2(\eta)(l_2\cos\eta - l_1\sin\eta)Y. \quad (54)$$

If we adopt

$$w_1 = -i \coth \xi, \quad w_2 = \frac{1}{l_1 \cos \eta + l_2 \sin \eta}, \tag{55}$$

we have

$$\phi(\xi, \eta) = [\sinh \xi (l_1 \cos \eta + l_2 \sin \eta)]^{-1/\nu}. \tag{56}$$

Now let us investigate the eigenvalue of the angular momentum operator L defined as

$$L = \epsilon_{ij} X_i P_j = i(fP_2 \partial_{P_1} - fP_1 \partial_{P_2}) = -i \cosh \xi \partial_\eta. \tag{57}$$

The eigenvalue equation reads

$$L\Phi_m = m\hbar \Phi_m, \quad (m \in Z). \tag{58}$$

Solving Eq. (58), we have

$$\Phi_m(\xi, \eta) = \frac{1}{\cosh \xi} e^{im\hbar\eta}. \tag{59}$$

Two-dimensional Harmonic oscillator

Now let us consider the isotropic harmonic Hamiltonian

$$\hat{H} = \frac{1}{2\mu}(P_1^2 + P_2^2) + \frac{1}{2}\mu w^2 (X_1^2 + X_2^2). \tag{60}$$

Using Eq. (47) and Eq. (50), we obtain the expression of H:

$$\hat{H} = -\frac{\mu w^2 \nu^2}{2}(\partial_\xi^2 + \coth \xi \partial_\xi + \coth^2 \xi \partial_\eta^2) + \frac{1}{2\mu\nu^2} \sinh^2 \xi. \tag{61}$$

Then, the Schrödinger equation reads

$$\left[-\frac{\mu w^2 \nu^2}{2}(\partial_\xi^2 + \coth \xi \partial_\xi + \coth^2 \xi \partial_\eta^2) + \frac{1}{2\mu\nu^2} \sinh^2 \xi\right] \psi = E\psi. \tag{62}$$

If we set $\psi = R(\xi)e^{im\eta}$, we have

$$R'' + \coth \xi R' + (\epsilon - A \sinh^2 \xi - m^2 \coth^2 \xi)R = 0, \tag{63}$$

where

$$\epsilon = \frac{2E}{\mu w^2 \nu^2}, \quad A = \frac{1}{\mu^2 w^2 \nu^4}.$$

Now consider the case that ξ is sufficiently small. In this case, we have

$$\sinh \xi \approx \xi, \quad \coth \xi \approx \frac{1}{\xi}.$$

Then, Eq. (63) reduces to

$$R'' + \frac{1}{\xi}R' + \left(\epsilon - A\xi^2 - \frac{m^2}{\xi^2}\right)R = 0. \tag{64}$$

If we set $\xi^2 = \mu w \nu^2 z$, we have

$$zR'' + R' + \left(\frac{1}{4}\mu w \nu^2 \epsilon - \frac{1}{4}z - \frac{m^2}{4z}\right)R = 0. \tag{65}$$

Solving Eq. (65), we get

$$R = e^{-\frac{\xi^2}{2\mu w \nu^2}} \left(\frac{\xi^2}{\mu w \nu^2}\right)^{\frac{m+1}{2}} L_n^m \left(\frac{\xi^2}{\mu w \nu^2}\right) \tag{66}$$

and

$$E_{nm} = \hbar w(2n + m + 1), \quad n = 0, 1, \ldots, \quad m = 0, 1, \ldots, n. \tag{67}$$

The ground state energy is given by

$$E_{00} = \hbar w, \tag{68}$$

which corresponds to the classical result.

Conclusion

Recently, Nowicki and Tkachuk [16] considered a one-dimensional deformed nonlinear Heisenberg algebra with function of deformation $f(P)$, namely $[X, P] = if(P)$. They discussed the relation of the nonlinear Heisenberg algebras with linear ones. We introduced the variable $\xi = \sinh^{-1}(\sqrt{\beta}p)$ to solve the one-dimensional harmonic problem for the small value of ξ. We extended Nowicki and Tkachuk's work to the two-dimensional case. We obtained the linearized algebra $\mathcal{E}$ from the two-dimensional nonlinear Heisenberg algebras by adding two generators. Introducing two variables

$$\xi = \sinh^{-1}(\nu^2 p_1^2 + \nu^2 p_2^2)$$

$$\eta = \tan^{-1}\left(\frac{p_2}{p_1}\right),$$

we expressed all generators of the algebra $\mathcal{E}$ in terms of ξ and η. We also solved the eigenvalue equation for the position and angular momentum operator. Finally, we discussed two-dimensional isotropic harmonic oscillator problem and obtained the corresponding energy eigenvalue and wave function for the small value of ξ. We found that the ground state energy for this model corresponds to the classical result.

Acknowledgments This Work (GNUDFF- 2014-25) was supported by the Academy-oriented Research Funds of Development Fund Foundation, Gyeongsang National University, 2014.

Appendix

More general solution of Eq. (3) is

$$X = if(p)\partial_p + ig(p). \tag{69}$$

We assume that the norm of ϕ takes the form

$$||\phi||^2 = \int_{-a}^{a} \mathrm{d}p\mu(p)|\phi(p)|^2, \tag{70}$$

where the measure function obeys $\mu(-p) = \mu(p)$. For the self-adjointness of X, we have

$$\mu' f + (f' - 2g)\mu = 0. \tag{71}$$

Solving Eq. (71), we get

$$\mu(p) = \frac{1}{f} e^{2\int \frac{g}{f}\mathrm{d}p}. \tag{72}$$

Ex.1 For $g = 0$, we have $\mu = \frac{1}{f}$.
Ex.2 For $f = 1 + \beta p^2, g = \gamma p$, we have $\mu = f^{\gamma/\beta - 1}$.
Ex.3 For $f = \sqrt{1 + \beta p^2}, g = \gamma p$, we have $\mu = \frac{1}{\sqrt{1+\beta p^2}} e^{\frac{2\gamma}{\beta}\sqrt{1+\beta p^2}}$.

References

1. Kempf, A., Mangano, G., Mann, R.B.: Hilbert space representation of the minimal length uncertainty relation. Phys. Rev. D **52**, 1108 (1995)
2. Kempf, A.: Noncommutative geometric regularization. Phys. Rev. D **54**, 5174 (1996)
3. Quesne, C., Tkachuk, V.M.: Deformed algebras, position-dependent effective masses and curved spaces: an exactly solvable Coulomb problem. J. Phys. A **36**, 10373 (2003)
4. Quesne, C., Tkachuk, V.M.: More on a SUSY QM approach to the harmonic oscillator with nonzero minimal uncertainties in position and/or momentum. J. Phys. A **37**, 10095 (2004)
5. Chang, L.N., Minic, D., Okamura, N., Takeuchi, T.: Exact solution of the harmonic oscillator in arbitrary dimensions with minimal length uncertainty relations. Phys. Rev. D **65**, 125027 (2002)
6. Dadic, I., Jonke, L., Meljanac, S.: Harmonic oscillator with minimal length uncertainty relations and ladder operators. Phys. Rev. D **67**, 087701 (2003)
7. Quesne, C., Tkachuk, V.M.: Dirac oscillator with nonzero minimal uncertainty in position. J. Phys. A **38**, 1747 (2005)
8. Quesne, C., Tkachuk, V.M.: Lorentz-covariant deformed algebra with minimal length and application to the (1 + 1)-dimensional Dirac oscillator. J. Phys. A **39**, 10909 (2006)
9. Fityo, T.V., Vakarchuk, I.O., Tkachuk, V.M.: One-dimensional Coulomb-like problem in deformed space with minimal length. J. Phys. A **39**, 2143 (2006)
10. Bouaziz, D., Bawin, M.: Regularization of the singular inverse square potential in quantum mechanics with a minimal length. Phys. Rev. A **76**, 032112 (2007)
11. Bouaziz, D., Bawin, M.: Singular inverse square potential in arbitrary dimensions with a minimal length: application to the motion of a dipole in a cosmic string background. Phys. Rev. A **78**, 032110 (2008)
12. Brau, F.: Minimal length uncertainty relation and hydrogen atom. J. Phys. A **32**, 7691 (1999)
13. Benczik, S., Chang, L.N., Minic, D., Takeuchi, T.: Hydrogen-atom spectrum under a minimal-length hypothesis. Phys. Rev. A **72**, 012104 (2005)
14. Stetsko, M.M., Tkachuk, V.M.: Perturbation hydrogen-atom spectrum in deformed space with minimal length. Phys. Rev. A **74**, 012101 (2006)
15. Stetsko, M.M.: Corrections to the ns levels of the hydrogen atom in deformed space with minimal length. Phys. Rev. A **74**, 062105 (2006)
16. Nowicki, A., Tkachuk, V.: Relation of deformed nonlinear algebras with linear ones. J. Phys. A: Math. Theor. **47**, 025207 (2014)

8

Conformally invariant spin-$\frac{3}{2}$ field equation in de Sitter space–time

N. Fatahi[1]

Abstract In the previous paper (Behroozi et al., Phys Rev D 74:124014, 2006; Dehghani et al., Phys Rev D 77:064028, 2008), conformal invariance for massless tensor fields (scalar, vector and spin-2 fields) was studied and the solutions of their wave equations and two-point functions were obtained. In the present paper, conformally invariant wave equation for massless spinor field in de Sitter space–time has been obtained. For this propose, we use Dirac's six-cone formalism. The solutions of massless spin-$\frac{1}{2}$ and -$\frac{3}{2}$ equations, in the ambient space notation, have been calculated.

Introduction

Today, latest discoveries of the modern astrophysics data show that our universe in first approximation is in a de Sitter (dS) phase. Therefore, it is important to find a formulation of de Sitter quantum field theory with the same level of completeness and rigor as for its Minkowskian counterpart. The various physical stimulants for studying quantum field in dS space are:

The de Sitter space–time is a curved space–time manifold with maximum symmetry and it is a solution of the Einstein equation with nonzero cosmological constant. It is one of the simple curved space–time that one can find unique vacuum state [3] and one can replace the usual spectral condition by a certain geodesic spectral condition [4]. For large separated points, the graviton propagator on dS space has a pathological behavior (infrared divergence) [5–7]. Therefore, one expectancy to discover inside the dynamics of the space–time is some natural mechanism to end inflation and to find a description for the cosmological constant [8–10]. Covariant quantum field theory in de Sitter space for the "massive" and "massless" conformally coupled scalar field [11, 12] and for the "massless" minimally coupled scalar field in de Sitter space [13] previously made. It has been shown that conformally coupled scalar field, vector field and spin-2 field in massive and massless cases correspond to the unitary irreducible representations (UIR) of de Sitter group [12, 14–18].

Conformal invariance was introduced for first time into physics by Cunningham and Bateman [19, 20] when they showed that Maxwell's equations are covariant under the larger 15-parameter conformal group rather than 10-parameter Poincare group. Equations of motion of charged particles with nonzero mass are not conformal invariant. The massless field in Minkowski space–time propagates on the light cone. These fields are invariant under the conformal group $SO(2,4)$. Fields with spin $s \geq 1$ are invariant under the gauge transformation as well. In dS space, for the set of observable transformations under the dS group $SO(1,4)$, mass is not an invariant parameter. However, concept of light-cone propagator does exist and leads to the conformal invariance. "Massless" is used in reference to the conformal invariance (propagation on the de Sitter light cone).

One of the original objectives of the quantum field theory is the quantization of gravitational field [21]. The quanta of gravitational field are massless particle with spin-2 that call "graviton" so that the equation of this particle "Einstein equation" is not conformal invariant. Therefore, conformal invariance may solve the problem of quantum gravity. The fermionic partner of the gravitational field is a

✉ N. Fatahi
n83fatahi@yahoo.com

[1] Department of Physics, Science and Research Branch, Islamic Azad University, Tehran, Iran

spin-$\frac{3}{2}$ field that call "gravitino". In the original formulation of supergravity, there was one real massless gravitino (accordingly $N = 1$) this means that one generator exists that change fermions to bosons $\delta B = \bar{\epsilon} F$. For unification between gravitational (spin-2) and nongravitational (spin-1) forces existence of this generator, that can change spin, is necessary. So, in this paper, we study conformal invariant equation spin-$\frac{3}{2}$ field in dS space–time.

In the previous papers [1, 2], the conformally invariant (CI) wave equations for scalar, vector and spin-2 fields in de Sitter space–time were obtained. We are interested in the conformal invariance properties of the massless spin-$\frac{1}{2}$ and spin-$\frac{3}{2}$ field in de sitter space–time. First, we introduce massless spinor field equations in de Sitter space. Afterwards explained the Dirac's six-cone formalism. With using this formalism, we write the conformal invariant spinor wave equations. Then, in the following sections, de Sitter field solutions can be obtained.

Massless spinor field equations in de sitter space

de Sitter space–time is visualized as the hyperboloid embedded in a five-dimensional Minkowski space–time:

$$X_H = \{x \in \mathrm{IR}^5; x^2 = \eta_\alpha \beta x^\alpha x^\beta = -H^{-2}\}, \quad \alpha, \beta = 0,1,2,3,4, \tag{2.1}$$

$$\eta_{\alpha\beta} = \mathrm{diag}(1, -1, -1, -1, -1).$$

where H is Habel constant (for simplicity, we can choose $H = 1$). The de Sitter metric is:

$$ds^2 = \eta_{\alpha\beta} dx^\alpha dx^\beta \,|_{x^2=-1} = g^{ds}_{\mu\nu} dX^\mu dX^\nu, \quad \mu, \nu = 0, 1, 2, 3, \tag{2.2}$$

where X^μ are the four space–time intrinsic coordinates in de Sitter hyperboloid and x^α are the five global coordinates in ambient space notation. Working in embedding space has two advantages: first it is close to the group theoretical language and second the equations are obtained in an easier way than they might be found in de Sitter intrinsic space.

Dirac [22] deduces wave equation for electron in de Sitter space–time. This equation, with using eigenvalue equation for second-order Casimir operator $Q^{(1)}$, previously is obtained [23]. There are two Casimir operators for dS group, these operators commute with all the action of the group generators and thus they are constant on each representation. Two Casimir operators for dS group are:

$$Q^{(1)} = -\frac{1}{2} L^{\alpha\beta} L_{\alpha\beta}, \tag{2.3}$$

$$Q^{(2)} = -W_\alpha W^\alpha, \tag{2.4}$$

where $W_\alpha = \frac{1}{8} \epsilon_{\alpha\beta\gamma\delta\eta} L^{\beta\gamma} L^{\delta\eta}$ and $\epsilon_{\alpha\beta\gamma\delta\eta}$ is the antisymmetric tensor in the ambient space notation with $\epsilon_{01234} = 1$. The generator of de Sitter group is

$$L_{\alpha\beta} = M_{\alpha\beta} + S_{\alpha\beta},$$

where the "orbital" part is:

$$M_{\alpha\beta} = -i(x_\alpha \partial_\beta - x_\beta \partial_\alpha) = -i(x_\alpha \bar{\partial}_\beta - x_\beta \bar{\partial}_\alpha), \tag{2.5}$$

and the "spinoral" part $S_{\alpha\beta}$ is [24]:

$$S_{\alpha\beta} = -\frac{i}{4}[\gamma_\alpha, \gamma_\beta], \tag{2.6}$$

the transverse derivative is $\bar{\partial}_\beta = \partial_\beta + x_\beta x.\partial$.

In this case, five γ matrices needed instead of the four ones of the flat Dirac theory. These matrices are found within the Clifford algebra issued from the metric $\eta^{\alpha\beta}$:

$$\gamma^\alpha \gamma^\beta + \gamma^\beta \gamma^\alpha = 2\eta^{\alpha\beta}, \quad \gamma^{\alpha\dagger} = \gamma^0 \gamma^\alpha \gamma^0, \tag{2.7}$$

an explicit representation is provided by:

$$\gamma^0 = \begin{pmatrix} I1 & 0 \\ 0 & I1 \end{pmatrix}, \quad \gamma^4 = \begin{pmatrix} 0 & I1 \\ -I1 & 0 \end{pmatrix}$$

$$\gamma^1 = \begin{pmatrix} 0 & i\sigma^1 \\ i\sigma^1 & 0 \end{pmatrix}, \quad \gamma^2 = \begin{pmatrix} 0 & -i\sigma^2 \\ -i\sigma^2 & 0 \end{pmatrix}, \quad \gamma^3 = \begin{pmatrix} 0 & i\sigma^3 \\ i\sigma^3 & 0 \end{pmatrix},$$

where I1 is unit 2×2 matric and σ^i are Pauli matrices.

By the eigenvalue of this two Casimir operator, the unitary irreducible representation of de sitter group SO(1,4) is characterized "the principal, complementary and discrete series" [23]. The UIR's $U_{s,\nu}$ in the principal series correspond to the:

$$Q_s^{(1)} = \left(\frac{9}{4} + \nu^2 - s(s+1)\right), \tag{2.8}$$

$$Q_s^{(2)} = \left(\left(\frac{1}{4} + \nu^2\right) s(s+1)\right), \tag{2.9}$$

where s is spin and ν is real positive parameter. The second-order field equations can be written as:

$$\langle Q_s^{(1)} - \langle Q_s^{(1)} \rangle\rangle \psi = 0. \tag{2.10}$$

For massive field with $s = \frac{1}{2}$, we have:

$$\left(Q_{\frac{1}{2}}^{(1)} - \frac{3}{2}\right)\psi(x) = \nu^2 \psi(x), \tag{2.11}$$

where $\psi(x)$ is a 4-component spinor with arbitrary degree of homogeneity ($x \cdot \partial\psi = \sigma\psi$). The second-order Casimir operator for spin-$\frac{1}{2}$ is given by

$$Q_{\frac{1}{2}}^{(1)} = -\frac{1}{2} M_{\alpha\beta} M^{\alpha\beta} - \frac{1}{2} S_{\alpha\beta} S^{\alpha\beta} - S_{\alpha\beta} M^{\alpha\beta}, \tag{2.12}$$

where one can show

$$Q_0^{(1)} = -\frac{1}{2}M_{\alpha\beta}M^{\alpha\beta}, \quad \frac{1}{2}S_{\alpha\beta}S^{\alpha\beta} = \frac{5}{2},$$
$$S_{\alpha\beta}M^{\alpha\beta} = -\frac{i}{2}\gamma_\alpha\gamma_\beta M^{\alpha\beta} = -\not{x}\,\not{\partial}^T, \tag{2.13}$$

note that $\gamma_\alpha x^\alpha \equiv \not{x}$. The massless elementary system with spin s is described by the discrete series $\prod_{p,s}^{\pm}$ with $\Delta = (s,s)$ for which [23]:

$$Q_s^{(1)} = 2(1-s^2), \quad Q_s^{(2)} = s^2(1-s^2).$$

In the representation of parameters $\Delta = (\frac{1}{2},\frac{1}{2})$, the eigenvalues of the Casimir operators become:

$$\left\langle Q_{\frac{1}{2}}^{(1)}\right\rangle = \frac{3}{2}, \quad \left\langle Q_{\frac{1}{2}}^{(2)}\right\rangle = \frac{3}{16}. \tag{2.14}$$

Then, Eq. (2.11) is substituted by

$$\left(Q_{\frac{1}{2}}^{(1)} - \frac{3}{2}\right)\psi(x) = 0, \tag{2.15}$$

after using the relations (2.13) this equation is written as:

$$(Q_0 + \not{x}\,\not{\partial}^\top - 4)\psi = 0. \tag{2.16}$$

Now, consider the de Sitter–Dirac operator D that define by:

$$D \equiv -\frac{i}{2}\gamma_\alpha\gamma_\beta M^{\alpha\beta} + 2 = -\not{x}\bar{\not{\partial}} + 2. \tag{2.17}$$

With using this operator, Eq. (2.11) can be written as follow:

$$(iD - \nu)(iD + \nu)\psi(x) = 0, \tag{2.18}$$

this relation is similar to the $(i\not{\partial} + m)(i\not{\partial} - m)\psi(x) = 0$, in Minkowski space. Then, the first order of field equation for particle with spin-$\frac{1}{2}$ and nonzero mass in de Sitter space introduced as follow:

$$(iD + \nu)\psi(x) = 0, \tag{2.19}$$

where $\nu \epsilon R$ and $\nu \neq 0$. Solutions of the de Sitter dirac first-order equation are also the solutions of Eq. (2.18). For massless case we have:

$$iD\psi(x) = 0. \tag{2.20}$$

Therefore, with using the above equation, the de Sitter massless spin-$\frac{1}{2}$ field equation (2.15) will be:

$$(Q_0 - 2)\psi = 0. \tag{2.21}$$

Now let us consider de Sitter massless spin-$\frac{3}{2}$ field equation. As previously mentioned [25]. The spinor–vector unitary irreducible representations can be classified using the eigenvalues of $Q^{(1)}$ and the field equation can be written as:

$$\left(Q_{\frac{3}{2}}^{(1)} - \left\langle Q_{\frac{3}{2}}^{(1)}\right\rangle\right)\kappa_\alpha(x) = 0. \tag{2.22}$$

For spin-$\frac{3}{2}$ field, according to the possible values of p and q, two types of the unitary irreducible representations are distinguished for the de Sitter group SO(1,4) namely, the principal and discrete series.

1. The unitary irreducible representations $U^{\frac{3}{2},\nu}$ in the principal series where $p = s = \frac{3}{2}$ and $q = \frac{1}{2} + i\nu$ correspond to the Casimir spectral values:

   $$\left\langle Q_{\frac{3}{2}}^{(1)}\right\rangle = \nu^2 - \frac{3}{2}, \quad \nu \in \mathrm{IR} \quad \nu > \frac{3}{2},$$

   Note that $U^{\frac{3}{2},\nu}$ and $U^{\frac{3}{2},-\nu}$ are equivalent.
2. The unitary irreducible representations $\Pi^{\pm}_{\frac{3}{2},q}$ of the discrete series, where $p = s = \frac{3}{2}$ correspond to:

   $$\left\langle Q_{\frac{3}{2}}^{(1)}\right\rangle = -\frac{5}{2}, \quad q = \frac{3}{2}, \quad \Pi^{\pm}_{\frac{3}{2},\frac{3}{2}}, \tag{2.23}$$

   $$\left\langle Q_{\frac{3}{2}}^{(1)}\right\rangle = -\frac{3}{2}, \quad q = \frac{1}{2}, \quad \Pi^{\pm}_{\frac{3}{2},\frac{1}{2}}, \tag{2.24}$$

where the sign $\pm$ stands for the helicity. The "massless" spin-$\frac{3}{2}$ field in dS space corresponds to the discrete series $\Pi^{\pm}_{\frac{3}{2},\frac{3}{2}}$ and the field equation is:

$$\left(Q_{\frac{3}{2}}^{(1)} + \frac{5}{2}\right)\kappa_\alpha(x) = 0, \tag{2.25}$$

where

$$Q_{\frac{3}{2}}^{(1)}\kappa_\alpha(x) = \left(-\frac{1}{2}M_{\alpha\beta}M^{\alpha\beta} + \frac{i}{2}\gamma_\alpha\gamma_\beta M^{\alpha\beta} - 3 - \frac{5}{2}\right)\kappa_\alpha(x)$$
$$- 2\partial x.\kappa(x) + 2x\partial.\kappa(x) + \gamma\gamma.\kappa(x). \tag{2.26}$$

However, with the subsidiary condition $\partial.\kappa = \partial^\top.\kappa = \gamma.\kappa(x) = 0$, the "massless" spin-$\frac{3}{2}$ field is singular. This type of singularity is actually due to the divergencelessness condition needed to associate this field with a specific unitary irreducible representation of dS group. To solve this problem, the subsidiary condition must be dropped. Then, the field equation (2.25) replaced with the following equation

$$\left(Q_{\frac{3}{2}}^{(1)} + \frac{5}{2}\right)\kappa_\alpha(x) - D_{\frac{3}{2}\alpha}\bar{\partial}.\kappa(x) = 0, \tag{2.27}$$

where

$$D_{\frac{3}{2}\alpha} = -\bar{\partial}_\alpha - \bar{\gamma}_\alpha\not{x}.$$

One can show that this equation is invariant under the gauge transformation:

$$\kappa_\alpha(x) \longrightarrow \kappa'_\alpha(x) = \kappa_\alpha(x) + D_{\frac{3}{2}\alpha}\zeta. \tag{2.28}$$

where ζ is an arbitrary spinor field. By introducing a gauge fixing parameter c, the wave equation now reads as:

$$\left(Q^{(1)}_{\frac{3}{2}} + \frac{5}{2}\right)\kappa(x) - cD_{\frac{3}{2}\alpha}\bar{\partial}.\kappa(x) = 0, \tag{2.29}$$

The role of c is just to fix the gauge field ζ. The first-order field equation is now simply introduced as follows:

$$\not x \bar{\not\partial}\kappa_\alpha(x) - \kappa_\alpha(x) - x_\alpha \not x \not\kappa + D_{\frac{3}{2}\alpha} \not x \not\kappa = 0, \tag{2.30}$$

where this equation is invariant under the gauge transformation as follow:

$$\kappa_\alpha(x) \longrightarrow \kappa'_\alpha(x) = \kappa_\alpha(x) + \partial^\top_\alpha \zeta. \tag{2.31}$$

The solutions of this equation obviously satisfied the field equation (2.29).

There exists another first-order field equation:

$$\not x \bar{\not\partial}\kappa_\alpha(x) - 3\kappa_\alpha(x) - x_\alpha \not x \not\kappa - \partial^\top_\alpha \not x \not\kappa = 0, \tag{2.32}$$

this equation is invariant under the gauge transformation $\kappa_\alpha(x) \longrightarrow \kappa'_\alpha(x) = \kappa_\alpha(x) + D_{\frac{3}{2}\alpha}\zeta$.

UIRs of the conformal group

In the Minkowski space, the massless field equations are conformally invariant. For every massless representation of Poincare group, there exists only one corresponding representation in the conformal group [26, 27]. The massless field with spin-$\frac{1}{2}$ associated with discrete series $\prod^{\pm}_{p,s}$ and their unitary irreducible representation are $\prod^{\pm}_{\frac{1}{2},\frac{1}{2}}$, where $p = q = s = \frac{1}{2}$ correspond to $\langle Q^{(1)}_{\frac{1}{2}} \rangle = \frac{3}{2}$ where these two representations have a Minkowskian interpretation. For spin-$\frac{3}{2}$ field, the two unitary irreducible representations $\Pi^{\pm}_{\frac{3}{2},\frac{3}{2}}$ have a Minkowskian interpretation. The direct sum of two UIR's $\mathcal{C}(j+1, j, 0)$ and $\mathcal{C}(-(j+1), j, 0)$ of the conformal group SO(2,4), with positive and negative energy and $j = \frac{1}{2}$ for spinor field $j = \frac{3}{2}$ for spin-$\frac{3}{2}$ field, is a unique extension of representation $\prod^{\pm}_{j,j}$. The massless Poincare UIR's $P^{>}(0, j)$ and $P^{<}(0, j)$ with positive and negative energies, respectively, and positive helicity. The following diagrams illustrate these relations:

$$\begin{array}{ccccccc} & & \mathcal{C}(j+1,j,0) & & \mathcal{C}(j+1,j,0) & \longleftrightarrow & \mathcal{P}^{>}(0,j) \\ \Pi^{+}_{j,j} & \hookrightarrow & \oplus & \xrightarrow{H=0} & \oplus & & \oplus \\ & & \mathcal{C}(-(j+1),j,0) & & \mathcal{C}(-(j+1),j,0) & \longleftrightarrow & \mathcal{P}^{<}(0,j), \end{array} \tag{3.1}$$

$$\begin{array}{ccccccc} & & \mathcal{C}(j+1,0,j) & & \mathcal{C}(j+1,0,j) & \longleftrightarrow & \mathcal{P}^{>}(0,-j) \\ \Pi^{-}_{j,j} & \hookrightarrow & \oplus & \xrightarrow{H=0} & \oplus & & \oplus \\ & & \mathcal{C}(-(j+1),0,j) & & \mathcal{C}(-(j+1),0,j) & \longleftrightarrow & \mathcal{P}^{<}(0,-j), \end{array} \tag{3.2}$$

where the arrows $\hookrightarrow$ designate unique extension. It is important to note that the representations $\Pi^{\pm}_{\frac{3}{2},\frac{1}{2}}$ do not have corresponding flat limit.

Diracs six-cone and the projective of the six-cone

Diracs six-cone is a five-dimensional super-surface in IR^6 that

define by:

$$u^2 = \eta_{ab}u^a u^b = u_0^2 - \overrightarrow{u}^2 + u_5^2 = 0, \tag{4.1}$$

where

$$\eta_{ab} = \mathrm{diag}(1, -1, -1, -1, -1, 1), \quad a, b = 0, 1, 2, 3, 4, 5$$

$$\overrightarrow{u} \equiv (u_1, u_2, u_3, u_4).$$

If operator $\hat{A}$ that act on the field ϕ in IR^6 satisfy the relation

$$\hat{A}u^2\phi = u^2\hat{A}\phi,$$

for any ϕ, it is to be intrinsic. Wave equations, subsidiary conditions, etc., must be expressed in terms of operators that are defined intrinsically on the cone. One of the intrinsic operators is powers of d'Alembertian $(\partial_a\partial^a)^n$ which acts intrinsically on fields of conformal degree $(n-2)$. For tensor or spinor fields of degree $-1, 0, 1, \ldots$, the intrinsic wave operators are $\partial^2, (\partial^2)^2, (\partial^2)^3, \ldots$ respectively [28]. Thus, the following CI system of equations, on the cone, has been used [1]:

$$\begin{cases} (\partial_a\partial^a)^n\Psi = 0, \\ \hat{N}_5\Psi = (n-2)\Psi, \end{cases} \tag{4.2}$$

where Ψ is tensor or spinor field of a definite rank and symmetry. Other examples of the intrinsic operators are:

1. Fifteen generators of the conformal group $\mathrm{SO}_0(2,4)$
   $$M_{ab} = i(u_a\partial_b - u_b\partial_a),$$
2. The conformal-degree operator N_5
   $$N_5 \equiv u^a\partial_a,$$
3. The intrinsic gradient
   $$\mathrm{Grad}_a \equiv u_a\partial_b\partial^b - (2N_5 + 4)\partial_a,$$
4. The powers of d'Alembertian

$$(\partial_a\partial^a)^n,$$

which acts intrinsically on fields of conformal degree (n-2).

With CI conditions that added to the above system, the space of solutions restricted. The following conditions are introduced by:

1. transversality

$$u_a\Psi^{ab\ldots}=0,$$

2. divergencelessness

$$\mathrm{Grad}_a\Psi^{ab\ldots}=0,$$

3. tracelessness

$$\Psi^a_{\ ab\ldots}=0.$$

Now, with regard to the following relations:

$$\begin{cases} x^\alpha=(u^5)^{-1}u^\alpha,\\ x^5=u^5,\end{cases}\tag{4.3}$$

the coordinates on the cone $u^2=0$ projected to the $4+1$ de sitter space. Note that in the projective coordinate, x^5 becomes superfluous. So that intrinsic operators that introduced read as:

1. The ten $SO_0(1,4)$ generators

$$M_{\alpha\beta}=i(x_\alpha\partial_\beta-x_\beta\partial_\alpha),\tag{4.4}$$

2. The conformal-degree operatorN_5

$$N_5=x_5\frac{\partial}{\partial x_5},\tag{4.5}$$

3. The conformal gradient [29]

$$\mathrm{Grad}_\alpha=-x_5^{-1}x_\alpha[Q^0-N_5(N_5-1)]+2\bar{\partial}_\alpha(N_5+1),\tag{4.6}$$

4. The powers of d'Alembertian$(\partial_a\partial^a)^n$ which acts intrinsically on field of conformal degree (n-2)

$$(\partial_a\partial^a)^n=-x_5^{-2n}\prod_{j=1}^{n}[Q_0+(j+1)(j-2)]\,.\tag{4.7}$$

In the next section, we use this formalism to write conformal invariant wave equation of massless field.

Conformal invariant spinor field equation in de sitter space

For spinor field, the simplest conformally invariant system is obtained from (4.2) with $n=1$:

$$\begin{cases}(\partial_a\partial^a)\Psi=0,\\ N_5\Psi=-\Psi.\end{cases}\tag{5.1}$$

We introduce $\psi=x_5\Psi$ where Ψ is spinor field on the cone and ψ is spinor field on the de Sitter space. With using (4.7) the conformally invariant equation derived as:

$$(Q_0-2)\psi=0,\tag{5.2}$$

which is a massless conformally coupled spinor field in de Sitter space. After making use of (2.16), the first-order field equation in this case becomes as follows:

$$(\not{x}\,\not{\partial}^\top-2)\psi(x)=0,\tag{5.3}$$

therefore, the field $\psi(x)$ associates with the UIR of dS group $\Pi^{\pm}_{\frac{1}{2},\frac{1}{2}}$ and propagates on the dS light cone.

Now, we apply this method to the spin-$\frac{3}{2}$ field. In this case, six degrees of freedom of spinor field on the cone are classified as:

$$\kappa_\alpha=x_5(\Psi_\alpha+x_\alpha x.\Psi),\quad \psi_1=x_5\Psi_5,\quad \psi_2=x_5x.\Psi,\tag{5.4}$$

where ψ_1,ψ_2 are spinor fields on dS space. The condition $x.\kappa=0$ satisfied by the above definition, therefore, κ_α is a spin-$\frac{3}{2}$ field that lives on de Sitter hyperboloid. With using $n=1$ in (4.2) conformal invariant wave equation for Ψ_α obtained as follows (Appendix 2):

$$(Q_0-2)\Psi_\alpha=0,\tag{5.5}$$

where Ψ_α is a spin-$\frac{3}{2}$ field on the cone.

After doing some calculations, following CI system of field equations is obtained (see Appendix 2):

$$(Q_0-2)\kappa_\alpha+2x_\alpha\bar{\partial}.\kappa+\bar{\partial}_\alpha\bar{\partial}.\kappa=0,\tag{5.6}$$

$$(Q_0-2)\psi_1=0,\quad (Q_0-2)\psi_2=0,\tag{5.7}$$

that indicates ψ_1 and ψ_2 are both conformal invariant massless spinor fields. Using the transversality condition on the cone we obtain:

$$\psi_2=\frac{1}{2}\bar{\partial}.\kappa,$$

$$(Q_0-2)\bar{\partial}.\kappa=0.\tag{5.8}$$

However, one can use (2.26), to write the CI equation (5.6) as:

$$\left(Q^{(1)}_{\frac{3}{2}}-\not{x}\,\bar{\not{\partial}}+\frac{7}{2}\right)\kappa_\alpha+\bar{\partial}_\alpha\bar{\partial}.\kappa-\gamma_\alpha\gamma.\kappa=0,\tag{5.9}$$

with using Eqs. (2.25) and (2.30) the above field equation can be rewritten as:

$$\left(Q^{(1)}_{\frac{3}{2}}-\frac{5}{2}\right)\kappa_\alpha-(\not x\,\not\bar\partial^\top\kappa_\alpha-\kappa_\alpha-x_\alpha\not x\,\not\kappa-\bar\partial_\alpha\bar\partial.\kappa+\gamma^\top_\alpha\not\kappa)=0, \tag{5.10}$$

the fields should be projected to the de Sitter space, the transverse projection implies the transversality of fields, $x.\kappa=0$, so that from the homogeneity condition, one obtains $x^\alpha x.\partial\kappa_\alpha=0$. These two conditions impose the following constrains on the projected fields: $x.\Psi=0=\Psi^5$, and consequently $\kappa_\alpha=\Psi_\alpha$. In Appendix 2, it is shown that the CI divergencelessness condition on the cone, namely $\nabla_a\Psi^a=0$, results in $\partial^T{}_\alpha\kappa^\alpha=0=\partial^T{}_\alpha\Psi^\alpha$, which indicates the divergenceless fields are only mapped from the cone on dS hyperboloid..

For simplicity and irreducibility of vector–spinor field representation, the CI condition $\gamma^a\Psi_a=0$ on the cone is imposed, this leads to $\gamma^\beta\kappa_\beta=\not\kappa=0$, which is the conformally invariant condition on the de Sitter hyperboloid. Imposing this condition and irreducibility (see (2.25)), one receives the following first- and second-order CI field equations

$$(\not x\,\not\partial^T-1)\kappa_\alpha(x)=0,\quad\text{and}\quad\left(Q^{(1)}_{\frac{3}{2}}+\frac{5}{2}\right)\kappa_\alpha=0. \tag{5.11}$$

In this case, κ_α associates with the UIR of dS group, namely $\Pi^\pm_{\frac{3}{2},\frac{3}{2}}$, and note that it propagates on the dS light cone. In the following sections, we find the solution of this vector–spinor field.

The solutions of conformally invariant field equation

Using the de Sitter plane waves [30], the de Sitter–Dirac plane wave for spinor field calculated [23]. In this section, the solution of the conformally invariant wave equation in terms of the de Sitter–Dirac plan wave calculated. A general solution of Eq. (5.11) can be written in terms of spinor fields ψ_1,ψ_2,ψ_3 as follows:

$$\kappa_\alpha(x)=Z^\top_\alpha\psi_1+D_{\frac{3}{2}\alpha}\psi_2+\gamma^\top_\alpha\psi_3, \tag{6.1}$$

where Z is an arbitrary five-component constant vector field

$$Z^\top_\alpha=\theta_{\alpha\beta}Z^\beta=Z_\alpha+H^2x_\alpha x\cdot Z,\quad x\cdot Z^\top=0.$$

If we want the field κ_α to obey simultaneously the second- and first-order field equation (5.11) we find that, the spinor fields ψ_1,ψ_2 and ψ_3 must obey the following equations:

$$(Q_0+\not x\,\not\partial^\top-3)\psi_1=0, \tag{6.2}$$

$$2(x.Z)\psi_1+(Q_0+\not x\,\not\partial^\top)\psi_2=0, \tag{6.3}$$

$$[\not x(Z.x)+\not Z]\psi_1+(Q_0+\not x\,\not\partial^\top)\psi_3=0, \tag{6.4}$$

and

$$(\not x\,\not\partial^\top-1)\psi_1=0, \tag{6.5}$$

$$(\not x-2)\psi_2-2\not x\psi_3=0, \tag{6.6}$$

$$\not x\,\not\partial^\top\psi_3-(4\not x+1)\psi_2-\not x.Z\psi_1=0, \tag{6.7}$$

with using (6.2) and (6.5) we find that:

$$(Q_0-2)\psi_1=0, \tag{6.8}$$

that indicates ψ_1 is massless conformally coupled spinor field with homogeneity degree of $\sigma=-1$ and -2 [23].

In Eq. (6.1), all sentences must have the same degree of homogeneity. The homogeneity degree of $Z^\top_\alpha$ and $\gamma^\top_\alpha$ is zero. Therefore, the homogeneity degree of ψ_3 must be equal with ψ_1. So, the homogeneity degree of ψ_2 must be $\sigma=0,-3$.

Eq. (6.6) results in

$$\psi_2=\frac{2}{5}(1-2\not x)\psi_3,\quad\text{or}\quad\psi_3=\frac{1}{2}(1+2\not x)\psi_2. \tag{6.9}$$

Now let us multiply $\not\partial^T$ from the left on Eqs. (6.6) and (6.7), after doing some calculations, this yields

$$Q_0\psi_2=\frac{2}{5}(1-2\not x)\psi_2+2[3x.Z-\not x\,\not Z^T]\psi_1, \tag{6.10}$$

$$Q_0\psi_3=-\frac{2}{5}(2\not x+7)\psi_3+[3x.Z-\not x\,\not Z^T]\psi_1. \tag{6.11}$$

Inserting these results in Eqs. (6.2), (6.3), (6.4) and after making use of (6.9), one can write ψ_2 and ψ_3 in term of ψ_1 as follows

$$\psi_2=\frac{1}{2}(1+3\not x)[\not Z+(1+3\not x)x\cdot Z]\psi_1, \tag{6.12}$$

$$\psi_3=-\frac{5}{4}(1-\not x)[\not Z+(1+3\not x)x\cdot Z]\psi_1. \tag{6.13}$$

Using the divergencelessness condition and the above relations, one obtains

$$(Z\cdot\partial^T+3x\cdot Z)\psi_1+\frac{1}{2}(\not x-3)[\not Z+(1+3\not x)x\cdot Z]\psi_1=0, \tag{6.14}$$

and making use of $\not\kappa=0$ leads one to write

$$\not Z^T\psi_1-(4\not x+\not\partial^T)\psi_2+4\psi_3=0,$$

and in terms of ψ_1, this equation becomes

$$\not Z^T\psi_1+2(\not x+5)[\not Z+(1+3\not x)x\cdot Z]\psi_1 -\frac{2}{5}(1+2\not x)x\cdot Z\psi_1=0. \tag{6.15}$$

Actually what we have obtained is that by dealing with ψ_1, the other two fields will be established as well. Consequently by gathering all the results, κ can be written as follows

$$\kappa_\alpha(x) = \mathbf{D}_\alpha(\mathbf{x}, \partial^\mathbf{T}, \mathbf{Z})\psi_\mathbf{1}, \tag{6.16}$$

where we have defined

$$\mathbf{D}_\alpha(\mathbf{x}, \partial^\mathbf{T}, \mathbf{Z}) \equiv Z_\alpha^T + \left[\frac{1}{2}\mathcal{D}_\alpha^{(\frac{3}{2})}(1+3\not{x}) - \frac{5}{4}\gamma_\alpha^T(1-\not{x})\right] [\not{Z} + (1+3\not{x})x\cdot Z].$$

In this formalism, a given spin-$\frac{3}{2}$ field could be constructed from the multiplication of the polarization vector $\mathbf{D}_\alpha$ or Z_α (5 degrees of freedom) with the spinor field ψ_1 (2 degrees of freedom), which appears naturally ten polarization states. After making use of (6.14) and (6.15), the degrees of freedom are indeed reduced to the usual four polarization states $m_{\frac{3}{2}} = \frac{3}{2}, \frac{1}{2}, -\frac{1}{2}$ and $-\frac{3}{2}$, where two of them are the physical states $\pm\frac{3}{2}$ [31].

Now, ψ_1 should be identified. Making use of the relation between the $s = 1/2$ and spin-zero Casimir operators, Eq. (6.8) can be written as

$$\left(Q_{\frac{1}{2}}^{(1)} - \frac{1}{2}\right)\psi_1(x) = 0. \tag{6.17}$$

This means that ψ_1 and its related two-point functions can in fact be extracted from a massive spinor field in the principal series representation given by (2.11) by setting $\nu = -i$.

Therefore, the solutions of (6.8) are found to be [23, 32]:

$$\psi_{1\xi,\mathcal{V}}(x) = \mathcal{V}(x,\xi)(Hx.\xi)^{-1}, \tag{6.18}$$

$$\psi'_{1\xi,\mathcal{U}}(x) = \mathcal{U}(\xi)(Hx.\xi)^{-3}, \tag{6.19}$$

where $\mathcal{V}(x,\xi) = \frac{\not{x}\not{\xi}}{x.\xi}\mathcal{V}(\xi)$ and

$$\xi \in C^+ = \{\xi; \eta_{\alpha\beta}\xi^\alpha\xi^\beta = (\xi^0)^2 - \vec{\xi}.\vec{\xi} - (\xi^4)^2 = 0,\ \xi^0 > 0\}.$$

The two spinors $\mathcal{V}(\xi)$ and $\mathcal{U}(\xi)$ are

$$\mathcal{U}^a(\xi) = \frac{\xi^0 - \vec{\xi}.\vec{\gamma}\gamma^0 + 1}{\sqrt{2(\xi^0+1)}}\mathcal{U}^a(\overset{o}{\xi}_+),$$
$$\mathcal{V}^a(\xi) = \frac{1}{\sqrt{2(\xi^0+1)}}\mathcal{U}^a(\overset{o}{\xi}_-), \quad a = 1,2, \tag{6.20}$$

where

$$\mathcal{U}_1(\underset{+}{\overset{o}{\xi}}) = \frac{1}{\sqrt{2}}\begin{pmatrix}\alpha\\ \alpha\end{pmatrix}, \mathcal{U}_2(\overset{o}{\xi}_+) = \frac{1}{\sqrt{2}}\begin{pmatrix}\beta\\ \beta\end{pmatrix}, \tag{6.21}$$

$$\mathcal{U}_1(\underset{-}{\overset{o}{\xi}}) = \frac{1}{\sqrt{2}}\begin{pmatrix}\alpha\\ alpha\end{pmatrix}, \mathcal{U}_2(\overset{o}{\xi}_-) = \frac{1}{\sqrt{2}}\begin{pmatrix}\beta\\ -\beta\end{pmatrix}, \tag{6.22}$$

with $\alpha = \begin{pmatrix}1\\0\end{pmatrix}$, $\beta = \begin{pmatrix}0\\1\end{pmatrix}$ and $\xi = \underset{\pm}{\overset{o}{\xi}} \equiv (1, \vec{0}, \pm 1)$.

Eventually, two solutions for $\kappa_\alpha(x)$ are

$$\kappa_{1\alpha}(x) = \left[Z_\alpha^T + \left[\frac{1}{2}\mathcal{D}_\alpha^{(\frac{3}{2})}(1+3\not{x}) - \frac{5}{4}\gamma_\alpha^T(1-\not{x})\right] [\not{Z} + (1+3\not{x})x\cdot Z]\right]\mathcal{V}(\xi)(Hx.\xi)^{-1}$$
$$\equiv \mathcal{V}_\alpha(x,\xi,Z)(Hx.\xi)^{-1}, \tag{6.23}$$

and

$$\kappa_{2\alpha}(x) = \left[Z_\alpha^T + \left[\frac{1}{2}\mathcal{D}_\alpha^{(\frac{3}{2})}(1+3\not{x}) - \frac{5}{4}\gamma_\alpha^T(1-\not{x})\right] [\not{Z} + (1+3\not{x})x\cdot Z]\right]\mathcal{U}(\xi)(Hx.\xi)^{-3}$$
$$\equiv \mathcal{U}_\alpha(x,\xi,Z)(Hx.\xi)^{-3}. \tag{6.24}$$

By taking the derivative of plan wave $(x.\xi)^\sigma$, the explicit forms of $\mathcal{U}_\alpha$ and $\mathcal{V}_\alpha$ are obtained in terms of ξ as follows:

$$\mathcal{V}_\alpha(x,\xi,Z) = \left[Z_\alpha^T + \left[\frac{1}{2}\mathcal{D}_\alpha^{(\frac{3}{2})}(1+3\not{x}) - \frac{5}{4}\gamma_\alpha^T(1-\not{x})\right] [\not{Z} + (1+3\not{x})x\cdot Z]\right]\frac{\not{x}\not{\xi}}{x.\xi}\mathcal{V}(\xi), \tag{6.25}$$

and

$$\mathcal{U}_\alpha(x,\xi,Z) = \left[Z_\alpha^T + \left[\frac{1}{2}\mathcal{D}_\alpha^{(\frac{3}{2})}(1+3\not{x}) - \frac{5}{4}\gamma_\alpha^T(1-\not{x})\right] [\not{Z} + (1+3\not{x})x\cdot Z]\right]\mathcal{U}(\xi). \tag{6.26}$$

Conclusion

Conformal transformations and conformal techniques have been used in general relativity for a long time [33]. As one knows, the quantum theory of gravity based on Einstein equation is not renormalizable [34]. Also, it is proved that the conformal theories of gravity are better to renormalize [35, 36]. The gravitational field is long range and propagates with the speed of light, thus in the linear approximation, expected the equations governing its dynamic must be conformally invariant, whereas, the Einstein equation is not conformally invariant equation.

In the linear approximation, gravitational field resembles as a massless particle with spin-2 that propagates on the background space–time. In this paper, we study fermionic partner of gravitational field, massless spin-$\frac{3}{2}$ field, in de Sitter space. We used Dirac's six-cone formalism to obtain CI massless spin-$\frac{3}{2}$ wave equation in dS space which corresponds to UIRs of the dS group. We obtain the solutions in ambient space notation.

Appendix 1: Some useful relations

In this appendix, some useful relations are given.

$$\bar{\partial}_\alpha = \partial_\alpha + x_\alpha x.\partial = \partial_\alpha - x_\alpha + x.\partial x_\alpha \quad (7.1)$$

$$Q_0\bar{\partial}_\alpha = \bar{\partial}_\alpha Q_0 + 2\bar{\partial}_\alpha + 2x_\alpha Q_0 \quad (7.2)$$

$$Q_0 x_\alpha = x_\alpha Q_0 - 4x_\alpha - 2\bar{\partial}_\alpha \quad (7.3)$$

$$\bar{\gamma}_\alpha = \Theta^\beta_\alpha \gamma_\beta = \gamma_\alpha + x_\alpha x.\gamma \quad (7.4)$$

$$D_{\frac{3}{2}\alpha} = -\bar{\partial}_\alpha - \bar{\gamma}_\alpha \not{x} \quad (7.5)$$

$$Q^{(1)}_{\frac{3}{2}} D_{\frac{3}{2}} = D_{\frac{3}{2}} Q^{(1)}_{\frac{1}{2}} \quad (7.6)$$

$$Q_0 = -\bar{\partial}^2 \quad (7.7)$$

$$\gamma^\top_\alpha \not{x} = -\not{x}\gamma^\top_\alpha \quad (7.8)$$

$$Q_{\frac{3}{2}}\partial^\top_\alpha = \partial^\top_\alpha (Q_0 + \not{x}\not{\partial}^\top) \quad (7.9)$$

$$\not{\partial}^\top D_{\frac{3}{2}\alpha} = D_{\frac{3}{2}\alpha}\not{\partial}^\top - \not{x}\partial^\top_\alpha - x_\alpha\not{\partial}^\top - 4x_\alpha\not{x} + 3\gamma^\top_\alpha \quad (7.10)$$

$$Q_0 D_{\frac{3}{2}\alpha} = D_{\frac{3}{2}\alpha} Q_0 - 4\partial^\top_\alpha + 4\gamma^\top_\alpha\not{x} - 2\gamma^\top_\alpha\not{\partial}^\top - 2x_\alpha\not{x}\not{\partial}^\top \quad (7.11)$$

$$\not{x}\not{\partial}^\top x_\alpha = x_\alpha\not{x}\not{\partial}^\top + \not{x}\gamma^\top_\alpha \quad (7.12)$$

$$Q_0\not{x}\not{\partial}^\top = \not{x}\not{\partial}^\top Q_0 \quad (7.13)$$

$$Z^\top_\alpha = Z_\alpha + x_\alpha x.Z \quad (7.14)$$

$$Q_0 Z^\top_\alpha = Z^\top_\alpha Q_0 - 2x_\alpha Z.\partial^\top - 4x_\alpha x.Z \quad (7.15)$$

$$\partial^\top.Z^\top = Z.\partial^\top + 4x.Z \quad (7.16)$$

$$\not{Z}^\top = \not{Z} + \not{x}x.Z \quad (7.17)$$

$$\not{\partial}^\top\not{Z}^\top = \not{Z}^\top\not{\partial}^\top + \not{x}\not{Z}^\top + 4x.Z \quad (7.18)$$

$$Q_0\not{x}(x.Z) = \not{x}x.ZQ_0 - 8\not{x}x.Z - 2\not{x}(Z.\partial^\top) - 2\not{\partial}^\top(x.Z) \quad (7.19)$$

$$Q_0\not{x}(Z.\partial^\top) = \not{x}(Z.\partial^\top)Q_0 - 2\not{x}(Z.\partial^\top) + 2\not{x}(x.Z)Q_0 - 2\not{\partial}^\top(Z.\partial^\top) \quad (7.20)$$

$$\not{\partial}^\top(Z.\partial^\top) = -2\not{x}(Z.\partial^\top) + \not{x}(x.Z) \quad (7.21)$$

$$\not{\partial}^\top(x.Z) = -\not{x}(x.Z) + \not{Z}^\top \quad (7.22)$$

Appendix 2: CI wave equation

By considering (4.7) and (5.1) we can write:

$$(Q_0 - 2)\Psi_\alpha = 0 \quad (8.1)$$

then using transversality condition we have:

$$(Q_0 - 2)x.\Psi = 0 \quad (8.2)$$

If we multiply the relation (2.1) this appendix in x_α from left we obtain:

$$Q_0 x_\alpha\Psi^\alpha + 2x_\alpha\Psi^\alpha + 2\bar{\partial}_\alpha\Psi^\alpha = 0$$

then

$$\bar{\partial}.\Psi = -2x.\Psi \quad (8.3)$$

Now we get divergence from κ_α in (5.4) as follows:

$$\bar{\partial}.\kappa = 2x_5 x.\Psi \quad (8.4)$$

where the relation (2.3) in this appendix has been used to obtain the above relation. Also, it is known that

$$(Q_0 - 2)\bar{\partial}.\kappa = 0$$

Now the operator $(Q_0 - 2)$ can work on relationship κ_α in (5.4):

$$(Q_0 - 2)\kappa_\alpha = x_5[-8x_\alpha x.\Psi - 2x_\alpha\bar{\partial}.\Psi - 2\bar{\partial}_\alpha(x.\Psi)]$$

then

$$(Q_0 - 2)\kappa_\alpha = x_5[-4x_\alpha x.\Psi - 2\bar{\partial}_\alpha(x.\Psi)]$$

Finally, we obtain CI spin-$\frac{3}{2}$ equation as follows:

$$(Q_0 - 2)\kappa_\alpha + 2x_\alpha\bar{\partial}.\kappa + \bar{\partial}_\alpha\bar{\partial}.\kappa = 0 \quad (8.5)$$

References

1. Behroozi, S., Rouhani, S., Takook, M.V., Tanhayi, M.R.: Conformally invariant wave-equationsand massless fields in de Sitter space time. Phys. Rev. D **74**, 124014 (2006)
2. Dehghani, M., Rouhani, S., Takook, M.V., Tanhayi, M.R.: Conformally invariant massless spin-2 field in the de Sitter universe. Phys. Rev. D **77**, 064028 (2008)
3. Allen, B.: Vacuum states in de Sitter space. Phys. Rev. D **32**, 3136 (1985)
4. Bros, J., Moschella, U.: Two-point functions and quantum field in the de Sitter Universe. Rev. Math. Phys. **8**, 327 (1996)
5. Allen, B.: An evaluation of the graviton propagator in de Sitter space. Turyn M. Nucl. Phys. B **292**, 813 (1987)
6. Floratos, E.G., Iliopoulos, J., Tomaras, T.N.: Tree-level scattering amplitudes in de Sitter space diverge. Phys. Lett. B **197**, 373 (1987)
7. Antoniadis, I., Mottola, E.: Graviton fluctuation in de Sitter space. J. Math. Phys. **32**, 1037 (1991)
8. Ford, H.L.: Quantum instability of de Sitter spacetime. Phys. Rev. D **31**, 710 (1985)
9. Antoniadis, I., Iliopoulos, J., Tomaras, T.N.: Quntum Instability of de Sitter Space. Phys. Rev. Lett. **56**, 1319 (1986)
10. Folacci, A.: Toy model for the zero-mode problem in the conformal sector of de Sitter quantum gravity. Phys. Rev. D **53**, 3108 (1996)
11. Bros, J., Gazeau, J.P., Moschella, U.: Quantum field theory in the de Sitter universe. Phys. Rev. Lett. **73**, 1746 (1994)
12. Bros, J., Moschella, U.: Two-Point functions and quantum fields in the de Sitter universe. Rev. Math. Phys. **8**, 327 (1996)
13. Gazeau, J.P., Renaud, J., Takook, M.V.: Gupta-Bleuler quantization for minimally coupled scalar field in de Sitter space. Class. Quant. Grav. **17**, 1415 (2000)
14. Gazeau, J.P., Takook, M.V.: Massive vector field in the de Sitter space. J. Math. Phys. **41**, 5920 (2000)

15. Garidi, T., Gazeau, J.P., Takook, M.V.: Massless vector field in de Sitter universe. J. Math. Phys. **43**, 6379 (2002)
16. Gabriel, C., Spindel, P.: Massive spin-2 propagator on de Sitter space. J. Math. Phys. **38**, 622 (1997)
17. Garidi, T., Gazeau, J.P., Takook, M.V.: Massive spin-2 field in De Sitter space. J. Math. Phys. **44**, 3838 (2003)
18. Laoues, M.: Massless particles in arbitrary dimensions. Rev. Math. Phys. **10**, 1079 (1998)
19. Cunningham, E.: The principle of relativity in electrodynamics and an extension thereof. Proc. Math. Soc. London **8**, 77 (1909)
20. Bateman, H.: The principle of relativity in electrodynamics and an extension thereof. Proc. Math. Soc. London **8**, 223 (1910)
21. Van Nieuwenhuizen, P.: Supergravity. Phys.Rept. **68**, 189 (1981)
22. Dirac, P.A.M.: The electron wave equation in the de Sitter space. Ann. Math. **36**, 657 (1935)
23. Bartesaghi, P., Gazeau, J.P., Moschella, U., Takook, M.V.: Dirac fields and thermal effects in the de Sitter universe. Class. Quant. Grav. **17**, 4373 (2001)
24. Moylan, J.P.: Unitary representations of the (4+1)-de Sitter group on irreducible representation spaces of the Poincar group. Math. Phys. **24**, 2706 (1983)
25. Takook, M.V., Azizi, A., Babaian, E.: Covariant quantization of "massive" spin-$\frac{3}{2}$ fields in the de Sitter Space. Eur. Phys. J. C **72**, 2026 (2012)
26. Barut, A.O., Bohm, A.: Reduction of a class of O(4,2) representation with respect to SO(4,1) and SO(3,2). J. Math. Phys. **11**, 2938 (1970)
27. Angelopoulos, E., Laoues, M.: Massless in n dimensions. Rev. Math. Phys. **10**, 271 (1998)
28. Fronsdal, C.: Singletons and massless integral-spin fields on de Sitter space. Phys. Rev. D **20**, 848 (1979)
29. Gazeau, J.P., Hans, M., Murenzi, R.: Invariant bilinear forms on 2+3 de Sitter space. Class. Quant. Grav. **6**, 329 (1989)
30. Bros, J., Gazeau, J.P., Moschella, U.: Quantum field theory in the de Sitter universe. Phys. Rev. Lett. **73**, 1746 (1994)
31. Takook, M.V.: Quantum Field Theory in de Sitter Universe: Ambient Space Formalism. arXiv:1403.1204v2 [gr-qc]
32. Takook, M.V.: Tèorie quantique des champs pour des systèmes èl'ementaires "massifs" ET de "masse nulle" sur l'espace-tempsde de Sitter. Thèse de l'université Paris VI (1997)
33. Faraoni, V., Gunzing, E., Nardone, P.: conformal transformations in classical gravitational theories and in cosmology. Fund. Cosmicphys. **20**, 121 (1999)
34. Birrel, N.D., Davies, P.C.W.: Quantum Fields in Curved Space. Cambridge University Press, Cambridge (1982)
35. Fronsdal, C.: Ghost-free nonlinear spin-two conformal gauge theory. Phys. Rev. D **30**, 2081 (1984)
36. Mannheim, P.D.: Are galactic rotation cureves really flat. Astrophys. J. **479**, 659 (1997)

The structural properties of boron and nitrogen adsorption on benzene molecule: a density functional study

Samaneh S. Hoseini · Edris Faizabadi

Abstract The structural properties of boron and nitrogen atoms added on benzene (Bz) molecule are studied using density functional theory within Gaussian 03 program package. The adsorption energy, HOMO–LUMO energy gap (Δ_{H-L}) and also the optimized bond lengths (C–C and C–H bond lengths) of the structures are evaluated. In this work, three adsorption sites for both boron and nitrogen were selected, hollow site (H), middle site (M) and top site (T), as their initial positions. It is found that for boron adsorption on Bz molecule, the relaxed middle site configuration has the most stable geometry, while in NBz, we obtained similar positions after optimization process. We have also illustrated that the relaxed NBz positions are in higher stability than the relaxed BBz positions. As a consequence, it is found that the stability of an isolated benzene molecule increases by adding boron or nitrogen on top of it.

Keywords Density functional theory · Electronic structure · Configuration · Adsorption energy · HOMO–LUMO energy gap · Optimized bond length

Introduction

Benzene, one of the organic molecules, because of its special properties has long been on the focus of both theoretical and experimental research. In chemical studies, benzene is used as a useful substance including as a solvent or also as a precursor of many synthetic materials, and so on [1]. While some works studied mono-benzene or multi-benzene molecules solely [2, 3], yet others considered benzene rings with hetero-atoms or hetero-molecules [4, 5]. Among all hetero-atoms, we have selected boron and nitrogen as two atoms on both sides of carbon atom in periodic table with difference of one electron less and more than the number of valence electrons in carbon, respectively. Some experimental and theoretical studies were performed on benzene with boron and nitrogen atoms. Kostic et al. [6] have theoretically studied borabenzene anion (BBz^-) and its sandwich complexes [$(BBz)_2Fe$ and $(BBz)_2Co$] to obtain their bonding, electronic structure and substitution reactions properties. A quantum mechanical ab initio calculations of benzene–BX_3(X=H, F, Cl) interaction at the level of Mp2/6-311++G**, was performed by Tarakeshwar et al. [7]. They also carried out some single-point calculations at a much higher level of correlation with larger basis set (MP4/6-311++G(2df,p) + diffuse(d,p)). Their binding energy calculations suggest that all of these complexes have weakly bound van der Waals and the weakest one is for benzene-BF_3. Another investigation on boron with benzene was carried out by Zhang et al. [8]. In their experimental work, the reactions of ground-state were studied using the crossed molecular beam technique. Venturo et al. [9] have done some theoretical calculations on neutral solute-(solvent)$_n$ clusters for the benzene-$(N_2)_n$ case, where $n = 1$–32. The size evolution of N_2 in benzene-(N_2) clusters was measured by mass-selective ionization-loss stimulated Raman spectroscopy in this experimental work. A theoretical study on the complexes of benzene with isoelectronic nitrogen was performed by Wang et al. [10] at the MP2 and DFT-SAPT calculations. They have calculated the C–H bond length and binding energies and have found that all energy components increase with the number of nitrogen atoms.

S. S. Hoseini (✉) · E. Faizabadi
School of Physics, Iran University of Science and Technology, Tehran 16846, Iran
e-mail: s.s.hoseini2288@gmail.com

Srivastava et al. [11] have investigated the influence of adding boron and nitrogen to benzene as impurities instead of the last carbon atom or the last hydrogen atom. They have analyzed their data based on single electron transition (SET) using density functional theory. Their calculations have indicated that at zero gate potential in SET environment, C_5H_6N has more stability in its positively charged state, whereas C_6H_6, C_5H_6B and C_6H_5N are more stable in their neutral state. Besides, C_6H_5B shows more stability in both neutral and positively charged states.

In this work, we have theoretically investigated the stable positions of benzene with boron (BBz) and with nitrogen (NBz) using density functional theory with the Gaussian 03 program package. In addition, the adsorption energies, HOMO–LUMO gaps and optimized bond lengths have been evaluated.

The paper is organized as follows; in the next section the method and numerical approach are presented. In "Results and Discussion", we have discussed our results. Finally, we end the paper with a brief conclusion.

Methods

All of our electronic structure calculations in this work were carried out using density functional theory (DFT) [12] within Gaussian 03 program package [13]. The selected structures of adsorbing boron and nitrogen on benzene were fully optimized at the B3LYP [14, 15] level using 6-31++G(d, p) basis set. In the present work for both boron and nitrogen three adsorption sites were selected, namely hollow site (H), middle site (M) and top site (T) as initial geometric configurations which can be seen in Fig. 1. The most stable structures and optimized bond lengths have been obtained via optimizing these positions. Moreover, the highest occupied and lowest unoccupied molecular orbital (HOMO and LUMO) energy gaps and the adsorption energies (E_{ads}) $\{E_{ads} = E_{Bz} + E_{adatom} - E_{complex}\}$ have been calculated, where E_{Bz} and E_{adatom} are the energy of the isolated benzene and adatom, respectively, and $E_{complex}$ is the energy of the fully relaxed benzene with adatom.

Results and discussion

It should be mentioned that the position of relaxed structure depends on its initial geometric configuration. Thus if we change the primary configuration, we could obtain another final position. We discuss the results as follows;

Boron on benzene

As it is clear in Figs. 1 and 2, all of the initial structures are deformed after optimization process. In hollow site position we have a curved benzene molecule with a boron atom on top of it, whereas in middle site case the boron atom with benzene molecule has formed a flat stable structure. In the third position an approximately plane benzene molecule with a boron atom on it can be seen. According to our calculations the structure with middle site initial configuration has the maximum value of adsorption energy, so it is the most stable structure. Another important quantity that was investigated in this work is HOMO–LUMO energy gap ($\Delta_{H\text{–}L}$) as an indicator of chemical reactivity for a structure, in which a larger $\Delta_{H\text{–}L}$ means less reactivity. It must be noted that, the most sustainable structure in this process has minimum $\Delta_{H\text{–}L}$ and so maximum chemical reactivity. Our results are shown in Table 1.

Comparing the results of boron on benzene molecule with an isolated benzene (see Table 1) shows that the isolated benzene is less stable than complex forms with lower chemical reactivity. The optimized bond lengths are shown in Table 2. By comparing the results of these structures with an isolated benzene molecule after

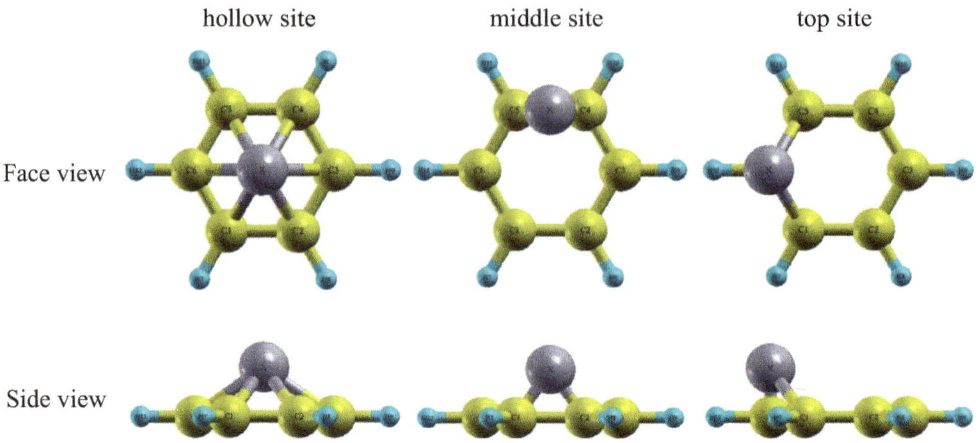

Fig. 1 The initial positions of adsorbing X (X=B, N) on benzene molecule

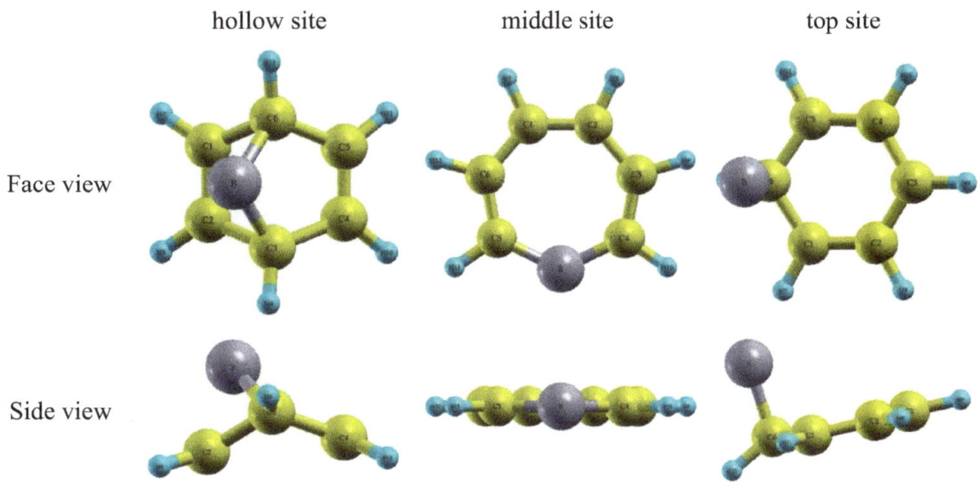

Fig. 2 The relaxed structures of adsorbing a boron atom on benzene molecule

Table 1 The total energies (E_t), the adsorption energies (E_{ads}) and the HOMO–LUMO gaps (Δ_{H-L}) for all of the relaxed structures

	E_t	E_{ads}	Δ_{H-L}
Bz	−6,320.3494	–	0.242
BBz(H)	−6,992.8105	1.440	0.172
BBz(M)	−6,993.8674	2.497	0.083
BBz(T)	−6,991.7476	0.377	0.103
NBz(H)	−7,806.8440	3.864	0.185
NBz(M)	−7,806.8440	3.864	0.185
NBz(T)	−7,806.8440	3.864	0.184

The energies are in (eV)

optimization process, it may be concluded that the C–H bond lengths are approximately similar, whereas some of the C–C bond lengths are different from the value in an isolated benzene which is approximately 1.398 Å. Based on our calculations, the difference between the maximum C–C bond length in complex structures and the C–C bond length in the isolated benzene is 0.104 Å.

Nitrogen on benzene

As it is shown in Fig. 3 the optimized positions are almost similar to each other, so there is no significant difference between their E_{ads} and Δ_{H-L} values. The results are shown in Table 1. Similar to previous stage the isolated relaxed benzene molecule, because of its absolute value of energy, is more unstable and has less chemical reactivity (see Table 1). As it is evident in Table 2, the optimized bond lengths are similar in these three relaxed complex structures and only slight difference is observable. Our finding shows that the C–H bond lengths in complex forms are approximately similar to those in the isolated benzene after optimization process. Whereas the C–C bond lengths are changed and their maximum value is 1.539 Å.

Finally, as a comparative study on adsorption energies between BBz and NBz, we have found that the NBz is more stable than BBz in per position, whereas the chemical reactivity of BBz is higher than NBz. Therefore, as a consequence, our calculation indicates that we can obtain

Table 2 The optimized bond lengths of the relaxed structures

	Bz	BBz(H)	BBz(M)	BBz(T)	NBz(H)	NBz(M)	NBz(T)
C_1–C_2	1.398	1.384	1.379	1.377	1.472	1.451	1.473
C_2–C_3	1.398	1.501	1.426	1.416	1.539	1.355	1.355
C_3–C_4	1.398	1.501	1.380	1.417	1.472	1.473	1.452
C_4–C_5	1.398	1.343	–	1.376	1.355	1.539	1.355
C_5–C_6	1.398	1.502	1.380	1.492	1.451	1.473	1.472
C_6–C_1	1.398	1.500	1.426	1.493	1.355	1.355	1.539
C_1–H_7	1.086	1.085	1.089	1.085	1.086	1.086	1.090
C_2–H_8	1.086	1.085	1.089	1.087	1.090	1.086	1.086
C_3–H_9	1.086	1.086	1.092	1.084	1.090	1.086	1.086
C_4–H_{10}	1.086	1.086	1.091	1.087	1.086	1.090	1.086
C_5–H_{11}	1.086	1.086	1.091	1.085	1.086	1.090	1.086
C_6–H_{12}	1.086	1.086	1.092	1.097	1.086	1.086	1.090

The bond lengths are in Å

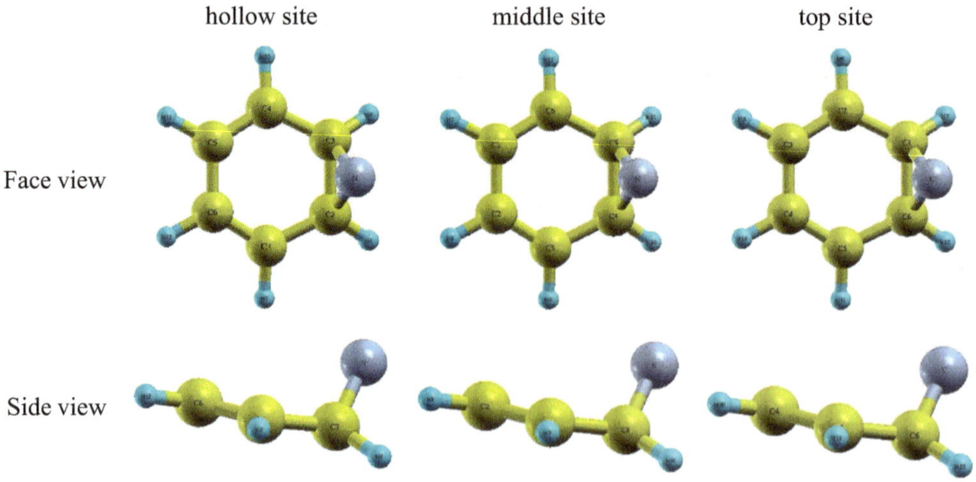

Fig. 3 The relaxed structures of adsorbing a nitrogen atom on benzene molecule

more stable structures by adsorbing nitrogen atom on a benzene molecule.

Conclusions

We studied the structural properties of adsorbing boron and nitrogen atoms on benzene molecule in three site positions using density functional theory method at the level of B3LYP/6-31++G(d, p). Via optimization process, the stable geometrical structures and optimized bond lengths have been obtained. We have investigated the most stable structures and their chemical reactivity by calculating the adsorption energies and HOMO–LUMO energy gaps, respectively. According to our results, the relaxed middle site position in the case of adsorbing boron atom is the most stable structure, whereas we obtained three similar configurations by relaxing the case of adsorbing nitrogen atom. Comparing these two groups of structures, we found that the NBz structures with higher adsorption energies are more stable, however, they are in less chemically reactive. Finally, we have realized that by adding boron and nitrogen, we are able to obtain the structures with more stability and chemical reactivity compared to isolated benzene.

Acknowledgments This work was supported by Iran University of Science and Technology Grant.

References

1. Foo, S.C.: Benzene pollution from gasoline usage. Sci. Total Environ. **103**, 19–26 (1991)
2. Pavone, M., Rega, N., Barone, V.: Implementation and validation of DFT-D for molecular vibrations and dynamics: the benzene dimer as a case study. Chem. Phys. Lett. **452**, 333–339 (2008)
3. Bludský, O., Rubeš, M., Soldán, P., Nachtigall, P.: Investigation of the benzene-dimer potential energy surface: DFT/CCSD(T) correction scheme. J. Chem. Phys. **128**, 114102 (2008)
4. Rungsirisakun, R., Jansang, B., Pantu, P., Limtrakul, J.: The adsorption of benzene on industrially important nanostructured catalysts (H-BEA, H-ZSM-5, and H-FAU): confinement effects. J. Mol. Struct. **733**, 239–246 (2005)
5. Hansen, N., Kerber, T., Sauer, J., Bell, A.T., Keil, F.J.: Quantum chemical modeling of benzene ethylation over H-ZSM-5 approaching chemical accuracy: a hybrid MP2:DFT study. J. Am. Chem. Soc. **132**, 11525–11538 (2010)
6. Kostić, N.M., Fenske, R.F.: Molecular orbital study of bonding, structure, and substitution reactions of Bis(borabenzene)iron. Organometallics **2**, 1319–1325 (1983)
7. Tarakeshwar, P., Lee, S.J., Lee, J.Y., Kim, K.S.: Ab initio study of benzene-BX_3 (X=H, F, Cl) interactions. J. Phys. Chem. B. **103**, 184–191 (1999)
8. Zhang, F., Guo, Y., Gu, X., Kaiser, R.I.: A crossed molecular beam study on the reaction of boron atoms, $B(^2P_j)$, with benzene, $C_6H_6(X^1A_{1g})$, and D6-benzene $C_6D_6(X^1A_{1g})$. Chem. Phys. Lett. **440**, 56–63 (2007)
9. Venturo, V.A., Maxton, P.M., Felker, P.M.: Size evolution of solvent vibrational structure in neutral solute-(solvent)$_n$ clusters: benzene-$(N_2)_n$, n = 1–32. J. Phys. Chem. **96**, 5234–5237 (1992)
10. Wang, W., Hobza, P.: Theoretical study on the complexes of benzene with isoelectronic nitrogen-containing heterocycles. ChemPhysChem **9**, 1003–1009 (2008)
11. Srivastava, A., Santhibhushan, B., Dobwal, P.: Performance analysis of impurity added benzene based single-electron transistor. Appl. Nanosci. **4**, 263–269 (2014)
12. Hohenberg, P., Kohn, W.: Inhomogeneous electron gas. Phys. Rev. B. **136**, 864 (1964)
13. Foresman, J.B., Frisch, A.: Exploring chemistry with electronic structure methods. Gaussian Inc., Wallingford (1996)
14. Becke, A.D.: Density-functional thermochemistry. III. The role of exact exchange. J. Chem. Phys. **98**, 5648 (1993)
15. Lee, C., Yang, W., Parr, R.G.: Development of the Colle-Salvetti correlation-energy formula into a functional of the electron density. Phys. Rev. B. **37**, 785 (1988)

10

Eigensolutions of the Schrödinger equation with a class of Yukawa potentials via supersymmetric approach

C. A. Onate[1] · J. O. Ojonubah[2]

Abstract Using the basic concept of the supersymmetric shape invariance approach and formalism, we obtained an approximate solution of the Schrödinger equation with an interaction of inversely quadratic Yukawa potential, Yukawa potential and Coulomb potential which we considered as a class of Yukawa potentials. By varying the potential strengths, we obtained a solution for Hellmann potential, Yukawa potential, Coulomb potential and inversely quadratic Yukawa potential. The numerical results we obtained show that the interaction of these potentials is equivalent to each of the potential.

Keywords Schrödinger equation · Eigensolutions · Class of Yukawa potentials

Introduction

In the recent time, physicists have developed much interest in searching for the exponential-type potentials. The reason is that most of the exponential-type potentials play an important role in physics, e.g. Yukawa potential is used in plasma, solid-state and atomic physics [1]. As a result, many authors have solved both relativistic and non-relativistic wave equations with these potentials. For instance, Zhang et al. [2] obtained approximate solutions of the Schrödinger equation with the Generalized Morse potential model including a centrifugal term. Jia et al. [3] solved six-parameter exponential-type potentials. Onate [4] obtained relativistic and non-relativistic solutions of the inversely quadratic Yukawa potential. Falaye et al. [5] obtained bound state solutions of the Schrödinger equation with Manning–Rosen potential. Hassanabadi et al. [6, 7] obtained Actual and general Manning–Rosen potential under spin and pseudospin symmetries of the Dirac equation; approximate solutions of the Schrödinger equation under Manning–Rosen potential in arbitrary dimensions via SUSY QM. Hamzavi et al. [8] obtained approximate spin and pseudospin solutions of the Dirac equation for inversely quadratic Yukawa potential and tensor interaction. Maghsoodi et al. [1] solved Dirac particles in the presence of Yukawa potential plus a tensor interaction in SUSY QM frame work. Ikhdair [9] obtained on the bound state solutions of the Manning–Rosen potential including an improved approximation to the orbital centrifugal term.

The solutions of the wave equations (either Schrödinger, Klein–Gordon, Dirac or D.K.P.) with any of these exponential-type potentials are obtained using different methods which include: asymptotic iteration method (AIM) [10–16], Nikiforov–Uvarov (N.U) method [17–20], exact/proper quantization rule [21], supersymmetric method [22–27], $1/N$ shifted expansion method [28], etc.

Motivated by the success in the exponential-type potentials, we attempt to investigate the solutions of the radial Schrödinger equation with a class of Yukawa potentials given as

$$V(r) = \frac{-br + rce^{-\delta r} - ae^{-2\delta r}}{r^2}, \qquad (1)$$

where a, b and c are potential strength and δ is the screening parameter. The potential is obtained by the addition of Hellmann potential and inversely quadratic

✉ C. A. Onate
oaclems14@physicist.net; onateca12@gm.us

[1] Theoretical Physics Section, Department of Physics, University of Benin, Benin, Nigeria

[2] Mathematics Department, Federal College of Education Okene, Okene, Nigeria

Yukawa potential which is equivalent to the interaction of inversely quadratic Yukawa potential, Coulomb potential and Yukawa potential. This potential has its application where its components are useful. It is noted that the exact solution of the radial Schrödinger equation with potential (1) is not possible due to the presence of the inverse square term. Therefore, to obtain an approximate solutions, we employ a suitable approximation scheme. It is found that such approximation proposed by Greene and Aldrich [29]

$$\frac{1}{r^2} = \frac{\delta^2}{\left(1 - e^{-\delta r}\right)^2}. \tag{2}$$

is a good approximation to the centrifugal/inverse square term which is valid for $\delta \ll 1$ for a short potential range.

Our work is arranged as follows: in the following section, we obtain the bound state solutions. We present discussion and concluding remarks at the end of the article.

Bound state solutions

To study any quantum physical system, we solve the original Schrödinger equation given as [9, 30, 31]

$$\left(\frac{P^2}{2\mu} + V(r)\right)\psi_{n,\ell,m}(r) = E_{n,\ell}\psi_{n,\ell,m}(r), \tag{3}$$

where $E_{n,\ell}$ is the energy, μ is the particle mass and $V(r)$ is the potential. Setting the wave functions $\psi_{n,\ell,m}(r) = \frac{R_{n,\ell}(r)Y_{\ell,m}(\theta,\varphi)}{r}$, we obtained the following radial Schrödinger equation

$$\left[\frac{d^2}{dr^2} + \frac{2\mu}{\hbar^2}\left(E_{n,\ell} - V(r)\right) - \frac{\ell(\ell+1)}{r^2}\right]R_{n,\ell}(r) = 0. \tag{4}$$

With Eqs. (1) and (2), (4) becomes

$$\frac{d^2 R_{n,\ell}(r)}{dr^2} = [V_{\text{eff}} - E_{\text{eff}}]R_{n,\ell}(r), \tag{5}$$

where $V_{\text{eff}} = \left[\frac{2\mu\delta}{\hbar^2}(a\delta - b + c) + \ell(\ell+1)\delta^2\right]\frac{e^{-\delta r}}{1-e^{-\delta r}} - \left[\frac{2\mu a\delta^2}{\hbar^2} - \ell(\ell+1)\delta^2\right]\frac{e^{-\delta r}}{(1-e^{-\delta r})^2}$ and $-E_{\text{eff}} = -\frac{2\mu\delta}{\hbar^2}\left(E_{n,\ell} + b\right) + \ell(\ell+1)\delta^2$. Since we are dealing with a Schrödinger-like equation that we solved by means of SUSY QM [6, 32–35], the first step in the SUSY approach is finding the solution of the Riccati equation [6]. Using the shape invariance formalism, it can be easily seen that

$$W^2(r) - \frac{dW(r)}{dr} = V_{\text{eff}} - E_{\text{eff}}, \tag{6}$$

whose solution is given as

$$W(r) = A - \frac{Be^{-\delta r}}{1 - e^{-\delta r}}, \tag{7}$$

The ground state wave function $R_{0,\ell}(r)$ as

$$R_{0,\ell}(r) = \exp\left(-\int W(r)dr\right), \tag{8}$$

where $W(r)$ is called the superpotential function in supersymmetric quantum mechanics [36–38]. Substitute Eq. (7) into Eq. (6), we deduce the following relations

$$A^2 = -E_{\text{eff}}, \tag{9}$$

$$B = \frac{2\mu a\delta}{\hbar^2} - (\ell+1)\delta, \tag{10}$$

$$A = \frac{2\mu\delta(b - c - a\delta) - \ell(\ell+1)\delta^2 - B^2}{2B}. \tag{11}$$

With Eq. (7), we can construct the two partner potentials as follows:

$$\begin{aligned} U_+ &= W^2(r) + \frac{dW(r)}{dr} \\ &= A^2 - \frac{B(B+2A)e^{-\delta r}}{1-e^{-\delta r}} + \frac{B(B+\delta)e^{-\delta r}}{(1-e^{-\delta r})^2}, \end{aligned} \tag{12}$$

$$\begin{aligned} U_+ &= W^2(r) + \frac{dW(r)}{dr} \\ &= A^2 - \frac{B(B+2A)e^{-\delta r}}{1-e^{-\delta r}} + \frac{B(B-\delta)e^{-\delta r}}{(1-e^{-\delta r})^2}. \end{aligned} \tag{13}$$

Using the shape invariance technique [6, 39–42], it can readily be shown that the two partner potentials are shape invariant. Therefore, their relationship is written as

$$U_+(r, a_0) = U_-(r, a_1) + T(a_1), \tag{14}$$

where a_0 is an old set of parameters in which the new set of parameters a_1 is obtained from and $T(a_1)$ is a remainder that is independent of the variable r. Here, $B = a_0$ via mapping of the form $a_1 \to a_0 - \delta,\ a_2 \to a_0 - 2\delta,\ a_3 \to a_0 - 3\delta$. Thus, a generalization is drawn as $a_n \to a_0 - n\delta$. In terms of the parameters of the problem, we obtain the following relations:

$$\begin{aligned} T(a_1) = &\left[\frac{2\mu\delta(b-c-a\delta) - \ell(\ell+1)\delta^2 - a_0^2}{2a_0}\right] \\ &- \left[\frac{2\mu\delta(b-c-a\delta) - \ell(\ell+1)\delta^2 - a_1^2}{2a_1}\right], \end{aligned} \tag{15}$$

$$\begin{aligned} T(a_2) = &\left[\frac{2\mu\delta(b-c-a\delta) - \ell(\ell+1)\delta^2 - a_1^2}{2a_1}\right] \\ &- \left[\frac{2\mu\delta(b-c-a\delta) - \ell(\ell+1)\delta^2 - a_2^2}{2a_2}\right], \end{aligned} \tag{16}$$

$$\begin{aligned} T(a_3) = &\left[\frac{2\mu\delta(b-c-a\delta) - \ell(\ell+1)\delta^2 - a_2^2}{2a_2}\right] \\ &- \left[\frac{2\mu\delta(b-c-a\delta) - \ell(\ell+1)\delta^2 - a_3^2}{2a_3}\right], \end{aligned} \tag{17}$$

$$T(a_n) = \left[\frac{2\mu\delta(b-c-a\delta) - \ell(\ell+1)\delta^2 - a_{n-1}^2}{2a_{n-1}}\right] - \left[\frac{2\mu\delta(b-c-a\delta) - \ell(\ell+1)\delta^2 - a_n^2}{2a_n}\right], \tag{18}$$

The energy spectral can then be determine as follows:

$$E_{0,\ell}^- = 0, \tag{19}$$

$$E_{n,\ell}^- = E_{\text{eff}} + E_{0,\ell}^- = \sum_{k=1}^{n} T(a_k) = \left[\frac{2\mu\delta(b-c-a\delta) - \ell(\ell+1)\delta^2 - a_n^2}{2a_n}\right]. \tag{20}$$

This gives energy equation as

$$E_{n,\ell} = \delta\left(\frac{\delta\hbar^2\ell(\ell+1)}{2\mu} - b\right) - \frac{\delta^2\hbar^2}{2\mu}\left[\frac{\frac{2\mu}{\delta\hbar^2}(b-c-a\delta) - (n+\ell+1)^2 - \ell(\ell+1)}{2(\ell+n+1)}\right]^2. \tag{21}$$

Wave function

To obtain the un-normalized wave function, we define a variable of the form $y = e^{-\delta r}$ and substitute it into Eq. (5) to have

$$\frac{d^2R_{n,\ell}(r)}{dy^2} + \frac{1}{y}\frac{dR_{n,\ell}(r)}{dy} + \frac{Ny^2Qy + P}{(y(1-y))^2}R_{n,\ell}(r) = 0, \tag{22}$$

where

$$P = \frac{2\mu(b\delta + E_{n,\ell})}{\delta^2\hbar^2} - \ell(\ell+1),$$
$$Q = -\frac{2\mu(b\delta + c\delta + E_{n,\ell})}{\delta^2\hbar^2},$$
$$N = \frac{2\mu(a\delta + c\delta^2 + 2E_{n,\ell})}{\delta^2\hbar^2} \tag{23}$$

Analyzing the asymptotic behavior of Eq. (22) at origin and at infinity, it can be tested that when $r \to 0 (y \to 1)$ and when $r \to \infty (y \to 0)$ Eq. (22) has a solution

$$R_{n,\ell}(y) = y^{\eta}(1-y)^{\varepsilon}, \tag{24}$$

where

$$\eta = \sqrt{-\frac{2\mu(b\delta + E_{n,\ell})}{\delta^2\hbar^2} + \ell(\ell+1)}, \tag{25}$$

$$\varepsilon = \frac{1}{2} + \frac{1}{2}\sqrt{1 - \frac{2\mu a}{\delta\hbar^2} + \ell(\ell+1)}. \tag{26}$$

Now, taking a trial wave function of the form $R_{n,\ell}(y) = y^{\eta}(1-y)^{\varepsilon}f(y)$ and substitute it into Eq. (22), we have

$$f''(y) + f'(y)\frac{2\eta + 1 - y(2\eta + 2\varepsilon + 1)}{y(1-y)} + f(y)\frac{(\eta+\varepsilon)^2 + N}{y(1-y)} = 0. \tag{27}$$

Equation (27) is a differential equation satisfied by the hypergeometric function. Thus, its solution is obtained as:

$$f(y) = {}_2F_1(-n, n + 2(\eta+\varepsilon); 2\eta + 1, y). \tag{28}$$

Replacing the function $f(y)$ with the hypergeometric function and write a complete radial wave function as:

$$R_{n,\ell}(y) = y^{\eta}(1-y)^{\varepsilon}{}_2F_1(-n, n + 2(\eta+\varepsilon); 2\eta + 1, y). \tag{29}$$

Results and discussion

Some special cases of interest are studied here, when $a = 0$, our potential (1) reduces to Hellmann potential

$$V(r) = \frac{-b + ce^{-\delta r}}{r}, \tag{30}$$

which reduces the energy Eq. (21) to

$$E_{n,\ell} = \delta\left(\frac{\delta\hbar^2\ell(\ell+1)}{2\mu} - b\right) - \frac{\delta^2\hbar^2}{2\mu}\left[\frac{\frac{2\mu}{\delta\hbar^2}(b-c) - (n+\ell+1)^2 - \ell(\ell+1)}{2(\ell+n+1)}\right]^2. \tag{31}$$

When $a = c = 0$, the potential (1) reduces to Coulomb potential

$$V(r) = \frac{-b}{r}, \tag{32}$$

and the energy Eq. (21) turns to

$$E_{n,\ell} = \delta\left(\frac{\delta\hbar^2\ell(\ell+1)}{2\mu} - b\right) - \frac{\delta^2\hbar^2}{2\mu}\left[\frac{\frac{2\mu b}{\delta\hbar^2} - (n+\ell+1)^2 - \ell(\ell+1)}{2(\ell+n+1)}\right]^2. \tag{33}$$

When $a = b = 0$, the potential (1) reduces to Yukawa potential

$$V(r) = \frac{-ce^{-\delta r}}{r}, \tag{34}$$

and the energy Eq. (21) turns to

Table 1 Bound states energy for the class of Yukawa potential $E_{n,\ell}^{\mathrm{CYP}}$, Hellmann potential $E_{n,\ell}^{\mathrm{HP}}$, Coulomb potential $E_{n,\ell}^{\mathrm{CP}}$ and Yukawa potential $E_{n,\ell}^{\mathrm{YP}}$

State	δ	$E_{n,\ell}^{\mathrm{CYP}}$	$E_{n,\ell}^{\mathrm{HP}}$	$E_{n,\ell}^{\mathrm{CP}}$	$E_{n,\ell}^{\mathrm{YP}}$
2p	0.001	−2.2474553	−2.2475002	−2.2485002	−2.2455002
	0.005	−2.2374615	−2.2375062	−2.2425062	−2.2275062
	0.010	−2.2249805	−2.2250250	−2.2350250	−2.2050250
3p	0.001	−0.9983214	−0.9983347	−0.9993347	−0.9963347
	0.005	−0.9916874	−0.9917007	−0.9967007	−0.9817007
	0.010	−0.9834563	−0.9834694	−0.9934694	−0.9634694
3d	0.001	−0.9969870	−0.9970003	−0.9980003	−0.9950003
	0.005	−0.9849930	−0.9850063	−0.9900063	−0.9750063
	0.010	−0.9700120	−0.9700250	−0.9800250	−0.9500250
4p	0.001	−0.5611224	−0.5611281	−0.5621281	−0.5591281
	0.005	−0.5556960	−0.5557016	−0.5607016	−0.5457016
	0.010	−0.5490507	−0.5490563	−0.5590563	−0.5290563
4d	0.001	−0.5603710	−0.5603766	−0.5613766	−0.5583766
	0.005	−0.5519085	−0.5519141	−0.5569141	−0.5419141
	0.010	−0.5414008	−0.5414063	−0.5514063	−0.5214063
4f	0.001	−0.5592447	−0.5592503	−0.5602503	−0.5572503
	0.005	−0.5462508	−0.5462563	−0.5572563	−0.5362563
	0.010	−0.5300196	−0.5300250	−0.5400250	−0.5100250

$$E_{n,\ell} = \frac{\delta^2\hbar^2\ell(\ell+1)}{2\mu} - \frac{\delta^2\hbar^2}{2\mu}\left[\frac{\frac{-2\mu c}{\delta\hbar^2} - (n+\ell+1)^2 - \ell(\ell+1)}{2(\ell+n+1)}\right]^2. \quad (35)$$

When $b = c = 0$, potential (1) reduces to inversely quadratic Yukawa potential

$$V(r) = \frac{-ae^{-2\delta r}}{r^2}, \quad (36)$$

and the energy Eq. (21) turns to

$$E_{n,\ell} = \frac{\delta^2\hbar^2\ell(\ell+1)}{2\mu} - \frac{\delta^2\hbar^2}{2\mu}\left[\frac{\frac{-2\mu a}{\hbar^2} - (n+\ell+1)^2 - \ell(\ell+1)}{2(\ell+n+1)}\right]^2. \quad (37)$$

In Table 1, we numerically reported the energy eigenvalues for a class of Yukawa potential $(a = 1 \times 10^{-5}, b = 2 \times 10^0, \quad c = -1 \times 10^0)$, Hellmann potential $(a = 0, b = 2 \times 10^0, \ c = -1 \times 10^0)$, Coulomb potential $(a = 0 = c = 0, \quad b = 3)$ and Yukawa potential $(a = b = 0, \ c = -3 \times 10^0)$ for 2p, 3p, 3d, 4p, 4d and 4f. In Table 2, we have reported the energy eigenvalues of these potentials for $n = 0, \quad \in \ell = 0$; $n = 1, \quad \ell = 0, 1$; $n = 2, \ \ell = 0, 1, 2$; $n = 3, \ \in \ell = 0, 1, 2, 3$ and $n = 4, \ \ell = 0, 1, 2, 3, 4$. In Table 1, energy increases as the screening parameter increases. In Table 2, energy increases as n increases. In Table 3, we compared our result for the Hellmann potential with the result from two other methods.

Table 2 Bound states energy for the class of Yukawa potential $E_{n,\ell}^{\mathrm{CYP}}$, Hellmann potential $E_{n,\ell}^{\mathrm{HP}}$, Coulomb potential $E_{n,\ell}^{\mathrm{CP}}$ and Yukawa potential $E_{n,\ell}^{\mathrm{YP}}$

n	ℓ	$E_{n,\ell}^{\mathrm{CYP}}$	$E_{n,\ell}^{\mathrm{HP}}$	$E_{n,\ell}^{\mathrm{CP}}$	$E_{n,\ell}^{\mathrm{YP}}$
0	0	−8.9896656	−8.9900250	−9.0000250	−8.9700250
1	0	−2.2400552	−2.2401000	−2.2501000	−2.2201000
	1	−0.9834563	−0.9834694	−0.9934694	−0.9634694
2	0	−0.9902117	−0.9902250	−1.0002250	−0.9702250
	1	−0.5490507	−0.5490563	−0.5590563	−0.5290563
	2	−0.3431582	−0.3431610	−0.3531610	−0.3231610
3	0	−0.5528944	−0.5529000	−0.5629000	−0.5329000
	1	−0.3481262	−0.3481290	−0.3581290	−0.3281290
	2	−0.2356234	−0.2356250	−0.2456250	−0.2156250
	3	−0.1670240	−0.1670250	−0.1770250	−0.1470250
4	0	−0.3506222	−0.3506250	−0.3606250	−0.3306250
	1	−0.2391345	−0.2391361	−0.2491361	−0.2191361
	2	−0.1709424	−0.1709434	−0.1809434	−0.1509434
	3	−0.1260556	−0.1260563	−0.1360563	−0.1060563
	4	−0.0948517	−0.0948522	−0.1048522	−0.0748522

Table 3 Ro-vibrational energy spectrum $(-E_{n,\ell})$ for the Hellmann potential with $2m = \hbar = 1$, $b = 2$ and $c = 1$

State	δ	SUSY	NU [43]	AP [43]
2p	0.001	0.063999	0.064000	0.063495
	0.005	0.069975	0.070000	0.067377
	0.010	0.077400	0.077500	0.072020
3p	0.001	0.029499	0.029279	0.028765
	0.005	0.036356	0.035309	0.032480
	0.010	0.044869	0.042903	0.036645
3d	0.001	0.029274	0.029388	0.028767
	0.005	0.035184	0.035817	0.032526
	0.010	0.042403	0.043825	0.036814
4p	0.001	0.017436	0.017128	0.016602
	0.005	0.024652	0.023200	0.020100
	0.010	0.033606	0.030925	0.023711
4d	0.001	0.017308	0.017180	0.016604
	0.005	0.023952	0.023464	0.020142
	0.010	0.032056	0.031256	0.023857
4f	0.001	0.017117	0.017311	0.016607
	0.005	0.022925	0.024027	0.020206
	0.010	0.029825	0.032356	0.024072

Conclusions

We have obtained approximate solutions of the Schrödinger equation by combining inversely quadratic Yukawa potential, Yukawa potential and Coulomb potential. We deduced that from the energy equation of these combined potentials, the energy equation of Hellmann potential as well as these individual potential can be obtained. In Tables 1 and 2, we have numerically reported the equivalence of the energy of these potentials.

References

1. Maghsoodi, E., Hassanabadi, E., Aydoğdu, O.: Dirac particles in the presence of the Yukawa potential plus a tensor interaction in SUSYQM framework. Phys. Scr. **86**, 015005 (2012)
2. Zhang, L.H., Li, X.P., Jia, C.S.: Approximate solutions of the Schrödinger equation with the generalized Morse potential model including the centrifugal term. Int. J. Quantum Chem. **111**, 1870 (2011)
3. Jia, C.S., Zeng, X.L., Li, S.C., Sun, L.T., Yang, Q.B.: Six parameter exponential-type potential and the identity for the exponential-type potentials. Commun. Theor. Phys. **37**, 523 (2002)
4. Onate, C.A.: Relativistic and Non-relativistic Solutions of the Inversely Quadratic Yukawa potential. Afr. Rev. Phys. **8**, 0046 (2013)
5. Falaye, B.J., Oyewumi, K.J., Ibrahim, T.T., Punyasena, M.A., Onate, C.A.: Bound state solutions of the Manning–Rosen potential. Can. J. Phys. **91**, 98 (2013)
6. Hassanabadi, H., Maghsoodi, E., Zarrinkamar, S., Rahimov, H.: Actual and general Manning-Rosen potentials under spin and pseudospin symmetries of the Dirac equation. Can. J. Phys. **90**, 633 (2012)
7. Hassanabadi, H., Lu, L.L., Zarrinkamar, S., Liu, G., Rahimov, H.: Appropriate solutions of Schrodinger equation under Manning-Rosen potential in arbitrary dimensions via SUSYQM. Acta Phys. Pol. A **122**, 650 (2012)
8. Hamzavi, M., Ikhdair, S.M., Ita, B.I.: Approximate spin and pseudospin solutions to the Dirac equation for the inversely quadratic Yukawa potential and tensor interaction. Phys. Scr. **85**, 045009 (2012)
9. Ikhdair, S.M.: On the bound state solutions of the Manning-Rosen potential including an improved approximation to the orbital centrifugal term. Phys. Scr. **83**, 015010 (2011)
10. Bayrak, O., Boztosun, I.: Bound state solutions of the Hulthen potential by using the asymptotic iteration method. Phys. Scr. **76**, 92 (2007)
11. Soylu, A., Bayrak, O., Boztosun, I.: κ state solutions of the Dirac equation for the Eckart potential with pseudospin and spin symmetry. J. Phys. A: Math. Theor. **41**, 065308 (2008)
12. Bayrak, O., Boztosun, I.: The pseudospin symmetric solution of the Morse potential for any κ state. J. Phys. A: Math. Theor. **40**, 11119 (2007)
13. Soylu, A., Bayrak, O., Boztosun, I.: An approximate solution of Dirac-Hulthen problem with pseudospin and spin symmetry for any κ state. J. Math. Phys. **48**, 082302 (2007)
14. Oyewumi, K.J., Falaye, B.J., Onate, C.A., Oluwadare, O.J., Yahya, W.A.: Thermodynamic properties and approximate solutions of the Schrodinger equation with the shifted Deng-Fan potential model. Mol. Phys. doi:10.1080/00268976
15. Oyewumi, K.J., Falaye, B.J., Onate, C.A., Oluwadare, O.J., Yahya, W.A.: k state solutions for the fermionic massive spin-1/2 particles interacting with double ring-shaped Kratzer and Oscillator potentials. Int. J. Mod. Phys. E **23**, 1450005 (2014)
16. Behere, S.H.: Anharmonicity constants of diatomic molecules for external Rydberg and Dimitreva-Zenevich potential functions. Indian J. Phys. **55B**, 272 (1982)
17. Zarrinkamar, S., Rajabi, A.A., Hassanabadi, H., Rahimov, H.: Analytical treatment of the two-body spinless Salpeter equation with the Hulthen potential. Phys. Scr. **84**, 065008 (2011)
18. Satare, M.R., Haidari, S.: Spin symmetry of the Dirac equation with the Yukawa potential. Phys. Scr. **81**, 065201 (2010)
19. Ikhdair, S.M., Sever, R.: Two approximation schemes to the bound states of the Dirac-Hulthen problem. J. Phys. A Math. Theor. **44**, 355301 (2011)
20. Onate, C.A.: Approximate Solutions of the Non-Relativistic Schrodinger Equation with an Interaction of Coulomb and Hulthen potentials. SOP Trans. Theor. Phys. **1**, 118 (2014)
21. Ikhdair, S.M., Hasna, J.A.: Quantization rule solutions to the Hulthen potential in arbitrary dimension with a new approximation scheme for the centrifugal term. Phys. Scr. **83**, 025002 (2011)
22. Wei, G.F., Dong, S.H.: Pseudospin symmetry in the relativistic Manning–Rosen potential including a Pekeris-type approximation to the pseudo-centrifugal term. Phys. Lett. B **686**, 288 (2010)
23. Hassanabadi, H., Zarrinkamar, S., Rahimov, H.: Approximate solution of D-dimensional Klein–Gordon equation with Hulthen-type potential via SUSYQM. Commun. Theor. Phys. **56**, 423 (2011)
24. Jia, C.S., Chen, T., He, S.: Bound state solutions of the Klein-Gordon equation with the improved expression of the Manning-Rosen potential energy model. doi:10.1016/physleta.2013.01.016
25. Hassanabadi, H., Maghsoodi, E., Zarrinkamar, S., Rahimov, H.: Actual and general Manning-Rosen potentials under spin and pseudospin symmetries of the Dirac equation. Can. J. Phys. **90**, 633 (2012)

26. Hassanabadi, H., Lu, L.L., Zarrinkamar, S., Liu, G., Rahimov, H.: Approximate Solutions of Schrodinger Equation under Manning-Rosen Potential in Arbitrary Dimension via SUSY MQ. Acta Phys. Pol., A **122**, 1111 (2012)
27. Onate, C.A.: Bound state solutions of Duffin-Kemmer-Petaiu equation with Yukawa potential. Afr. Rev. Phys. **9**, 0033 (2015)
28. Hammed, R.H.: Approximate Solution of the Schrodinger equation with Manning-Rosen potential in Two Dimensions by using the shifted 1/N expansion method. J. Basrah Res. **38**, 51 (2012)
29. Greene, R.L., Aldrich, C.: Variational wave functions for a screened Coulomb potential. Phys. Rev. A **14**, 2363 (1976)
30. Schiff, L.I.: Quantum Mechanics, 3rd edn. Mc Graw-Hill, New York (1968)
31. Landau, L.D., Lifshitz, E.M.: Quantum Mechanics: Non-Relativistic Theory, 3rd edn. Pergamon, New York (1977)
32. Zhang, L.H., Li, X.P., Jia, C.S.: Few-Body Syst. **52**, 11 (2011)
33. Gendenstein, L.E.: Derivation of Exact Spectra of the Schrodinger Equation by means of supersymmetry. J. Exp. Theor. Phys. Lett. **38**, 356 (1983)
34. Hassanabadi, H., Maghsoodi, E., Zarrinkamar, S., Rahimov, H.: An Approximate solution of the Dirac equation for Hyperbolic scalar and vector potentials and a Coulomb tensor interaction by SUSY QM. Mod. Phys. Lett. A **26**, 2703 (2011)
35. Zarrinkamar, S., Rajabi, A.A., Hassanabadi, H.: Dirac equation for the harmonic scalar and vector potentials and linear plus coulomb like tensor potential : the SUSY approach. Ann. Phys. **325**, 2522 (2010)
36. Cooper, F., Freedman, B.: Aspects of supersymmetric quantum mechanics. Ann. Phys. **146**, 262 (1983)
37. Onate, C.A.: An approximate solution of dirac equation for second poschl-teller like scalar and vector potentials with a coulomb tensor interaction. Afr. Rev. Phys. **8**, 0020 (2013)
38. Ikot, A.N., Maghsoodi, E., Ibanga, E.J., Zarrinkamar, S., Hassanabadi, H.: Spin and pseudospin symmetries of the Dirac equation with shifted Hulthen potential using supersymmetric quantum mechanics. Chin. Phys. B **22**, 120302 (2013)
39. Onate, C.A., Ojonubah, J.O.: Relativistic and Non-relativistic solutions of the generalized Poschl-Teller and Hyperbolical potentials with some Thermodynamic properties. Int. J. Mod. Phys. E **24**, 1550020 (2015)
40. Onate, C.A., Oyewumi, K.J., Falaye, B.J.: Approximate Solutions of the Schrodinger with the Hyperbolical potential: Supersymmetric Approach. Few-Body Syst. **55**, 61 (2014)
41. Hassanabadi, H., Maghsoodi, E., Zarrinkamar, S.: Dirac equation with vector and scalar cornell potentials and an external magnetic field. Ann. der Physik. Ann. der Phys. **525**, 944 (2013)
42. Hassanabadi, H., Maghsoodi, E., Zarrinkamar, S.: Relativistic symmetries of Schioberg and general Manning-Rosen potentials. Commun. Theor. Phys. **58**, 807 (2012)
43. Hamzavi, M., Thylwe, K.E., Rajabi, A.A.: Approximate Bound states solution of the Hellmann potential. Commun. Theor. Phys. **60**, 1 (2013)

11

Quantum analytical modeling and simulation of CNT on insulator (COI) and CNT on nothing (CON) FET: a comparative analysis

Sudipta Mukherjee[1] · Dipan Bandyopadhyay[1] · Pranab Kishore Dutta[1] · Subir Kumar Sarkar[1]

Abstract A comprehensive performance analysis by quantum analytical modeling of CNT on insulator (COI) and CNT on nothing (CON) FET having channel length 20 nm has been proposed and investigated on the basis of 2D Poisson's Equation and solution of 1-D Schrodinger's Equation and validated using ATLAS 2D simulator. As classical approximations fail to describe carrier quantization, charge inversion and potential profile of a device at sub-100 nm regime, here for the first time an analytical model in quantum mechanical aspect for COI/CON devices has been derived. Effects of high-k dielectrics in place of conventional SiO_2 over the device characteristics have been thoroughly discussed. Moreover, all noticeable benefits of our device to the so called SOI/SON architecture have also been vividly justified.

Keywords ATLAS 2D simulator · CNT · COI/CON · High-k dielectric · Inversion charge · Quantum threshold voltage

Introduction

VLSI/ULSI industry is continually being grown up over the platform of rigorous downsizing of devices [1]. This needed an optimization between ultra high speed and ultra low power consumption in making denser circuits [2–6]. But lowering dimensions indeed lowers the control of gate over the channel as lateral electric field turns to be significantly larger than the vertical electric field which in consequence invokes degradation of device performance including DIBL, gate tunnelling, threshold voltage roll-off, hot carrier effect [2–5, 7] etc. Therefore to suppress those undesired phenomena as well as to design the device in such away it can operate faster, researchers moved forward with the discovery of non-classical structures. Fully depleted silicon on insulator (FDSOI) was a device [6] with buried oxide layer to minimize electrostatic coupling, better subthreshold behaviour, and reduced junction capacitance gaining higher speed of operation. But FDSOI [2, 3] suffered a lot from accumulation of positive charges at the buried layer. That is why a little rectification was done to the device structure by replacing thick oxide with embedded air gap to form silicon on nothing (SON) MOSFET [2, 5, 8] keeping all the advantages of SOI structure unaffected.

In our present work, we have designed an implanted CNT channel on weakly inverted SOI and SON devices for higher speed of operation, diminishing SCE and improved subthreshold characteristics. From the very invention in 1991 by S. Iijima [9], CNT being a rolled graphite sheet in tubular form [10–16] originating Quasi-1 dimensional approach [17, 18] in nano-science has been considered attractive by scientists to be used in non-classical devices as it holds exceptional mechanical and electrical properties [9, 18] like very high tensile strength, high flexibility and elasticity, low thermal expansion coefficient, high electrical conductivity and can be found with significantly high aspect ratio. Conductivity of CNT sincerely depends upon its chirality [17, 18], diameter of nanotube and degree of twist [10]. Having strong carbon–carbon bond in the planar honeycomb structure of graphite [9], the modulus of elasticity at its basal plane is superior than most of the materials known has proved itself trustworthy as per their fabrication is concerned. Near ballistic transport [14, 15] of mobile carriers through the CNT channel can be observed

✉ Sudipta Mukherjee
sudipta.conference@gmail.com

[1] Department of E.T.C.E, Jadavpur University, 188, Raja S. C. Mallick Road, Kolkata 700032, West Bengal, India

with the absence of dangling bonds. In this paper, keeping channel length 20 nm and taking diameter of nanotube 1.5 nm for optimum transport, we have analytically modeled COI/CON devices [19] influenced with quantum mechanical [16, 20–23] effects as dimensions can be compared with the De-Broglie wavelength. Novelty of this paper lies on simplicity of our mathematical model for most of the general characteristics of COI/CON devices.

Device structure

Figure 1 schematically shows the cross sectional view of weakly inverted carbon-nanotube on insulator (WICOI) or carbon-nanotube on nothing (WICON) FET. Here source and drain are heavily doped where semiconducting zigzag CNT channel [18] is lightly doped. t_f, t_{cnt}, $t_{ox/air}$, and t_{sub} are front gate oxide thickness, thickness of channel, buried oxide/air gap thickness and substrate thickness, respectively. '*L*' is defined as channel length. Contacts are provided with voltages accordingly.

Analytical model

As potential distribution along the channel is absolutely two dimensional in nature for this kind of devices, Poisson's Equation can be applied [8, 15] at CNT channel along its length and thickness. Assuming comparable acceptor and hole concentration for zigzag (*n*, 0) CNT, the normalized density of states per unit length can be expressed as [17]: $DOS = \sum_{q=1}^{2n} g(E,q)$ where, $g(E,q) = \frac{1}{\pi}\left|\frac{\partial E}{\partial K}\right|^{-1} = \frac{1}{\pi}\left|\frac{\partial K}{\partial E}\right|$ where, *q* is an integer ranging from 1 to 2*n* and *K* is lattice wave vector. Here wave vector being $K_{zz}(E,q) = \pm\frac{2}{\sqrt{3}a} \times \cos^{-1}\left[\frac{1}{4}\sec\left(\frac{\pi q}{n}\right)\left(\frac{E^2}{\gamma_0^2} - 3 - 2\cos\left(\frac{2\pi q}{n}\right)\right)\right]\gamma_0$ is nearest neighbour overlap energy, *a* is bravais lattice constant of CNT $\approx$ 2.46 Å. We may write

$$DOS_{zz}(E)\partial E = \frac{2\alpha}{\sqrt{3}a\pi} \times \sum_{q=1}^{2n} \tan^{-1}\left(\frac{2E^2 - E_{vh1}^2 - E_{vh2}^2}{2\sqrt{(E^2 - E_{vh1}^2)(E_{vh2}^2 - E^2)}}\right)\Bigg|_E^{E+\partial E}$$

for total number of states between *E* and $E + \partial E$ where *E* can be $E_{cb} \le E \le E_{ct}$ for conduction band and $E_{vb} \le E \le E_{vt}$ for valance band implying '*b*' and '*t*' suffixes are used to define bottom and top of the band. Again, E_{vh1} and E_{vh2} being Van Hove Singularity [9–18, 24] energies where DOS is real and finite to be: $E_{vh1} = \pm\gamma_0\left|\left(1 + 2\cos\left(\frac{\pi q}{n}\right)\right)\right|$ and $E_{vh2} = \pm\gamma_0\left|\left(1 - 2\cos\left(\frac{\pi q}{n}\right)\right)\right|$.

Now band gap is determined with the expression: $E_g \approx 2\gamma_0\left(\frac{2n\pi+\pi}{\sqrt{3}n} - \frac{2\pi}{\sqrt{3}}\right) = 2\gamma_0\left(\frac{\pi}{\sqrt{3}n}\right) = 2\gamma_0\left(\frac{a_{cc}}{d_t}\right)$ where $d_t = a_{cc}n\sqrt{3}/\pi$ and a_{cc} is carbon–carbon length $\approx$1.42 Å. Therefore, hole concentration can be written with valance band DOS [17] as: $p = 2 \times \mathrm{DOS}_{zzv}(E) \times e^{\frac{E_{vh1}-E_F}{KT}}$ where, $E_{vh1} = \frac{E_g}{2}$. If drain to source voltage is kept low, dependence of potential on lateral direction can be parabolic in nature and expressed as: $\phi(x,y) = \phi_f(y) + C1(y)x + C2(y)x^2$ where $\phi_f(y)$ is the surface potential, $C1(y)$ and $C2(y)$ are arbitrary and functions of *y* only. *Y* and *X* are considered as horizontal and vertical positional coordinates. If $\varepsilon ox/air$ is the dielectric constant of oxide/air implantation, $\phi_f(y)$ and $\phi_b(y)$ stands for front side oxide-CNT interface and back side CNT-oxide/air interface so that $\phi_f(y) = \phi(0,y)$ and $\phi_b(y) = \phi(t_{cnt}, y)$. Moreover, *tf* is the gate oxide thickness and $t_{ox/air}$ is the buried oxide/air gap thickness. Coefficients and other variables can be found in detail at Appendix: [A].

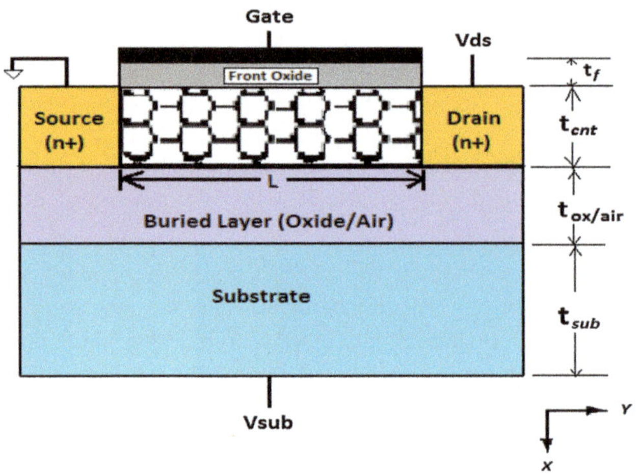

Fig. 1 Cross sectional view of WICOI/WICON FET structure

Surface potential

Now these constants are put on Poisson's Equation and using the parabolic approximation, we get expression of surface potential as:

$$\phi_f(y) = C_1 \times e^{\sqrt{S_f}\cdot y} + C_2 \times e^{-\sqrt{S_f}\cdot y} - \frac{D_f}{S_f} \quad (1)$$

Electric field

The lateral electric field can be obtained by differentiating the surface potential with respect to *y*:

$$E(y) = \frac{d\phi_f(x,y)}{dy}\bigg|_{x=0} = C_1 \times \sqrt{S_f} \times e^{\sqrt{S_f}\cdot y} - C_2 \times \sqrt{S_f} \times e^{-\sqrt{S_f}\cdot y} \quad (2)$$

Subthreshold Slope

To calculate the subthreshold slope following equation must be taken under consideration:

$$S = 2.3KT \frac{\partial \phi_f(y)}{\partial V_{gs}}\bigg|_{y=y_0} \quad (3)$$

where expression for $\frac{\partial \phi_f(y)}{\partial V_{gs}}$ and $\phi_{f\min}(y)$ are shown in detail at Appendix: [B].

Moreover, dependency of y_0 over ϕ_{bi} is also taken care of.

Classical threshold voltage model

As threshold voltage is the value of gate to source voltage for which,

$$\phi_{f\min}(y) = 2 \times \phi_b = 2 \times V_T \times \ln\left(\frac{N_a}{n_i}\right) \quad (4)$$

where V_T is the thermal voltage and n_i is intrinsic doping concentration, then classical part of threshold voltage can be found as:

$$V_{th,\text{classical}} = \frac{-B - 2P \pm 2\left[P^2 - \left(\frac{\left\{V_T \ln\left(\frac{N_a}{n_i}\right)\right\}^2 + \frac{TCD}{4}}{1-m^2}\right)\right]^{\frac{1}{2}}}{-2 \times A} \quad (5)$$

Appendix: [C], different parameters like A, B, C, D, P, T and m, as used in Eq. (5), are defined.

Formulation of inversion charge and quantum threshold voltage model

Carrying out investigation over inversion charge along the channel, first the nature of potential well is to be determined. We found it 'Triangular' as depicted in Fig. 2. Now total threshold voltage can be derived as $V_{th}^q = V_{th}^{cl} + \Delta V_{th}^q$. Shift of quantum threshold voltage [8] can be derived to be: $\Delta V_{th}^q = \frac{S}{(kT/q)\ln(10)} \Delta\phi_s^q(x_m)$. Refer Appendix: [D] for elucidated derivations.

Results and discussion

WICOI/WICON structure with gate oxide thickness, t_f to be 2 nm, channel thickness, t_{cnt} 1.5 nm, source/drain doping of 5×10^{24} m^{-3}, Effective mass of electron $0.053 \times 9.1 \times 10^{-31}$ is considered here for mathematical model as well as for simulation with ATLAS-2D simulator to validate the model. Semiconducting zigzag CNT (19, 0) of diameter 1.5 nm has been taken as channel material having carbon–carbon length of 1.42 Å, nearest neighbour overlap energy being 2.7 eV and band gap of 0.51 eV.

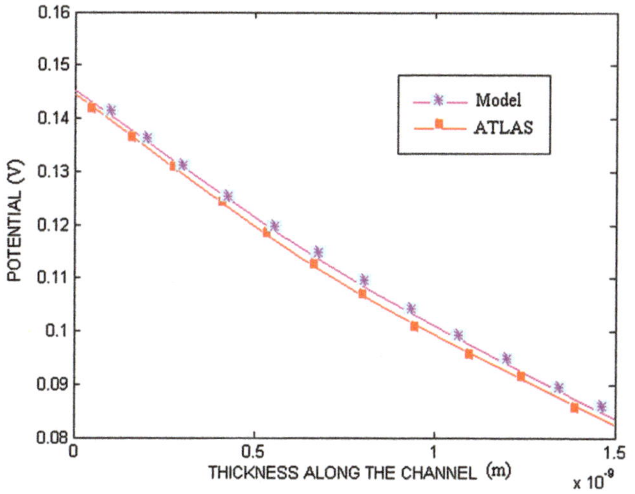

Fig. 2 Potential well formation along film thickness at $y = 10$ nm for WICOI/WICON FET structure keeping $V_{gs} = 0.2$ V and $V_{ds} = 0.35$ V, respectively

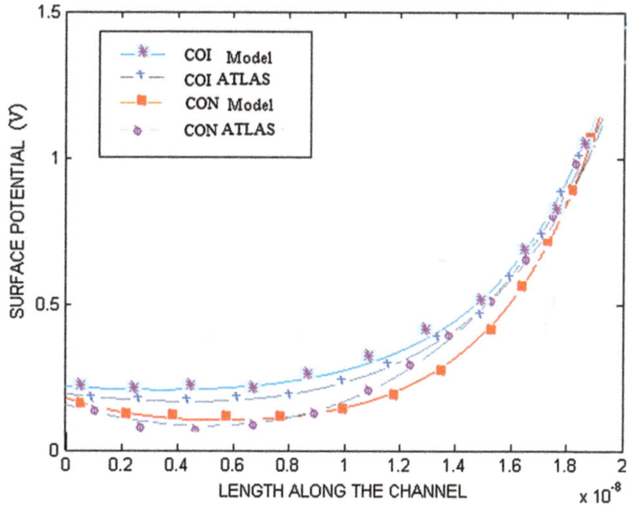

Fig. 3 Surface potential profile along channel length for WICOI/WICON FET structure keeping $V_{gs} = 0.2$ V and $V_{ds} = 0.35$ V, respectively

Figure 3 shows asymmetric surface potential distribution for gate to source voltage 0.2 V and drain to source voltage 0.35 V. Shift of minima of the potential parabola towards source side is the consequence of different channel potential. The downward shift of the minima for CON structure implies better immunity to common SCE's and DIBL also. Figure 4 tells us about the effect of change in the permittivity of gate dielectric as channel length downsizes from 200 nm. Application of high-k material simply creates higher value of C_f contributing to lower surface potential for both COI and CON devices.

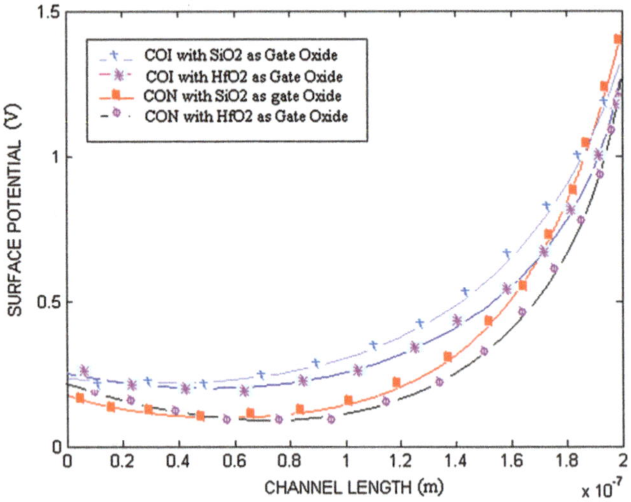

Fig. 4 Surface potential profile with increasing channel length for WICOI/WICON FET structure with comparison for high-k dielectric as gate oxide

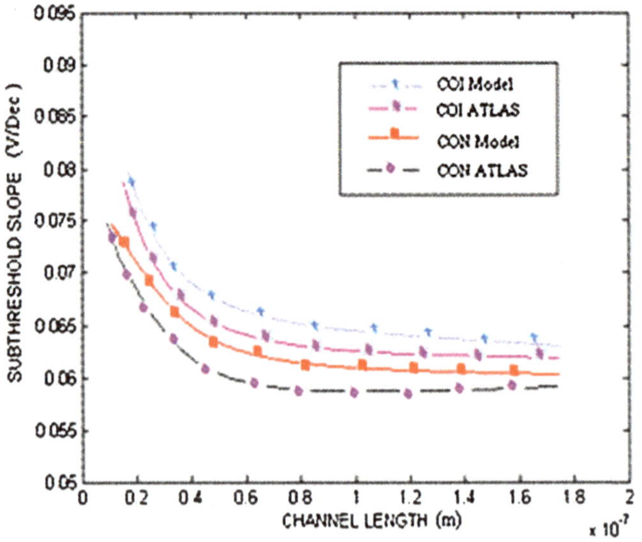

Fig. 6 Subthreshold slope for WICOI/WICON FET structure expressed in (V/Dec.)

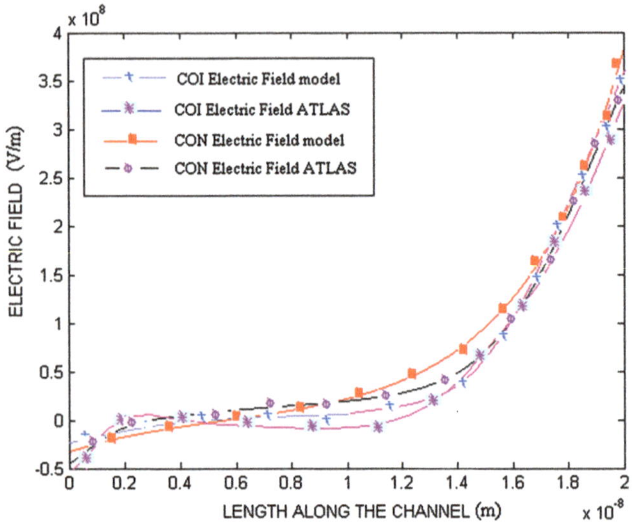

Fig. 5 Electric field along the channel for WICOI/WICON FET structure keeping $V_{gs} = 0.2$ V and $V_{ds} = 0.35$ V, respectively

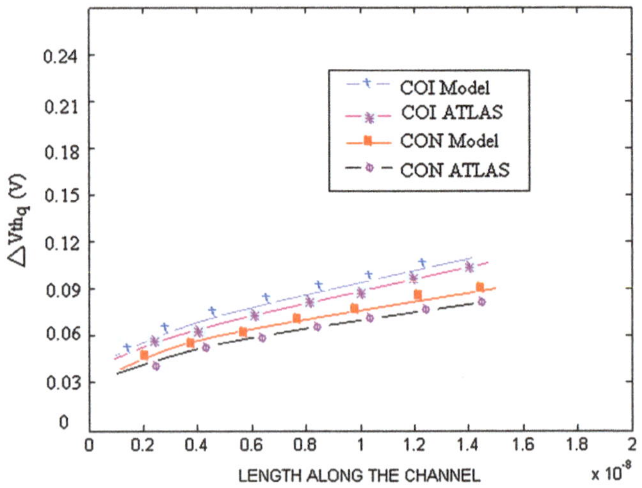

Fig. 7 Variation of quantum threshold voltage shift along the channel for WICOI/WICON FET structure

Electric field, as found in Fig. 5, increases gradually towards drain with small negative initial values at source side. At middle of the channel, electric field for COI places more downward deviation than CON justifying better carrier transport for CON devices. Figure 6 is about comparison of subthreshold slope where CON experiences lower slope value than COI due to its reduced front to back potential coupling ratio.

Figure 7 expresses the variation of quantum threshold voltage shift along the channel differed by device structure where COI has relatively steep fall than CON device signifying less effect of channel length variation over CON. Figure 8 defines the effect of high-k dielectric to vary quantum threshold voltage on COI/CON device. Threshold voltage drop is more with the application of low-k material as gate dielectric, making it prone to undesired performance complications.

Figure 9 depicts the reduction in quantum threshold voltage shift with increment of channel thickness. This can be elucidated as potential-well previously formed, experiences increment when diameter of implanted CNT is of higher value. Relaxation of quantized nature occurs gradually in terms of increased channel thickness as separation of any two energy levels possesses minimum value.

Figure 10 is of true importance as it justifies CON to possess immense improvement over SOI, SON and COI as per DIBL is concerned. Here DIBL is calculated in terms of the difference between linear and saturation threshold

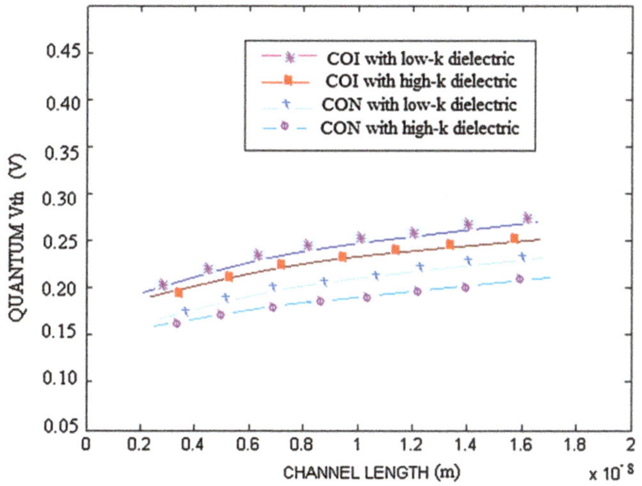

Fig. 8 Effect of high-k material on quantum threshold voltage for WICOI/WICON FET structure

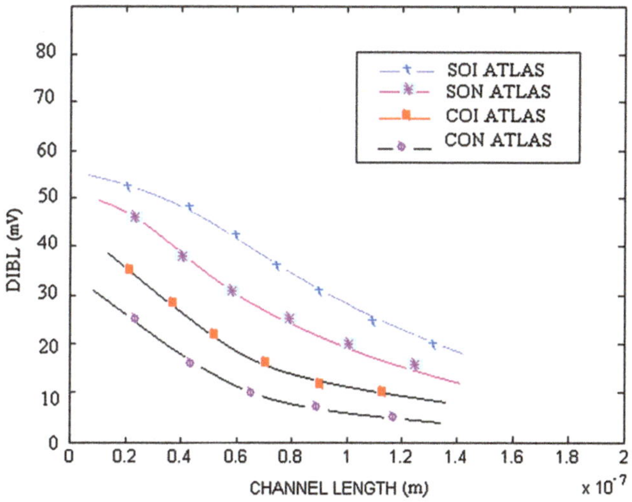

Fig. 10 DIBL effect on SOI/SON/WICOI/WICON FET structure for different channel lengths up to 200 nm

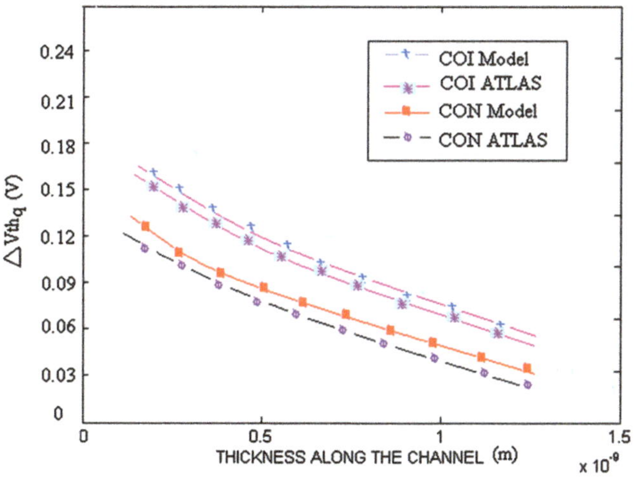

Fig. 9 Variation of quantum threshold voltage shift along channel thickness for WICOI/WICON FET structure keeping $V_{gs} = 0.2$ V and $V_{ds} = 0.35$ V, respectively

voltage. We have validated our mathematical model with those found by simulation. Control of gate over channel thus can be kept intact by implementing buried air gap as well as implanting CNT as channel material. Starting from about 30 mV, it tends to be lowered up to 5 mV at 140 nm.

Conclusion

Present analysis reveals most of the performance related characteristics of COI/CON structure being sincerely considered for modeling in quantum mechanical aspect. Quantum threshold voltage has been successfully derived on the basis of the formation of inversion charge as well as quantum wells with discrete energy levels. Proposed model has its ultimate immunity against major SCE's including DIBL as validated with good agreement with the data found by ATLAS in "Results and discussion". Reduced subthreshold slope implying strong gate to channel coupling with application of high-k gate dielectric recommends the device to fulfil the ever increasing demand of nano-technological industry today. This structure, if fabricated precisely, can be implemented in device-circuit mutual integration platform to deliver ultra-high speed with its nearly ballistic transport. Moreover, this simple model can further be improvised with the application of pocket implantation and work function engineered binary metal alloy gate as future scopes.

Acknowledgments The work has been supported by UGC_UPE phase-II with Ref. No: R-11/43/2013. The authors also thankfully acknowledge the valuable technical discussions with Gargee Bhattacharyya.

Appendix: [A]

Coefficients are: $C_1 = \frac{e^{-\sqrt{S_f \cdot L}}}{(1-e^{-2\cdot\sqrt{S_f \cdot L}})} \times \left[\phi_{bi} - \phi_{bi} \times e^{-\sqrt{S_f \cdot L}} + V_{ds} + \frac{D_f}{S_f} - \frac{D_f}{S_f} \times e^{-\sqrt{S_f \cdot L}}\right]$, $C_2 = \frac{1}{(1-e^{-2\cdot\sqrt{S_f \cdot L}})} \cdot \left[\phi_{bi} - \phi_{bi} . e^{-\sqrt{S_f} . L} - V_{ds} . e^{-\sqrt{S_f} . L} + \frac{D_f}{S_f} - \frac{D_f}{S_f} . e^{-\sqrt{S_f} . L}\right]$ where $D_f = \frac{q \times N_a}{\varepsilon_{cnt}} - 2 \times$

$\frac{\left(\frac{C_f}{C_{cnt}}+\frac{C_f}{C_{ox/air}}\right)\times V'_{gs}+2\times V'_{sub}}{t_{cnt}^2\times\left(1+2\times\frac{C_{cnt}}{C_{ox/air}}\right)}$ and $S_f=2\times\frac{\left(1+\frac{C_f}{C_{cnt}}+\frac{C_f}{C_{ox/air}}\right)}{t_{cnt}^2\times\left(1+2\times\frac{C_{cnt}}{C_{ox/air}}\right)}$ with $C_f=\frac{\varepsilon_{ox}}{t_f}$ to be front channel oxide capacitance and $C_{ox/air}=\frac{\varepsilon_{ox/air}}{t_{ox/air}}$ to be back channel oxide/air capacitance.

Appendix: [B]

In determining $\frac{\partial\phi_f(y)}{\partial V_{gs}}$, we can take, $w=\frac{e^{-\sqrt{S_f}\cdot y}}{(1-e^{-2\cdot\sqrt{S_f}\cdot L})}$, $A=\frac{\left(\frac{C_f}{C_{cnt}}+\frac{C_f}{C_{ox/air}}\right)}{t_{cnt}^2\times\left(1+2\times\frac{C_{cnt}}{C_{ox/air}}\right)S_f}$, $Y=\frac{\left(\frac{C_f}{C_{cnt}}+\frac{C_f}{C_{ox/air}}\right)e^{-\sqrt{S_f}\cdot L}}{t_{cnt}^2\times\left(1+2\times\frac{C_{cnt}}{C_{ox/air}}\right)S_f}$ and $Z=e^{\sqrt{S_f}\cdot y}$

Thus, we have

$$\frac{\partial\phi_f(y)}{\partial V_{gs}}=[W\times(-2\times X+2\times Y)\times Z]+[(-2\times X)-2\times Y]\times W+(2\times X) \tag{6}$$

Similarly,

If $N=e^{-\sqrt{S_f}\cdot L}$, $K=\frac{D_f}{S_f}$, $Q=\frac{e^{-\sqrt{S_f}\cdot L/2}}{(1-e^{-2\cdot\sqrt{S_f}\cdot L})}$, $H=\phi_{bi}$.

Then

$$\phi_{f\min}(y)=-K+Q\times[\{H-(H\times N)+V_{ds}+K-(K\times N)\}\times\{H-(H\times N)-(V_{ds}\times N)+K-(K\times N)\}]^{1/2} \tag{7}$$

Here, y_0 is undoubtedly, position in the channel where surface potential is minimum and if $J=\frac{L}{2}+\frac{1}{2\sqrt{S_f}}$ then it is obtained as:

$$y_0=\mathrm{J}\times\ln[\{H-(\mathrm{H}\times\mathrm{N})-(V_{ds}\times\mathrm{N})+\mathrm{K}-(\mathrm{K}\times\mathrm{N})\}/\{\mathrm{H}-(\mathrm{H}\times\mathrm{N})+V_{ds}+\mathrm{K}-(\mathrm{K}\times\mathrm{N})\}].$$

Appendix: [C]

Now, to find the classical threshold voltage, we assume $B=D_f=\frac{q\times N_a}{\varepsilon_{cnt}S_f}+2\times\frac{\left(\frac{C_f}{C_{cnt}}+\frac{C_f}{C_{ox/air}}\right)\times V'_{gs}-2\times V'_{sub}}{t_{cnt}^2\times\left(1+2\times\frac{C_{cnt}}{C_{ox/air}}\right)S_f}$, $C=\phi_{bi}-\phi_{bi}e^{-\sqrt{S_f}\cdot L}+V_{ds}$, $D=\phi_{bi}-\phi_{bi}e^{-\sqrt{S_f}\cdot L}-V_{ds}e^{-\sqrt{S_f}\cdot L}$, as well as considering $T=4\frac{e^{-2\sqrt{S_f}\cdot L}}{(1-e^{-2\sqrt{S_f}\cdot L})^2}$, $m=1-e^{-\sqrt{S_f}\cdot L}$ and $P=\frac{V_T\ln\left(\frac{N_a}{n_i}\right)-\frac{\mathrm{Tm}(C+D)}{4}}{1-m^2}$. This makes simply understandable classical threshold voltage.

Appendix: [D]

Schrodinger's equation can be solved for triangular potential well which is given by:

$$\frac{\partial^2\psi}{\partial x^2}+\frac{2qm^*}{\hbar^2}(E-Fx)\psi=0. \tag{8}$$

F can be treated as electric field across film thickness. At surface, $F|_{x=0}=F_S=-\frac{\partial}{\partial x}(\phi_s(x)+C_1(y)x+C_2(y)x^2|_{x=0}=-C_1(y)$

Solution of Eq. (8) can invoke quantized energy levels as: $E^j=\left(\frac{\hbar}{2qm*}\right)^{1/3}+\left(\frac{3\pi|F_S|}{2}(j-\frac{1}{4})\right)^{2/3}$

Now, critical charge is derived with the help of Boltzmann's Statistics as:

$$\begin{aligned}Q_t&=qn_i\exp\left(\frac{\phi_s^{cl}(y_m)}{V_T}\right)\frac{e^{\frac{C_1^2}{4V_T|C_2|}}\sqrt{\pi V_T}}{2\sqrt{|C_2|}}\\&\quad\times\left(erf\left(\frac{-C_1-2t_{cnt}|C_2|}{2\sqrt{V_T|C_2|}}\right)-erf\left(\frac{-C_1}{2\sqrt{V_T|C_2|}}\right)\right)\\&=\exp\left(\frac{\phi_s^q(y_m)}{V_T}\right)\times qkT\sum_j DOS_{zz}(E)\times g\\&\quad\times\ln\left[1+\exp\left(-\frac{1}{V_t}\left(E^j+\frac{E_g}{2}-\phi_b(y,t_{cnt})\right.\right.\right.\\&\quad\left.\left.\left.-\phi_s(x)+E_F(y)\right)\right)\right]\end{aligned}$$

Thus shift of classical and quantum potential distributions can be found as follows:

$$\begin{aligned}\phi_s^{cl}-\phi_s^q=&\ln\left(\frac{n_i\pi^{1/2}}{2\sqrt{kTq|C_2|}}\right)+\frac{C_1^2}{4V_T|C_2|}+\ln\left(erf\left(\frac{-C_1-2t_{cnt}|C_2|}{2\sqrt{V_T|C_2|}}\right)\right.\\&\left.-erf\left(\frac{-C_1}{2\sqrt{V_T|C_2|}}\right)\right)-\ln\left(\sum_j DOS_{zz}(E)\times g\right.\\&\left.\times\ln\left[1+\exp\left(-\frac{1}{V_t}\left(E^j+\frac{E_g}{2}-\phi_b(y,t_{cnt})-\phi_s(y)+E_F(y)\right)\right)\right]\right)\end{aligned}$$

Now, effective mass of CNT can be written as: $m*(E,q)=\frac{16\hbar^2}{3a^2}\frac{E^3}{(E_{vh1}^2E_{vh2}^2-E^4)}$, which can be simplified as: $m_{1-D}\propto\frac{E_g}{2\gamma_0+E_g}\propto\frac{E_g}{2\gamma_0}$ with the help of fermi-dirac distribution, total inversion charge per unit valley becomes as:

$$\begin{aligned}Q_{inv,q}=&\,qkT\sum_j DOS_{zz}(E)\times g\ln\\&\times\left[1+\exp\left(-\frac{1}{V_t}\left(E^j+\frac{E_g}{2}-\phi_b(y,t_{cnt})-\phi_s(x)+E_F(y)\right)\right)\right]\end{aligned}$$

References

1. Emerging Research Devices, ITRS. London, UK (2011)
2. Manna, B., Sarkhel, S., Islam, N., Sarkar, S., Sarkar, S.K.: Spatial composition grading of binary metal alloy gate electrode for

shortchannel SOI/SON MOSFET application. IEEE Trans. Electron Devices **11**(3), 472–478 (2012)
3. Deb, S., Singh, N.B., Islam, N., Sarkar, S.K.: Work function engineering with linearly graded binary metal alloy gate electrode for short-channel SOI MOSFET. IEEE Trans. Nanotechnol. **59**(12), 3280–3287 (2012)
4. JagadeshKumar, M., Orouji, A.A., Dhakad, H.: New dual-material SG nanoscale MOSFET: analytical threshold-voltage model. IEEE Trans. Electron Devices **53**(4), 920–923 (2006)
5. Deb, S., Singh, N.B., Das, D., De, A.K., Sarkar, S.K.: Analytical model of threshold voltage and sub-threshold slope of SOI and SON MOSFETs: a comparative study. J. Electron Devices **8**, 300–309 (2010)
6. Young, K.K.: Short-channel effect in fully depleted SOI-MOSFETs. IEEE Trans. Electron Devices **36**(2), 399–402 (1989)
7. Samudra, Rajendran, K.: Comparative analysis of minimum surface potential and location of barrier peaks in various Si MOSFET devices. Int. J. Electron. (Taylor and Francis) **87**(5), 513–530 (2000)
8. Shee, S., Bhattacharyya, G., Sarkar, S.K.: Quantum analytical modeling for device parameters and I–V characteristics of nanoscale dual-material double-gate silicon-on-nothing MOSFET. IEEE Trans. Electron Devices **61**(8), 2697–2704 (2014)
9. Iijima, S.: Helical microtubules of graphitic carbon. Nature **354**, 56–58 (1991)
10. Saito, R., Dresselhaus, M.S., Dresselhaus, G.: Physical Properties of Carbon Nanotubes. Imperial College, London (1998)
11. Arefinia, Z., Orouji, A.A.: Quantum simulation study of a new carbon nanotube field-effect transistor with electrically induced source/drain extension. IEEE Trans. Device Mater. Reliab. **9**(2), 237–243 (2009)
12. Tu, J.R., Farmer, D.B., Guo, J., Gordon, R.G., Dai, H.J.: High performance n-type carbon nanotube field-effect transistors with chemically doped contacts. Nano Lett. **5**(2), 345–348 (2005)
13. Datta, S.: Quantum transport: atom to transistor. Cambridge Univ. Press, Cambridge (2005)
14. Kazmierski, T.J., Zhou, D., Al-Hashimi, B.M., Ashburn, P.: Numerically efficient modeling of CNT transistors with ballistic and nonballistic effects for circuit simulation. IEEE Trans. Nanotechnol. **9**(1), 99–107 (2010)
15. Akturk, A., Pennington, G., Goldsman, N.: Quantum modeling and proposed designs of CNT-embedded nanoscale MOSFETs. IEEE Trans. Electron Devices **52**(4), 577–584 (2005)
16. Guo, J., Javey, A., Dai, H., Lundstrom, M. Performance analysis and design optimization of near ballistic carbon nanotube field-effect transistors. IEDM Tech. Dig. 703–706 (2004)
17. Akinwande, D., Nishi, Y., Philip Wong. H.S. An analytical derivation of the density of states, effective mass, and carrier density for achiral carbon nanotubes. IEDM. 753–756 (2007)
18. Sathyakam, P.U., Karthikeyan, A., Mallick, P.S.: Role of semiconducting carbon nanotubes in crosstalk reduction of CNT interconnects. IEEE Trans. Nanotechnol. **12**(5), 662–664 (2013)
19. Alam, K., Lake, R.: Performance metrics of a 5 nm, planar, top gate, carbon nanotube on insulator (COI) transistor. IEEE Trans. Nanotechnol. **6**(2), 186–190 (2007)
20. Shankar, R.: Principles of Quantum Mechanics. Plenum, New York (1994)
21. Kreyszig, : Advanced Engineering Mathematics. Wiley, New York (2011)
22. Davies, J.H.: The Physics of Low Dimensional Semiconductors. Plenum, New York (1998)
23. Naskar, S., Sarkar, S.K.: Quantum analytical model for inversion charge and threshold voltage of short-channel dual-material double-gate SON MOSFET. IEEE Trans. Electron Devices **60**(9), 2734–2740 (2013)
24. Bushmaker, A.W., Amer, M.R., Cronin, S.B.: Electrical transport and channel length modulation in semiconducting carbon nanotube field effect transistors. IEEE Trans. Nanotechnol. **13**(2), 176–181 (2014)

12

DFT study of hydrogen fluoride and sulfur trioxide interactions on the surface of Pt-decorated graphene

Ali Shokuhi Rad[1]

Abstract In this study, we investigate the adsorption properties of hydrogen florid (HF) and sulfur trioxide (SO_3) on the surface of platinum decorated graphene (PtG) using density functional theory. We found one optimized configuration for HF and two ones for SO_3 upon adsorption on the surface of PtG. Our result show significant adsorption on PtG with calculated energy adsorption of −73.6 (−54.2 BSSE) kJ/mol for HF at its only position and −172.4 (−144.8 BSSE) and −62.7 (−53.7 BSSE) kJ/mol for SO_3 at its two positions; P1 and P2, respectively), whereas there is weak physisorption of these analytes on pristine graphene (PG). Results of charge analyses reveled interesting net charge transfer; while the direction of charge is from HF to PtG, reverse direction is found for SO_3 for its two configurations. To deep understand the concept of adsorption properties, we used orbital analyses including density of states for interaction of mentioned analytes on the surface of PtG.

Keywords Pt-decorated graphene · Hydrogen florid · Sulfur trioxide · Gas sensor

Introduction

Graphene has connected carbons in the array of two-dimensional sp^2. It has excellent properties such as high surface-to-volume fraction [1]. Due to its marvelous electronic structure, it has been used for numerous times as gas sensor for different analytes [2, 3]. All graphene-made electronic devices benefit from its exceptional charge transferring and its high steadiness. Along with carbon nanotubes [4, 5], there are many experimental and theoretical studies focused on graphene for sensing materials [6–8]. Some research works on graphene as gas sensor were listed and discussed by Mao et al. [9]. They disputed on the opportunities of utilizing these materials for gas sensor purpose.

The properties of graphene could be improved particularly by modification of its surface by doping and decoration of different hetro atoms [10–13]. In our recent study, we have reported the potential of N-doped graphene towards its interaction with boron [14], and SOx [15] and CO [16]. We documented that this modified surface is able to increase the adsorption properties of these compounds. Moreover, our group reported different application of Al and B-doped graphene sheets [17–24]. It has been proved in these papers that metal doping by Al or B increases the adsorbent potential of graphene for practical application.

Besides doping process, metal decoration is another way to intensify the adsorption property of graphene. Upon decoration, single atom could be placed on high surface area carbons. It will induce to addition of several re-bonding per metal atom with releasing significant energy, which consequents more steadiness of single atom decorated structure in comparison with bulk atoms decorated structure [25]. There are diverse experimentally techniques for single atom decoration on nanostructures. As an example, Wang et al. [26] showed that a two-step process is a well-ordered technique to stabilize single atoms on graphene. Diverse metals including Pt, Co, and In, have been successfully incorporated in graphene in the single-atom form. As an example, Baby et al. [27] built an amperometric glucose biosensor by decoration of Pt and

✉ Ali Shokuhi Rad
a.shokuhi@gmail.com; a.shokuhi@qaemiau.ac.ir

1 Department of Chemical Engineering, Qaemshahr Branch, Islamic Azad University, Qaemshahr, Iran

Au nano-particles on graphene. Based on their research, decorated particles could able to immobilize glucose oxidase by physical adsorption follow-on an enhancement in the performance of biosensor. Moreover, our group searched on the adsorption of NO molecule on the surface of Pt-decorated graphene [28]. We found very high adsorption on this modified surface while there was weak adsorption in the case of using pristine graphene. In parallel we used Pt-decorated graphene as an ideal adsorbent for C_2H_2 and C_2H_4 molecules [29], MeOH and EtOH [30], and SO_2 and O_3 [31].

Hydrogen fluoride (HF) is a colorless gas and main industrial source of fluorine, frequently as an aqueous solution named hydrofluoric acid. It is the forerunner for different significant compounds such as polymers (for example Teflon). HF is extensively utilized in the petrochemical industry as a part of super acids. The adsorption of HF on the surface of pristine graphene has been deeply investigated by Sun et al. [32]. They used various configurations for adsorption of HF using DFT and found that at its most energetically favorable site, the amount of adsorption energy is very weak (physisorption).

On the other hand, sulfur trioxide (SO_3) in the gaseous form is a vital pollutant, which is the primary agent in acid rain. It has been produced on an industrial scale as a precursor to sulfuric acid. Adsorption of SO_3 on pristine graphene has been investigated previously by our group [15]. We found there is very weak adsorption of SO_3 on pristine graphene.

In this study, we select PtG sheet to have more investigation since its potential as adsorbent is well proven. Here we aim to establish the potential of PtG as a new adsorbent for HF and SO_3 molecules by using first-principles simulation. The gas molecules HF and SO_3 are of pollution for industrial as well as environmental applications. They are very poison gases and so there is much require to their removal at industrial situations. To the best of our knowledge, there is no reported study on the adsorption properties of HF and SO_3 molecules on the surface of PtG. The inspiration of this study is to attain basic insights to the impact of adsorbed HF and SO_3 molecules on the electronic structure of PtG.

Computational method

All relaxation for PG and PtG sheets in the present and absence of HF and SO_3 molecules were done using B3LYP density functional with split basis set (for all atom excluding Pt the basis set was 6-31G (d,p), for Pt atom the basis set was lanl2dz) as implemented in Gaussian 09 suite of program [33]. To save the time, initial optimizations were carried by using 3-21G(d)/lanl2dz basis set (for all possible configurations) to distinguish the most stable configurations. Then, among all relaxed structures achieved by this initial functional/basis set, those having highest stability (judging from their calculated adsorption energy) were selected as input files to do next optimization at stronger basis set/functional (6-31G(d,p)/B3LYP) to achieve more precise values of parameters (adsorption energy, equilibrium distance, charge transfer,...).

The 6-31G(d,p) basis set is good for general calculations, besides the B3LYP density functional has been known appropriate for nano-structure studies [28–31].

All Calculations including charge analysis, density of states (DOS), the energy of lowest unoccupied molecular orbital (LUMO), the energy of highest occupied molecular orbital (HOMO), and the HOMO–LUMO energy gap (E_g) have been done using above-mentioned level of theory.

The values of adsorption energy (E_{ads}) upon adsorption of analyte on PtG were specified using Eq. (1):

$$E_{ads}(PtG) = E_{PtG-A} - (E_{PtG} + E_A) \quad (1)$$

where the $E_{PtG\text{-}A}$ corresponds to the adsorbed system of PtG, E_{PtG} correspond to the isolated PtG and E_A corresponds to the isolated analyte (HF or SO_3), respectively. For all adsorbed systems, calculated adsorption energies were modified based on the Eq. (2).

$$E_{ads,CP} = E_{ads} - E_{BSSE} \quad (2)$$

which $E_{ads,\ CP}$ is counterpoise corrected adsorption energy of related complexes and E_{BSSE} is basis set superposition errors energy.

Result and discussions

A supercell of 4 × 4 graphene (12.30 × 12.30 × 16 Å) (left side of Fig. 1) has been used as surface of interaction. There is no interaction between graphene sheets of neighboring supercells since the z axis (16 Å) of this supercell is big enough.

As can be seen in the right side of Fig. 1, there is three potential sites for Pt decoration on graphene: the bridge site (considering the middle of a C–C bond), the top site (directly above a C atom) and the hollow site (at the middle of a hexagon). As we reported in our recent publications [28–31], the bridge site of graphene is the most energetically position for the decoration of Pt compared to the two other ones. Therefore, we placed the Pt atom on this site and let this initial configuration to be optimized at the above-mentioned basis set/functional to made PtG.

As mentioned in the section of introduction, the adsorption properties of both HF and SO_3 on the surface of pristine graphene were investigated before [15, 32] and their relaxed structure were well discussed. For both

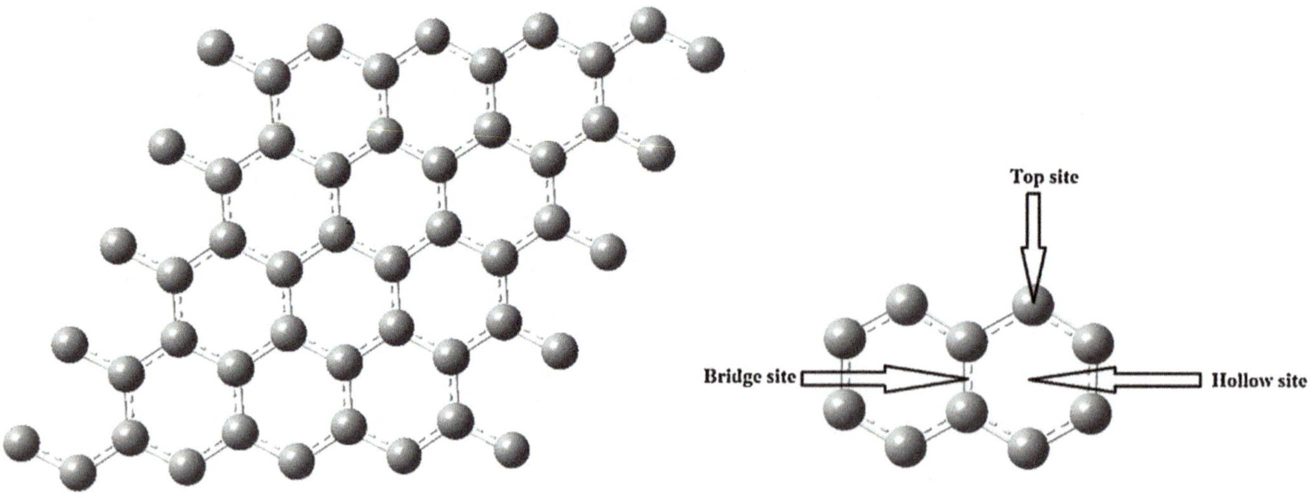

Fig. 1 A supercell of graphene (*left*) and three possible adsorption sites for decoration of Pt (*right*)

analytes, pristine graphene revealed very low adsorption (−7.1 kJ/mol for HF and −9.1 kJ/mol for SO_3, respectively). For this reason, we disregarded to repeat the study of adsorption of these analytes on pristine graphene.

To find the most stable configurations of HF and SO_3 on Pt-decorated graphene, first we made all possible configurations as input files. For HF, it was placed on top of Pt atom so that the H-F axis in one state has parallel configuration and in other states has perpendicular configurations (one, from the H-side and another, from the F-side) than the surface. Moreover, we placed SO_3 on top of Pt from the O-side (perpendicular to the surface) and the S-side (parallel to the surface). All these initial configurations were subjected to relaxation at minimum basis set of 3-21G(d) (lanl2dz for Pt) using DFT to primary finding of relaxed structures. We found that all three input structures for HF turn to only one position upon optimization, whereas there are two relaxed structures for output of SO_3-including systems. Additionally, we used these initial optimized structures (one for HF and two for SO_3) as input files for next optimization at the stronger basis set of 6-31G (d,p)/(B3LYP, lanl2dz for Pt) to get more precise structure properties. Figure 2 represents the relaxed structures of these adsorbed systems.

Relaxation of HF on PtG accomplishes with releasing −73.6 kJ/mol with closest equilibrium distance of 2.27 Å (see Fig. 2). These values are considerable compared to those reported for HF adsorption on pristine graphene (the adsorption energy of −7.1 kJ/mol and adsorption distance of 3.66 Å) [32]. Moreover, fully optimization of SO_3 on PtG corresponds to releasing energies of −172.4 and −62.7 kJ/mol for position 1 and 2, respectively (P1 and P2). The adsorption value of SO_3 on PtG in both configuration are much higher compared to that reported on pristine graphene (−9.1 kJ/mol) [15]. The closest adsorption distances are achieved 2.16 and 2.00 Å for P1 and P2, respectively, which are completely shorter compared to its adsorption on pristine graphene (3.25 Å) [15].

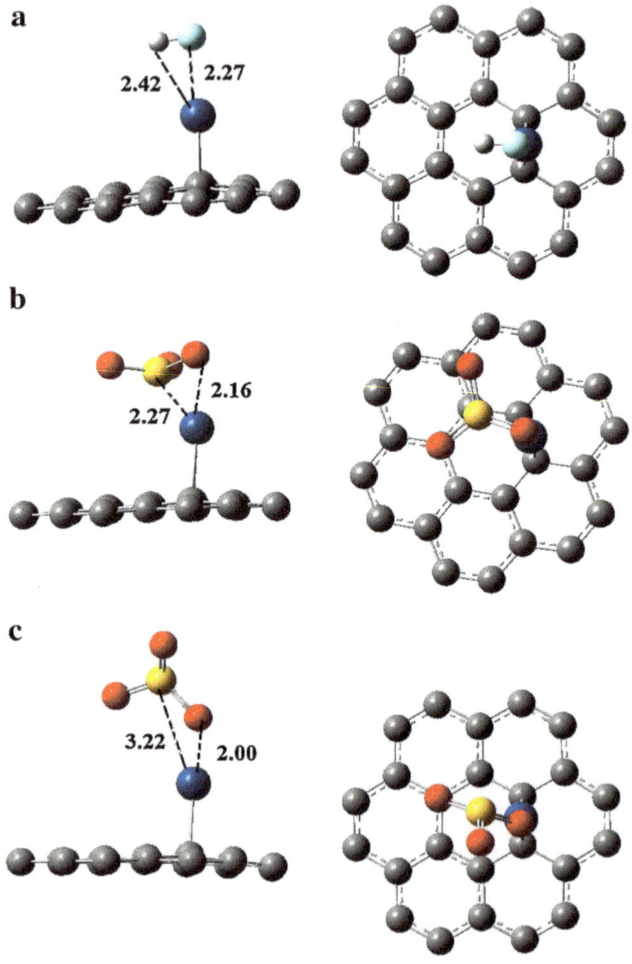

Fig. 2 Side views (*left*) and top views (*right*) of some parts of relaxed structure of adsorbed HF (**a**) and SO_3 [P1 (**b**) and P2 (**c**)] on PtG

Despite P1 represents much higher adsorption energy compared to P2, however, the closest adsorption distance of the later is relatively lower than that of P1. This is because in P1, all atoms of SO_3 involve in formation of bond with Pt, while in P2, only one atom (O) interacts with Pt.

We can conclude that interactions of HF and SO_3 (in P1 and P2) on PtG could be categorized in chemisorption region. These extensive higher values of HF and SO_3 adsorption on PtG than their adsorption on PG is a result of high interaction between Pt…F and Pt…O upon their interaction on the earlier surface.

Then, we investigate on the net charge transfer for adsorption of HF and SO_3 (P1 and P2) on the surface of PtG using natural bond orbital (NBO). The data are listed in Table 1. The net charge transfer for adsorption of HF and SO_3 were calculated +0.171, −0.337 (P1), and −0.364 (P2) e, respectively. Reverse direction of charge transfer for HF compared to SO_3 points to the ability of PtG to act as n-type and p-type semiconductors at the same time. This can be confirmed by considering the local charge of Pt upon adsorption of these analytes (−0.101 e for HF, +0.236 e for SO3 (P1), and +0.258 e for SO_3 (P2), as can be seen in Table 1).

These results accentuate that the adsorption of HF and SO_3 on PtG significantly change its electronic property.

To examine the change in the electronic structure of graphene resulted by decoration of Pt, the HOMO and LUMO distributions of pristine and Pt-decorated graphene sheets are depicted in Fig. 3. It can be seen in Fig. 3, that the HOMO of PG is mostly restricted on the C–C bonds whereas the LUMO is situated on the conflicting site. After Pt is decorated on graphene, major parts of both HOMO and LUMO relocate on Pt atom which result more reactivity of PtG compared to PG. Additionally, the result of NBO analysis showed that the C-atoms neighboring Pt attract charges owing to their high electron affinity and then result a reduce (+0.132 e) in the electron density of Pt.

Table 1 Adsorption energy (E_{ads}) and orbital characteristic: E_{HOMO}, E_{LUMO}, energy of Fermi level (E_{FL}), HOMO–LUMO energy gap (E_g), the nearest equilibrium distance (d_e) for all systems

System	E_{HOMO} (eV)	E_{FL} (eV)	E_{LUMO} (eV)	E_g (eV)	Q_{NBO} (e) of adsorbate	Q_{NBO} (e) of Pt	d_e (Å)	E_{ads} (kJ/mole)
HF	−10.24		1.68	11.92	–	–	–	–
SO_3	−9.46		−2.96	6.50	–	–	–	–
Isolated PtG	−5.00	−3.79	−2.58	2.42	–	+0.132	–	–
PtG-HF	−4.92	−3.38	−1.84	3.08	+0.171	−0.101	2.27	−73.6 (54.2)[a]
PtG-SO_3 (1)	−5.85	−4.37	−2.89	2.96	−0.337	+0.236	2.16	−172.4 (144.8)[a]
PtG-SO_3 (2)	−5.49	−4.76	−4.04	1.45	−0.364	+0.258	2.00	−62.7 (53.7)[a]

[a] Counterpoise corrected energy

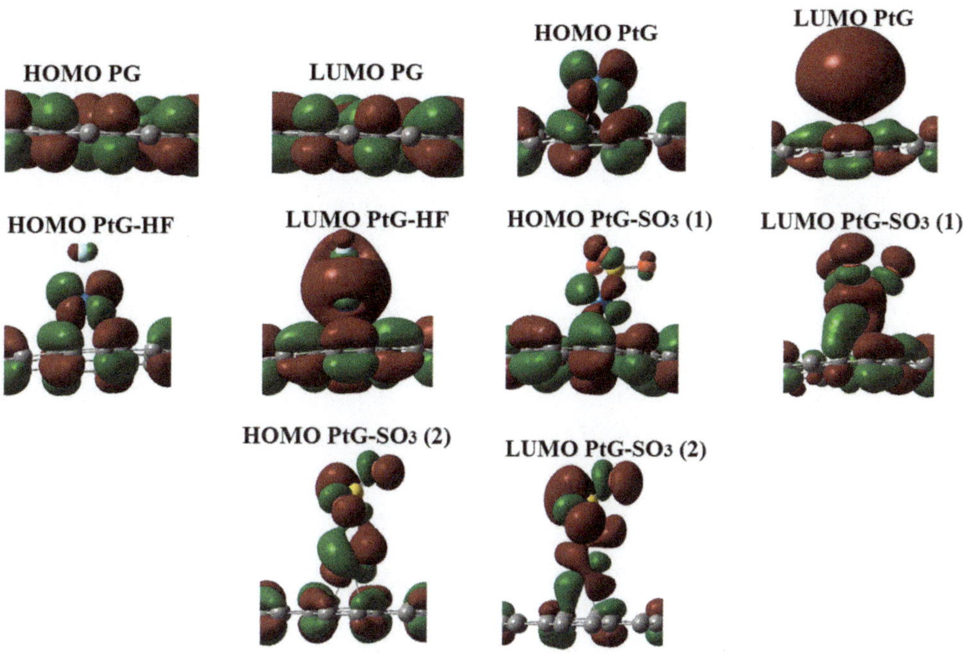

Fig. 3 HOMO and LUMO distributions of different systems

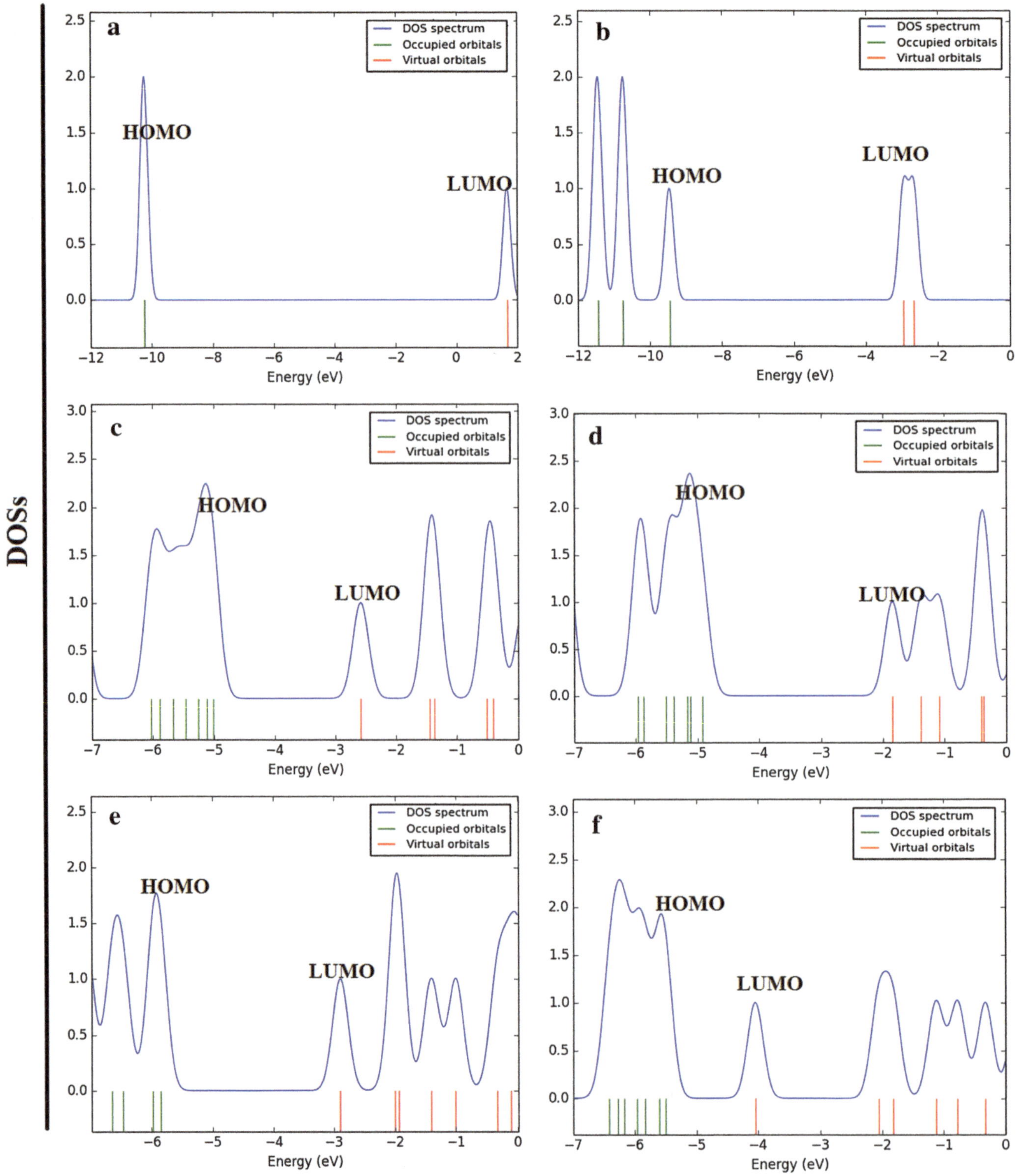

Fig. 4 DOSs for free HF (**a**), free SO_3 (**b**), free PtG (**c**), PtG-HF complex (**d**), PtG-SO_3 (position 1) (**e**), and PtG-SO_3 (position 2) (**f**)

The HOMO and LUMO of adsorbed system for complexed PtG are portrayed in Fig. 3 and the data are listed in Table 1. By comparing the HOMO–LUMO distributions of free PtG with those of adsorbed ones, it can be found that the distribution of both HOMO and LUMO are effected by adsorbate. These effects are more pronounced in the case of SO_3 rather than HF as can be expected by their difference in the calculated values of adsorption energy. By considering the data of Table 1, we can say that upon adsorption of HF, increase in the energies of HOMO and LUMO of system from −5.00 and −2.58 eV to −4.92 and −1.84 eV will happen, respectively. This increase associates with transfer of charge from HF to PtG (p-type semiconductor). On the other hand, for SO_3 adsorption, the energies of HOMO and LUMO of system decrease to −5.85 and −2.89 eV for P1 and −5.49 and −4.04 eV from their initial values (listed in Table 1). These changes accomplishes with transfer of charge from PtG to SO_3 (n-type semiconductor).

To deep understanding of the electronic property of PtG upon adsorption of HF and SO_3, the DOSs for all systems in free (Fig. 4a, b) and complexed forms were depicted close to the Fermi level (see Fig. 4). Comparing the DOSs of free PtG to that of PtG-HF complex (Fig. 4c, d) we can find significant changes in the location of HOMO and LUMO, in which the E_g ($E_{HOMO} - E_{LUMO}$) of system increase from 2.42 to 3.08 eV. This increase in the E_g corresponds to increase in the stability of system upon adsorption [34–36].

In the other hand, by comparing the DOSs of free PtG (Fig. 4c) with PtG-SO_3 (Fig. 4c, e, f) important change confirmation of hybridization between SO_3 and PtG (in both positions) could be find upon adsorption which introduce this modified surface as excellent adsorbent for SO_3.

Conclusion

In this study, we search the adsorption of HF and SO_3 molecules on the surfaces of PtG. First, we calculated the energy of any relaxed system followed by calculation of orbital descriptions, DOSs and NBO analyses. It was found that the interaction of HF and SO_3 molecules on PtG changes its electronic structure. The reason for our investigation on PtG was based on knowing this matter of fact that pristine graphene has very weak interaction with these molecules. Decoration of graphene by Pt considerably enhances the potential of graphene to be interacted. The NBO charge analysis reveals reverse direction of charge transfer for SO_3 compared to HF, pointing PtG is n-type and p-type semiconductors at the same time.

Acknowledgments I highly acknowledge financial support from Iran Nanotechnology Initiative Council, Iran.

References

1. Gadipelli, S., Xiao, G.Z.: Graphene-based materials: synthesis and gas sorption, storage and separation. Prog. Mater. Sci. **69**, 1–60 (2015)
2. Schedin, F., Geim, A.K., Morozov, S.V., Hill, E.W., Blake, P., Katsnelson, M.I., Novoselov, K.S.: Detection of individual gas molecules adsorbed on graphene. Nat. Mater. **6**, 652–655 (2007)
3. Barbolina, I.I., Novoselov, K.S., Morozov, S.V., Dubonos, S.V., Missous, M., Volkov, A.O., Christian, D.A., Grigorievam, I.V., Geim, A.K.: Submicron sensors of local electric field with single-electron resolution at room temperature. Appl. Phys. Lett. **88**, 013901 (2006)
4. Majzlíkova, P., Sedlacek, J., Prasek, J., Pekarek, J., Svatos, V., Bannov, A.G., Jasek, O., Synek, P., Elias, M., Zajickova, L., Hubalek, J.: Sensing properties of multiwalled carbon nanotubes grown in MW plasma torch: electronic and electrochemical behavior, gas sensing, field emission, ir absorption. Sensors **15**, 2644–2661 (2015)
5. Liu, S.F., Petty, A.R., Sazama, G.T., Swager, T.M.: Single-walled carbon nanotube/metalloporphyrin composites for the chemiresistive detection of amines and meat spoilage. Angew. Chem. Int. Ed. **54**, 6554–6557 (2015)
6. Leenaerts, O., Partoens, B., Peeters, F.M.: Adsorption of H_2O, NH_3, CO, NO_2, and NO on graphene: a first-principles study. Phys. Rev. B **77**, 125416 (2008)
7. Ao, Z.M., Yang, J., Li, S., Jiang, Q.: Enhancement of CO detection in Al doped graphene. Chem. Phys. Lett. **461**, 276–279 (2008)
8. Kang, H.S.: Theoretical study of binding of metal-doped graphene sheet and carbon nanotubes with dioxin. J. Am. Chem. Soc. **127**, 9839–9843 (2005)
9. Mao, S., Lu, G., Chen, J.: Nanocarbon-based gas sensors: progress and challenges. J. Mater. Chem. A **2**, 5573–5579 (2014)
10. Zacharia, R., Kim, K.Y., Kibria, A.K.M.F., Nahm, K.S.: Enhancement of hydrogen storage capacity of carbon nanotubes via spill-over from vanadium and palladium nanoparticles. Chem. Phys. Lett. **412**, 369–375 (2005)
11. Kim, H.S., Lee, H., Han, K.S., Kim, J.H., Song, M.S., Park, M.S., Lee, J.Y., Kang, J.K.: Hydrogen storage in Ni nanoparticle-dispersed multiwalled carbon nanotubes. J. Phys. Chem. B **109**, 8983–8986 (2005)
12. Li, Y.W., Yang, R.T.: Significantly enhanced hydrogen storage in metal-organic frameworks via spillover. J. Am. Chem. Soc. **128**, 726–727 (2006)
13. Gao, S., Ren, Z., Wan, L., Zheng, J., Guo, P., Zhou, Y.: Density functional theory prediction for diffusion of lithium on boron-doped graphene surface. Appl. Surf. Sci. **257**, 7443–7446 (2011)
14. ShokuhiRad, A., Shadravan, A., Soleymani, A.A., Motaghedi, N.: Lewis acid-base surface interaction of some boron compounds with N-doped graphene; first principles study. Curr. Appl. Phys. **15**, 1271–1277 (2015)
15. ShokuhiRad, A., Esfahanian, M., Maleki, S., Gharati, G.: Application of carbon nanostructures towards SO_2 and SO_3 adsorption: a comparison between pristine graphene and N-doped graphene by DFT calculations. J. Sulfur Chem. **37**, 176–188 (2016)
16. ShokuhiRad, A., Shabestari, S.S., Jafari, S.A., Zardoost, M.R., Mirabi, A.: N-doped graphene as a nanostructure adsorbent for carbon monoxide: DFT calculations. Mol. Phys. **114**, 1756–1762 (2016)
17. ShokuhiRad, A., Kashani, O.R.: Adsorption of acetyl halide molecules on the surface of pristine and Al-doped graphene: ab initio study. Appl. Surf. Sci. **355**, 233–241 (2015)
18. ShokuhiRad, A.: Al-doped graphene as sensitive nanostructure sensor for some ether molecules: ab initio study of adsorption. Synth. Met. **209**, 419–425 (2015)
19. ShokuhiRad, A.: First principles study of Al-doped graphene as

nanostructure adsorbent for NO_2 and N_2O: DFT calculations. Appl. Surf. Sci. **357**, 1217–1224 (2015)
20. ShokuhiRad, A., Foukolaei, V.P.: Density functional study of Al-doped graphene nanostructure towards adsorption of CO, CO_2 and H_2O. Synth. Met. **210**, 171–178 (2015)
21. ShokuhiRad, A.: Al-doped graphene as a new nanostructure adsorbent for some halomethane compounds: DFT calculations. Surf. Sci. **645**, 6–12 (2016)
22. ShokuhiRad, A., Jouibary, Y.M., Foukolaei, V.P., Binaeian, E.: Study on the structure and electronic property of adsorbed guanine on aluminum doped graphene: first principles calculations. Curr. Appl. Phys. **16**, 527–533 (2016)
23. ShokuhiRad, A., Shabestari, S.S., Mohseni, S., Aghouzi, S.A.: Study on the adsorption properties of O_3, SO_2, and SO_3 on B-doped graphene using DFT calculations. J. Solid State Chem. **237**, 204–210 (2016)
24. ShokuhiRad, A.: Adsorption of mercaptopyridine on the surface of Al- and B-doped graphenes: theoretical study. J. Alloy. Compd. **682**, 345–351 (2016)
25. Contescu, C.I., van Benthem, K., Li, S., Bonifacio, C.S., Pennycook, S.J., Jena, P., Gallego, N.C.: Single Pd atoms in activated carbon fibers and their contribution to hydrogen storage. Carbone **49**, 4050–4058 (2011)
26. Wang, H., Wang, Q., Cheng, Y., Li, K., Yao, Y., Zhang, Q., Dong, C., Wang, P., Schwingenschlögl, U., Yang, W., Zhang, X.X.: Doping monolayer graphene with single atom substitutions. Nano Lett. **12**, 141–144 (2012)
27. Baby, T.T., Aravind, S.S.J., Arockiadoss, T., Rakhi, R.B., Ramaprabhu, S.: Metal decorated graphene nanosheets as immobilization matrix for amperometric glucose biosensor. Sensor Actuat. B Chem. **145**, 71–77 (2010)
28. ShokuhiRad, A., Abedini, E.: Chemisorption of NO on Pt-decorated graphene as modified nanostructure media: a first principles study. Appl. Surf. Sci. **360**, 1041–1046 (2016)
29. ShokuhiRad, A.: Adsorption of C_2H_2 and C_2H_4 on Pt-decorated graphene nanostructure: ab initio study. Synth. Met. **211**, 115–120 (2016)
30. ShokuhiRad, A.: Density functional theory study of the adsorption of MeOH and EtOH on the surface of Pt-decorated graphene. Physica E **83**, 135–140 (2016)
31. ShokuhiRad, A., Zareyee, D.: Adsorption properties of SO_2 and O_3 molecules on Pt-decorated graphene: a theoretical study. Vacuum **130**, 113–118 (2016)
32. Sun, Y., Chen, L., Zhang, F., Li, D., Pan, H., Ye, J.: First-principles studies of HF molecule adsorption on intrinsic graphene and Al-doped graphene. Solid State Commun. **150**, 1906–1910 (2010)
33. Frisch, M.J., et al. Gaussian 09, Revision D.01, Gaussian, Inc., Wallingford, CT (2009)
34. ShokuhiRad, A., Ayub, K.: A comparative density functional theory study of guanine chemisorption on $Al_{12}N_{12}$, $Al_{12}P_{12}$, $B_{12}N_{12}$, and $B_{12}P_{12}$ nano-cages. J. Alloy. Compd. **672**, 161–169 (2016)
35. ShokuhiRad, A., Ayub, K.: Detailed surface study of adsorbed nickel on $Al_{12}N_{12}$ nano-cage. Thin Solid Films **612**, 179–185 (2016)
36. ShokuhiRad, A., Ayub, K.: Ni adsorption on $Al_{12}P_{12}$ nano-cage: DFT study. J. Alloy. Compd. **678**, 317–324 (2016)

13

First principle study of the effect of defects on performance of single-molecule pentacene field effect transistors

Bahniman Ghosh[1,2] · **Akash Gramin**[2]

Abstract In this work, we have performed first principle study on a single-molecule pentacene field effect transistor and studied various oxygen- and hydrogen-induced defects in the same device configuration. Further, we have investigated the effect of these defects on the various electronic transport properties of the device and compared them with those of the original device along with reporting the negative differential region window and the peak-to-valley ratio in different cases. For this purpose, we have applied the density functional theory in conjugation with non-equilibrium green's function (NEGF) formalism on a 14.11 Å pentacene device to obtain the I–V characteristics, conductance curves and transmission spectra in various device scenarios.

Keywords Pentacene · Field effect transistors · Non-equilibrium green's function · Density functional theory

Introduction

In the recent times, organic semiconductors have generated significant interest in the research and academic community. Due to the extensive application, possibilities of these materials in various areas such as organic flexible displays, smart devices and cards, solar cells and light-emitting diodes (OLED) among many others as well as the advantage of having low-cost, simpler processing and manufacturing techniques have led to an immense pace of development of these materials [1–5]. Moreover, consistent and rapid improvements in both device function and efficiency of organic-based devices have been promising in realizing "completely organic" (i.e., inorganic less) solutions [6]. So much so, that it would not be surprising to see organic-based mainstream commercial products in the coming few years. With this, organic electronics is already dubbed as being the next big revolution in the field of microelectronics.

Out of these, aromatic hydrocarbons have gained much attention with regard to their application in molecular devices [7]. Pentacene has been one of the most encouraging materials in realizing organic transistor, as it offers better reliability, mobility, stability and on–off ratio than many other organic materials [8]. Pentacene ($C_{22}H_{14}$) is an aromatic organic molecule with planar geometry having five benzene rings fused alongside each other as shown in Fig. 1a. It generally occurs in a herringbone pattern to form a triclinic structure having two molecules in a single unit cell. There are many different types of pentacene polytypes [9–11]. The electronic and structural properties of pentacene have been studied in some detail for single molecule and molecular solid systems [12]. In [13], authors study the physics of electron transport in single molecule. In [14], authors perform theoretical investigations of transport behavior in single molecule covalently bonded to gold electrodes. Although the effects of defects and impurities on mobility in pentacene crystal have been studied earlier in some detail, there has not been any study on pentacene single molecule field effect transistor (FET) [15].

In this work, we have applied non-equilibrium green's function (NEGF) formalism in conjugation with density

✉ Bahniman Ghosh
bghosh@utexas.edu

1 Microelectronics Research Center, University of Texas at Austin, 10100 Burnet Road, Bldg. 160, Austin, TX 78758, USA

2 Department of Electrical Engineering, Indian Institute of Technology Kanpur, Kanpur 208016, India

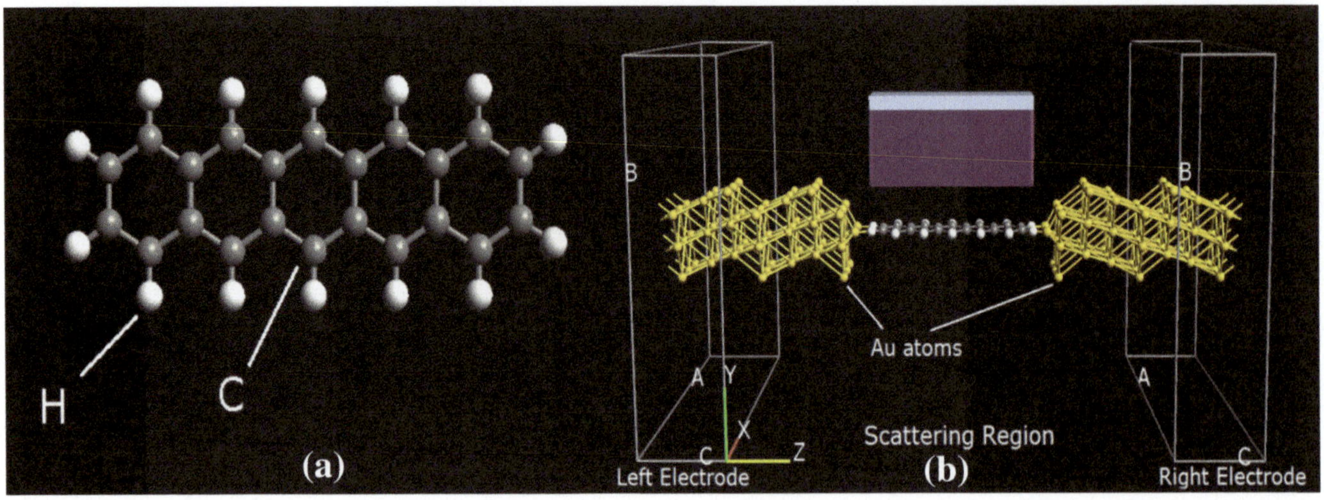

Fig. 1 **a** An isolated pentacene ($C_{22}H_{14}$) molecule. **b** Single-molecule pentacene FET with a pentacene molecule between the two gold electrodes

functional theory (DFT) to carry out first principle studies of the effect of oxygen- and hydrogen-induced defects on the electronic transport properties of a single pentacene molecule-based FET. We have also investigated additional properties in the I–V characteristics such as negative differential resistance (NDR) behavior, which is defined as the region on the I–V curve where an increase in the applied potential results in a decrease in the corresponding current (negative slope) and peak-to-valley ratio (PVR), which is the ratio of maximum peak current to the minimum valley current through the device, under different device scenarios.

Model

In our model, we have sandwiched a pentacene molecule between two gold electrodes thus making a FET configuration as shown in Fig. 1b. The nature of chemical bonding of a pentacene to gold is a van der Waals type. We find a supercell that can hold pentacene crystal on the Au(111) surface. A very good match with a mean strain below 2 percent is obtained through repetition of the Au(111) into a $(2 \times 3\sqrt{3})$ structure. For pentacene on the Au(111) surface, we obtained work function difference of 0.97 eV. The difference in work function arises from the charge transfer from gold to pentacene which gives rise to an interface dipole. This shows up in the electrostatic difference potential. The gate voltage (V_g) is applied only on top of the pentacene molecule. The entire device has three parts: the right electrode (R), central scattering region (SR) and left electrode (L). The gold electrodes are extracted along (111) direction from bulk gold, [16–18]. The pentacene molecule is attached to the two electrodes on either side. The bond lengths between the molecule and the electrodes along with the overall geometry of the configuration are determined by total energy optimizations in which the distance between electrodes is fixed. In this manner, the configuration is completely optimized such that the maximum atomic force between the atoms in less than 0.05 eV/Å, without disturbing the electrode gold atoms. Device channel length is 14.11 Å and width is 4.93 Å. The size of the supercell in the *y*-direction is taken as 26.66 Å. Also, HfO_2 ($\varepsilon_r = 25$) is used as the gate dielectric material with oxide thickness equal to 5.5 Å.

We have studied three types of defects in the pentacene ($C_{22}H_{14}$) molecule device which are namely, C–H_2 defect, C–O defect and C–HOH defect. The C–H_2 defect is created by adding one H atom to any one of the C–H units thus making it C–H_2 unit and the $C_{22}H_{14}$ molecule, a $C_{22}H_{15}$ one. Although the H atom can be connected to any of the corner C atoms, the most stable configuration is formed when it is attached as shown in Fig. 2a [19]. Similarly, the C–O defect is created by replacing an H atom with an O atom thus forming a double bond with that C atom and making it fourfold coordinated. In the same fashion, the C–HOH defect is formed by adding an OH group to the same C as mentioned above and further shown in the Fig. 2d. All the three defect configurations are optimized for geometry and bond relaxation. Next after the geometrical optimizations, all the three configurations are studied for electronic transport properties.

For all the electronic transport and geometry optimization study-related calculations of the configuration, we have used Atomistix ToolKit (ATK) simulator which is based on the combined theories of NEGF and DFT (NEGF + DFT) [20–22]. The exchange–correlation

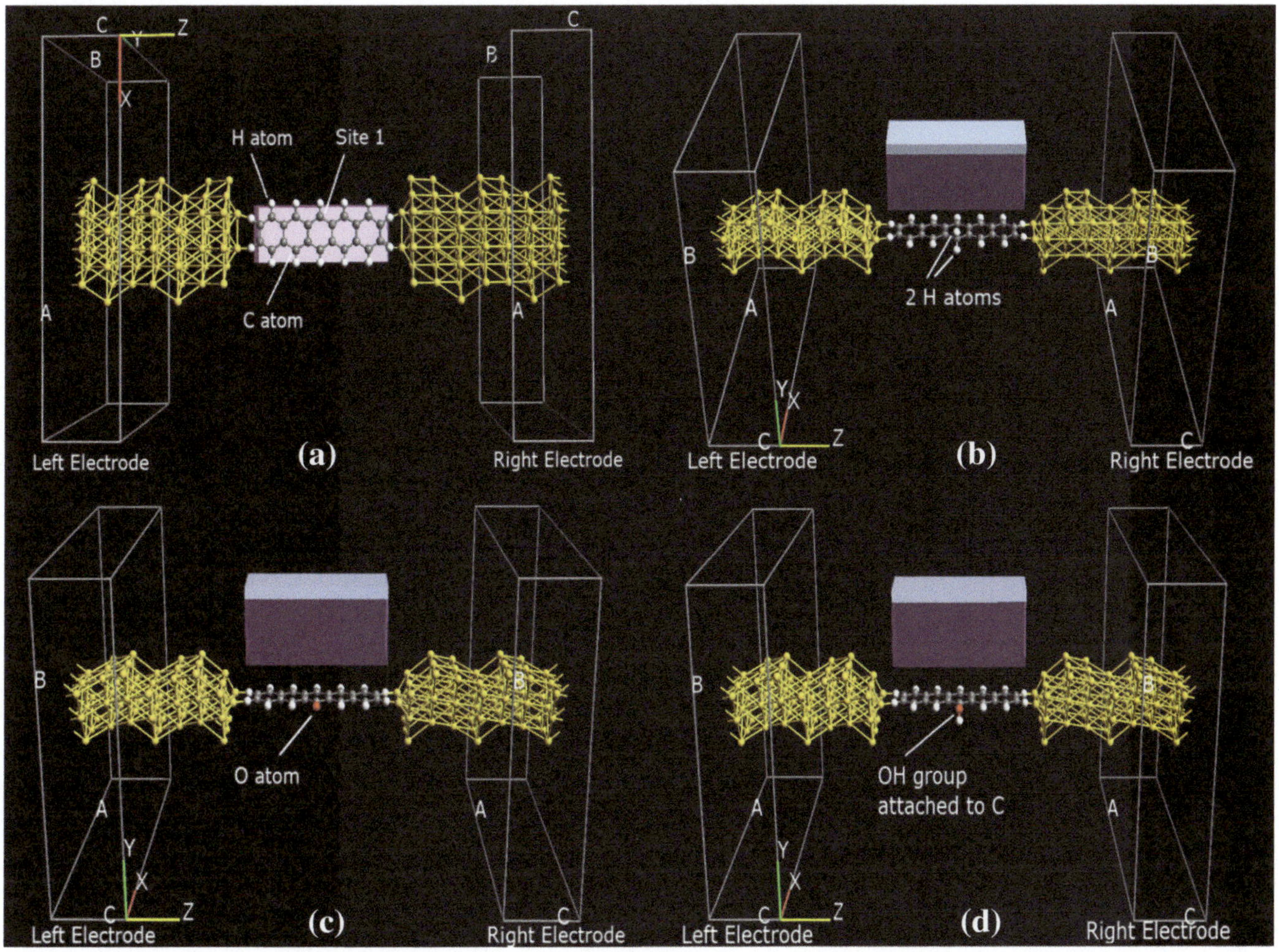

Fig. 2 **a** *Bottom view* of the device showing the C atom (site 1) where the most stable defects are formed. **b** device with C–H_2 defect. ($C_{22}H_{15}$) **c** device with C–O defect ($C_{22}H_{13}O$) **d** device with C–HOH defect ($C_{22}H_{15}O$)

potential in our model is described by the Perdew–Zunger functional of the local density approximation (LDA.PZ). The basis set used is a double zeta polarized (DZP) basis for all the atoms. First principle norm-conserving Fritz-Haber-Institute (FHI) pseudo-potentials are employed and the energy correlation functional is spin independent [23]. The electrode Poisson solver calculations are carried out with boundary conditions as periodic and the Brillouin zone integration is carried out with 1 × 10 × 100 *k*-point sampling. A real space grid is used for computing electrostatic potentials, with mesh cutoff energy equal to 75 Hartree. The electron temperature is taken as 300 K. Moreover, a large supercell dimension is taken in the perpendicular direction of electronic transport so that there is minimal or no interaction among the pentacene molecule and its images.As per the NEGF formalism, the current $I_{s\text{-}d}$ flowing between the source and drain terminals of the device resulting from applying a source–drain bias $V_{s\text{-}d}$ is given by the following formula [22]:

$$I_{s-d}=\frac{2e}{h}\int T(E,V_{s-d})[f(E-\mu_L)-f(E-\mu_R)]\mathrm{d}E, \quad (1)$$

where e and h are electron charge and Planck's constant, respectively, $\mu_{L(R)}=E_F\pm eV/2$ represents the electrochemical potential of the two electrodes and E_F denotes the Fermi level of the configuration which is taken to be zero. The electron Fermi distribution function of the electrodes is represented by $f(E-\mu_{L(R)})$. The total transmission probability is denoted by $T(E,V_{s\text{-}d})$ at energy E and source–drain bias $V_{s\text{-}d}$, and can be evaluated using:

$$T(E,V_{s-d})=\mathrm{Tr}[\Gamma_L G^R \Gamma_R G^A], \quad (2)$$

where $G^{R(A)}$ denotes the retarded (advanced) Green's function, and $\Gamma_{L(R)}=i\left[\sum^R_{L(R)}-\sum^A_{L(R)}\right]$ denotes the contact broadening function with $\sum\limits_{L(R)}^{R}$ and $\sum\limits_{L(R)}^{A}$ being the self-energy terms containing the effect of the electrodes. By shifting the SR terms of the Hamiltonian along with V_g, the effect of gate voltage is accounted for. Further details on such a theoretical arrangement can be studied in Refs. [20, 22].

Results and discussions

In all the cases, the applied gate voltage is taken 0.4 V while the drain voltage is taken over a range of 0–0.8 V.

In Fig. 3a, the I–V characteristic of the original pentacene device without any defect is shown. The device exhibits square-law-type characteristics, with saturated output curves, negative differential resistance regions and a low on-current. As can be clearly seen from the plot, the channel current shows a rising edge at zero bias voltage. The maximum current through the channel is of magnitude 2350 nA which is achieved at 0.42 V bias voltage. Also, the minimum channel current is observed at 0.74 V bias voltage and is equal to 1200 nA. Another important

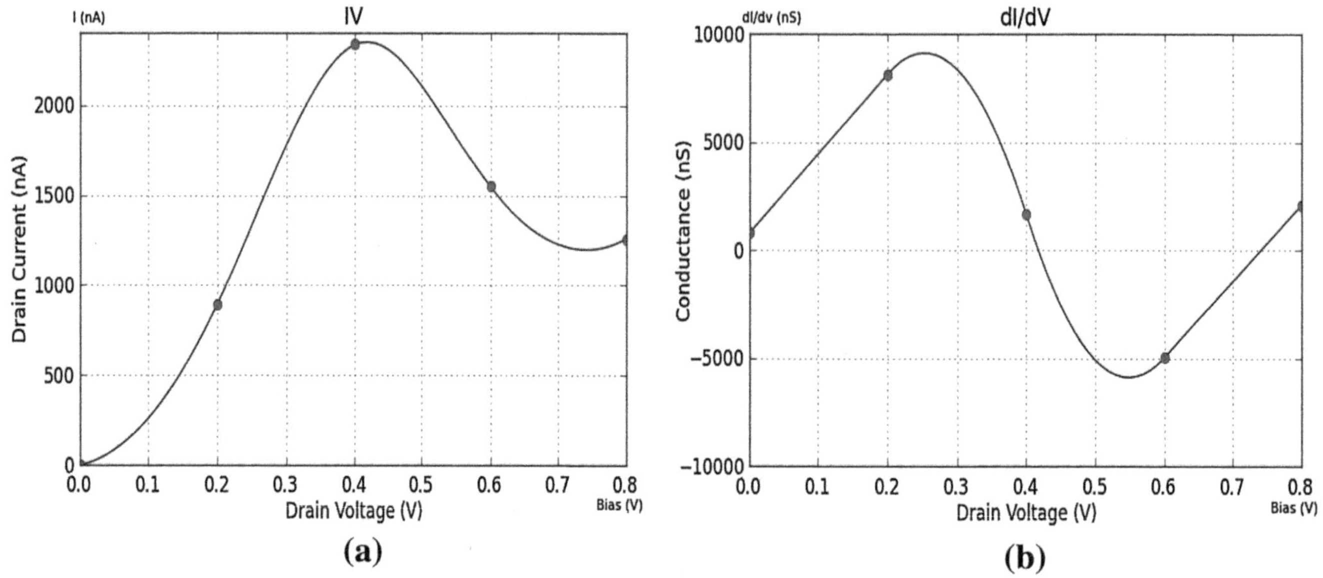

Fig. 3 **a** I–V characteristics of normal device ($V_g = 0.4$ V) **b** conductance of normal device ($V_g = 0.4$ V)

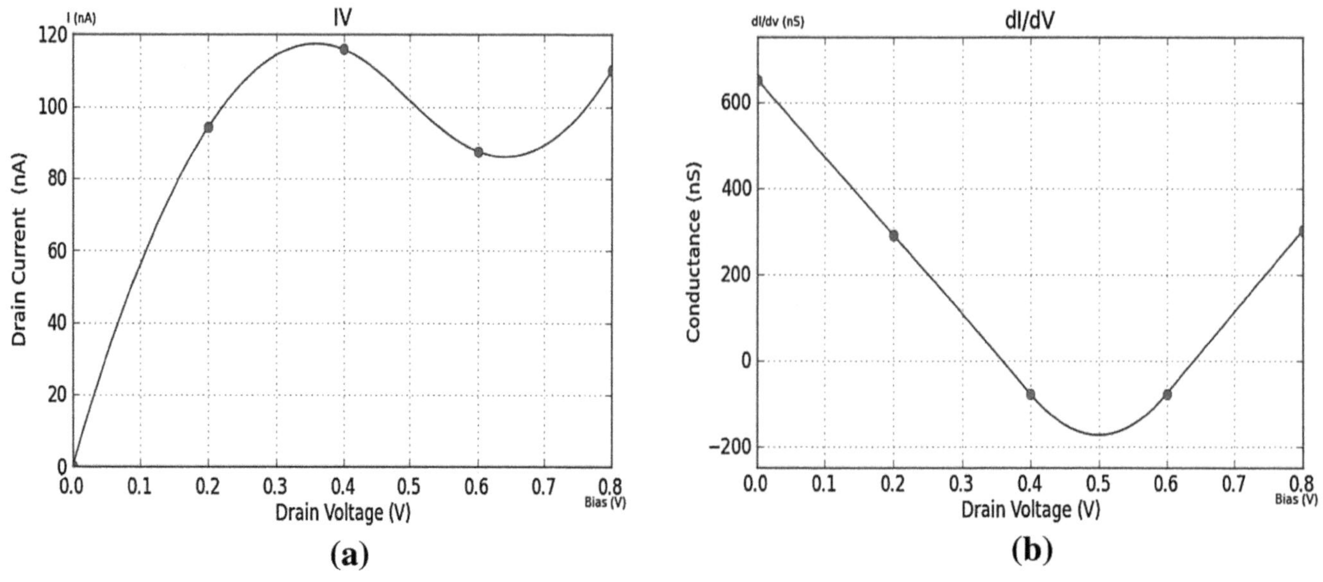

Fig. 4 **a** I–V characteristics of device with C–H_2 defect ($V_g = 0.4$ V). **b** Conductance of device with C–H_2 defect ($V_g = 0.4$ V)

observation is to note the NDR region which extends from 0.42 to 0.74 V drain voltage with PVR window of 1150 nA. The conductance vs bias voltage plot is shown in Fig. 3b, in which the peak conductance is seen at 0.25 V bias voltage which then decreases until it attains minima at 0.55 V bias voltage and increases afterwards.

Figure 4a shows the I–V curve of the device with C–H_2 defect. As can be seen from the plots, the channel current shows a rising edge at zero bias voltage. In this case, the maximum current is achieved at a bias voltage of 0.36 V, of magnitude 116 nA. The channel current then decreases with increasing bias voltage until achieving a minimum at 0.64 V bias voltage with corresponding channel current of 86 nA and increases afterwards. The NDR region is seen to extend from 0.36 to 0.64 V bias voltage and PVR windows of 30 nA. As the drain voltage is increased beyond 0.36 V, some of the transmission modes close to the Fermi level are suppressed. This leads to NDR behavior in I–V curves. Also, conductance decreases till it reaches 0.5 V at which point it starts to increase as shown in Fig. 4b.

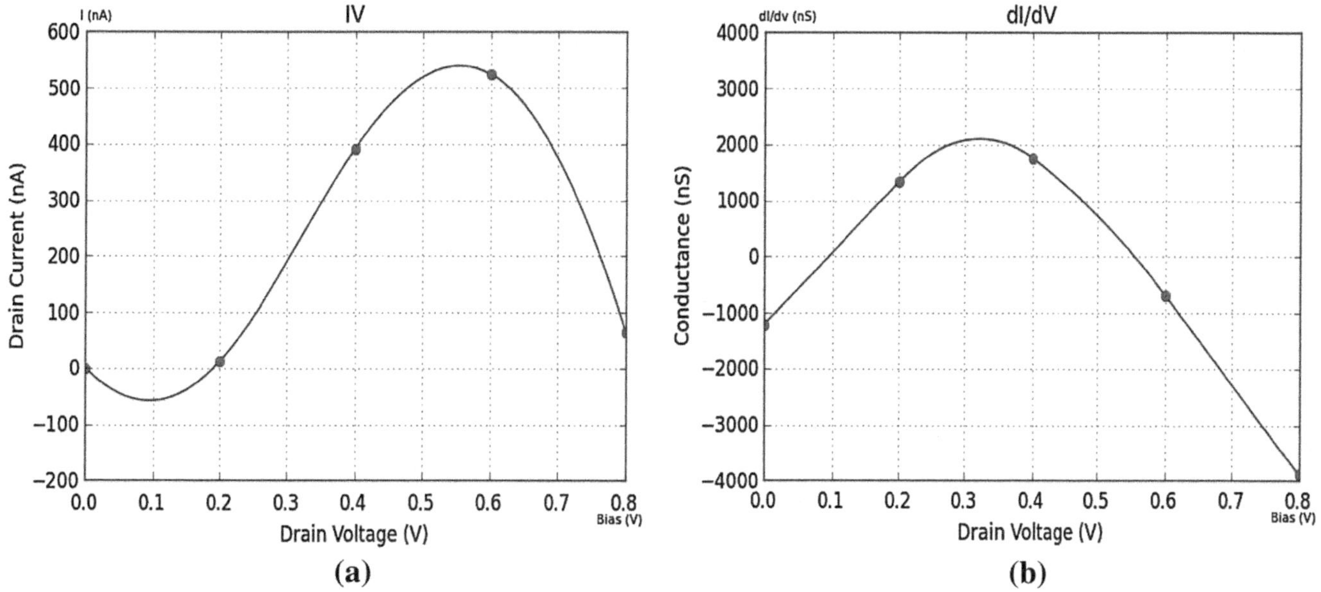

Fig. 5 **a** I–V characteristics of device with **b** conductance plot of the device with C–O defect ($V_g = 0.4$ V) C–O defect ($V_g = 0.4$ V)

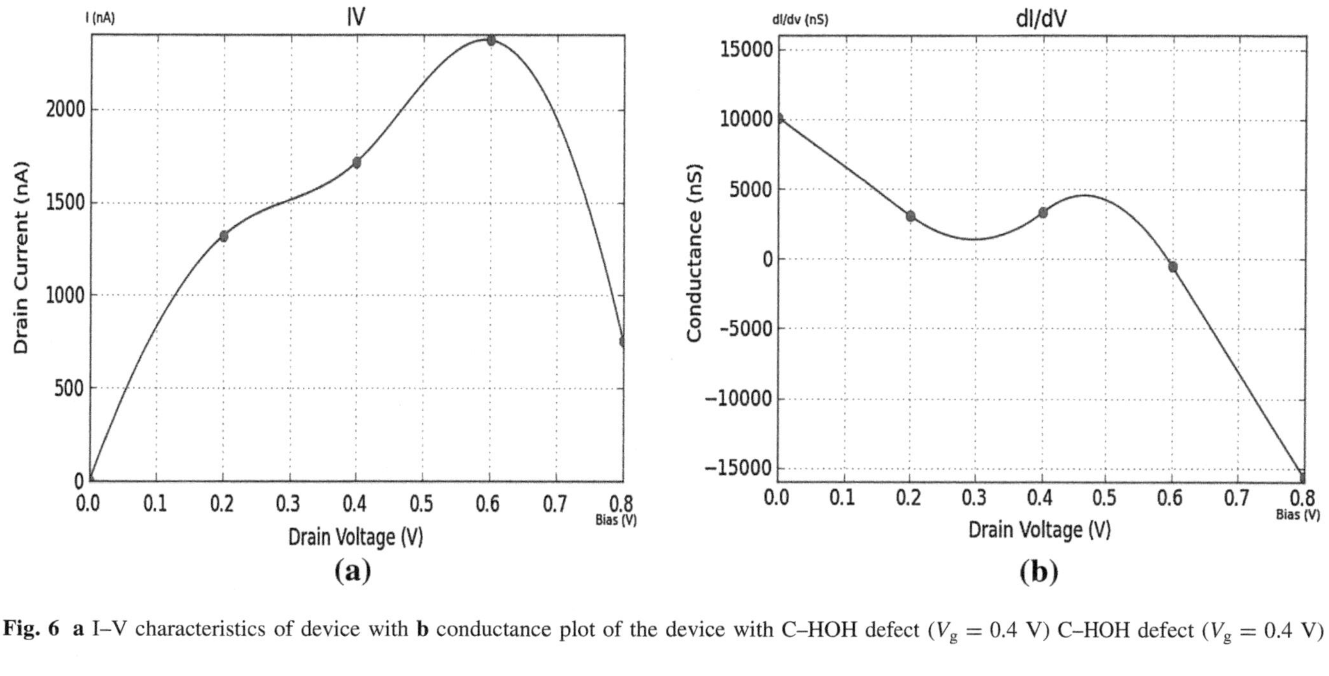

Fig. 6 **a** I–V characteristics of device with **b** conductance plot of the device with C–HOH defect ($V_g = 0.4$ V) C–HOH defect ($V_g = 0.4$ V)

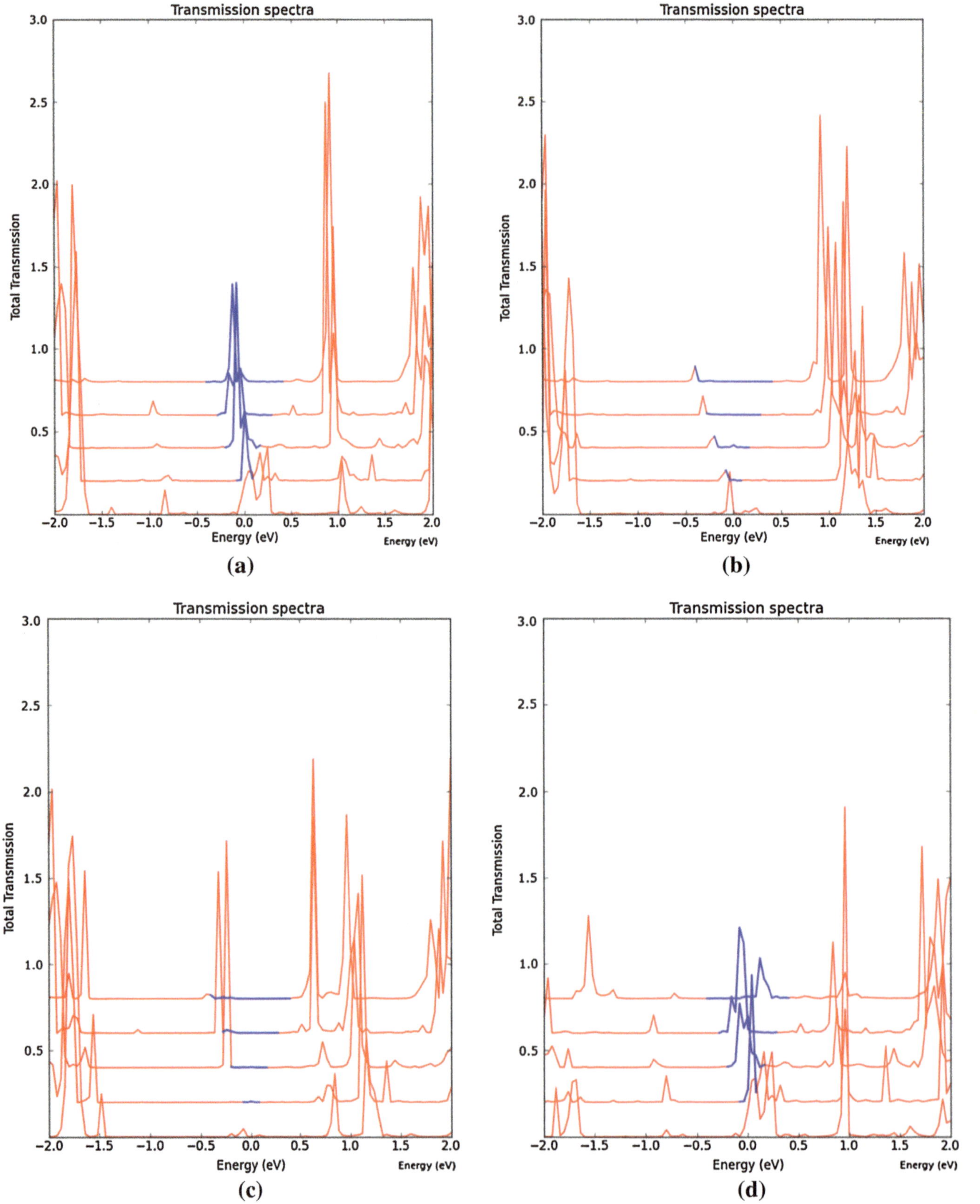

Fig. 7 Transmission spectra corresponding to all the four cases: **a** normal device with no defect **b** device with C–H_2 defect **c** device with C–O defect **d** device with C–HOH defect

In Fig. 5a, the I–V characteristic of the device with C–O defect is shown. As the plot shows, the channel current shows a falling edge at zero bias voltage. In this case, the channel current first decreases till 0.1 V at which point its magnitude is 60 nA in the reverse direction and then increases till it reaches the bias voltage of 0.55 V where it attains a maximum of 530 nA and then decreases further on. The NDR region in this case can be seen to exist in two intervals, i.e., from 0 to 0.1 V and from 0.55 V onwards with PVR window of 590 nA. Figure 5b shows the conductance curve which attains a maximum of 2100 nA at bias voltage of 0.33 V.

In Fig. 6a, the I–V characteristic of the device with C–HOH defect is shown. As is evident from the plot, the curve shows a rising edge at zero bias voltage. In this case, the channel current increases till it achieves a maximum at 0.6 V bias voltage with corresponding current of 2475 nA after which it decreases further right being 550 nA at 0.8 V. The NDR region can be seen to start from this point (0.6 V) onwards with PVR window of 1925 nA. Also, Fig. 6b shows the conductance plot which decreases until it reaches 0.3 V where it attains minima and then increases slightly till 0.46 V and then decreases steadily. As the drain voltage is increased beyond 0.6 V, some of the transmission modes close to the Fermi level get suppressed and as explained earlier this results in drop in the transmission transparency of the channel, thus exhibiting NDR behavior.

The transmission spectra corresponding to all the four cases are shown in Fig. 7.

Conclusion

We have studied the effect of various hydrogen (H)- and oxygen (O)-induced defects, namely, C–H_2 defect, C–O defect and C–HOH defect on the I–V characteristics, conductance and transmission spectrum of single-molecule pentacene FET and compared them with the results of the device with no defects. We found that the channel current shows a significant decrease in the case of C–H_2 and C–O defect, while the results of C–HOH defect are comparable to those of the normal device with no defects. We have also reported the NDR region in all the four cases and learned that its range is comparable in the case of no defect and C–H_2 defect, while that of C–O defect is comparable to C–HOH defect along with reporting the PVR window. These properties of the single-molecule FET can be used in various applications such as in the development of SRAMs, high-frequency oscillators and sensors.

Acknowledgments The authors would like to thank the Department of Science and Technology of the government of India for partially funding this work.

References

1. Sun, Y., Liu, Y., Zhu, D.: J. Mater. Chem. **15**, 53 (2005)
2. Murphy, A.R., Fréchet, J.M.J.: Chem. Rev. **107**, 1066 (2007)
3. Anthony, J.E.: AngewandteChemie Int. Ed. Engl **47**, 452 (2008)
4. Allard, S., Forster, M., Souharce, B., Thiem, H., Scherf, U.: AngewandteChemie Int. Ed. Engl. **47**, 4070 (2008)
5. Yamashita, Y.: Sci. Technol. Adv. Mater. **10**, 024313 (2009)
6. Koch, Norbert: Chem Phys Chem **8**, 1438–1455 (2007)
7. Horowitz, G.: Adv. Mater. **10**, 365 (1998)
8. Volkman, S.K., Molesa, S., Mattis, B., Chang, P.C., Subramanian V.: Mater. Res. Soc. Symp. Proc. 769, (2003)
9. Siegrist, T., et al.: AngewandteChemie Int. Ed. Engl. **40**, 1732 (2001)
10. Campbell, R.B., Robertson, J.M., Trotter, J.: Acta Crystallogr. **15**, 289 (1962)
11. Tiago, M.L., Northrup, J.E., Louie, S.G.: Phys Rev B **67**, 115212 (2003)
12. Endres, R.G., Fong, C.Y., Yang, L.H., Witte, G., Woll, C.: Materials Science (cond-mat.mtrl-sci)arXiv:cond-mat/0211420. (2002)
13. Chen, F., Tao, N.J.: Electron transport in single molecules: from benzene to graphene. Acc. Chem. Res. **42**, 429–438 (2009)
14. Li, X.L., Xu, B.Q., Xiao, X.Y., Hihath, J., Tao, N.J.: Measurement of electron transport Properties of single molecules. J. Jpn. Appl. Phys. **44**, 5344–5347 (2005)
15. Jurchescu, O.D., Baas, J., Palstra, T.T.M.: Appl. Phys. Lett. **84**, 3061 (2004)
16. Zheng, X.H., Shi, X.Q., Dai, Z.X., Zeng, Z.: Phys. Rev. B **74**, 085418 (2006)
17. Taylor, J., Brandbyge, M., Stokbro, K.: Phys. Rev. B **68**, 121101 (2003)
18. Ke, S.-H., Baranger, H.U., Yang, W.: Phys. Rev. B **71**, 113401 (2005)
19. Northrup, J.E., Chabinyc, M.L.: Phys. Rev. B **68**, 041202(R) (2003)
20. Taylor, J., Guo, H., Wang, J.: Physical Review B **63**, 245407 (2001)
21. Brandbyge, M., Mozos, J.L., Ordejón, P., Taylor, J., Stokbro, K.: Phys. Rev. B **65**, 165401 (2002)
22. Atomistix Virtual Nanolab, QuantumWise Inc
23. Troullier, N., Martins, J.: Phys. Rev. B **43**, 10222 (1991)
24. Buttiker, M., Landauer, R.: Phys. Rev. B **31**, 6207 (1985)

14

Generalized ghost dark energy in Horava–Lifshitz cosmology

Bharat Borah · M. Ansari

Abstract Purpose of this paper is to study generalized quantum chromodynamics ghost dark energy (GDE) in the frame work of Horava–Lifshitz cosmology. Considering interacting and non-interacting scenario of GDE with dark matter in a spatially non-flat universe, we investigate the cosmological implications of this model in detail. We obtain equation of state parameter, deceleration parameter and the evolution of dark energy density to explain the expansion of the universe. Also, we show that the results we calculate have a good compatibility with previous work and restore it in limiting case. Further, we investigate validity of generalized second law of thermodynamics in this scenario. Finally, we find out a cosmological application of our work by evaluating a relation for the equation of state of dark energy for law redshifts.

Keywords Ghost dark energy (GDE) · Modified gravity · Horava–Lifshitz cosmology · Redshift

Introduction

The discovery that universe is in an accelerating expansion phase is one of the most remarkable advancement in modern cosmology. Several cosmological observations clearly indicate that the present day universe is experiencing a phase of accelerated expansion [1–4]. This current accelerating expansion of the universe is dominated by an unknown energy termed as dark energy with negative pressure. Numerous models have been proposed to understand the dark energy phenomena and evolution of the universe. Comprehensive reviews of these models have been widely reported [5, 6]. However, nature and cosmological origin of dark energy still remain enigmatic at present.

In most of recent dark energy models, normally we need to consider a new degree of freedom or a new parameter, to explain the cosmic acceleration of Universe. However, our prior choice to handle the dark energy problem without introducing new degrees of freedom beyond what are already known. An interesting and successful attempt, in this direction is the so called "ghost dark energy" (GDE), which has been motivated from Veneziano ghost of chromodynamics. The Veneziano ghost field was proposed to resolve the $U(1)$ problem in quantum chromodynamics (QCD) [7–9]. It is well known that QCD ghost fields have positive and negative norms and cancel with each other, leaving no trace in the physical subspace, but it was argued that they have small contribution to the vacuum energy in the curved space or time-dependent background [10]. The ghost field has no contribution in the vacuum energy in Minkowski space-time but it has a non-vanishing contribution to the vacuum energy, in a dynamical background such as our universe, is proportional to $\Lambda QCD^3 H$, where H is Hubble parameter and ΛQCD^3 is QCD mass scale [10]. Taking into account the fact that $\Lambda QCD \sim 100\,\text{MeV}$ and $H \sim 10^{-33}\,\text{eV}$ for present time, this gives the right order of magnitude $\rho D \sim (3 \times 10^{-3}\,\text{eV})^4$ for the ghost energy density [10]. Thus GDE model can solve the fine tuning problem [10–12]. Energy density of ghost dark energy reads as [10–12]

$$\rho D = \alpha H, \tag{1}$$

where H is Hubble parameter $H = \frac{\dot{a}}{a}$ and α is constant parameter of the model, which should be determined.

B. Borah (✉) · M. Ansari
Department of Mathematics, North-Eastern Hill University, Shillong 793022, India
e-mail: bharatborah@gmail.com

Author of Ref. [13] discussed that the contribution of the Veneziano QCD ghost field to the vacuum energy is not exactly of order H and a sub-leading term H^2 appears due to the fact that the vacuum expectation value of the energy–momentum tensor is conserved in isolation [14]. It was argued that the vacuum energy of the ghost field can be written as $H + O(H^2)$, where the sub-leading term H^2 in the GDE model might play a crucial role in the early stage of the universe evolution, acting as the early DE [15]. It was shown [15, 16] that taking the second term into account can give better agreement with observational data compared to the usual GDE. The density of this generalized ghost dark energy (GGDE) reads [15]

$$\rho D = \alpha H + \beta H^2 \tag{2}$$

with β another constant parameter of the model with [energy]^2.

The cosmological implications of GDE model in a flat universe were considered in [17]. The generalization to the non-flat Friedmann–Robertson–Walker (FRW) universe and its consistency check with recent observational data were studied in [18]. The GDE model in the framework of Brans–Dicke cosmology and its stability against perturbations were also investigated in [19, 20]. The correspondence between this model and scalar field models of dark energy has been widely extended by many authors in the recent years [21]. Furthermore GDE has been studied and extended in the frame work of modified gravity [22, 23].

It is a well-known fact that the biggest challenge towards a theory of quantum gravity is the question of renormalizability whose lack, in effect, would imply loss of theoretical control and predictability at high energies. General relativity although performs fairly well at the solar system scale but fails at the cosmic scales at which it is unable to explain the origin of dark energy or presents accelerated cosmic acceleration properly. Therefore the classical generalization of general relativity is considered as the gravitational alternative for a unified description of the early-time inflation with late-time cosmic acceleration. Thus, to study the structure and cosmological properties of modified theories, the cosmological reconstruction of different modified gravities is provided in detail. In other words, the modified gravity, which represents a classical generalization of general relativity, becomes reasonable alternative. Scalar tensor theories are also modified theories of gravity. Lifshitz and Eksp [24] introduced a scalar theory in which the temporal dimension has weight equal to three if the space dimension has weight equal to one. Recently Horava [25–28] proposed a new quantizable theory of gravity, using the ideas from solid state physics, which is similar to Lifshitz theory, thus this theory is called Horava–Lifshitz gravity theory. This is a power-counting renormalizable theory with consistent ultraviolet (UV) behavior and the theory has one fixed point in the infrared limit (IR), namely general relativity [25–28]. Thus, this theory leads to a modification of the Einstein's general relativity at high energies producing interesting features in cosmology. Furthermore Horava–Lifshitz gravity has been studied and extended in detail [29–36] and applied as a cosmological frame work of the universe [37–46]. Recently, different dark energy models have been studied in the framework of Horava–Lifshitz cosmology using different infrared cut-offs. For comprehensive reviews on the scenario where the cosmological evolution is governed by Horava–Lifshitz cosmology one can go through the references [47–53]. Moreover, the ghost energy density belongs to a dynamical cosmological constant, therefore we need a dynamical frame to accommodate it instead of general relativity. Also the original GDE is unstable in all range of the parameter spaces in standard [20]. So it is worthwhile to investigate the GGDE model in the framework of the Horava–Lifshitz cosmology.

This paper is organized as follows: in Sect. 2, we discuss briefly some important features of Horava–Lifshitz Cosmology. In Sect. 3, we study GGDE in Horava–Lifshitz cosmology for interacting and non-interacting scenario of dark energy with dark matter. Section 4 is devoted for a cosmological application of our work, where we obtain the equation of state of dark energy at low redshifts. Since, study of thermodynamic properties of ghost dark energy in Horava–Lifshitz cosmology is also an important direction, in Sect. 5, we investigate the validity of generalized second law of thermodynamics. Finally in the final section we summarize our results.

Horava–Lifshitz cosmology

In this section, we briefly review Horava–Lifshitz cosmology. Under the projectability condition, the full metric in the 3 + 1-dimensional Arnowitt–Deser–Misner formalism is given by

$$ds^2 = -N^2dt^2 + g_{ij}(dx^i + N^i dt)(dx^j + N^j dt), \tag{3}$$

where dynamical variables N and N^i are the lapse and shift functions, respectively. The indices are raised and lowered by using the spatial metric g_{ij}. The scaling transformation of the co-ordinates for $z = 3$ is given by

$$t \longrightarrow l^3 t \quad \text{and} \quad \mathrm{x}^{\mathrm{i}} \longrightarrow \mathrm{l}\mathrm{x}^{\mathrm{i}}. \tag{4}$$

Decomposing the gravitational action into a kinetic and a potential part, action of Horava–Lifshitz cosmology can be written as

$$S_g = \int dtd^3x\sqrt{g}N(L_K + L_V). \tag{5}$$

Under the assumption of detailed balance [25] the full action of Horava–Lifshitz gravity is given by

$$S_g = \int dtd^3x\sqrt{g}N\left[\frac{2}{\kappa^2}(K_{ij}K^{ij} - \lambda K^2) + \frac{\kappa^2}{2\omega^4}C_{ij}C^{ij} + \frac{\kappa^2\mu^2}{8}R_{ij}R^{ij} - \frac{\kappa^2\mu\,\eta^{ijk}}{2\omega^2\sqrt{g}}R_{il}\nabla_j R_k^l + \frac{\kappa^2\mu^2}{8(3\lambda-1)}\left(\frac{1-4\lambda}{4}R^2 + \Lambda R - 3\Lambda^2\right)\right], \tag{6}$$

where

$$K_{ij} = \frac{1}{2N}\left(\dot{g}_{ij} - \nabla_i N_j - \nabla_j N_i\right), \tag{7}$$

$$C_{ij} = \frac{e^{ijk}}{\sqrt{g}}\nabla_k\left(R_i^j - \frac{1}{4}R\delta_i^j\right), \tag{8}$$

are extrinsic curvature and cotton tensor, respectively. Also, η^{ijk} is the totally antisymmetric unit tensor, λ is a dimensionless constant and Λ is a negative constant which is related to the cosmological constant in the IR limit. The variables, κ, ω and μ are constants with mass dimensions $-1, 0$ and 1, respectively. To include dark matter component in a universe governed by Horava gravity, two options can be considered: one is to introduce a scalar field ϕ with action [54] and the second is add a cosmological stress–energy tensor to the gravitational field equations [55]. Here in our work we follow the second one.

In the cosmological contents, we consider FRW metric,

$$N = 1, \quad g_{ij} = a^2(t)\gamma_{ij}, \quad N^i = 0, \tag{9}$$

with

$$\gamma_{ij}dx^i dx^j = \frac{dr^2}{1-kr^2} + r^2 d\Omega_2^2, \tag{10}$$

where, $k = -1, 0, +1$ refer to spatially open, flat and closed universe, respectively. Taking the variation of action with respect to the metric components N and g_{ij}, one can obtain the equation of motion as

$$H^2 = \frac{\kappa^2}{6(3\lambda-1)}\rho_m + \frac{\kappa^2}{6(3\lambda-1)}\left[\frac{3\kappa^2\mu^2k^2}{8(3\lambda-1)a^4} + \frac{3\kappa^2\mu^2\Lambda^2}{8(3\lambda-1)}\right] - \frac{\kappa^4\mu^2\Lambda k}{8(3\lambda-1)^2a^2}, \tag{11}$$

$$\dot{H} + \frac{3}{2}H^2 = -\frac{\kappa^2}{4(3\lambda-1)}\rho_m - \frac{\kappa^2}{4(3\lambda-1)}\left[\frac{\kappa^2\mu^2k^2}{8(3\lambda-1)a^4} - \frac{3\kappa^2\mu^2\Lambda^2}{8(3\lambda-1)}\right] - \frac{\kappa^4\mu^2\Lambda k}{16(3\lambda-1)^2a^2}, \tag{12}$$

where $H = \frac{\dot{a}}{a}$ is the Hubble parameter. Keeping in mind the Friedmann equations [given by Eqs. (11, 12)], we can define the energy density ρD and pressure density pD of dark energy as

$$\rho D \equiv \frac{3\kappa^2\mu^2k^2}{8(3\lambda-1)a^4} + \frac{3\kappa^2\mu^2\Lambda^2}{8(3\lambda-1)} \tag{13}$$

$$pD \equiv \frac{\kappa^2\mu^2k^2}{8(3\lambda-1)a^4} - \frac{3\kappa^2\mu^2\Lambda^2}{8(3\lambda-1)}. \tag{14}$$

The term proportional to a^{-4} is the "dark radiation term" in Horava–Lifshitz cosmology. Rewriting Friedmann Eqs. (11) and (12), using Eqs. (13) and (14), we have

$$H^2 + \frac{k}{a^2} = \frac{8\pi Gc}{3}(\rho m + \rho D) \tag{15}$$

$$\dot{H} + \frac{3}{2}H^2 + \frac{k}{2a^2} = -4\pi Gc(pm + pD) \tag{16}$$

with

$$Gc = \frac{\kappa^2}{16\pi(3\lambda-1)} \quad \text{and} \quad \frac{\kappa^4\mu^2\Lambda}{8(3\lambda-1)^2} = 1, \tag{17}$$

where Gc represents the Newton's cosmological constant. Additionally, we can define Newton's gravitational constant as

$$G_g = \frac{\kappa^2}{32\pi}. \tag{18}$$

It is interesting that in the IR limit ($\lambda = 1$), Gc and G_g are the same.

Generalized ghost dark energy in Horava–Lifshitz cosmology

To study generalized ghost dark energy in the frame work of Horava–Lifshitz cosmology, we consider a spatially non-flat FRW Universe containing DE and DM. We define the critical energy density, ρcr, and the energy density of the curvature, ρ_k, as

$$\rho cr = \frac{3H^2}{8\pi Gc}, \tag{19}$$

$$\rho_k = -\frac{3k}{8\pi Gca^2}. \tag{20}$$

Then the dimensionless density parameter can also be defined as usual

$$\Omega_m = \frac{\rho_m}{\rho cr} = \frac{8\pi Gc}{3H^2}\rho_m, \tag{21}$$

$$\Omega D = \frac{\rho D}{\rho cr} = \frac{8\pi Gc}{3H^2}\rho D = \frac{8\pi Gc}{3H^2}(\alpha H + \beta H^2), \tag{22}$$

$$\Omega_k = \frac{\rho_k}{\rho cr} = -\frac{k}{a^2H^2}, \quad (23)$$

Using Eqs. (21), (22) and (23), the first Friedmann Eq. (15), can be written as

$$1 - \Omega_k = \Omega D + \Omega_m \quad (24)$$

Non-interacting scenario

In this section, we extend our study to the case where PLECHDE and dark matter are interaction free. In non-interacting scenario the energy conservation laws for spatially non-flat FRW universe are as follows

$$\dot{\rho D} + 3H\rho D(1 + wD) = 0, \quad (25)$$

$$\dot{\rho_m} + 3H\rho_m = 0, \quad (26)$$

where $wD = \frac{pD}{\rho D}$ is the equation of state parameter of PLECHDE.

Differentiating Eqs. (2) and (15) with respect to time, we obtain

$$\dot{\rho}D = \dot{H}(\alpha + 2\beta H), \quad (27)$$

$$\frac{\dot{H}}{H^2} = -\Omega_k - \frac{3}{2}\Omega D(1 + u + wD), \quad (28)$$

where, $u = \frac{\rho_m}{\rho D} = \frac{\Omega_m}{\Omega D}$.

Combining Eqs. (27) and (28) with continuity Eq. (25), we have

$$wD = \frac{1}{\Omega D\left(2 - \Omega D - \frac{8\pi Gc\beta}{3}\right)}\left[\left(\Omega D + \frac{8\pi Gc\beta}{3}\right)\left(1 - \frac{\Omega_k}{3}\right) - 2\Omega D\right]. \quad (29)$$

here, it is easy to see that at late time where $\Omega D \longrightarrow 1$ (in limiting case $\Omega_k \longrightarrow 0$), we have $wD \longrightarrow -1$, which means that GGDE mimics a cosmological constant. Also if $\beta = 0$ or $\frac{8\pi Gc}{3} \longrightarrow 0$, Eq. (29) restores the EoS parameter of the original GDE model [18]

$$wD = -\frac{1}{2 - \Omega D}\left(1 + \frac{\Omega_k}{3}\right). \quad (30)$$

On the other hand, for flat universe $\Omega_k = 0$, the equation of state parameter of GGDE in Horava–Lifshitz cosmology is

$$wD = \frac{1}{\Omega D\left(2 - \Omega D - \frac{8\pi Gc\beta}{3}\right)}\left[\frac{8\pi Gc\beta}{3} - \Omega D\right] \quad (31)$$

It is clear from this relation also that if $\beta = 0$ or $\frac{8\pi Gc}{3} \longrightarrow 0$, it restores its respective one in GDE [18].

Imposing the present observational values $\Omega_\Lambda \approx 0.73$ and $\Omega_k \approx 0.02$ in Eq. (29), we can see that $wD < -1$ if $\frac{8\pi Gc\beta}{3} < \frac{2}{3}$. This implies that GGDE model in Horava–Lifshitz cosmology replicates the role of Phantom dark energy model at present time if $\frac{8\pi Gc\beta}{3} < \frac{2}{3}$.

We can also study the evolution behavior of the GGDE. For this, differentiating Eq. (22) with respect to time and using Eqs. (27), (28) and relation $\dot{\Omega D} = H\Omega D'$ we obtain

$$\frac{\Omega D'}{\Omega D} = \frac{3}{2}(1 + u + wD)\left(\Omega D + \frac{8\pi Gc\beta}{3}\right) + 2\Omega_k \quad (32)$$

Also using Eqs. (24) and (29), we have

$$1 + u + wD = \frac{1}{\Omega_d\left(2 - \Omega D - \frac{8\pi Gc\beta}{3}\right)}\left[2(1 - \Omega_k - \Omega D) + \frac{2}{3}\Omega_k\left(\Omega D + \frac{8\pi Gc\beta}{3}\right)\right]. \quad (33)$$

Using Eq. (33) in Eq. (32), finally we obtain the expression that describes the evolution behavior of the non-interacting GGDE in the frame work of Horava–Lifshitz cosmology as

$$\frac{\Omega D'}{\Omega D} = \left[\frac{\Omega_k(\Omega D + \frac{8\pi Gc\beta}{3}) + 3(1 - \Omega_k - \Omega D)}{\Omega D\left(2 - \Omega D - \frac{8\pi Gc\beta}{3}\right)}\right]\left[\Omega D + \frac{8\pi Gc\beta}{3}\right] + 2\Omega_k. \quad (34)$$

For completeness, we find the deceleration parameter 'q' which represents the decelerated or accelerated phase of the expansion of the universe. The positive value of deceleration parameter $(q > 0)$ indicates the decelerated phase of expansion and the negative value $(q<0)$ represents the accelerated phase of expansion of the universe. Thus using Eq. (28) we obtain the decelerated parameter as

$$q = -\frac{\ddot{a}}{aH^2} = -1 - \frac{\dot{H}}{H^2} = -1 + \Omega_k + \frac{3}{2}\Omega D(1 + u + wD). \quad (35)$$

Combining Eqs. (29) and (35), finally we have the decelerated parameter is

$$q = \frac{1}{2}(1 - \Omega_k) + \frac{3}{2\left(2 - \Omega D - \frac{8\pi Gc\beta}{3}\right)} \times \left[\left(\Omega D + \frac{8\pi Gc\beta}{3}\right)\left(1 - \frac{\Omega_k}{3}\right) - 2\Omega D\right]. \quad (36)$$

Interacting scenario

To be more general, in this section we consider an interaction between both the dark components, the pressure-less dark matter and the ghost dark energy . The recent observational evidence provided by the galaxy clusters also supports the interaction between dark energy and dark matter [56]. In this case, the energy densities of dark energy and dark matter no longer satisfy independent conservation laws. They obey instead

$$\dot{\rho D} + 3H\rho D(1 + wD) = -Q, \quad (37)$$

$$\dot{\rho_m} + 3H\rho_m = Q, \tag{38}$$

where Q represents the interaction term. Following [57], we consider the simplest form of Q as

$$Q = 3b^2H(\rho_m + \rho D), \tag{39}$$

where b^2 is a coupling constant. The positive b^2 is responsible for the transition from dark energy to matter and vice-versa for negative b^2. Note that the form of Q chosen is purely phenomenological, i.e. to obtain desirable cosmological findings although more general phenomenological interaction terms can be used. Following as above, we obtain expressions for wD and $\Omega D'$ as

$$wD = \frac{1}{\Omega D\left(2 - \Omega D - \frac{8\pi Gc\beta}{3}\right)} \times \left[\left(\Omega D + \frac{8\pi Gc\beta}{3}\right)\left(1 - \frac{\Omega_k}{3}\right) - 2\Omega D - 2b^2(1 - \Omega_k)\right]. \tag{40}$$

and

$$\frac{\Omega D'}{\Omega D} = \left[\frac{\Omega_k\left(\Omega D + \frac{8\pi Gc\beta}{3}\right) + 3(1 - \Omega_k - \Omega D) - 3b^2(1 - \Omega_k)}{\Omega D\left(2 - \Omega D - \frac{8\pi Gc\beta}{3}\right)}\right] \times \left[\Omega D + \frac{8\pi Gc\beta}{3}\right] + 2\Omega_k. \tag{41}$$

As in the non-interacting case here also if $\beta = 0$ or $\frac{8\pi Gc}{3} \longrightarrow 0$, Eq. (40) restores the EoS parameter of the original GDE model [18] in interacting case. Also for flat universe, where $\Omega_k = 0$, the equation of state parameter of GGDE in interacting case takes the form

$$wD = \frac{1}{\Omega D\left(2 - \Omega D - \frac{8\pi Gc\beta}{3}\right)}\left[\frac{8\pi Gc\beta}{3} - \Omega D - 2b^2\right]. \tag{42}$$

Using Eq. (40) in Eq. (35), we obtain the deceleration parameter in interacting case as

$$q = \frac{1}{2}(1 - \Omega_k) + \frac{3}{2\left(2 - \Omega D - \frac{8\pi Gc\beta}{3}\right)} \times \left[\left(\Omega D + \frac{8\pi Gc\beta}{3}\right)\left(1 - \frac{\Omega_k}{3}\right) - 2\Omega D - 2b^2(1 - \Omega_k)\right]. \tag{43}$$

Equation of state of generalized ghost dark energy at low redshift

This section is devoted for cosmological applications of our work, using the differential expression that we obtained above, which determine the evolution of dark energy parameter. We will set up the Ghost dark energy equation of state parameter for low redshift at present time. Since we have derived the expression for $\Omega D'$, hence we can calculate $\omega(z)$ for low redshift z. We can measure ω as in [58], due to ρD. In this regard, since $\rho D \sim \frac{1}{a^{3(1+\omega)}}$, we will use Taylor expansion for ρD

$$\ln\rho D = \ln\rho D^0 + \frac{d\ln\rho D}{d\ln a}\ln a + \frac{1}{2}\frac{d^2\ln\rho D}{d(\ln a)^2}(\ln a)^2 + \cdots. \tag{44}$$

Here derivatives are taken at present time $a_0 = 1$, therefore $\omega(\ln a)$ is

$$\omega(\ln a) = -1 - \frac{1}{3}\left[\frac{d\ln\rho D}{d\ln a} + \frac{1}{2}\frac{d^2\ln\rho D}{d(\ln a)^2}\ln a\right], \tag{45}$$

up to second order. Replacing $\ln a = -\ln(1 + z) = -z$, which is valid for small redshifts, we can calculate $\omega(z)$ as

$$\omega(z) = -1 - \frac{1}{3}\left(\frac{d\ln\rho D}{d\ln a}\right) + \frac{1}{6}\left(\frac{d^2\ln\rho D}{d(\ln a)^2}\right)z. \tag{46}$$

Using

$$\rho D = \frac{\Omega D\rho_M}{\Omega_M} = \frac{\rho_{m_0}\Omega D a^{-3}}{1 + \Omega_k - \Omega D}, \tag{47}$$

we can find out linear equation of state parameter $\omega(z)$ as

$$\omega(z) = \omega_0 + \omega_1(z), \tag{48}$$

where,

$$\omega_0 = -\frac{1}{3}\left[\frac{\Omega D'}{\Omega D} + \frac{\Omega_k' + \Omega D'}{1 - \Omega_k - \Omega D}\right], \tag{49}$$

and

$$\omega_1 = \frac{1}{6}\left[\frac{\Omega D''}{\Omega D} - \left(\frac{\Omega D'}{\Omega D}\right)^2 - \frac{\Omega_k'' + \Omega D''}{1 - \Omega_k - \Omega D} - \frac{(\Omega_k' + \Omega D')^2}{(1 - \Omega_k - \Omega D)^2}\right]. \tag{50}$$

Using Friedmann second Eq. (16), we can find Ω_k' and Ω_k'' from Eq. (23) as

$$\Omega_k' = 4\Omega_k(1 - \Omega_k), \tag{51}$$

and

$$\Omega_k'' = 4\Omega_k'(1 - 2\Omega_k), \tag{52}$$

Also from Eq. (34), we have

$$\Omega D'' = \Omega D'(\Omega D' - 2\Omega D\Omega_k)\left[\frac{1}{\left(\Omega D + \frac{8\pi Gc\beta}{3}\right)} - \frac{1}{\left(2 - \Omega D - \frac{8\pi Gc\beta}{3}\right)}\right] + \left[\Omega D'(\Omega_k - 3) + \Omega_k'\left(\Omega D + \frac{8\pi Gc\beta}{3} - 3\right)\right]\frac{\left(\Omega D + \frac{8\pi Gc\beta}{3}\right)}{\left(2 - \Omega D - \frac{8\pi Gc\beta}{3}\right)^2} + 2(\Omega D\Omega_k' + \Omega D'\Omega_k). \tag{53}$$

Inserting Eqs. (34) and (51) in the expression of ω_0 given in Eq. (49), we obtain

$$w_0 = -\frac{1}{3}\left[\left\{\frac{\Omega_k(\Omega D + \frac{8\pi Gc\beta}{3}) + 3(1-\Omega_k - \Omega D)}{\Omega D(2 - \Omega D - \frac{8\pi Gc\beta}{3})}\right\}\left(\Omega D + \frac{8\pi Gc\beta}{3}\right)\right.$$
$$\left. + 2\Omega_k\right]\left(1 + \frac{\Omega D}{1-\Omega_k-\Omega D}\right) - \left\{\frac{4\Omega_k(1-\Omega_k)}{3(1-\Omega_k-\Omega D)}\right\}. \quad (54)$$

Using Eqs. (34), (51), (52) and (53) into Eq. (50), we finally obtain the expression for ω_1 as

$$\omega_1 = -\frac{1}{6}\left[\left(\frac{\Omega D'}{\Omega D}\right)^2 + \left(\frac{\Omega_k' + \Omega D'}{1-\Omega_k-\Omega D}\right)^2\right] - \frac{2\Omega_k'(1-2\Omega_k)}{3(1-\Omega_k-\Omega D)}$$
$$+\frac{1}{6}\Omega D''\left(\frac{1}{\Omega D} - \frac{1}{1-\Omega_k-\Omega D}\right). \quad (55)$$

Equations (54) and (55) together determine the linear equation of state of dark energy for low redshift.

Validity of generalized second law of thermodynamics

In this section, we are going to investigate validity of generalized second law of thermodynamics of a universe bounded apparent horizon and governed by Horava–Lifshitz gravity. Bekenstein [59] introduced that, in view of the proportionality relation between entropy of black hole horizon and horizon area, the sum of black hole entropy and the background entropy must be an increasing quantity with time. Also, the thermodynamic interpretation of field equations can be applicable for any horizon, provided that the gravitational theory is diffeomorphism invariant [60], however, apparent horizon is widely used in the literature in FRW geometry [61–64]. In the case of Horava–Lifshitz gravity, the breaking of diffeomorphism invariance leaves us the apparent horizon as a reasonable choice, thus here we will use apparent horizon as black hole horizon. The dynamical apparent horizon, a marginally trapped surface with vanishing expansion can be determined by the equality $h^{ij}\partial_i RA \partial_j RA = 0$. We can find the radius of cosmological apparent horizon as [65],

$$RA = \frac{1}{\sqrt{H^2 + \frac{k}{a^2}}}. \quad (56)$$

Therefore the evolution of the apparent horizon follows,

$$\dot{R}A = 4\pi GcHRA^3[\rho_m + \rho D(1+wD)]. \quad (57)$$

The first law of thermodynamics gives

$$TdS_1 = dE_1 + PDdV, \quad (58)$$

where T, S_1, V, E_1 and PD represent temperature, entropy, volume, energy of the matter distribution and pressure of the system, respectively. We consider that temperature of the source inside horizon is in equilibrium with temperature associated with horizon, so starting with volume and energy of the system bounded by apparent horizon as

$$V = \frac{4}{3}\pi RA^3 \quad \text{and} \quad E_1 = \frac{4}{3}\pi RA^3 \rho D, \quad (59)$$

and using Eqs. (57), (58) and (59), we obtain

$$\frac{dS_1}{dt} = \frac{4}{T}(1+wD)\rho D\pi RA^3 H \times \left[\frac{\kappa^2}{4(3\lambda-1)}RA^2\{\rho_m + \rho D(1+wD)\} - 1\right]. \quad (60)$$

To connect the entropy of the horizon to its radius RA in Horava–Lifshitz cosmology, we consider the expression as in [66]

$$S_h = \frac{\kappa^2}{32\Lambda Gc^2}[\Lambda RA^2 + 2k\ln(\sqrt{\Lambda}RA)]. \quad (61)$$

Differentiating Eq. (61), we have

$$\frac{dS_h}{dt} = \frac{\kappa^2}{16Gc^2}RA\dot{R}A + \frac{\kappa^2 k}{16\Lambda Gc^2 RA}\dot{R}A. \quad (62)$$

Also, temperature of the horizon is related to its radius for spherical geometry [60], and we have,

$$T = \frac{1}{2\pi RA}. \quad (63)$$

In view of the above equations, we can find the rate of change of total entropy as

$$\frac{dS_{\text{total}}}{dt} = \frac{dS_1}{dt} + \frac{dS_h}{dt}$$
$$= \frac{\kappa^2}{4(3\lambda-1)}HRA^3\rho D\left[(1+wD)(8\pi RA^3\rho_m + 1 - 2\pi RA) + \frac{\kappa^2}{16Gc^2}\frac{\rho_m}{\rho D}\left(RA + \frac{k}{\pi RA}\right)\right]. \quad (64)$$

From the above equation, it is difficult to give definite conclusion regarding validity of the generalized second law of thermodynamics, but if generalized ghost dark energy is not of phantom nature then the second law of thermodynamics is valid in the generalized ghost dark energy model if $(8\pi RA^3\rho_m + 1 - 2\pi RA) > 0$.

Conclusion

In this paper, we have investigated the GGDE scenario in the frame work of Horava–Lifshitz cosmology. Since GDE density corresponds to a dynamical cosmological constant, we need a dynamical framework, instead of general relativity, to consistently accommodate it. There fore we investigate the GDE in the framework of Horava–Lifshitz cosmology. Experimental data have implied that our universe is not a perfectly flat universe and it possesses a small positive curvature ($\Omega_k \sim 0.02$) [47], so here we consider a non-flat FRW universe. We have extracted the exact differential equations that derive evolution of the dark energy parameter, which gives a suitable estimate for the state parameter of dark energy as well as the deceleration parameter to study an expansion of the universe for both interacting and non-interacting case. We have obtained that GGDE model in Horava–Lifshitz cosmology replicates the role of Phantom dark energy model at present time if $\frac{8\pi Gc\beta}{3} < \frac{2}{3}$. Furthermore, we have established that our work has a good compatibility with previous works and restores it in limiting cases. It is also interesting to see that we have found out a cosmological application of our work by evaluating a relation for the equation of state of dark energy for low redshifts. Moreover, we have obtained that generalized second law of thermodynamics valid in case of non-phantom nature of generalized ghost dark energy, but in general we can not give a definite conclusion regarding validity of generalized second law.

References

1. Riess, A.G., et al.: Observational evidence from supernovae for an accelerating universe and a cosmological constant. Astron. J. **116**, 1009 (1998)
2. Perlmutter, S.J., et al.: Measurements of Ω and Λ from 42 high-redshift supernovae. Astrophys. J. **517**, 565 (1999)
3. Tegmark, M., et al.: (SDSS Collaboration), Cosmological parameters from SDSS and WMAP. Phys. Rev. D **69**, 103501 (2004)
4. Spergel, D.N., et al.: [WMAP Collaboration], First year Wilkinson Microwave Anisotropy Probe observations : determination of cosmological parameters. Astrophys. J. Suppl. **148**, 175 (2003)
5. Copland, E.J., Sami, M., Tsujikawa, S.: Dynamics of dark energy. Int. J. Mod. Phys. D **15**, 1753 (2006)
6. Li, M., Li, X.D., Wang, S., Wang, Y.: Dark Energy. Commun. Theor. Phys. **56**, 525 (2011)
7. Witten, E.: Current Algebra theorem for the U(1) "Goldstone Boson". Nucl. Phys. B **156**, 269 (1979)
8. Veneziano, G.: U(1) without instantons. Nucl. Phys. B **159**, 213 (1979)
9. Rosenzweig, C., Schechter, J., Trahern, C.G.: Is the effective Lagrangian for quantum Chromodynamics a σ model. Phys. Rev. D **21**, 3388 (1980)
10. Ohta, N.: Dark Energy and QCD Ghost. Phys. Lett. B **695**, 41 (2011)
11. Urban, F.R., Zhitnitsky, A.R.: The cosmological constant from the QCD Veneziano Ghost. Phys. Lett. B **688**, 9 (2010)
12. Urban, F.R., Zhitnitsky, A.R.: The Cosmological constant from the ghost: A Toy model. Phys. Rev. D **80**, 063001 (2009)
13. Zhitnitsky, A.R.: Contact term, its holographic description in QCD and dark energy. Phys. Rev. D **86**, 045026 (2012)
14. Maggiore, M.: Zero-point quantum fluctuations and dark energy. Phys. Rev. D **83**, 063514 (2011)
15. Cai, R.G., Tuo, Z.L., Wu, Y.B., Zhao, Y.Y.: More on QCD Ghost Dark Energy. Phys. Rev. D **86**, 023511 (2012)
16. Ebrahimi, E., Sheykhi, A.: Interacting generalized ghost dark energy in a non-flat universe. Central Eur. J. Phys. **11**, 949 (2013)
17. Cai, R.G., Tuo, Z.L., Zhang, H.B.: Notes on Ghost Dark Energy. Phys. Rev. D **84**, 123501 (2011)
18. Sheykhi, A., Sadegh, M.: Movahed, Interacting ghost dark energy in non-flat universe. Gen. Relativ. Gravit. **44**, 449 (2012)
19. Ebrahimi, E., Sheykhi, A.: Interacting ghost dark energy in Brans-Dicke theory. Phys. Lett. B **705**, 19 (2011)
20. Ebrahimi, E., Sheykhi, A.: Instability of QCD ghost dark energy model. Int. J. Mod. Phys. D **20**, 2369 (2011)
21. Sheykhi, A., Bagheri, A.: Quintessence Ghost Dark Energy Model. Europhys. Lett. **95**, 39001 (2011)
22. Khodam-Mohammadi, A., Malekjani, M., Monshizadeh, M.: Reconstruction of modified gravity with ghost dark energy models. Mod. Phys. Lett. A **27**, 1250100 (2012)
23. Yousefi, L., Sheykhi, A.: QCD Ghost Dark Energy in RS II Braneworld with Bulk-Brane Interaction. Int. J. Theor. Phys. **53**, 1472 (2014)
24. Lifshitz, E.M.: On the theory of second order phase transitions I and II. Zh. Eksp. Teor. Fiz. **11**, 255 (1949)
25. Horava, P.: Membranes at quantum criticality. J. High Energy Phys. **3**, 20 (2009)
26. Horava, P.: Quantum Gravity at a Lifshitz Point. Phys. Rev. D **79**, 084008 (2009)
27. Horava, P.: Spectral Dimension of the Universe in Quantum Gravity at a Lifshitz Point. Phys. Rev. Lett. **102**, 161301 (2009)
28. Horava, P.: Quantum criticality and YangMills gauge theory. Phys. Lett. B **694**, 172 (2010)
29. Cai, R.G., Cao, L.M., Ohta, N.: Topological black holes in Hoava-Lifshitz gravity. Phys. Rev. D **80**, 024003 (2009)
30. Germani, C., Kehagias, A., Sfetsos, K.: Relativistic Quantum Gravity at a Lifshitz Point. Journal of High Energy Physics **9**, 60 (2009)
31. Bogdanos, C., Saridakis, E.N.: Perturbative instabilities in Hoava gravity. Classical and Quantum Gravity **27**, 075005 (2010)
32. Myung, Y.S.: Generalized uncertainty principle and Hoava-Lifshitz gravity. Physics Letters B **679**, 491 (2009)
33. Alexandre, J., Farakos, K., Pasipoularides, P., Tsapalis, A.: Schwinger-Dyson approach for a Lifshitz-type Yukawa model. Phys. Rev. D **81**, 045002 (2010)
34. Capasso, D., Polychronakos, A.P.: Particle kinematics in Hoava-Lifshitz gravity. Journal of High Energy Physics **2**, 68 (2010)
35. Koyama, K., Arroja, F.: Pathological behaviour of the scalar graviton in Hoava-Lifshitz gravity. Journal of High Energy Physics **3**, 61 (2010)
36. Li, M., Pang, Y.: A trouble with Hoava-Lifshitz gravity. Journal of High Energy Physics **8**, 15 (2009)
37. Kiritsis, E., Konas, G.: Hoava-Lifshitz cosmology. Nuclear Physics B **821**, 467 (2009)
38. Wang, A., Wu, Y.: Cosmology in nonrelativistic general covariant theory of gravity. J. Cosmol. Astropart. Phys. **7**, 12 (2009)
39. Sotiriou, T.P., Visser, M., Weinfurtner, S.: Quantum gravity without Lorentz invariance. Journal of High Energy Physics **10**, 33 (2009)
40. Mukohyama, S.: Scale-invariant cosmological perturbations from Horava-Lifshitz gravity without inflation. J. Cosmol. Astropart. Phys. **6**, 1 (2009)
41. Cai, Y.F., Zhang, X.: Primordial perturbations with a modified dispersion relation. Phys. Rev. D **80**, 043520 (2009)
42. Gao, X., Wang, Y., Xue, W., Brandenberger, R.: Fluctuations in a Hoava-Lifshitz Bouncing Cosmology. J. Cosmol. Astropart. Phys. **2**, 20 (2010)
43. Kim, S.S., Kim, T., Kim, Y.: Surplus solid angle as an imprint of

Hoava-Lifshitz gravity. Phys. Rev. D **80**, 124002 (2009)
44. Dutta, S., Saridakis, E.N.: Observational constraints on Hoava-Lifshitz cosmology. J. Cosmol. Astropart. Phys. **1**, 13 (2010)
45. Kehagias, A., Sfetsos, K.: The black hole and FRW geometries of non-relativistic gravity. Physics Letters B **678**, 123 (2009)
46. Greenwald, J., Papazoglou, A., Wang, A.: Black holes and stars in Horava-Lifshitz theory with projectability condition. Phys. Rev. D **81**, 084046 (2010)
47. Jamil, M., Saridakis, E.N., Setare, M.R.: The generalized second law of thermodynamics in. Horava Lifshitz cosmology. J. Cosmol. Astropart. Phys. **11**, 32 (2010)
48. Jamil, M., Saridakis, E.N.: New agegraphic dark energy in Hova-Lifshitz cosmology. J. Cosmol. Astropart. Phys. **7**, 28 (2010)
49. Setare, M.R., Jamil, M.: Holographic dark energy with varying gravitational constant in Horava-Lifshitz cosmology. J. Cosmol. Astropart. Phys. **2**, 10 (2010)
50. Pasqua, A., Chattopadhyay, S.: New Agegraphic Dark Energy model in chameleon Brans-Dicke cosmology for different forms of the scale factor. Astrophys. Space Sci. **348**, 541 (2013)
51. Elizalde, E., Nojiri, S., Odintsov, S.D., Saez-Gomez, D.: Unifying inflation with dark energy in modified F(R) Horava-Lifshitz gravity. European Physical Journal C **70**, 351 (2010)
52. Lepe, Samuel, Saavedra, Joel: On Horava-Lifshitz cosmology. Astrophys. Space Sci. **350**, 839 (2014)
53. Setare, M.R., Momeni, D., Moayedi, S.K.: Interacting dark energy in Hoava-Lifshitz cosmology. Astrophys. Space Sci. **338**, 405 (2012)
54. Calcagni, G.: Cosmology of the Lifshitz universe. J. High Energy Phys. **9**, 112 (2009)
55. Chaichian, M., Nojiri, S., Odintsov, S.D., Oksanen, M., Tureanu, A.: Modified F(R) Horava-Lifshitz gravity: a way to accelerating FRW cosmology. Class. Quantum Gravity **27**, 185021 (2010)
56. Bartolami, O., Gil, F.: Perdro and M. LeDelliou, Dark Energy-Dark Matter Interaction and putative violation of the Equivalence Principle from the Abell Cluster A586. Phys. Lett. B **654**, 165 (2007)
57. Pavon, D., Zimdahl, W.: Holographic dark energy and cosmic coincidence. Phys. Lett. B **628**, 206 (2005)
58. Riess, A.G., et al.: Type Ia Supernova Discoveries at $z > 1$ From the Hubble Space Telescope: Evidence for Past Deceleration and Constraints on Dark Energy Evolution. Astrophys. J. **607**, 665 (2004)
59. Bekenstein, J.D.: Black Holes and Entropy. Phys. Rev. D **7**, 2333 (1973)
60. Padmanabhan, T.: Thermodynamical Aspects of Gravity: New insights. Rept. Prog. Phys. **73**, 046901 (2010)
61. Frolov, A.V., Kofman, L.: Inflation and de Sitter thermodynamics. J. Cosmol. Astropart. Phys. **05**, 009 (2003)
62. Bousso, R.: Cosmology and the S matrix. Phys. Rev. D **71**, 064024 (2005)
63. Cai, R.G., Kim, S.P.: First Law of Thermodynamics and Friedmann Equations of Friedmann-Robertson-Walker Universe. JHEP **0502**, 050 (2005)
64. Akbar, M., Cai, R.G.: Thermodynamic behavior of the Friedmann equation at the apparent horizon of the FRW universe. Phys. Rev. D **75**, 084003 (2007)
65. Bak, D., Rey, S.J.: Cosmic Holography. Class. Quant. Grav. **17**, L83 (2000)
66. Jamil, M., Saridakis, E.N., Setare, M.R.: The generalized second law of thermodynamics in Horava-Lifshitz cosmology. J. Cosmol. Astropart. Phys. **11**, 1475 (2010)

15

Micromagnetic analysis of Heusler alloy-based perpendicular double barrier synthetic antiferromagnetic free layer MTJs

Bahniman Ghosh[1,2] · Kshitij Dwivedi[2]

Abstract We investigate spin transfer torque switching in a perpendicular double barrier synthetic antiferromagnetic free layer MTJ stack using micromagnetic simulations. For the material used in free layers, we use two different Cobalt-based Heusler alloys and compare their performance on the basis of switching speed, thermal stability and Tunnel magnetoresistance. We show that for Heusler alloys switching from one state to other is significantly faster but they suffer from the drawback of low thermal stability.

Keywords Magnetic tunnel junctions · Micromagnetic simulations · Perpendicular anisotropy · Synthetic antiferromagnet · Heusler alloys

Introduction

Spin transfer torque (STT)-based magnetoresistive random access memory (MRAM) is emerging as a promising memory technology for the next generation due to its scalability, high operation speed and unlimited endurance [1–16]. However, the challenge with STT-MRAM is that the current required for switching is relatively large (10^7–10^8 A/cm^2) and hence relatively large transistors are required to drive them and power consumption is also high, thus limiting the information storage density in this technology. Cobalt-based Heusler compounds are reported to have the advantage of low mismatch of lattice constant, high spin polarization factor and high tunneling magneto resistance (TMR) ratio. [17, 18]. In 1903, Heusler reported that Cu–Mn alloy can be turned into a ferromagnetic material by the addition of a *sp* element e.g., Al, In, Sn, Sb or Bi even though there is no ferromagnetic element in the alloy. The reason for the ferromagnetism in these alloys is the double-exchange mechanism between neighboring magnetic ions. They are characterized by rich electronic and magnetic properties such as shape memory and half metallic behavior [19]. They are interesting candidates for spintronic devices due to their large magnetization, high magnetic critical temperatures and half metallicity. It has been shown that for critical current required for switching in MTJs J_c can be reduced by employing materials with values of M_s and α lesser than that of CoFeB MTJs in free layers. In this work, we use Co_2MnSi (CMS) and $Co_2FeAl_{0.4}Si_{0.6}$ (CFAS) as the material in the free layer for the MTJ stack used in [20] with perpendicular anisotropy. Both these Heusler alloys have low M_s and α values and hence are expected to show faster switching. We show significant reduction in switching time for the same switching current densities. We also compare the TMRs of the structures using most general model proposed by Julliere.

Micromagnetic model

For the Heusler alloy-based MTJs, we expect significant reduction in the switching current density due to lower saturation magnetization (M_s) values, smaller Gilbert damping constant (α) and higher spin polarization factor (P). We have used micromagnetic simulation tool OOMMF for detailed analysis of switching dynamics.

✉ Bahniman Ghosh
bghosh@utexas.edu

[1] Microelectronics Research Center, University of Texas at Austin, 10100, Burnet Road, Bldg. 160, Austin, TX 78758, USA

[2] Department of Electrical Engineering, Indian Institute of Technology Kanpur, Kanpur 208016, India

Table 1 Material parameters for different alloys used in micromagnetic simulation

Material	M_s (A/m)	A	P	$K1$ (J/m^3)	A (J/m)
Cobalt	1000×10^3	0.014	0.4	5×10^3	3×10^{-11}
CMS	800×10^3	0.008	0.56	3×10^3	2×10^{-11}
CFAS	1000×10^3	0.01	0.6	9×10^3	2.3×10^{-11}

Material parameters for different alloys used are provided in Table 1 [21, 22]. We have used a perpendicular DBSAF-AP MTJ stack as used in earlier works [20] for simulations and varied parameters accordingly for the different alloys used. The stack consists of two (antiparallel) fixed layers on the top. Below the fixed layers are (two for the two fixed layers) MgO tunnel barrier layers and below the MgO tunnel barriers are (two) synthetic antiferromagnetic free layers separated by Ruthenium.

OOMMF uses a time evolver that integrates a Landau–Lifshitz–Gilbert ODE with a spin momentum term. For simplicity we have neglected perpendicular spin torque terms in our calculations.

$$\frac{dm}{dt} = -\gamma_0 \times m \times H_{\text{eff}} + \alpha \times m \times \frac{dm}{dt} + T_{\text{STT}} \quad (1)$$

where m *is* reduced magnetization M/M_s, α is damping constant, γ_0 is Gilbert gyromagnetic ratio, $H_{\text{eff}} = H_{\text{ext}} + H_{\text{exchange}} + H_{\text{demag}} + H_{\text{dip}} + H_{\text{anisotropy}}$, $\beta = |\frac{\hbar}{\mu e}|\frac{J}{d_f \times M_s}$. Here, H_{eff} is the effective field which includes contributions due to external applied magnetic field H_{ext} which is zero in this case, the exchange field H_{exchange}, the dipolar field due to fixed layers H_{dip}, the demagnetization field H_{demag} and the anisotropy induced field $H_{\text{anisotropy}}$. The Eq. (1) has three torque terms on the right hand side. First term is the torque on free layer due to sum total of all magnetic fields, the second term is the damping torque which acts as a friction force impeding the motion and is proportional to the velocity of the motion while the third term is the torque produced due to spin polarized current and is given by $T_{\text{STT}} = |\gamma|\beta\varepsilon(m \times m_p \times m)$ in which ε is the effective spin polarization efficiency factor and is given by $\varepsilon = \frac{P\wedge^2}{(\wedge^2+1)+(\wedge^2-1)(m.m_p)}$. Here, P is the spin polarization, J is the current charge density and m_p is the unit polarization direction of spin polarized current and d_f is the thickness of the free layer. As we can see that T_{STT} is inversely proportional to saturation magnetization (M_s) and directly proportional to spin polarization factor (P). So lesser switching time is expected from Heusler alloy-based MTJs for the same switching current applied. Also for a macro-spin model at zero temperature the critical current density which is required to reverse the magnetization in a MTJ can be expressed as :

$$J_c = \frac{2e \times \alpha \times M_s \times d_f \times (H_{\text{ext}} + H_{\text{anisotropy}} + 2\pi M_s)}{\hbar P} \quad (2)$$

From Eq. (2) it can be easily noticed that to reduce the critical current density a material which has lower value of M_s, and α and higher P values should be preferred. Reducing thickness of free layer can also reduce the J_c but after a certain thickness the thermal stability of STTRAM bit gets compromised. For a stable bit, an energy barrier of about 60 $k_B T$ (k_B is Boltzmann's constant and T is temperature in Kelvins) is required for 10 years retention. In this work, we have concentrated on the current densities and confirmed our assumptions using simulations. We have also calculated and compared the stability constants for the same free layer thickness for all three materials and showed its dependence on free layer volume.

Switching dynamics

Switching magnetization in the MTJ structure under study proceeds as follows. We inject perpendicularly polarized current in the top free layer (TFL). Initially the magnetization of the TFL is in the same direction ($+z$) as the spin polarization of the electrons injected. Up-spin electrons conduct thorough the metallic Ru layer and are injected in the bottom free layer (BFL). Initial magnetization of the BFL is in $-z$ direction due to RKKY coupling. The up-spin electrons injected exert torque on the BFL, also there are reflected up-spin electrons from the bottom pinned layer (BPL) which has its magnetization pinned in the $-z$ direction. If the current density is large enough the torque on the BFL is enough to flip the magnetization of the BFL in the $+z$ direction and enters in a metastable state where both the free layers have magnetization in the same direction $+z$ in this case. If we stop the current at this stage then magnetization of TFL switches in $-z$ direction by itself to reach in another stable state (AP). The time taken to switch from P to metastable state is t_r and the time taken to switch from metastable state to AP is t_f. However, the switching from metastable state to stable AP states is too slow ($t_f >> t_r$) as compared to switching from P to metastable state so to assist switching from metastable state to AP state we reverse the direction of the current after the metastable state is reached. If the bi-directional MTJ stack is exposed to negative polarity current while switching from metastable state to stable state for a period longer than the switching time, then the BFL eventually switches again to $-z$ direction leading to another metastable state. This leads to an unintentional switching event and a write failure. To avoid this, a timing margin is necessary. A design to significantly reduce this failure has been proposed. [20] For switching from AP to P current is injected from the opposite direction and hence the electrons injected this time have spin in $-z$ direction and they exert torque on the BFL to flip its magnetization in $-z$ direction leading to another metastable state in which both the free layers have their magnetization in $-z$ direction and then again we reverse the direction of current at metastable state to switch back to P state.

Micromagnetic simulation results and discussion

For comparing the switching dynamics two cobalt-based Heusler alloys are selected which have values of M_s and α less than that of cobalt and high spin polarization factor P. Co_2MnSi (CMS) and $Co_2Fe_{0.4}Al_{0.6}$ (CFAS) are chosen as the ferromagnetic material to design coupled free layers. We simulate the given device by applying current pulses of amplitudes ranging from 1×10^7 to 5×10^7 A/cm^2 in steps of 1×10^7 A/cm^2 to study the effect of STT and compare the dynamics of different materials. The dimensions of the free layers in device used are $40 \times 20 \times 2$ nm^3. The cell size chosen for simulation is $1 \times 1 \times 0.1$ nm^3. Figure 1 shows the magnetization switching from P to AP state for the Cobalt-based MTJ for current density of 2×10^8 A/cm^2. At the beginning device is in P state and since both the free layers have magnetizations in opposite direction the overall reduced magnetization m is equal to zero and when a spin polarized current is applied the device reaches the metastable state and at this stage m becomes equal to 1. Then the polarity is reversed to reach stable AP configuration and m again reduces to zero. However, as discussed earlier if the current is not stopped after the AP state the device goes into another metastable state as shown in the Fig. 1 and the overall reduced magnetization becomes equal to -1. Figure 2 compares the switching dynamics of the three materials used to design free layers at the same current density of 2×10^8 A/cm^2. As expected the material with lowest value of M_s and α (CMS) shows the fastest switching and the material with highest value of M_s and α shows slowest switching. To show the comparison more precisely, we have studied the switching of all the three devices by varying the current density and plotting the switching time versus current density for all of them.

As can be observed from Fig. 3 for the same current density applied switching time for CMS is lowest while it is highest for Cobalt. We expected similar results from theoretical analysis since the spin transfer torque is inversely proportional to M_s^2 and directly proportional to P and we can see from the Table 1 that the CMS has the lowest value of M_s.

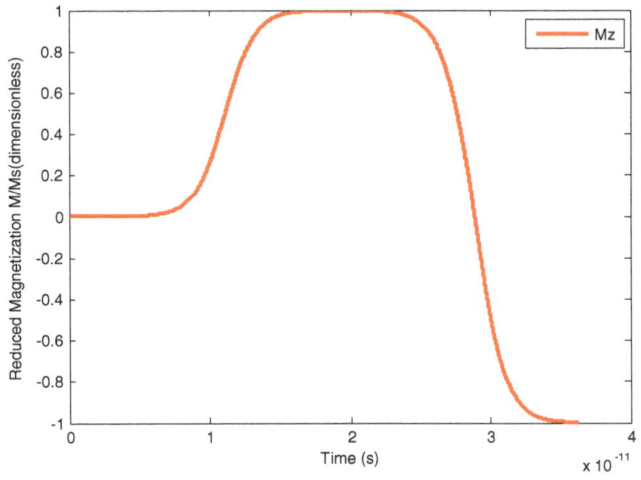

Fig. 1 Magnetization versus time plot to understand the switching dynamics

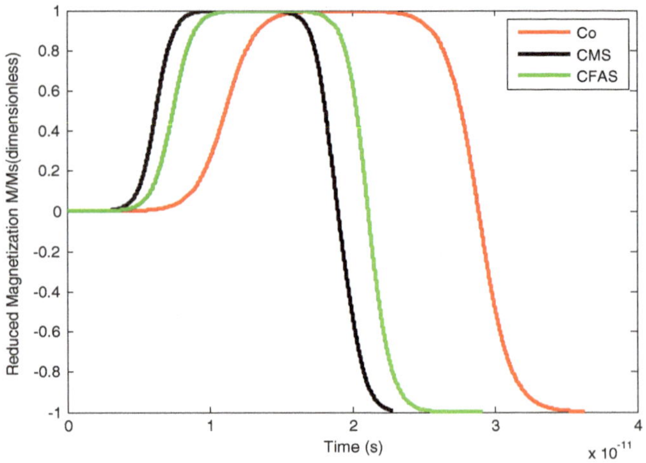

Fig. 2 Plot to compare the switching dynamics of different materials used for $J = 2 \times 10^8$ A/cm^2

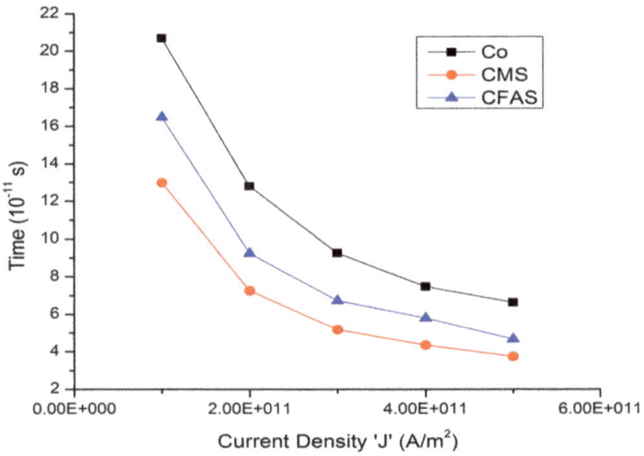

Fig. 3 Switching time versus current density plot to compare the switching speed in Cobalt, CMS and CFAS

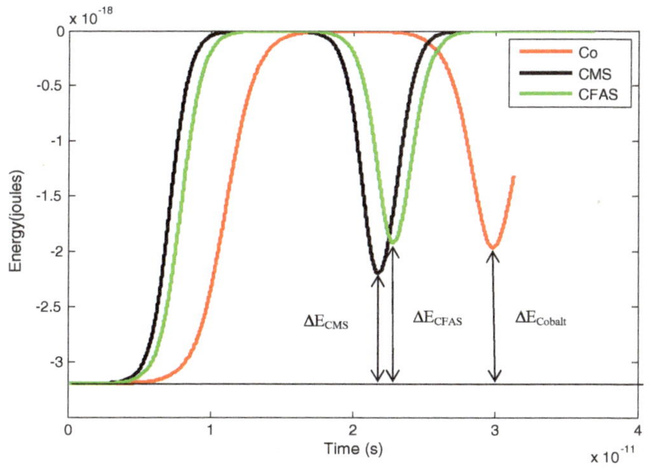

Fig. 4 Plot to compare stability by comparing ΔE of different materials used

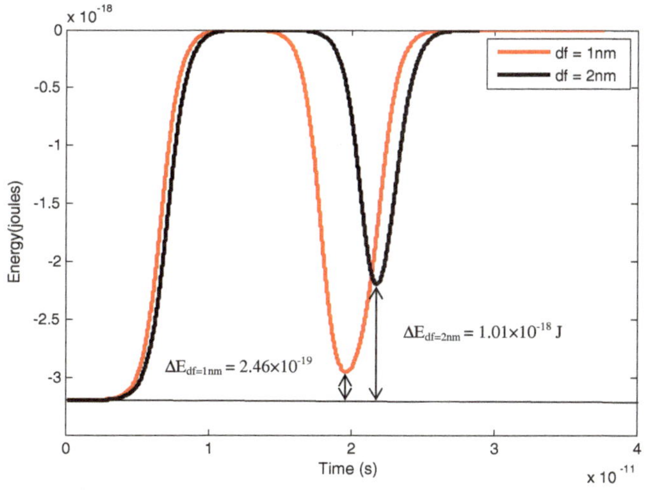

Fig. 5 Plot to study the dependence of ΔE on free layer thickness

Stability analysis was done for the three materials used by comparing the energy difference (ΔE) between two states P and AP. As mentioned earlier for 10-year data retention an energy difference of about 60 k_BT is required (Fig. 4) [23]. Thermal stability of STTRAM bit often sets trade-off between maximum storage density and data retention requirement. We have shown in Fig. 5 that ΔE decreases on decreasing the density of free layer by decreasing its thickness. To show the dependence of ΔE on the volume of the free layer we have compared ΔE for the thickness values of 1 and 2 nm and found that for 1 nm $\Delta E < k_BT$ for $T = 300$ K while ΔE for 2 nm thick free layer is almost 100 times k_BT and hence satisfies the requirement for data retention. In Figs. 5 the plots show variation of total energy with time while the switching takes place. The two minima, first one at $t = 0$ s and the second one after the maxima, correspond to stable P and AP states, respectively. We have also compared the ΔE values for Cobalt, CMS and CFAS free layers having thickness 2 nm and observed that ΔE is maximum for CFAS and minimum for CMS free layer. The following observation can be explained using Eq. (3).

$$\Delta E = \frac{\mu \times M_s \times V \times H_k}{2} \tag{3}$$

where H_k is the effective magnetic anisotropy field, μ is the permeability of free space, V is the volume of the free layer. Effective anisotropy field is directly proportional to uniaxial anisotropic constant $K1$. So the energy difference ΔE depends on the product of saturation magnetization M_s and $K1$. We can observe from the parameters given in Table 1 that the product has the maximum value for CFAS and minimum value for CMS, so CFAS should have the maximum ΔE for fixed volume of free layer and it is confirmed from the simulation results obtained. Though the switching is fastest in CMS-based MTJ due to lowest M_s and high P, CFAS-based MTJ shows the highest stability due to high value of anisotropic constant $K1$. For a better comparison between the two current density and stability constants ($\Delta E/k_BT$) curve can be plotted and one which has lower current density at same stability constant should be preferred.

TMR calculations

Resistance of a MTJ depends on the angle between the magnetization directions of the free layer and the pinned layer since in the transport between the majority and minority spin states the tunneling which occurs depends on the spin. In the tunneling process, the electrons maintain their spin direction and the probability that an electron with a certain spin will tunnel through the barrier from the pinned layer to free layer depends on the number of states with the same spin direction available in free layer. So the probability of tunneling is not equal for both parallel and antiparallel states since they correspond to different density of state. Tunnel magnetoresistance (TMR) is defined as:

$$\frac{\Delta R}{R} = \frac{R_{AP} - R_P}{R_P} \tag{4}$$

where R_P and R_{AP} are the resistance for parallel (P) and antiparallel (AP) magnetic configurations, respectively. In this study, we calculate the TMR of a simple MTJ structure with in-plane anisotropy. There are only two ferromagnetic layers with a tunneling barrier between them. One of the ferromagnetic layers has its magnetization fixed and is called reference layer or pinned layer while the

Table 2 TMR values for Cobalt, CMS and CFAS-based in-plane MTJs

	CMS	CFAS	Cobalt
TMR (%)	91.3	112.5	53.3

magnetization of other layer is not fixed and is free to rotate. We have used the simplest model proposed by Julliere to calculate the TMR of the materials used and compared them. Though we compared TMRs for a simple MTJ stack using a simple model we expect similar results will hold in case of multilayered MTJ stack such as DBSAF MTJ stack used for studying dynamics.

In the Julliere model to explain the TMR [24], it depends only on the spin polarization of the ferromagnetic material used in pinned and free layers and is given by,

$$\frac{\Delta R}{R} = \frac{2P_1P_2}{1 - P_1P_2} \tag{5}$$

where P_1 and P_2 are spin polarization of pinned and free layer, respectively. In this case, same ferromagnetic material is used in both pinned and free layers so P_1 is equal to P_2. TMR values are calculated for CoFeB-, CMS- and CFAS-based MTJs and are given in Table 2. As it can be observed from the table a higher P value means a higher TMR so we can conclude Heusler alloy-based MTJs are expected to show higher TMR values than the Co-based MTJs. These results for TMR are for making a comparison and based on the simplest model for calculation and so are not highly accurate. A high TMR is essential for high readability of MTJs, and hence from this observation, Heusler alloys are much better option to use as the ferromagnetic material in MTJs. It has also been reported that CMS produces high TMR values at low temperatures and its TMR value decreases with increasing temperature while CFAS is reported to have high TMR values for all temperature and is weakly dependent on temperature [25].

Conclusion

We performed detailed micromagnetic simulations to study the advantages and disadvantages of Heusler alloy-based perpendicular double barrier synthetic antiferromagnetic magnetic tunnel junctions. Their low saturation magnetization (M_s) values and high spin polarization (P) lead to faster switching as compared to Cobalt-based MTJs. However, due to low M_s thermal stability of the bit is compromised. In this study, we used CMS and CFAS Heusler alloys as the material to fabricate free layers and compared them with Cobalt-based free layers. Switching in CMS is fastest among the three while its thermal stability is lowest for the same free layer thickness. Switching of CFAS free layer is faster than Cobalt but is slow when compared to CMS free layer while its thermal stability is highest due to high value of crystalline anisotropy constant $K1$. We also compare TMR of in-plane simple MTJ based on Jullier model for all the three materials and observe that the TMR of CFAS MTJ is highest.

References

1. Slonczewski, J.: Current-driven excitation of magnetic multilayers. J. Magn. Magn. Mater. **159**, L1–L7 (1996)
2. Berger, L.: Low-field magnetoresistance and domain drag in ferromagnets. J. Appl. Phys. **49**, 2156 (1978)
3. Tsoi, M., Jansen, A.G.M., Bass, J., Chiang, W.-C., Seck, M., Tsoi, V., Wyder, P.: Excitation of a magnetic multilayer by an electric current. Phys. Rev. Lett. **80**, 4281–4284 (1998)
4. Myers, E.B., Ralph, D.C., Katine, J.A., Louie, R.N., Buhrman, R.A.: Current-induced switching of domains in magnetic multilayer devices. Science **285**, 867 (1999)
5. Kiselev, S.I., Sankey, J.C., Krivorotov, I.N., Emley, N.C., Schoelkopf, R.J., Buhrman, R.A., Ralph, D.C.: Microwave oscillations of a nanomagnet driven by a spin-polarized current. Nat Lond **425**, 380 (2003)
6. Tserkovnyak, Y., Brataas, A., Bauer, G.E.W., Halperin, B.I.: Nonlocal Magnetization dynamics in ferromagnetic hybrid nanostructures. Rev. Mod. Phys. **77**, 1375 (2005)
7. Albert, F.J., Katine, J.A., Buhrman, R.A., Ralph, D.C.: Spin-polarized current switching of a Co thin film nanomagnet. Appl. Phys. Lett. **77**, 3809 (2000)
8. Berger, L.: Emission of spin waves by a magnetic multilayer traversed by a current. Phys. Rev. B **54**, 9353–9358 (1996)
9. Slonczewski, J.: Currents, torques, and polarization factors in magnetic tunnel junctions. Phys. Rev. B **71**, 024411 (2005)
10. Kieselev, S.I., Sankey, J.C., Krivorotov, I.N., Emley, N.C., Schoelkopf, R.J., Buhrman, R.A., Ralph, D.C.: Microwave oscillations of a nanomagnet driven by a spin-polarized current. Nat Lond **425**, 380 (2003)
11. Katine, J.A., Albert, F.J., Buhrman, R.A., Myers, E.B., Ralph, D.C.: Current-driven magnetization reversal and spin-wave excitations in Co/Cu/Co pillars. Phys. Rev. Lett. **84**, 3149 (2000)
12. Al Haj Darwish, M., Kurt, H., Urazhdin, S., Fert, A., Loloee, R., Pratt Jr, W.P., Bass, J.: Controlled normal and inverse current induced magnetization switching and magnetoresistance in magnetic nanopillars. Phys. Rev. Lett. **93**, 157203 (2004)
13. Mangin, S., Henry, Y., Ravelosona, D., Katine, J.A., Fullerton, E.E.: Reducing the critical current for spin-transfer switching of perpendicularly magnetized nanomagnets. Appl. Phys. Lett. **94**, 012502 (2009)
14. Finocchio, G., Krivorotov, I., Carpentieri, M., Consolo, G., Azzerboni, B., Torres, L., Martinez, E., Lopez-Diaz, L.: Magnetization dynamics driven by the combined action of ac magnetic field and dc spin-polarized current. J. Appl. Phys. **99**(8), 08G507 (2006)
15. Carpentieri, M., Ricci, M., Burrascano, P., Torres, L., Finocchio, G.: Noise-like sequences to resonant excite the writing of a universal memory based on spin-transfer-torque MRAM. IEEE Trans. Magn. **48**(9), 2407–2414 (2012)
16. Cui, Y.-T., Sankey, J.C., Wang, C., Thadani, K.V., Li, Z.-P., Buhrman, R.A., Ralph, D.C.: Resonant spin-transfer-driven switching of magnetic devices assisted by microwave current pulses. Phys. Rev. B **77**(21), 214440 (2008)

17. Ebke, D.: Cobalt-based Heusler compounds in magnetic tunnel junctions. Ph.D. Thesis, Bielefeld University, Germany (2010)
18. Yakushiji, K., Saito, K., Mitani, S., Takanashi, K., Takahashi, Y.K., Hono, K.: Current-perpendicular-to-plane magnetoresistance in epitaxial Co2MnSi/Cr/Co2MnSi trilayers. Appl. Phys. Lett. **88**, 222504-1-3 (2006)
19. Graf, T., Felser, C., Parkin, S.S.P.: Simple rules for the understanding of Heusler compounds. Progr. Solid State Chem. **39**(1), 1–50 (2011)
20. Raychowdhury, A.: Numerical analysis of a novel MTJ stack for high readability and writability. In: Proceedings of the 41st European Solid-State Device Research Conference (ESSDERC), 347–350, Helsinki, Finland, 12–16 September (2011)
21. Vaz, C.A.F., Rhensius, J., Heidler, J., Wohlhüter, P., Bisig, A., Körner, H.S., Mentes, T.O., Locatelli, A., Le Guyader, L., Nolting, F., Graf, T., Felser, C., Heyderman, L.J., Kläui, M.: Spin configurations in $Co_2FeAl_{0.4}Si_{0.6}$ Heusler alloy thin film elements. Appl. Phys. Lett. **99**, 182510 (2011)
22. Huang, H.B., Ma, X.Q., Liu, Z.H., Zhao, C.P., Chen, L.Q.: Micromagnetic simulation of high-power spin-torque oscillator in half-metallic Heusler alloy spin valve nanopillar. AIP Adv **3**, 032132 (2013)
23. Apalkov, D., Watts, S., Driskill-Smith, A., Chen, E., Diao, Z., Nikitin, V.: Comparison of scaling of in-plane and perpendicular spin transfer switching technologies by micromagnetic simulation. IEEE Trans. Magn. **46**, 2240–2243 (2010)
24. Julliere, M.: Tunneling between ferromagnetic films. Phys. Lett. A **54**(3), 225–226 (1975)
25. Shan, R., Sukegawa, H., Wang, W.H., Kodzuka, M., Furubayashi, T., Ohkubo, T., Mitani, S., Inomata, K., Hono, K.: Demonstration of half-metallicity in fermi-level-tuned Heusler alloy Co_2-$FeAl_{0.5}Si_{0.5}$ at room temperature. Phys. Rev. Lett. **102**, 246601 (2009)

16

Metallic and semi-conducting resistivity behaviour of $La_{0.7}Ca_{0.3-x}K_xMnO_3$ ($x = 0.05$, 0.1) manganites

Dinesh Varshney · Neha Dodiya

Abstract The temperature dependence of electrical resistivity, ρ, of ceramic $La_{0.7}Ca_{0.3-x}K_xMnO_3$ ($x = 0.05$, 0.1) is investigated in metallic and semi-conducting phase. The metallic resistivity is attributed to be caused by electron–phonon, electron–electron and electron–magnon scattering. Substitutions affect average mass and ionic radii of *A*–site resulting in an increase in Debye temperature θ_D attributed to hardening of lattice with K doping. The optical phonon modes shift gradually to lower mode frequencies leading to phonon softening. Estimated resistivity compared with reported metallic resistivity, accordingly $\rho_{diff.} = [\rho_{exp.} - \{\rho_0 + \rho_{e-ph}\ (=\rho_{ac} + \rho_{op})\}]$, infers electron–electron and electron–magnon dependence over most of the temperature range. Semi-conducting nature is discussed with variable range hopping and small polaron conduction model. The decrease in activation energies and increase in density of states at the Fermi level with enhanced Ca doping is consistently explained by cationic disorder and Mn valence.

Keywords Magnetic materials · Ab initio calculations · Electrical properties · Phonons

Introduction

The unique properties of mixed-valence manganites as $La_{1-x}A_xMnO_3$ (A = Na, K, Rb) depend mainly on the relative amount of Mn^{3+} and Mn^{4+} ions. Of particular interest are the monovalent-doped manganites as the substitution with alkali metal ions can induce twice the increase in Mn^{4+} ions as compared to the alkaline-earth metal substitution. Most of the literature is devoted to the interplay between magnetic and transport properties [1, 2]. While discussing the correlation between the magnetic and transport properties, ideas such as double-exchange (DE) mechanism and the Jahn–Teller (JT) formation have been proposed. The metallic behaviour is usually described in terms of electron–phonon interaction, and polaronic conduction mechanism is effective in describing semiconducting behaviour above T_C [3].

Raman, infrared spectroscopy and neutron inelastic scattering measurements had obtained the most convincing evidence for the phonons and structural properties related to phase transition. Raman spectroscopy measurement of $La_{0.7}Ca_{0.3-x}K_xMnO_3$ ($x = 0.05$, 0.1) confirmed the structural lattice disorder with orthorhombic to rhombohedral structure and identified three peaks. The spectral maxima of the above are mainly located in the three intervals at about 150–200, 600–700 and 1,050–1,100 cm^{-1}, respectively [4].

The $La_{0.7}Ca_{0.3-x}K_xMnO_3$ manganites show an orthorhombic structure with space group *Pnma* for ($x = 0.05$) and rhombohedral structure with space group $R\ \bar{3}C$ for ($x = 0.1$) and illustrate a metallic behaviour below a temperature referred as metal–insulator transition temperature (T_{MI}) and behave like a semiconductor above that temperarture. For $La_{0.7}Ca_{0.3-x}K_xMnO_3$, T_{MI} enhances with the increased K doping concentration and is 269 (275) K for $x = 0.05$ (0.1) [5].

The disordered Holstein double-exchange model is solved to investigate the disorder effect at strong electron–phonon coupling and notice that even weak disorder enormously enhances the resistivity ρ at $T = 0$ K,

D. Varshney (✉) · N. Dodiya
Materials Science Laboratory, School of Physics,
Vigyan Bhawan, Devi Ahilya University,
Khandwa Road Campus, Indore 452001, India
e-mail: vdinesh33@rediffmail.com

simultaneously suppressing the density of states at the Fermi level [6]. The conductivity behaviour of doped manganites in the low-temperature ferromagnetic regime ($T < T_{MI}$) documents metallic behaviour, and the data have been fitted with different theoretical expressions to find the best fit. No systematic effort has been made to understand the electrical resistivity in the ferromagnetic metallic and paramagnetic semi-conducting phase of monovalent-doped manganites as $La_{0.7}Ca_{0.3-x}K_xMnO_3$.

The electrical resistivity of the pristine $Pr_{2/3}(Ba_{1-x}Cs_x)_{1/3}MnO_3$ (PBMO) documents two metal–insulator transitions which are systematically shifted to lower temperatures with Cs doping. An upturn in resistivity behaviour below 50 K in PBMO manganites were explained due to the combined effect of weak localization, electron–electron and electron–phonon scattering [7]. An analogous analysis at low temperature well below T_{MI} was also put forward in the $La_{2/3}Ca_{1/3}MnO_3$ manganites [8].

Rozenberg [9] commented on resistivity fitting of manganites at low temperature by Chen et al. [8] as the claim seems to be doubtful due to the fundamental inapplicability of the existence of low temperature minimum resulting from the existence of the intrinsic spin disorder. It stressed that the spin-dependent tunnelling and scattering of carriers between grains (grain-boundary effect) account reasonably for the resistivity minima [9]. The ferromagnetic transition in $La_{0.75}Ca_{0.25}MnO_3$ manganites is driven by an exchange interaction of polaronic carriers with localized spins [10]. An additional contribution arising from the electron–electron contribution in $La_{0.91}Rb_{0.06}Mn_{0.94}O_3$ must analyse the ferromagnetic metallic resistivity behaviour [11–13].

In the high-temperature semi-conducting regime ($T > T_{MI}$), the electrical conduction is explained either by the spin polaronic conduction (SPC) model [14] or by the Mott's variable range hopping (VRH) model [15]. The doping at main site influences the polaronic transport as it causes change in the Polaron hopping distance and also the polaron concentration [16]. For $T > T_{MI}$, the resistivity data of the Co-doped bilayer manganites $LaSr_2Mn_2O_7$ could be fit well using the VRH model and the SPC model and also using thermally activated conduction (TAC) law [17].

Looking to earlier experimental and theoretical effort neither temperature-independent (<20 K) nor temperature-dependent (>50 K) resistivity in divalent alkaline-earth metals (Ca, Sr, and Ba) or monovalent alkali metals like Na, K, Rb, there is a clear need to properly understand the resistivity behaviour. It is worth commenting that for doped monovalent alkali metals like K at Ca site in $La_{0.7}Ca_{0.3-x}K_xMnO_3$, no theoretical efforts have been made so far. In view of the electron–electron (T^2) and electron–magnon ($T^{4.5}$) contributions apart from electron–phonon scattering to resistivity, the time-tested strategy is to seek their roles in $La_{0.7}Ca_{0.3-x}K_xMnO_3$ manganites.

We first discuss the interaction potential, which includes the long-range Coulomb interaction, van der Waals (vdW) interaction and the short-range repulsive interaction up to second-neighbour ions [18–24]. It makes us to compute the Debye and Einstein temperatures of $La_{0.7}Ca_{0.3-x}K_xMnO_3$. Later on, we shall use the Bloch–Gruneisen method to estimate the independent contributions of acoustic and optical phonons and both. We further compute resistivity behaviour for semi-conducting region, with both Mott's variable range hopping (VRH) model and small polaron conduction (SPC). The technical details of the numerical analysis and its results are discussed in Sect. 3.

The main findings for resistivity behaviour in $La_{0.7}Ca_{0.3-x}K_xMnO_3$ [$x = 0.05$ (0.1)] include the following:

(i) The estimated Debye and Einstein temperatures from the EIoIP are consistent with the specific heat measurements and Raman spectroscopy results.
(ii) The Debye temperatures increase with enhanced K doping concentration results in lattice hardening.
(iii) The optical phonon modes shift gradually to lower mode frequencies with increased doping concentration.
(iv) The classical electron–phonon model of resistivity, for example., the Bloch–Gruneisen (BG) model retraces the reported metallic resistivity behaviour in the temperature range [$T < T_{MI} \cong 269$ (275) K].
(v) The quadratic temperature (T^2) dependence of $\rho_{diff.}$ [$=\rho_{exp.}-(\rho_0 + \rho_{e-ph})$] is interpreted in terms of 3D electron–electron inelastic scattering and is essential for the description of resistivity.
(vi) The $T^{4.5}$ temperature dependence of $\rho_{diff.}$ [$=\rho_{exp.}-(\rho_0 + \rho_{e-ph} + \rho_{e-e})$] accounting for electron–magnon interactions is also essential for the correct description of resistivity.
(vii) The $La_{0.7}Ca_{0.3-x}K_xMnO_3$ [$x = 0.05$ (0.1)] is a good metal in the temperature region $T < T_{MI}$ and Mott–Ioffe–Regel criterion for metallic conductivity is valid, $k_F l > 1$, $\varepsilon_F \tau > 1$.
(viii) The VRH model is inappropriate for the description of resistivity behaviour in the high-temperature region, $T > T_{MI}$.
(ix) The small Polaron conduction model consistently explains the higher temperature resistivity behaviour ($T > \theta_D/2$).

To our knowledge, the importance of electron–phonon, electron–electron and electron–magnon interactions and small polaron conduction for the metallic and semi-conducting resistivity behaviour of $La_{0.7}Ca_{0.3-x}K_xMnO_3$ [$x = 0.05$ (0.1)] manganites are analysed for the first time. We end up with a summary and conclusions in Sect. 4.

The model

The Raman spectroscopy is an impressive technique that probes the lattice vibrational properties, which not only identifies the spin–phonon interaction but also accounts for the electron–phonon interaction and is effective in mixed-valent manganites. The ideal perovskite ABO_3 of cubic structure presents around 15 normal modes of vibration. The lowest frequency (external mode) corresponds to a vibration of the *A* ions against the rigid BO_6 octahedra. The intermediate frequency (bending mode) corresponds to a vibration where the B ion and two apical oxygen molecules move against the other four oxygen molecules of the octahedron. At the highest frequency (stretching mode), the *B* ion moves against the rigid oxygen octahedron [25]. We can thus model the phonon spectrum consisting of an acoustic branch of Debye type and a separated optical peak with the characteristic Einstein temperature.

The dynamic properties of materials are inhibited in an effective interionic potential. This is particularly important in $La_{0.7}Ca_{0.3-x}K_xMnO_3$, as there is much disagreement as to whether long- or short-range interactions are at the origin. For simplification, we assume that the change in force constants is small, the short-range interactions are effective up to the second-neighbour ions and harmonic elastic forces hold the atoms together without any internal strains within the crystal. We thus express the crystal energy for a particular lattice separation (r) as

$$U(r) = U_C(r) + U_R(r) + U_V(r). \tag{1}$$

The Coulomb energy is the first term:

$$U_C(r) = -\sum_{ij} \frac{Z_i Z_j e^2}{r_{ij}} = -\frac{\alpha_m Z^2 e^2}{r}. \tag{2}$$

α_m is the Madelung constant [18] and r_{ij} being the separation distance between i and j ions.

The second term in Eq. (1) refers to the short-range overlap repulsive energy and is

$$U_R(r) = nb\beta_{ij}\exp\left(\frac{r_i + r_j - r_{ij}}{\rho}\right) + n'b\beta_{ii}\exp\left(\frac{2r_i - kr_{ij}}{\rho}\right) + n'b\beta_{jj}\exp\left(\frac{2r_j - kr_{ij}}{\rho}\right). \tag{3}$$

The short-range overlap repulsive energy is written as suggested by Hafemeister and Flygare approach [18–24]. The ionic radii are r_i and r_j, k is the structure factor, and $n(n')$ is the number of nearest (next nearest) ions, respectively. Further, the notations b and ρ denote the hardness and range parameters, respectively.

The Pauling coefficients, β_{ij}, are defined in terms of valence [Z_i (Z_j)] and the number of outermost electrons [n_i (n_j)] in the anions (cations), respectively, as

$$\beta_{ij} = 1 + (Z_i/n_i) + (Z_j/n_j). \tag{4}$$

The last term in Eq. (1) is the van der Waals (vdW) energy and is

$$U_V(r) = -\left(\sum_{ij}\frac{c_{ij}}{r_{ij}^6} + \sum_{ij}\frac{d_{ij}}{r_{ij}^8}\right) \tag{5}$$

$$= -\left(\frac{C}{r^6} + \frac{D}{r^8}\right). \tag{6}$$

The van der Waals (vdW) energy is attributed to dipole–dipole (*d–d*) and dipole–quadruple (*d–q*) interactions. The overall vdW coefficients C and D in Eq. (1) are defined as [26]

$$C = c_{ij}S_6(r) + \frac{1}{2}(c_{ii} + c_{jj})S_6(0) \tag{7}$$

and

$$D = d_{ij}S_8(r) + \frac{1}{2}(d_{ii} + d_{jj})S_8(0). \tag{8}$$

c_{ij} and d_{ij} are the vdW coefficients due to *d–d* and *d–q* interactions. We follow the variational method [27] to derive as

$$c_{ij} = \frac{3}{2}\frac{e\hbar}{\sqrt{m_e}}\alpha_i\alpha_j\left[(\alpha_i/N_i)^{1/2} + (\alpha_j/N_j)^{1/2}\right]^{-1} \tag{9}$$

and

$$d_{ij} = \frac{27}{8}\frac{\hbar^2}{m}\alpha_i\alpha_j\left[(\alpha_i/N_i)^{1/2} + (\alpha_j/N_j)^{1/2}\right]^2 \left[(\alpha_i/N_i) + \frac{20}{3}(\alpha_i\alpha_j/N_iN_j)^{1/2} + (\alpha_j/N_j)\right]^{-1}. \tag{10}$$

Here, m_e is the electron mass, α_I is the electronic polarizability and N_i denotes the effective number of electrons of the ith ion. The values of the overall vdW coefficients are obtained using Eqs. (7) and (8) and weighted in terms of appropriate lattice sums [$S_6(0)$, $S_6(r)$, $S_8(0)$ and $S_8(r)$] [26]. The individual vdW coefficients c_{ij} and d_{ij} are obtained with certainty and accuracy as the excitation energies are ignored in Eqs. (9) and (10).

We shall seek the interionic interaction in between a pair such as Mn–O and La/Ca/K–O. It is clear from the above descriptions that the present effective interionic potential contains only two free parameters (b and ρ), which are determined from the equilibrium conditions:

$$\left(\frac{dU}{dr}\right)_{r=r_0} = 0 \tag{11}$$

and bulk modulus

$$B_T = \frac{1}{9kr_0}\left(\frac{d^2U}{dr^2}\right)_{r=r_0}. \tag{12}$$

The model parameters obtained from Eq's. (11) and (12) have been used to compute the second-order elastic constants (C_{11}, C_{12} and C_{44}) as [28–31]

$$C_{11} = \frac{e^2}{4r_0^4}\left[-5.112Z_m^2 + A_1 + \frac{(A_2 + B_2)}{2}\right], \quad (13)$$

$$C_{12} = \frac{e^2}{4r_0^4}\left[0.226Z_m^2 - B_1 + \frac{(A_2 - 5B_2)}{2}\right], \quad (14)$$

$$C_{44} = \frac{e^2}{4r_0^4}\left[2.556Z_m^2 + B_1 + \frac{(A_2 + 3B_2)}{4}\right]. \quad (15)$$

The short-range parameters for the nearest and the next nearest neighbours are (A_1, B_1) and (A_2, B_2), respectively. These are defined as

$$A_1 = \frac{4r_0^3}{e^2}\left[\frac{d^2}{dr^2}V_{ij}(r)\right]_{r=r_0}, \quad (16)$$

$$A_2 = \frac{4(r_0\sqrt{2})^3}{e^2}\left[\frac{d^2}{dr^2}V_{ii}(r) + \frac{d^2}{dr^2}V_{jj}(r)\right]_{r=r_0\sqrt{2}}, \quad (17)$$

$$B_1 = \frac{4r_0^3}{e^2}\left[\frac{d}{dr}V_{ij}(r)\right]_{r=r_0}, \quad (18)$$

$$B_2 = \frac{4(r_0\sqrt{2})^2}{e^2}\left[\frac{d}{dr}V_{ii}(r) + \frac{d}{dr}V_{jj}(r)\right]_{r=r_0\sqrt{2}}. \quad (19)$$

The short-range potentials between the ions is $V_{ij}(r)$ and is written as

$$V_{ij}(r) = b\beta_{ij}\exp\left(\frac{r_i + r_j - r_{ij}}{\rho}\right) + c_{ij}r_{ij}^{-6} + d_{ij}r_{ij}^{-8}. \quad (20)$$

The elastic force constant κ is derived at the equilibrium inter-ionic distance r_0 as

$$\kappa = \frac{r_0}{2}\left[\pi^2(C_{11} - C_{12})(C_{11} + C_{12} + 2C_{44})(C_{44})\right]^{\frac{1}{3}}. \quad (21)$$

We have thus computed the elastic force constants for a pair such as Mn–O and La/Ca/K–O and have the total elastic force constants of the K-doped $La_{0.7}Ca_{0.3}MnO_3$. This continuum model thus takes care of the clear physical binding in doped manganites. We stress that simpler models based on this potential can describe those cohesive properties of such solids that depend on van der Waals interactions. However, the true potential must recognize the correct charge distribution and the relative orientations of the interacting atoms in manganites, which is a complicated task.

We shall now estimate the acoustic Debye branch characterized by the Debye temperature θ_D and an optical peak defined by the Einstein temperature θ_E. The Debye frequency is characterized as a cutoff frequency at the Brillouin zone boundary. It can be expressed in terms of the effective value of ionic mass and elastic force constant for crystal lattices with two different kinds of atoms such as Mn–O and La/Ca/K–O, which we deal with. The acoustic-mode and optical-mode frequencies are estimated in an ionic model using a value of effective ion charge $Ze = -2e$.

For the sake of simplicity, we have chosen an acoustic mass $M = (2\,M_+ + M_-)$ [Mn(O) which is symbolized by M_+ (M_-)]; $\kappa^* = 2\kappa$ for each directional oscillation mode to get the acoustic phonon frequency as [32–37]

$$\omega_D = \sqrt{\frac{2\kappa^*}{M}}. \quad (22)$$

In this regard, for a pairwise potential, the photons are optic in origin and the frequency is determined by the reduced mass of the pair as $\mu^{-1} = M(A)^{-1} + M(B)^{-1}$, where A is the cation (La/Ca/K, Mn) and B is the anion (O)

$$\omega_{LO}^2 = \frac{\kappa + \alpha}{\mu}, \quad (23)$$

$$\omega_{TO}^2 = \frac{\kappa - \alpha}{\mu}. \quad (24)$$

Here, α is the force constant as

$$\alpha = \frac{8\pi}{3}\frac{(Ze)^2}{\Omega}, \quad (25)$$

ω_{LO} (ω_{TO}) symbolized for the longitudinal (transverse) optical phonon frequency and Ω for the volume of the unit cell.

It is known to us that in the metallic state, the electron–phonon, electron–electron, electron-magnon scattering and polaronic effects are the major proponents of various conceptions in electrical transport. We shall begin with the description of the scattering of electron–phonon for the resistivity in the ferromagnetic metallic state. The temperature-dependent part of the metallic resistivity, following the Debye model, is

$$\rho \approx \left(\frac{3}{\hbar e^2 v_F^2}\right)\frac{k_B T}{M v_s^2}\int_0^{2k_F}|F(q)|^2\left[\frac{xq^3dq}{[e^x - 1][1 - e^{-x}]}\right], \quad (26)$$

with $x = \hbar\omega/k_BT$. In the above equation, $F(q)$ is the Fourier transform of the potential associated with one lattice site, v_F being the Fermi velocity and v_s being the sound velocity. Equation (26) shows in terms of acoustic phonon contribution yields the Bloch–Gruneisen function of temperature dependence of resistivity:

$$\rho_{ac}(T, \theta_D) = 4A_{ac}T(T/\theta_D)^4\int_0^{\theta_D/T} x^5(e^x - 1)^{-1}(1 - e^{-x})^{-1}dx, \quad (27)$$

where A_{ac} is a constant of proportionality defined as

$$A_{ac} \cong \frac{3\pi^2 e^2 k_B}{k_F^2 v_s^2 L\hbar v_F^2 M}. \quad (28)$$

Since the resistivity is additive, the Matthiessen rule is obeyed, and resistivity is represented as a sum $\rho(T) = \rho_0 + \rho_{e-ph}(T)$, where ρ_0 is the residual resistivity that does not depend on temperature as electrons also scatter off impurities, defects and disordered regions. However, in case of the Einstein type of phonon spectrum, (an optical mode) $\rho_{op}(T)$ may be described as follows:

$$\rho_{op}(T, \theta_E) = A_{op}\theta_E^2 T^{-1}[e^{\theta_E/T} - 1]^{-1}[1 - e^{-\theta_E/T}]^{-1}, \quad (29)$$

A_{op} is defined analogously to Eq. (28).

We have thus conveniently modelled the phonon resistivity in the ferromagnetic metallic state by combining both terms arising from acoustic and optical phonons

$$\rho_{e-ph}(T) = \rho_{ac}(T, \theta_D) + \rho_{op}(T, \theta_E). \quad (30)$$

The total resistivity is given by

$$\begin{aligned}\rho(T, \theta_D, \theta_E) &= \rho_0 + \rho_{ac}(T, \theta_D) + \rho_{op}(T, \theta_E) \\ &= \rho_0 + 4A_{ac}(T/\theta_D)^4 T \\ &\quad \times \int_0^{\theta_D/T} x^5(e^x - 1)^{-1}(1 - e^{-x})^{-1} dx \\ &\quad + A_{op}\theta_E^2 T^{-1}[\exp(\theta_E/T) - 1]^{-1}[1 - \exp(-\theta_E/T)]^{-1}.\end{aligned} \quad (31)$$

In order to analyse the resistivity data of high-temperature region, $T > T_{MI}$, we have made computation following Model I: variable range hopping (VRH) and Model II: adiabatic small Polaron Conduction (SPC). In the high-temperature range, for example, in the paramagnetic semiconducting state, we have fitted the resistivity data using VRH model.

The expression as derived by Mott for conductivity is as follows:

$$\sigma = \sigma_{oh} \exp(-T_0/T)^{0.25}. \quad (32)$$

with σ_{oh} as a constant and T_0 defined in terms of the density of state in the vicinity of Fermi energy $N(\varepsilon_F)$ and the localization length a as

$$T_0 = \frac{18}{k_B N(\varepsilon_F) a^3}. \quad (33)$$

Here T_o describes hopping transport in doped manganites where the carriers are localized by random potential fluctuations and preferred hopping is between sites lying within a certain range of energy.

Let us now briefly sketch the description of resistivity in the temperature range $T > T_{MI}$, associated with Model II. It is worth mentioning that the most rapid motion of a small polaron occurs when the carrier hops each time the configuration of vibrating atoms in an adjacent site coincides with that in the occupied site. Henceforth, the charge carrier motion within the adiabatic regime is faster than the lattice vibrations, and the resistivity for SPC is as follows:

$$\rho = \rho_{os} T \exp\left(\frac{E_p}{k_B T}\right), \quad (34)$$

where E_p is the polaron formation energy and k_B being the Boltzmann constant. The resistivity coefficient ρ_{os} is given by

$$\rho_{os} = \frac{k_B}{n(1-x)e^2 D}. \quad (35)$$

Here, n is charge carrier density ($\sim 10^{20}$ cm^{-3}), x is the hole (Mn^{4+}) content, e, the electronic charge and D is the polaron diffusion constant. The polaron diffusion constant for a typical cubic coordination can be given explicitly as $D = a^2 v/6$, where a is the lattice constant and v, the characteristic frequency of the longitudinal optical phonon that carries polaron through the lattice.

Parent $LaMnO_3$ at high temperatures is an insulator with a cubic structure and becomes tetragonal due to distortions at low temperatures. Doping at a La site by Ca, Sr, Ba, Na and K and so forth leads to a decrease not only in the structural phase transition temperature but also in the overall behaviour of resistivity. However, for optimized doped $x \cong 0.3$, the material is still insulating (at about room temperature and higher) and the resistivity is much higher than the Mott limit. The substitution at the La site by divalent (monovalent) ions changes the valence of the Mn site as outer Mn d-orbital shows two-fold degeneration, and the result is Jahn–Teller (breathing type) distortion of the oxygen octahedra focused at each of the Mn-occupied (unoccupied) sites. The energy required for the formation of a local lattice distortion is about 0.6 eV per site. Identifying a strong electron–phonon coupling and hence polaronic transport is essential in the doped materials. At high temperatures, the Jahn–Teller distortions are decorated but do not disappear.

On the other hand, the double-exchange mechanism shows that the oxygen ion is closed shell and supports hopping. In a true sense, an electron jumps onto a Mn ion on the right simultaneously with an electron hopping onto the O ion from the Mn on the left. Henceforth, in the double hopping, both hopping electrons should have the same spin via the oxygen (O^{2-}) ions. Following Hund's rule, the Mn^{3+} and Mn^{4+} ions should have the electron in parallel spin moments for hopping. This mechanism connects the parallel alignment of men moments (ferromagnetism) with hopping of carriers (metallic conduction). Thus an effective positive exchange coupling induced by the carriers was known as double exchange.

Actually, the lattice vibration of O ions essentially pushes the electrons towards vacant states in the Mn ion inducing a local distortion of the lattice. Such self-trapping of the charge carriers is substantially above the Curie temperature. However, below the Curie temperature, the self-trapping disappears because the bandwidth broadens and the electrons are much more mobile. It is inferred that at high temperature, the effective positive exchange coupling induced by the carriers does not solely contribute to the resistivity and that a strong electron–phonon interaction arising from the Jahn–Teller splitting of the outer Mn *d* level is essentially required [38–40].

Discussion and analysis of results

Any discussion of the mixed-valent oxides necessitates the knowledge of the structural aspects, and this is particularly true of the calculations reviewed here. Also, applying the available information on the developed theory inevitably entails certain complications, and one has to find suitable data that vary from technique to technique. Special attention is paid in this approach to address the issue whether long-range or short-range interactions are at the origin of the substantial properties of the monovalent K-doped manganites. The formulated effective inter-ionic interaction potential (EIoIP) with the long-range Coulomb interaction, van der Waals (vdW) interaction and the short-range repulsive interaction up to second-neighbour ions within the Hafemeister and Flygare approach is employed to estimate temperature-dependent resistivity for temperatures less than metal–insulator transition. We first elaborate the main findings and motivate them by simple physical arguments.

The present investigation concerns with electron–phonon scattering and theoretically analysed within the framework of the classical electron–phonon model of resistivity, i.e. the Bloch–Gruneisen (BG) model, as a dominating process for electrical resistivity. We first retrace the experimental data with ρ_0 as the free parameter, and then contribution from acoustic and optical phonons together with the residual resistivity is compared to the reported data and the difference is analysed in the light of electron–electron and electron–magnon scattering mechanisms. The above discussion of manganites needs an extremely accurate knowledge of the structural and transport parameters involved in either the metallic or semi-conducting state and this is particularly true for the calculation documented below. For the analysis of transport properties, it is necessary to know the real value of some physical parameters, which actually govern resistivity behaviour, however, they vary from technique to technique as well as the probe employed for characterizations. The effective interionic potential is constructed in an easily generalizable manner.

We have thus studied structural properties in an ordered way. The values of Debye and Einstein temperatures have been computed using the values of the two model parameters, namely range (ρ) and hardness (b), for a pair such as La/Ca/K–O and Mn–O, which have been evaluated from the equilibrium distance [41–43] and bulk modulus [44]. We first deduced the individual and overall vdW coefficients from the variational method [27], which are listed in Table 1. The material parameters such as hardness (b) and range parameters (ρ) as functions of doping concentration (x) are obtained from the equilibrium distance and the bulk modulus using the equilibrium conditions. Within the continuum model, we notice that the hardness parameter (b_1 and b_2) increases and the range parameter (ρ_1 and ρ_2) decreases with the increased K doping concentration (x) in doped $La_{0.7}Ca_{0.3\text{-}x}K_xMnO_3$. The input data along with their relevant references and the model parameters are given in Table 2.

Deduced values of second-order elastic constants C_{11}, C_{12} and C_{44} are 9.655 (9.729), 4.816 (4.89) and 3.598 (3.667) 10^{12} Dyne cm^{-2}, respectively for $x = 0.05$ (0.1). Using Eq. (21), we find the force

Table 1 Van der Waals coefficients of $La_{0.7}Ca_{0.3-x}K_xMnO_3$ ($x = 0.05, 0.1$) (c_{ij} in units of 10^{-60}ergcm^6 and d_{ij} in unit of 10^{-76}ergcm^8)

		c_{ii}	c_{ij}	c_{jj}	C	d_{ii}	d_{ij}	d_{jj}	D
	Mn–O	120.1	180.3	270.8	1543	131.5	193.36	284.11	1355
$x = 0.05$	La/Ca/K–O	194.2	201.6	269.9	1750	547.6	446.0	284.10	3074
$x = 0.1$	La/Ca/K–O	197.5	203.0	269.1	1761	552.1	448.5	284.10	3091

Table 2 Input crystal data and model parameters for $La_{0.7}Ca_{0.3-x}K_xMnO_3$ ($x = 0.05, 0.1$)

		Input parameters				Model parameters	
		r_i (Å)	r_j (Å)	r_0 (Å)	B_T (GPa)	$b(10^{-12}$ erg)	$\rho(10^{-1}$ Å)
$x = 0.05$	Mn–O	0.660 [43]	1.32 [43]	1.946 [41]	179 [44]	3.712	0.336
	La/Ca/K–O	1.223 [43]	1.32 [43]	2.881 [42]	179 [44]	3.162	0.214
$x = 0.1$	Mn–O	0.660 [43]	1.32 [43]	1.942 [41]	179 [44]	3.692	0.335
	La/Ca/K–O	1.242 [43]	1.32 [43]	2.881 [42]	179 [44]	2.712	0.218

constant, κ, accounting for long-range Coulomb interaction , van der Waals (vdW) interaction and the short-range repulsive interaction up to second-neighbour ions as 3.755 (3.776) $\times$ 10^5 gs^{-2} for $x = 0.05$ (0.1), respectively. With these parameters, the Debye frequency is estimated as 41.5 (41.8) meV for $x = 0.05$ (0.1), respectively.

The Debye temperature plays an important role in the study of a large number of solid-state problems involving lattice vibrations because at this temperature, nearly all modes of vibrations in a solid are excited. It is clear that Debye frequencies are found to increase with x, as A-site radius varies and the observed behaviour might be due to structural transition and decrease in the Mn–O distance as a result of substitution of a K^+ ion. The size of the K^+ ion is larger than those of the La^+ and Ca^+ ions, which in turn creates more structural distortions in the lattice. Because of this, there is a change in Mn–O bond distance. This leads to affect the DE interaction resulting in increase in Debye frequency. We attempt to analyse our results in the framework of DE interaction with electron–lattice interaction. Therefore, when La site is substituted by Ca and K, the Mn–O distance decreases resulting in increase in θ_D i.e. hardening of lattice. Henceforth, it suggests that increased hole doping drives the system effectively towards the weak electron–phonon coupling region. A similar behaviour is found for $La_{1-x}A_xMnO_3$ when A-site is substituted for Na, K [45, 46].

We must mention that these computed values for $La_{0.7}Ca_{0.3-x}K_xMnO_3$ for $x = 0.05$ (0.1) could not be compared due to the lack of experimental data. However, the computed values of Debye and Einstein temperatures for $La_{1-x}Na_xMnO_3$ manganites following the EIoIP approach yield consistent results [47–49] with the available experimental results. Furthermore, the optical phonon mode is obtained as $\omega_{TO} \approx 65.8$ (65.7) meV and $\omega_{LO} \approx 68.8$ (68.7) meV for $x = 0.05$ (0.1), respectively, consistent with the Raman spectroscopy results [4]. Optical Phonon modes shift towards a lower side with doped element. The La/Ca/K–O and Mn–O bonds govern these modes. If mode frequency is governed by local factors, such as the force constant and ionic mass of the substitutive elements as well, it will be proportional to $(k/M)^{1/2}$, where k is the force constant and M is the reduced mass [50]. The valence and ionic radius of Ca^{2+}, K^+, ions are different to those of La^{3+} ion. Therefore, k force constant and the mass of substitutive ion may cause a change in mode frequency. Analogous to that of $Nd_{1-x}Sr_xMnO_3$ [37], substitution of K^+ ion shows phonon softening with x. The average mass of the A-site and ionic radii affect the normal modes related to the La–O and Mn–O bonds; phonon modes shift gradually to lower mode frequencies with the doping concentration.

It is true that the two-orbital model based on Wannier functions predicts the electronic states such as charge ordering in manganites. It is pointed that the Wannier function approach of the electronic problem is useful for the description of electron dynamics following semi-classical theory as well as the magnetic interactions in solids [52]. In the present investigation, we do not intend to discuss the electron dynamics as well as the magnetic interactions but focus on determining the acoustic (optical) phonon frequency to estimate the electron–phonon contribution of resistivity in the ferromagnetic metallic phase.

The developed effective interionic interaction potential takes care of the interactions between a pair such as Mn–O and La/Ca/K–O. The interactions thus are attractive Coulomb, and van der Waals (vdW) as well as short-range overlap repulsive interaction following Hafemeister and Flygare type potential. The advantage of using this potential is that it takes care of the number of nearest (next nearest) ions, the valence, and the number of outermost electrons in the anions (cations), respectively. Thus it takes care of the structural parameters that yield an approximately correct description of the interactions between a pair such as Mn–O and La/Ca/K–O. Henceforth, we are able to estimate the acoustic (optical) phonon frequency consistent with the Raman measurements to estimate the electron–phonon contribution of resistivity.

The elastic parameters are determined using a mean field approximation with emphasis on Mn–O bond lengths and evaluated Mn–Mn and Mn–O force constants for the lattice distortions [38–40]. On the other hand, in the present model, we have considered both Mn–O and La/Ca/K–O bond lengths to obtain the Mn–O, La/Ca/K–O and the total force constants for strong electron–phonon interaction. The interionic interaction between a pair such as Mn–O and La/Ca/K–O enables us to find the total force constant with consistent Debye and Einstein temperatures.

The transport mechanism of the $La_{0.7}Ca_{0.3-x}K_xMnO_3$ for ($x = 0.05$ 0.1) in the intermediate-temperature FM region below T_{MI} is studied with the electron–phonon interaction to investigate the metallic behaviour of $\rho(T)$. Figures 1 and 2 illustrate the results of temperature dependence of resistivity with the electron–phonon interaction from Eq. (31) with our earlier choice of $\theta_D = 485$ (489) K and $\theta_E = 805$ (808) K for $x = 0.05$ (0.1). The contributions of acoustic and optical phonon towards resistivity are clubbed, and the resultant resistivity is exponential at low temperatures and nearly linear at high temperatures until transition temperature. The coefficients A_{ac} (A_{op}) as functions of doping concentration are shown in Table 3.

Given the numbers for coefficients (A_{ac} and A_{op}) that emerge from the analysis, it seems fair to conclude that the contribution from acoustic and optical phonons are

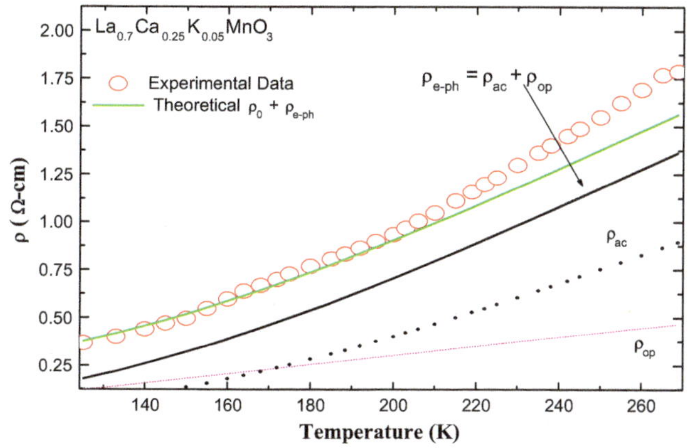

Fig. 1 Variation of ρ_{e-ph} with temperature, the contribution of acoustic phonon ρ_{ac} as well as of optical phonon ρ_{op} to the resistivity for sample $La_{0.7}Ca_{0.25}K_{0.05}MnO_3$. *Hollow circles* represent the experimental data [5]

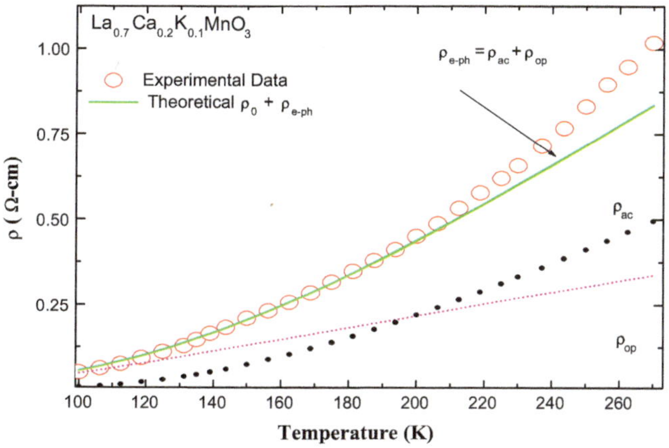

Fig. 2 Variation of ρ_{e-ph} with temperature, the contribution of acoustic phonon ρ_{ac} as well as of optical phonon ρ_{op} to the resistivity for sample $La_{0.7}Ca_{0.2}K_{0.1}MnO_3$. *Hollow circles* represent the experimental data [5]

Table 3 The fitting parameters obtained from the Bloch–Gruneisen model for $La_{0.7}Ca_{0.3-x}K_xMnO_3$ ($x = 0.05$, 0.1)

	ω_D (meV)	ω_{TO} (meV)	ω_{LO} (meV)	A_{ac}	A_{op}
$x = 0.05$	41.5	65.8	68.8	2.08×10^{-3}	6.7×10^{-3}
$x = 0.1$	41.8	65.7	68.7	1.5×10^{-3}	3.69×10^{-3}

significant in both samples, while in $La_{0.7}Ca_{0.25}K_{0.05}MnO_3$, contribution of acoustic phonon is more for electron–phonon resistivity and in the case of $La_{0.7}Ca_{0.2}K_{0.1}MnO_3$, there is large contribution of optical phonon. In spite of the difference in contribution from both phonons, both samples show electron–phonon interaction up to nearly same temperature range. Therefore, this electron–phonon interaction inherent in K-doped $La_{0.7}Ca_{0.3}MnO_3$ sample may be due to possible larger lattice distortions provoked by the K ions. It

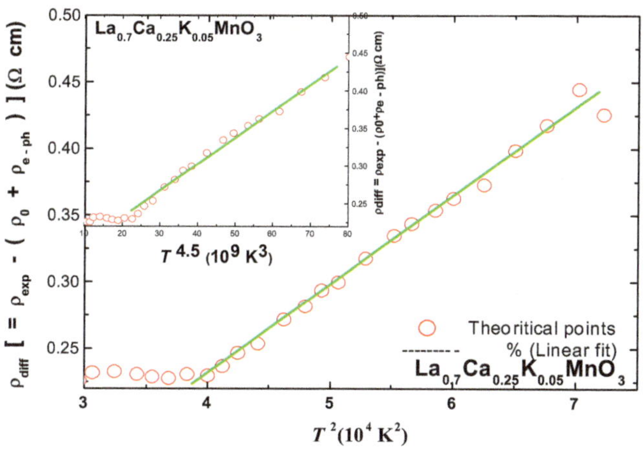

Fig. 3 Variation of $\rho_{diff.}$ [$= \rho_{exp.} - \{\rho_0 + \rho_{e-ph} (= \rho_{ac} + \rho_{op})\}$] with T^2 of $La_{0.7}Ca_{0.25}K_{0.05}MnO_3$. The *inset* shows variation of $\rho_{diff.}$ [$= \rho_{exp.} - \{\rho_0 + \rho_{e-ph} (= \rho_{ac} + \rho_{op})\}$] with $T^{4.5}$

is noticed that the electron–phonon along with residual resistivity partially retraces the metallic resistivity behaviour of $La_{0.7}Ca_{0.25}K_{0.05}MnO_3$ and $La_{0.7}Ca_{0.2}K_{0.1}MnO_3$ manganites, and other temperature-dependent mechanisms such as electron–electron and electron–magnon scattering should also be invoked.

We have discussed only how electron–phonon interaction retraces approximately the resistivity behaviour, yet our goal is to determine the manner in which other interactions modify the electrical resistivity. The additional term due to electron–electron contribution was required in understanding the resistivity behaviour, as extensive attempts to fit the data with residual resistivity and phonon resistivity were unsuccessful. It is noteworthy to comment that in conventional metals, the electron–electron contribution to the resistivity can at best be seen only at very low temperatures, due to its small magnitude in comparison with the phonon contribution.

The temperature-dependent resistivity, $\rho_{diff.}$[$= \rho_{exp.} - \{\rho_0 + \rho_{e-ph} (= \rho_{ac} + \rho_{op})\}$], is proportional to power temperature (T^2) from 198 to 245 K for $La_{0.7}Ca_{0.25}K_{0.05}MnO_3$and 200 to 250 K for $La_{0.7}Ca_{0.2}K_{0.1}MnO_3$ as shown in Figs. 3 and 4. This infers that electron–electron scattering is also a dominant process for resistivity of $La_{0.7}Ca_{0.25}K_{0.05}MnO_3$ and $La_{0.7}Ca_{0.2}K_{0.1}MnO_3$. The fitting results indicate that K doping significantly decreases the electron–phonon interaction, which favours the movement of the charge carriers. Slight deviations need an additional scattering mechanism for consistent explanation of the metallic behaviour of resistivity. At high temperatures, electron scattering is isotropic due to the predominance of electron–phonon scattering, which is nearly isotropic at high temperatures.

The inset of Figs. 3 and 4 shows the fitting in terms of $T^{4.5}$ behaviour. It is noticed that $T^{4.5}$ behaviour

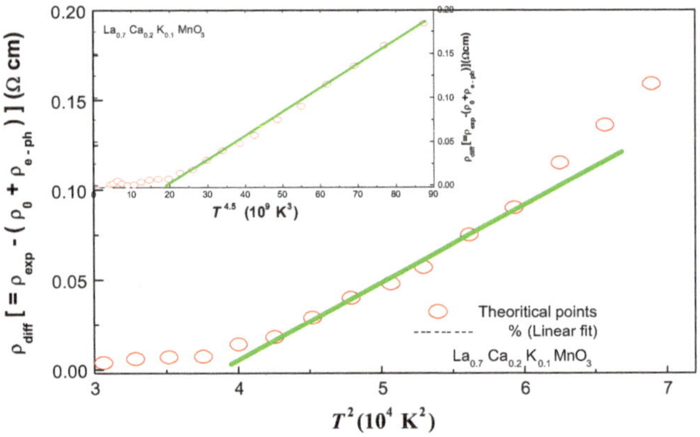

Fig. 4 Variation of $\rho_{diff.}$ [$= \rho_{exp.} - \{\rho_0 + \rho_{e-ph} (= \rho_{ac} + \rho_{op})\}$] with T^2 of $La_{0.7}Ca_{0.2}K_{0.1}MnO_3$. The *inset* shows variation of $\rho_{diff.}$ [$= \rho_{exp.} - \{\rho_0 + \rho_{e-ph} (= \rho_{ac} + \rho_{op})\}$] with $T^{4.5}$

remarkably increases with increasing K doping level, implying that the electron–magnon scattering enhances with the increase in K doping. Thus, the transport mechanism in the near-transition temperature region can be attributed to the electron–electron and electron–magnon scattering, which further demonstrates that the metallic region lies in the FM phase. For the present temperature region, the contribution of a phonon is usually ignored due to the weak vibration of crystal lattice.

For hole-doped manganites, an additional $T^{4.5}$ contribution as a result of electron–magnon scattering process is argued to be essential [53]. In the recent past, we have also noticed that the spin wave scattering in the FM phase is important in discussing the electrical resistivity behaviour of hole-doped $La_{0.91}Rb_{0.06}Mn_{0.94}O_3$ manganites [11–13]. We admit that in the higher temperature limit, the difference can be predicted linearly with $T^{4.5}$ in accordance with the electron–magnon scattering in the double-exchange process. The feature of $T^{4.5}$ temperature dependence of $\rho_{diff.}$ is consistent with the quantum theory of two-magnon scattering [53] and is valid for half-metallic ferromagnets. Consequently, besides electron–phonon and electron–electron interaction, another possibility for the changes in carrier density arose due to the presence of spin waves in the metallic system and is caused by spin wave scattering.

We shall now address the metallic behaviour of doped manganites. If the high-frequency phonon modes are indeed strongly coupling with charge carriers, the effective mass of the carriers should be substantially enhanced. The Fermi energy can be known from the reported values of thermoelectric power using $\varepsilon_F = -\pi^2 k_B^2 T/[3|e|S_c^{diff.}(T)]$, we find $\varepsilon_F \cong 0.677$ (0.897) eV for $La_{0.7}Ca_{0.25}K_{0.05}MnO_3$ ($La_{0.7}Ca_{0.2}K_{0.1}MnO_3$), respectively [2]. Deduced Fermi energy corresponds to the density of states $N(\varepsilon_F) \cong 4.019$ (6.371) $\times 10^{19}$ eV^{-1} cm^{-3} for doping concentration, $x = 0.05$ (0.1), respectively.

The density of states at Fermi level found to increase with K doping. The observed behaviour can be explained on the basis of Mn valence present in these samples. The K doping in $La_{0.7}Ca_{0.3}MnO_3$ showed increase in Mn^{4+} ions that favour double exchange between Mn ions. As a consequence, the increase in $N(\varepsilon_F)$ suggested an appreciable increase in the number of charge carriers accompanied by increase of conductivity of K doping, which is further supported by resistivity data [5]. The density of states in turn yields the electronic specific heat coefficient using $\gamma = (\pi k_B)^2 N(\varepsilon_F)/3$. We thus find the values of $\gamma = 3.523$ (5.585) $mJmol^{-1}$ K^{-2} consistent with the γ values in hole-doped divalent manganites [54]. These could not be compared due to the lack of experimental data on $La_{0.7}Ca_{0.3-x}K_xMnO_3$ monovalent manganites.

The effective mass of the carrier along the conducting Mn–O plane is deduced from the electronic specific heat coefficient γ, using $m^* = 3\hbar^2\pi\gamma d/k_B^2$. Using the MnO distance, d, of 7.7301 (5.4662) Å [41], we find $m^* = 2.031$ (2.514) m_e. The estimated electron parameters for $La_{0.7}Ca_{0.25}K_{0.05}MnO_3$ ($La_{0.7}Ca_{0.2}K_{0.1}MnO_3$) are the Fermi wave vector $k_F = 3.581$ (4.586) $\times 10^7$ cm^{-1}, the Fermi velocity $v_F = 2.035$ (2.105) $\times 10^7$ cms^{-1} and the plasma frequency $\omega_p = 1.406$ (1.831) eV. We stress that the effects induced by electron correlations and mass renormalizations by electron–electron interactions are crucial in magnetic systems such as doped manganites [55–65].

In usual conventional metals, electron–phonon scattering is mathematically identical to conventional impurity scattering and leads to a resistivity proportional to $(v_F^2)^{-1}$ where l is the mean free path. The mean free path in this approximation is usually related to the Fermi velocity and is estimated following $l = v_F\tau$. We find l to be about 3.058 (3.072) Å, for $x = 0.05$ (0.1), respectively. We follow the Drude relation, $\tau^{-1} = \rho_0\omega_p^2/4\pi$ to obtain $\tau^{-1} = 6.664$ (6.852) $\times 10^{15}$ s^{-1}. We must mention that the residual resistance obtained for nominally the same compositions may vary significantly for different groups of compounds. It remains unclear whether ρ_0 only characterizes the sample's quality or if there is an intrinsic component in the residual resistivity [66]. For the Sr-doped manganites, the resistivity data in the crystalline films yield ρ_0 as low as 10^{-5} Ω cm and are in the range of typical metallic conductors [67]. We notice that the $La_{0.7}Ca_{0.25}K_{0.05}MnO_3$ ($La_{0.7}Ca_{0.2}K_{0.1}MnO_3$) manganites are good metals as the product $\varepsilon_F\,\tau > 1$.

The electron correlations in view of the mass enhancement effect in narrow band materials are important in doped manganites. The phenomenon that mass enhanced with substituting ion can be interpreted as follows: substitution stimulate manganese content and these Mn^{4+}ions induced an increase in bending of Mn–O–Mn bond angle, in this manner narrowing band width and enhancing the effective mass [68]. We notice that enhanced electron mass $m^* = 2.031\ (2.514)\ m_e$ in $La_{0.7}Ca_{0.3-x}K_xMnO_3$ leads to the reduced plasma frequency, and hence the reduced zero temperature elastic scattering rate in comparison to conventional metals. It is perhaps worth noticing that, in hole-doped manganites, the scattering rate at low temperatures is of the order of $10^{15}\ s^{-1}$ [55–65]. Furthermore, the Mott–Ioffe–Regel criterion for metallic conductivity is valid, as the deduced mean free path is larger than the Mn–O bond length. The use of the Bloch–Gruneisen expression in estimating the electron–phonon contributions is thus validated as the product $k_F l > 1$.

The monovalent (K)-doped $La_{0.7}Ca_{0.3}MnO_3$ manganites thus illustrate a metallic behaviour below 269 (275) K a temperature T_{MI} for $x = 0.05$ (0.1), respectively. The metal–insulator transition temperature T_{MI} increases with the K doping concentration. With increased doping, $La_{0.7}Ca_{0.3-x}\ K_xMnO_3$ manganites behave as good metal and this is attributed to the fact that there are twice as many Mn^{4+} ions as there are K ions and each of this Mn^{4+} will contribute a hopping hole. Consequently, K doping will cause increase in the number of charge carriers and hence resistivity will be decreased.

The effects of the grain-boundary-induced lattice disorder on the resistivity in $Sm_{0.55}Sr_{0.45}MnO_3$ at temperatures near the metal–insulator transition have earlier been emphasized [69]. The low temperature resistivity data ($T \leq 75$ K) of the SmSrMnO were successfully fitted using the relation $\rho = \rho_0 + \rho_2 T^2 + \rho_5 T^5$ with ρ_0 as the residual resistivity and ρ_2 and ρ_5 are the electron–electron and the electron–phonon scattering coefficients, respectively. It is further noticed that in this temperature range, the disorder does not affect the temperature dependence of ρ; however, it causes an increase in the coefficients ρ_0, ρ_2 and ρ_5 by two orders of magnitude.

In contrast to electron–phonon scattering as the source of resistivity in ferromagnetic metallic (FM) state, the angle-resolved photoemission spectroscopy data for the bilayer manganite $La_{1.2}Sr_{1.8}Mn_2O_7$ identify a coherent polaronic metallic ground state below metal–insulator transition. The FM state is a polaronic metal with a strong anisotropic character of the electronic excitations, strikingly similar to the pseudogap phases in heavily under-doped cuprate high-temperature superconductors such as Bi 2212. A strong mass enhancement and a small renormalization factor are found to account for the metallic properties [70, 71]. The temperature dependence of resistivity in the metallic state is intimately related to polaronic metallic ground state and the insulator-to-metallic state can be attributed to the polaron coherence condensation process acting in concert with the double-exchange mechanism. We may comment that the present theory finds an enhanced mass of holes as carriers [$m^* = 2.031\ (2.514)\ m_e$

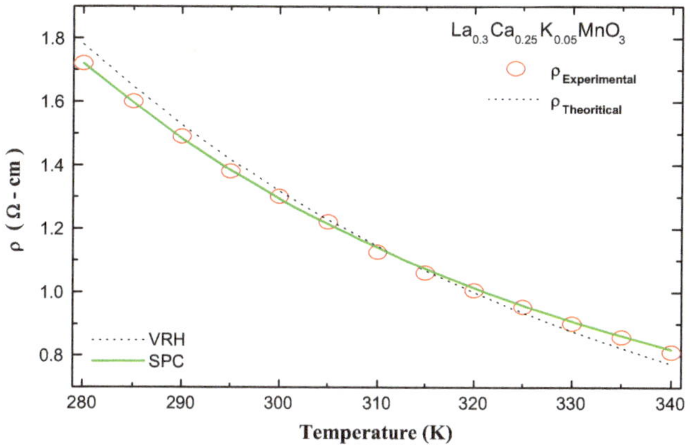

Fig. 5 Variation of ρ for high-temperature region, $T > T_{MI}$ Open circles are the experimental data taken from Sdiri et al. [5]. *Solid line* is the fitting by Small Polaron conduction model and *dashed line* is from the Variable range hopping model of $La_{0.7}Ca_{0.25}K_{0.05}MnO_3$

for $La_{0.7}Ca_{0.25}K_{0.05}MnO_3$ ($La_{0.7}Ca_{0.2}K_{0.1}MnO_3$)] from the residual resistivity to validate the Mott–Ioffe–Regel criterion. However, a detailed analysis is required to understand the polaronic metallic state in the ferromagnetic phase and the condensation process analogous to under-doped cuprate high-temperature superconductors.

Let us now discuss the semi-conducting behaviour of resistivity data in the high-temperature region and paramagnetic semi-conducting state ($T > T_{MI}$). We have computed the temperature-dependent resistivity of $La_{0.7}Ca_{0.3-x}K_xMnO_3$ ($x = 0.05,\ 0.1$) using variable range hopping (VRH) and adiabatic small polaron conduction (SPC) model. Keeping in mind that the charge carrier motion is faster than the lattice vibrations in the adiabatic regime, the nearest-neighbour hopping of a small polaron leads to mobility with a thermally activated form.

To understand the nature of electrical conduction behaviour in K-doped $La_{1-x}\ Ca_xMnO_3$ materials, we have analysed the resistivity data using Mott's variable range hopping VRH model [14]. Originally in the VRH theory, there is a competition between the potential energy and the hopping distance of electrons. Figures 5 and 6 show the plots of high-temperature resistivity ρ as a function of temperature T revealing good concurrence with VRH model [Eq. (32)]. The values of fitting parameters obtained from the fit of the high-temperature resistivity of $La_{0.7}Ca_{0.25}K_{0.05}MnO_3$ ($La_{0.7}Ca_{0.2}K_{0.1}MnO_3$) by the VRH model are $\rho_{oh} = 0.04069$ (0.7) $\times\ 10^{-6}\ \Omega$ cm and $T_0 = 2.683 \times 10^6$ (2.158×10^6) K. The fitted value of T_0 and the localization length or the average hopping distance {$a = 4.6$ Å [15]} yield the density of states at the Fermi level, $N(\varepsilon_F)$, of the order of 0.799 (9.943) $\times\ 10^{20}\ eV^{-1}$ - cm^{-3}. The above sets of values are consistent with the fitting parameters for $Nd_{0.33}Ln_{0.34}Sr_{0.33}MnO_3$ [72].

To have a cross check, we use the thermo electric power measurement to calculate experimental value of the density

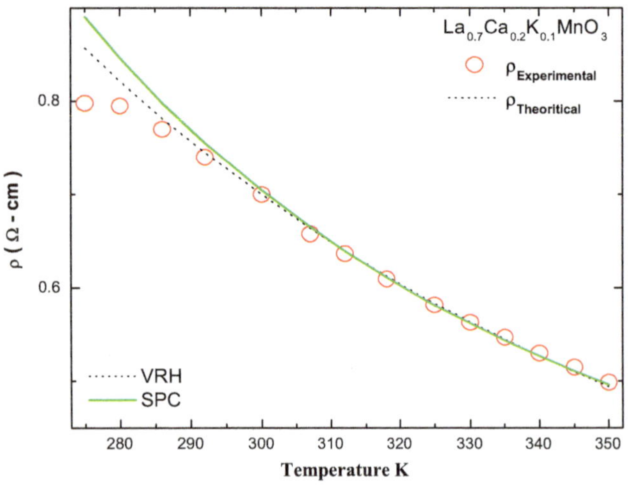

Fig. 6 Variation of ρ for high-temperature region, $T > T_{MI}$ Open circles are the experimental data taken from Sdiri et al. [5]. *Solid line* is the fitting by Small Polaron conduction model and *dashed line* is from the Variable range hopping model of $La_{0.7}Ca_{0.2}K_{0.1}MnO_3$

of states at the Fermi level [2]. Deduced density of states $N(\varepsilon_F) \cong 4.019$ (6.123) $\times$ 10^{19} eV^{-1} cm^{-3} for doping concentration, $x = 0.05$ (0.1), respectively. We comment that the value of density of states at the Fermi level obtained from semi-conducting resistivity fit using VRH is higher by an order as obtained from thermoelectric power measurements. This unphysical result leads us to argue that the VRH model is not a proper choice to describe resistivity behaviour in the high-temperature region, $T > T_{MI}$ for $La_{0.7}Ca_{0.3-x}K_xMnO_3$ ($x = 0.05, 0.1$). Earlier, Mott and Davies argued that the VRH model is applicable only for the temperature region ($T_{MI} < T < \theta_D/2$) in doped semi-conductors [73]. The above is true for K-doped manganites $La_{0.7}Ca_{0.3-x}$ K_xMnO_3 ($x = 0.05$, 0.1) with $T_{MI} \cong 269$ (275) K and $\theta_D \cong 485$ (486) K.

The above discrepancy in density of states at the Fermi level might be due to spin-dependent potential in doped manganites [15]. Quite generally, in strongly correlated electron systems such as mixed-valent manganites, the carrier hopping is always of a variable-range type at low temperature. As at low temperatures, the thermal energy (k_BT) is insufficient to allow electrons to hop to their nearest neighbours. However, electrons conveniently hop to farther neighbours to find a smaller potential difference, and there is a competition between the potential energy difference and the distance electrons can hop. On the other hand, at higher temperatures, the electron conduction is due to the activation above the mobility edge. The above results are consistent with previous work [11, 47–49].

We have further used the adiabatic small polaron conduction (SPC) model for the analysis of resistivity behaviour for high temperature. Charge carriers in the insulating region, above T_{MI}, are not itinerant and thermally activated

Table 4 Parameters obtained from the fitting of experimental data using VRH and SPC model

	T_0 10^7 (K)	$N(\varepsilon_F)$ 10^{20} (eV^{-1} cm^{-3})	D 10^{-2} (cm^2 s^{-1})	ρ_{os} 10^{-5} (Ω cm K^{-1})	E_p (meV)
$x = 0.05$	2.683	0.799	5.2	3.049	128
$x = 0.1$	0.215	9.943	5.3	6.872	91

carriers (polaron) controlled the transport mechanism. It is argued that the conductivity of the high-temperature ($T > T_{MI}$) PM insulating state is attributed to the hopping motion of self-trapped small polaron. To elucidate the nature of the charge transport mechanism in this high-temperature region, ρ has been plotted as a function of T in Figs. 5 and 6. The good linear fits reveal that the conduction in the PM semi-conducting region obeys the small polaron hopping model. The activation energy obtained from fitting of experimental data is given in Table 4. The decrease in E_p is attributed to the increase in the number of charged carrier caused by K addition.

We have first determined the diffusion coefficient using the lattice constant; the longitudinal optical phonon frequency and polaron formation energy are enlisted in Table 4. The decrease in E_p is also attributed to the effect of cationic disorder by the larger size of the monovalent ion K^+ compared to La^{3+} and Ca^{2+} ions. In addition to the effect of cationic disorder, change in Mn valance effectively leads to structural change. K doping provoked Mn^{4+} ions consequently with increasing x there is decrease in Mn–O–Mn bond angle substantial increases density of states at Fermi level and enhancing effective mass of charge carrier due to this effective band gap decreases. Therefore, activation energy values decrease with x. It is interesting to note that the sample with the higher Mn^{4+} content shows the lower E_P value.

The computed values of E_p for $La_{0.7}Ca_{0.3-x}K_xMnO_3$ for ($x = 0.05$ 0.1) could not be compared due to the lack of experimental data. However, the polaron formation energy ranges from 110 to 160 meV in $La_{1-x}Ca_xMnO_3$ ($0 \leq x \leq 0.5$) and ranges from 50 to 110 meV in $La_{1-x}K_xMnO_3$ ($0 \leq x \leq 0.33$) [74–76]. The deduced value of the E_p lies between above range and it reveals that addition of K increases the carrier density. The fact that E_p gradually decreases for higher doping is due to the phonon softening effect on carrier transport owing to Jahn–Teller distortions in the K-doped $La_{0.7-}Ca_{0.3}MnO_3$. Henceforth, above T_{MI}, the Jahn–Teller-type distortion present in the unit cell traps the charge carriers and gives rise to polarons. Hence, above T_{MI}, resistivity shows polaronic-type behaviour. We end up by commenting that the SPC model with realistic physical parameters successfully retraces the reported experimental behaviour ($T > \theta_D/2$; $T > T_{MI}$) of resistivity in doped manganites.

Conclusion

The present investigations report the analysis of experimentally known behaviour of electrical resistivity in monovalent (K) -oped $La_{0.7}Ca_{0.3}MnO_3$ manganites for metallic and semi-conducting states. For $T < T_{MI}$ in the ferromagnetic metallic state, we have employed the electron–phonon interaction with the model phonon spectrum consisting of two parts: an acoustic branch of Debye type and optical mode with characteristic Einstein temperature. Debye and Einstein temperatures from the EIoIP within the Hafemeister and Flygare approach are deduced.

The proposed model may, in spite of its simple structure, provide a consistent account for the experimental fact, such as Debye and Einstein temperatures consistent with the specific heat and Raman spectroscopy measurements. It is emphasized that conclusions have been established only within the framework of single (longitudinal and transverse) optical phonon mode with a flat dispersion relation. It is due to the fact that substitution of a K^+ ion creates more structural distortions in the lattice as a result of decrease in Mn–O bond distance, consequently affecting the DE interaction and thus resulting in increase in Debye frequency. Einstein temperatures were found to decrease with doping. This reduction is ascribed by the blue-shift of Optical Phonon modes, which are governed by the La/Ca/K–O and Mn–O bonds. The average mass of the *A*–site and ionic radii affects normal modes related to the La/Ca/K–O and Mn–O bonds; phonon modes are blue-shifted gradually with enhanced alkali metal K doping.

Signatures of different types of interaction terms as electron–phonon, electron–electron and electron–magnon are found to govern the transport mechanism in the low-temperature phase ($T < T_{MI}$). The temperature-dependent resistivity data in this region, where the local ferromagnetic order is almost complete, have been successfully fitted with $\rho = \rho_0 + \rho_{\text{e-ph}} + \rho_{\text{e-e}}\ (T^2) + \rho_{\text{e-mag}}\ (T^{4.5})$. It is thus argued that apart from the resistivity due to domain, grain-boundary and the electron–phonon scattering, the electron–electron and electron–magnon scattering are essential for a complete description of the metallic behaviour of $La_{0.7}Ca_{0.3-x}K_xMnO_3$ for ($x = 0.05\ 0.1$) manganites. However, the contribution of electron–phonon scattering appears similar in both samples and only contributions from acoustic and optical phonons are different. In this way, K doping impelled lattice distortion confirmed through electron–phonon contribution. The mean free path is larger than the Mn–O bond length, and the product $k_F l \geq 1$ and $\varepsilon_F \tau > 1$ favours metallic conduction. Hence, it is appropriate to use the Bloch–Gruneisen expression in estimating the electron–phonon contributions at $T < T_{MI}$, and this is associated with the dynamic Jahn–Teller distortion, arising from the local lattice distortion due to the strong electron–phonon coupling.

The resistivity data of semi-conducting state at high-temperature region, $T > T_{MI}$, is analysed using both the VRH and adiabatic SPC models. Deduced values of $N(\varepsilon_F)$ are found to increase and this phenomenon is associated with Mn valence induced by K doping. The small polaron conduction model with realistic physical parameters consistently retraces the semi-conducting behaviour. The nearest-neighbour hopping of a small polaron leads to a mobility with a thermally activated form and successfully retraced the reported experimental curve in the paramagnetic phase. E_p gradually decreases which may be due to decreasing charge localization with doping and also from phonon softening effect on carrier transport owing to the Jahn–Teller distortions in the $La_{0.7}Ca_{0.3-x}K_xMnO_3$. Above T_{MI}, the Jahn–Teller-type distortion present in the unit cell traps the charge carriers and gives rise to polarons and for $T > T_{MI}$, the resistivity behaviour is indeed polaronic-type.

To an end, we comment that although we have provided a simple explanation of these effects, there is a clear need for good theoretical understanding of the resistivity behaviour in view of the formation of small polarons that may be of magnetic origin in manganites.

Acknowledgments The authors are thankful to UGC, New Delhi for financial assistance.

References

1. Malavasi, L., Mozzati, M.C., Ghigna, P., Azzoni, C.B., Flor, G.: Lattice disorder, electric properties, and magnetic behavior of $La_{1-x}Na_xMnO_{3+\delta}$ manganites. J. Phys. Chem. B **107**, 2500–2505 (2003)
2. Bhattacharya, S., Pal, S., Mukherjee, R.K., Chaudhuri, B.K., Neeleshwar, S., Chen, Y.Y., Mollah, S., Yang, H.D.: Development of pulsed magnetic field and study of magnetotransport properties of K-doped $La_{1-x}Ca_{x-y}K_yMnO_3$ CMR materials. J. Magn. Magn. Mater. **269**, 359–371 (2004)
3. Salamon, M.B., Jaime, M.: The physics of manganites: structure and transport. Rev. Mod. Phys. **73**, 583–628 (2001)
4. Bejar, M., Feki, H., Dhahri, E., Ellouze, M., Balli, M., Hlil, E.K.: Effects of substituting divalent by monovalent ion on the physical properties of $La_{0.7}Ca_{0.3-x}K_xMnO_3$ compounds. J. Magn. Magn. Mater **316**, 707–709 (2007)
5. Sdiri, N., Jemai, R., Bejar, M., Hussein, M., Khirouni, K., Dhahri, E., Mazen, S.: Electrical conductivity and dielectric analysis of the perovskite $La_{0.7}Ca_{0.3-x}K_xMnO_3$ ($x =$ 0:00, 0.05 and 0.10). Solid State Commun. **148**, 577–581 (2008)
6. Kumar, S., Majumdar, P.: Singular effect of disorder on electronic transport in strongly coupled electron–phonon systems. Phys. Rev. Lett. **94**, 136601–136604 (2005)

7. Panwar, N., Pandya, D.K., Rao, A., Wu, K.K., Kaurav, N., Kuo, Y.K., Agarwal, S.K.: Electrical and thermal properties of $Pr_{2/3}(Ba_{1-x}Cs_x)_{1/3}MnO_3$ manganites. Eur. Phys. J. B **65**, 179–186 (2008)
8. Chen, Z., Xu, Y., Su, Y., Cao, S., Zhang, J.: Resistivity minimum behavior and weak magnetic disorder characteristics in $La_{2/3}Ca_{1/3}MnO_3$ manganites. J. Supercond. Nov. Magn **22**, 465–469 (2009)
9. Rozenberg, E.: Comment on 'Resistivity Minimum Behavior and Weak Magnetic Disorder Characteristics in $La_{2/3}Ca_{1/3}MnO_3$ Manganites'. J. Supercond. Nov. Magn. **23**, 183–186 (2010)
10. Alexandrov, A.S., Kornilovitch, P.B.: Mobile small polaron. Phys. Rev. Lett. **82**, 807–810 (1999)
11. Varshney, D., Dodiya, N.: Interpretation of metallic and semiconducting temperature dependent resistivity of $La_{0.91}Rb_{0.06}Mn_{0.94}O_3$ manganites. Solid State Sci **13**, 1623–1632 (2011)
12. Varshney, D., Dodiya, N.: Structural and magnetotransport studies of magnetic ion doping for monovalent-doped $LaMnO_3$ manganites. J. Mater. Res. **29**, 1183–1198 (2014)
13. Varshney, D., Shaikh, M.W.: Substitutional effects on structural and magnetotransport properties of $La_{0.85-x}Sm_xK_{0.15}MnO_3$ ($x = 0.05$, 0.1 and 0.15). J. Alloys Compnds. **589**, 558–567 (2014)
14. Snyder, G.J., Hiskes, R., Dicarolis, S., Beasley, M.R., Geballe, T.H.: Intrinsic electrical transport and magnetic properties of $La_{0.67}Ca_{0.33}MnO_3$ and $La_{0.67}Sr_{0.33}MnO_3$ MOCVD thin films and bulk material. Phys. Rev. B **53**, 14434–14444 (1996)
15. Viret, M., Ranno, L., Coey, J.M.D.: Magnetic localization in mixed-valence manganites. Phys. Rev. B **55**, 8067–8070 (1997)
16. Jaime, M., Hardner, H.T., Salamon, M.B., Rubinstein, M., Dorsey, P., Emin, D.: Hall-effect sign anomaly and small-polaron conduction in $(La_{1-x}Gd_xd)_{0.67}Ca_{0.33}MnO_3$. Phys. Rev. Lett. **78**, 951–954 (1997)
17. Ang, R., Lu, W.J., Zhang, R.L., Zhao, B.C., Zhu, X.B., Song, W.H., Sun, Y.P.: Effects of Co doping in bilayered manganite $LaSr_2Mn_2O_7$: resistivity, thermoelectric power, and thermal conductivity. Phys. Rev. B **72**(184417), 1–8 (2005)
18. Hafemeister, D.W., Flygare, W.H.: Outer-shell overlap integral as a function of distance for halogen–halogen, halogen–alkali, and alkali–alkali ions in the alkali halide lattices. J. Chem. Phys. **43**(1965), 795–800 (1965)
19. Varshney, D., Joshi, G., Varshney, M., Shriya, S.: Pressure dependent elastic and structural (B3–B1) properties of Ga based monopnictides. J. Alloys Compd. **495**, 23–32 (2009)
20. Varshney, D., Joshi, G., Varshney, M., Shriya, S.: Pressure induced structural phase transition and elastic properties in BSb, AlSb, GaSb and InSb compounds. Phys B **405**, 1663–1676 (2010)
21. Varshney, D., Shriya, S., Varshney, M.: Study of pressure induced structural phase transition and elastic properties of lanthanum pnictides. Eur. Phys. J. B **85**, 241–272 (2012)
22. Varshney, D.: Mechanical and elastic properties of Europium mono-oxides and mono-chalcogenides (EuX; $X =$ O, S, Se, Te), Europium: synthesis, characteristics and potential applications. In: Attia, M.S. (ed.) pp. 247–292. Nova Science Publication, New York (2013)
23. Varshney, D., Shriya, S.: Elastic, mechanical and thermodynamic properties at high pressures and temperatures of transition metal monocarbides. Int. J. Refract. Metals Hard Mater. **41**, 375–401 (2013
24. Varshney, D., Joshi, G., Varshney, M., Shriya, S.: Pressure dependent elastic and structural (B3–B1) properties of Ga based monopnictides. J Alloys Compds. **495**, 23–32 (2010)
25. Dodiya, N., Varshney, D.: Structural properties and Raman spectroscopy of rhombohedral $La_{1-x}Na_xMnO_3$ with ($0.075 \leq x \leq 0.15$). J. Mol. Struct. **2013**, 104–109 (1031)
26. Tosi, M.P.: Cohesion of ionic solids in the born model. Solid State Phys. **16**, 1–120 (1964)
27. Slater, J.C., Kirkwood, J.G.: The van der Waals forces in gases. Phys. Rev. **37**, 682–697 (1931)
28. Varshney, D., Rathore, V., Kinge, R., Singh, R.K.: High-pressure induced structural phase transition in alkaline earth CaX ($X =$ S, Se and Te) semiconductors: NaCl-type (B1) to CsCl-type (B2). J. Alloys Compd. **484**, 239–245 (2009)
29. Varshney, D., Dagaonkar, G., Varshney, M.: Pressure and doping dependent elastic and thermodynamical properties of $Ga_{1-x}In_xP$ mixed valent compounds. Mater. Res. Bull. **45**, 916–926 (2010)
30. Varshney, D., Joshi, G., Varshney, M., Shriya, S.: Pressure induced mechanical properties of boron based pnictides. Solid State Sci. **12**, 864–872 (2010)
31. Varshney, D., Kaurav, N., Sharma, U., Singh, R.K.: Pressure induced structural phase transition and elastic behavior of Y and Sc antimonides. J. Alloys Compds. **448**, 250–256 (2008)
32. Varshney, Dinesh: Dodiya, N.: Electrical resistivity of the hole doped $La_{0.8}Sr_{0.2}MnO_3$ manganites: Role of electron–electron/phonon/magnon interactions. Mater. Chem. Phys. **129**, 896–904 (2011)
33. Varshney, D., Dodiya, N.: Electrical resistivity of alkali metal doped manganites $La_xA_yMn_wO_3$ ($A =$ Na, K, Rb): Role of electron–phonon, electron–electron and electron–magnon interactions. Curr. Appl. Phys. **13**, 1188–1198 (2013)
34. Varshney, D., Dodiya, N., Shaikh, M.W.: Structural properties and electrical resistivity of Na-substituted lanthanum manganites: $La_{1-x}Na_xMnO_3$ ($x = 0.1$, 0.125 and 0.15). J. Alloys Compds. **509**, 7447–7457 (2011)
35. Shaikh, M., Varshney, D.: Structural properties and electrical resistivity behavior of $La_{1-x}K_xMnO_3$ ($x = 0.1$, 0.125 and 0.15) manganites. Mater. Chem. Phys. **134**, 886–898 (2012)
36. Varshney, D., Choudhary, D., Khan, E.: Electrical transport in the ferromagnetic and paramagnetic state of potassium substituted manganites $La_{1-x}K_xMnO_3$ ($x = 0.05$, 0.1 and 0.15), J. Mater. Sci. **48**, 5904–5916 (2013)
37. Shaikh, M.W.: Varshney, D.: Structural and electrical properties of $Pr_{1-x}Sr_xMnO_3$ ($x = 0.25$, 0.3, 0.35 and 0.4) manganites. Mater. Sci. Semicond. Process. **27**, 418–426 (2014)
38. Millis, A.J.: Cooperative Jahn-Teller effect and electron–phonon coupling in $La_{1-x}A_xMnO_3$. Phys. Rev. B **53**, 8434–8441 (1996)
39. Millis, J., Littlewood, P.B., Shraiman, B.I.: Double exchange alone does not explain the resistivity of $La_{1-x}Sr_xMnO_3$. Phys. Rev. Lett. **74**, 5144–5147 (1995)
40. Ederer, C., Lin, C., Millis, A.J.: Structural distortions and model Hamiltonian parameters: From LSDA to a tight-binding description of $LaMnO_3$. Phys. Rev. B **76**(155105), 1–11 (2007)
41. Bejar, M., Dhahri, E., Hlil, E.K., Heniti, S.: Influence of A-site cation size-disorder on structural, magnetic and magnetocaloric properties of $La_{0.7}Ca_{0.3-x}K_xMnO_3$ compounds. J. Alloys Compds. **440**, 36–42 (2007)
42. Radaelli, P.G., Cox, D.E., Marezio, M., Cheong, S.W.: Charge, orbital, and magnetic ordering in $La_{0.5}Ca_{0.5}MnO_3$. Phys Rev. B. **55**, 3015–3023 (1997)
43. Lide D.R. (Ed.) CRC Handbook of Chemistry and Physics, 79th edn, CRC Press, New York (1999)
44. Buch, J.J.U., Pathak, T.K., Lakhani, V.K., Vasoya, N.H., Modi, K.B.: High temperature thermoelectric power study on calcium substituted lanthanum manganites. J. Phys. D: Appl. Phys. **40**, 5306–5312 (2007)
45. Das, S., Dey, T.K.: Structural and magnetocaloric properties of $La_{1-y}Na_yMnO_3$ compounds prepared by microwave processing. J. Phys. D: Appl. Phys. **40**, 1855–1863 (2007)
46. Das, S., Dey, T.K.: Thermoelectric power of potassium doped lanthanum manganites at low temperatures. J. Magn. Magn. Mater. **311**, 714–723 (2007)
47. Varshney, D., Choudhary, D., Shaikh, M.W.: Electrical resistivity behavior of sodium substituted manganites: electron–phonon,

Electron–electron and Electron–magnon interactions. Eur. Phys. J. B **76**, 327–338 (2010)
48. Varshney, D., Choudhary, D., Shaikh, M.W.: Interpretation of metallic and semiconducting temperature dependent resistivity of $La_{1-x}Na_xMnO_3$ ($x = 0.07, 0.13$) manganites. Compd. Mater. Sci. **47**, 839–847 (2010)
49. Varshney, D., Choudhary, K.K., Singh, R.K.: Interpretation of temperature-dependent resistivity of electron-doped cuprates. Supercond. Sci. Technol. **15**, 1119–1126. (2002)
50. Wu, D., Deng, Y., Mak, C.L., Wong, K.H., Li, A.D., Zhang, M.S., Min, N.B.: Raman scattering study of La-, Nd and Sm-substituted $Bi_4Ti_3O_{12}$. Appl. Phys. A: Mater. Sci. Process. **80**, 607–610 (2005)
51. Jandl, S., Mukhin, A.A., Ivanov, V.Y., Nekvasil, V., Sadowski, M.L.: Raman-active phonons and Nd^{3+} crystal-field studies of weakly doped $Nd_{1-x}Sr_xMnO_3$. Phys. Rev. B **72**(024423), 1–7 (2005)
52. Marzari, N., Vanderbilti, D.: Maximally localized generalized Wannier functions for composite energy bands. Phys. Rev. B **56**, 12847–12865 (1997)
53. Kubo, K., Ohata, N.: A quantum theory of double exchange. J. Phys. Soc. Jpn. **33**, 21–32 (1972)
54. Ghivelder, L., Castillo, I.A., Alford, N.M., Tomka, G.T., Riedi, P.C., MacManus- Driscoll, J., Akther Hossain, A.K.M., Cohen, L.F.: Specific heat of $La_{1-x}Ca_xMnO_3$. J. Magn. Magn. Mater. **189**, 274–282 (1998)
55. Varshney, D., Kaurav, N.: Analysis of low temperature specific heat in the ferromagnetic state of the Ca-doped manganites. Eur. Phys. J. B. **37** 301–309 (2004)
56. Varshney, D.: Effect of impurity scatterers on phonon, electron and magnon thermal transport in electron doped cuprate superconductors. Sup. Sci. Technol. **19** 433–444 (2006)
57. Varshney, D., Kaurav, N.: Electrical resistivity in the ferromagnetic metallic state of La-Ca-MnO_3: Role of electron-phonon interaction. Eur. Phys. J. B **40** 129–136 (2004)
58. Varshney, D., Kaurav, N.: Interpretation of temperature-dependent resistivity of La–Pb–MnO_3: role of electron–phonon interaction. J. Low Temp. Phys. **141** 165–178 (2005)
59. Varshney, D., Mansuri, I., Kaurav, N.: Effect of electron/hole doping on the transport properties of lanthanum manganites $LaMnO_3$. J. Phys. Condens. Matter **19**, 246211 1–13 (2007)
60. Varshney, D., Shaikh, M. W., Mansuri, I.: Interpretation of temperature-dependent resistivity of $La_{0.7}Ba_{0.3}MnO_3$ manganites. J. Alloys Compd. **486**, 726–732 (2009)
61. Varshney, Dinesh., Mansuri, I.: Influence of Ce doping on structural and transport properties of $Ca_{1-x}Ce_xMnO_3$ ($x = 0.2$) manganite. J. Low Temp. Phys. **162**, 52–61 (2011)
62. Mansuri, I., Varshney, D., Kaurav, N., Lu, C.L., Kuo, Y.K.: Effects of A-site disorder on magnetic, electrical and thermal properties of $La_{0.5-x}Ln_xCa_{0.5-y}Sr_yMnO_3$ manganites. J. Magn. Magn. Mater. **323**(2011), 316–323 (2011)
63. Mansuri, I., Varshney, D.: Structure and electrical resistivity of $La_{1-x}Ba_x$MnO3 (0.25 & #x2264; x & #x2264; 0.35) perovskites. J. Alloys Compd. **513**, 256–265 (2012)
64. Varshney, D., Mansuri, I., Kaurav, N., Lung, W.Q., Kuo, Y.K.: Influence of Ce doping on electrical and thermal properties of $La_{0.7-x}Ce_xCa_{0.3}MnO_3$ ($0.0 \leq x \leq 0.7$) manganites. J. Magn. Magn. Mater. **324**, 3276–3285 (2012)
65. Varshney, D., Mansuri, I., Shaikh, M.W., Kuo, Y.K.: Effect of Fe and Co doping on electrical and thermal properties of $La_{0.5}$-$Ce_{0.5}Mn_{1-x}(Fe,Co)_xO_3$ manganites. Mater. Res. Bull. **48**, 4606–4613 (2013)
66. Quijada, M., Cerne, J., Simpson, J.R., Drew, H.D., Ahn, K.H., Millis, A.J., Shreekala, R., Ramesh, R., Rajeswari, M., Venkatesan, T.: Optical conductivity of manganites: Crossover from Jahn-Teller small polaron to coherent transport in the ferromagnetic state. Phys. Rev. B **58**, 16093–16102 (1998)
67. Urushibara, A., Moritomo, Y., Arima, T., Asamitsu, A., Kido, G., Tokura, Y.: Insulator-metal transition and giant magnetoresistance in $La_{1-x}Sr_xMnO_3$. Phys. Rev. B. **51**, 14103–14109 (1995)
68. Malavasi, L., Mozzati, M.C., Azzoni, C.B., Chiodelli, G., Flor, G.: Role of oxygen content on the transport and magnetic properties of $La_{1-x}Ca_xMnO_{3+\delta}$ manganites. Solid State Commun. **123**, 321–326 (2002)
69. Egilmez, M., Chow, K.H., Jung, J., Fan, I., Mansour, A.I., Salman, Z.: Metal-insulator transition, specific heat, and grain boundary-induced disorder in $Sm_{0.55}Sr_{0.45}MnO_3$. Appl. Phys. Lett. **92**(132505), 1–3 (2008)
70. Mannella, N., Yang, W.L., Tanaka, K., Zhou, X.J., Zheng, H., Mitchell, J.F., Zaanen, J., Devereaux, T.P., Nagaosa, N., Hussain, Z., Shen, Z.X.: Polaron coherence condensation as the mechanism for colossal magnetoresistance in layered manganites. Phys. Rev. B **76**, 233102 (2007)
71. Mannella, N., Yang, W.L., Zhou, X.J., Zheng, H., Mitchell, J.F., Zaanen, J., Devereaux, T.P., Nagaosa, N., Hussain, Z., Shen, Z.X.: Nodal quasiparticle in pseudogapped colossal magnetoresistive manganites. Nature **438**, 474–478 (2005)
72. Padmavathi, K., Venkataiah, G., Venugopal Reddy, P.: Electrical behavior of some rare-earth-doped Nd0.33Ln0.34Sr0.33MnO3 manganites. J. Magn. Magn. Mater. **309**, 237–243 (2007)
73. Mott, N.F., Davis, E.A.: Electronic, Processes in Non-crystalline Material, 2nd Edn, Oxford, New York (1979)
74. Mahendiran, R., Tiwary, S.K., Raychaudhuri, A.K., Ramakrishnan, T.V., Mahesh, R., Rangavittal, N., Rao, C.N.R.: Structure, electron-transport properties, and giant magnetoresistance of hole-doped $LaMnO_3$ systems. Phys. Rev. B **53**, 3348–3358 (1996)
75. Das, S., Dey, T.K.: Temperature dependence of the thermoelectric power of $La_{1-x}K_xMnO_3$ compounds in light of a two phase model. Phys. B **381**, 280–288 (2006)
76. Lakshmi, Y.K., Venkataiah, G., Vithal, M., Reddy, P.V.: Magnetic and electrical behavior of $La_{1-x}A_xMnO_3$ (A = Li, Na, K and Rb) manganites. Phys. B **403**, 3059–3066 (2008)

17

Theoretical analysis on nonlinear vibration of fluid flow in single-walled carbon nanotube

P. Valipour[1] · S. E. Ghasemi[2] · Mohammad Reza Khosravani[3] · D. D. Ganji[4]

Abstract In this study, the concept of nonlocal continuum theory is used to characterize the nonlinear vibration of an embedded single-walled carbon nanotube. The Pasternak-type model is employed to simulate the interaction of the SWNTs. The parameterized perturbation method is used to solve the corresponding nonlinear differential equation. The effects of the vibration amplitude, flow velocity, nonlocal parameter, and stiffness of the medium on the nonlinear frequency variation are presented. The result shows that by increasing the Winkler constant, the nonlinear frequency decreases, especially for low vibration amplitudes. In addition, it is resulted that influence of the nonlocal parameter is greater at higher flow velocities in comparison with lower flow velocities.

Keywords Theoretical analysis · Single-walled carbon nanotube (SWCNT) · Parameterized perturbation method · Fluid flow · Nonlinear vibration

✉ S. E. Ghasemi
s.ebrahim.ghasemi@gmail.com

1 Department of Textile and Apparel, Qaemshahr Branch, Islamic Azad University, Qaemshahr, Iran

2 Young Reseachers and Elite Club, Sari Branch, Islamic Azad University, Sari, Iran

3 Chair of Solid Mechanics, University of Siegen, Paul-Bonatz-Str. 9-11, 57068 Siegen, Germany

4 Department of Mechanical Engineering, Babol University of Technology, Babol, Iran

Introduction

Carbon nanotubes (CNTs) are an allotrope of carbon. They take the form of cylindrical carbon molecules and have novel properties that make them potentially useful in a wide variety of applications in nanotechnology, electronics, optics, and other fields of materials science. In the recent few years, carbon nanotubes (CNTs) have been one of the most promising studies in the field of mechanics, physics, chemistry, materials science, and so on. They exhibit extraordinary strength and unique electrical properties, and are efficient conductors of heat.

The discovery of CNTs by Iijima [1], especially the discovery of the single-wall nanotube (SWNT) and the successful composition of the CNT in the macrography scale, has received considerable attention in recent years. At the present, CNT has been the chief research subject in the area of the fullerene, and it has been one of the most promising researches in the field of mechanics, physics, chemistry, materials science, and so on. Because of their novel electronic, mechanical, and other physical and chemical properties, CNT holds substantial promise as building blocks for nanoelectronics, nanodevices, and nanocomposites [2–4]. For example, the stiffness of CNT is 100 times as that of the steel, but the weight is one-sixth times as that of the steel. It is foreseen to be the most promising one-dimensional nanophase materials in the 21st century. Therefore, it is necessary and significant to study its mechanical property.

The classic elastic continuum models have been presented to investigate the vibration behavior of CNTs. Yoon et al. [5] examined the effect of moving internal fluid on free vibration and flow induced flutter instability of cantilever CNTs according to the continuum elastic model. It was resulted that the resonance frequencies were related to the fluid-flow velocity. Structural instability of the CNTs could happen at a critical flow velocity.

A theoretical approach based on a continuum elastic model is studied by Natsuki et al. [6] to investigate the sound wave propagation in both single- and double-walled CNTs filled with fluids and found that there exist several critical frequencies, and free wave speed of the fluids influences the wave propagation in fluid-filled CNTs.

Reddy et al. [7] examined the influence of fluid flow on the free vibration and instability of fluid-conveying single-walled CNTs (SWCNTs). Wang et al. [8] studied the Timoshenko beam model and the DQ method for free vibration analysis of multi-walled CNTs (MWCNTs). Zhang et al. [9]. used a double-elastic beam model for transverse vibrations of double-walled carbon nanotubes under compressive axial load according to the Euler–Bernoulli beam theory. The effect of internal moving fluid and compressive axial load on the nonlinear vibration and stability of conveying CNTs is presented by Rasekh and Khadem [10].

The nonlocal Euler–Bernoulli elasticity theory [11] as a modifying theory for nanostructers and nanoscale models have been greatly used to model CNTs due to the nanoscale effects. In this theory, the small scale effects are captured by assuming that the stress at a point is a function of the strains at all other points of the domain. Several references are found related to the nonlocal elasticity theory [12, 13].

Most of the studies are limited to linear vibration at present; however, the deformations of CNTs are nonlinear in nature. Only when considering the nonlinearities in geometry and physics, the more precise dynamic properties of CNTs can be obtained, and then, the nanostructures can receive more wide applications. Fortunately, more and more researchers have begun to realize these ideas, and the nonlinear mechanical behaviors of CNTs have received considerable attentions in recent years. Pantano and Boyce investigated the effect of the characteristic wavelike or wrinkles on the bending mode of CNTs under considering the geometric nonlinearity and explained the phenomenon that the curve modes of CNTs decrease with the increase in the diameter of CNTs. Fu et al. [14] analyzed the nonlinear vibration for embedded CNTs and got the amplitude frequency response curves of the nonlinear free vibration for the SWCNTs and DWCNTs. Yan et al. [15] showed that the radial vibration modes of the inner and outer tubes of simply supported DWCNTs and concluded that the system had twice dynamical mode transitions as the frequency increased.

MWCNTs composite nanofibers with various MWCNT contents were fabricated by electrospinning process, and their microwave absorption properties were evaluated in the frequency range of 8–12 GHz at room temperature [16].

Mathematical modeling is a vantage point to reach a solution in an engineering problem, so the accurate modeling of nonlinear engineering problems is an important step to obtain accurate solutions [17–21]. Most differential equations of engineering problems do not have exact analytic solutions, so approximation and numerical methods must be used. Recently, some different methods have been introduced to solving these equations, such as the variational iteration method (VIM) [22, 23], homotopy perturbation method (HPM) [24, 25], parameterized perturbation method (PPM) [26], differential transformation method (DTM) [27, 28], modified homotopy perturbation method (MHPM) [29], least square method (LSM) [30–32], collocation method (CM) [33, 34], galerkin method (GM) [35], optimal homotopy asymptotic method (OHAM) [36, 37], and differential quadrature method (DQM) [38].

In this study, the nonlocal continuum theory is utilized to simulate the nonlinear vibration of an SWCNT conveying fluid, employing Pasternak-type elastic foundation. To solve the governing equations of the problem, one of the newest analytical methods named the PPM is used. This method is one of the strong and effect method for solving nonlinear problems and is investigated and developed by some authors to solve nonlinear equations arising in engineering problems.

This approach can be used as an efficient and practical technique for solving such nonlinear problems in vibration field. Comparisons revealed that PPM is a powerful approach for solving these problems. In addition, the results showed that the main attributions of this method are very straightforward calculations and low computational burden compared with previous analytical and numerical approaches.

It can be found that PPM is very efficient but straightforward technique, while you just employ a transformation equation and solve the original equation without the separation of linear and nonlinear terms. This method can be easily extended to other strongly nonlinear vibration equations and can be found widely applicable in engineering and science.

The obtained results show good agreement with the numerical simulation and also with a previous study. The effects of the vibration amplitude, flow velocity, nonlocal parameter, and stiffness of the medium on the nonlinear frequency variation are illustrated.

Mathematical modeling

Figure 1 shows an SWCNT conveying fluid, as a hollow cylindrical tube in a Pasternak-type elastic medium. The nanotube is assumed to be simply supported at both ends, and the effect of gravity is negligible.

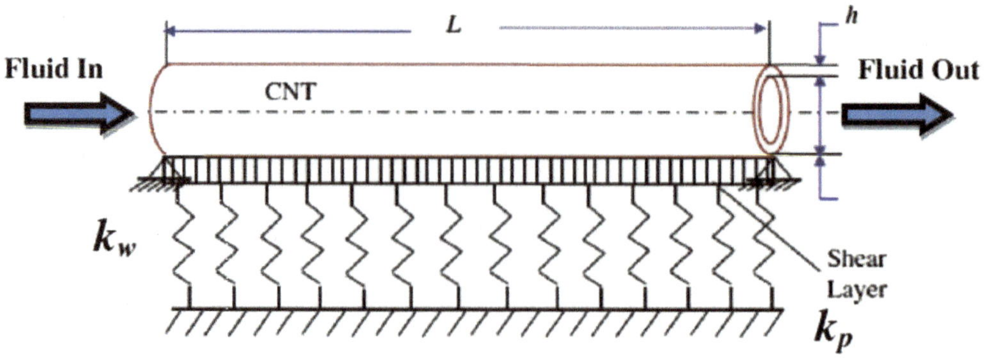

Fig. 1 Single-walled carbon nanotube embedded in a Pasternak-type foundation model

For the Euler–Bernoulli beam theory, the relationships among the transverse shear force Q, the bending moment of the model M, and the longitudinal force N are [39]:

$$\frac{\partial Q}{\partial x}=\frac{\partial^2 M}{\partial^2 x}+N\frac{\partial^2 w}{\partial^2 x} \tag{1}$$

N and M are the stress resultants, defined as follows:

$$M=\int z.\sigma_{xx}\mathrm{d}A_c=\int z.E\varepsilon_{xx}\mathrm{d}A_c,\quad N=\int \sigma_{xx}\mathrm{d}A_c=\int E\varepsilon_{xx}\mathrm{d}A_c \tag{2}$$

where E is Young's modulus of the SWCNT.

The nonlocal continuum theory, presented by Eringen in 1983, shows a more precise constitutive rule for small-scale structures in comparison with the common local elastic theories. This definition of nonlocal elasticity is based on lattice dynamics and observations on phonon dispersion. The nonlocal constitutive equation for the uniaxial bending stress state forms as:

$$\sigma_{xx}=E\varepsilon_{xx}+(e_0a)^2\frac{\partial^2\sigma_{xx}}{\partial x^2} \tag{3}$$

The parameter (e_0a) shows the small-scale effect, which is called the nonlocal parameter, in which the parameter e_0 is estimated such that the relations of the nonlocal elasticity model could provide a good approximation of atomic dispersion curves of plane waves with those of atomic lattice dynamics, and a represents an internal length, such as lattice parameter and granular size

$$M-(e_0a)^2\frac{\partial^2 M}{\partial x^2}=\int zE\varepsilon_{xx}\mathrm{d}A_c \tag{4}$$

Based on the Euler–Bernoulli continuum theory, the displacement field of the model is expressed as:

$$u(x,z,t)=u(x,t)-z.\frac{\partial w}{\partial x},\quad w(x,z,t)=w(x,t) \tag{5}$$

In addition, the von Karman strain based on the displacement field is approximately expressed [40]:

$$\varepsilon_{xx}\equiv\frac{\partial u(x,z,t)}{\partial x}+\frac{1}{2}\left(\frac{\partial w(x,z,t)}{\partial x}\right)^2=\frac{\partial^2 u}{\partial x^2}-z.\frac{\partial^2 w}{\partial x^2}+\frac{1}{2}\left(\frac{\partial w(x,z,t)}{\partial x}\right)^2 \tag{6}$$

From Eqs. (3) and (6), the nonlocal stress resultant can be defined as

$$M-(e_0a)^2\frac{\partial^2 M}{\partial x^2}=EI\frac{\partial^2 w}{\partial x^2} \tag{7}$$

where the following relation has been used

$$\int z\mathrm{d}A_c=0,\quad \int z^2\mathrm{d}A_c=I \tag{8}$$

The equations of motion can now be expressed in terms of displacements. Substituting for the second derivative of M from Eq. (1) into Eq. (8), we obtain

$$M=(e_0a)^2\left[\frac{\partial Q}{\partial x}-N\frac{\partial^2 w}{\partial x^2}\right]+EI\frac{\partial^2 w}{\partial x^2} \tag{9}$$

Now, substituting for M from Eq. (9) into Eq. (7), the governing equation of motion is readily identified as

$$EI\frac{\partial^4 w}{\partial x^4}+\frac{\partial Q}{\partial x}-N\frac{\partial^2 w}{\partial x^2}-(e_0a)^2\left[\frac{\partial^3 Q}{\partial x^3}-N\frac{\partial^4 w}{\partial x^4}\right]=0 \tag{10}$$

Hence, the governing equations for a fluid-conveying SWCNT can be written as

$$\begin{aligned}&m_c\frac{\partial^2 w}{\partial t^2}+EI\frac{\partial^4 w}{\partial x^4}+k_e w-k_p\frac{\partial^2 w}{\partial x^2}+F\frac{\partial^2 w}{\partial x^2}\\&+m_f\left(2v\frac{\partial^2 w}{\partial x\partial t}+v^2\frac{\partial^2 w}{\partial x^2}+\frac{\partial^2 w}{\partial t^2}\right)-(e_0a)^2\\&\times\left[m_c\frac{\partial^4 w}{\partial t^2\partial x^2}+k_e\frac{\partial^2 w}{\partial x^2}-k_p\frac{\partial^4 w}{\partial x^4}+F\frac{\partial^4 w}{\partial x^4}\right.\\&+m_f\left(2v\frac{\partial^2 w}{\partial x^3\partial t}+v^2\frac{\partial^4 w}{\partial x^4}+\frac{\partial^4 w}{\partial x^2\partial t^2}\right)-\frac{EA_c}{2L}\\&\left.\cdot\frac{\partial^4 w}{\partial x^4}\cdot\int_0^L\left(\frac{\partial w}{\partial x}\right)^2\mathrm{d}x\right]-\frac{EA_c}{2L}\frac{\partial^2 w}{\partial x^2}\int_0^L\left(\frac{\partial w}{\partial x}\right)^2\mathrm{d}x=0\end{aligned} \tag{11}$$

The deflection of the nanotube is subjected to the following boundary conditions:

$$w(0,t)=\frac{\partial^2 w(0,t)}{\partial x^2}=0 \quad \text{at } x=0$$
$$w(L,t)=\frac{\partial^2 w(L,t)}{\partial x^2}=0 \quad \text{at } x=\text{L} \tag{12}$$

W(x,t) can be expanded as:

$$w(x,t)=q(t).\phi_1(x) \tag{13}$$

Φ_1 performs as the normalized mode functions of the nanotube from the linear vibration analysis due to the specified boundary conditions.

Substituting Eq. (13) in Eq. (11) leads to:

$$\ddot{q}(t)+\frac{\left[1+e^2(K_e+K_p-T-U^2)+K_e+K_p-T-U^2\right].\omega_0^2}{1+e^2}.q(t)+\frac{\omega_0^2}{4r^2}.q^3(t)=0 \tag{14}$$

These equations can be made dimensionless using the following definitions

$$\omega_0=\frac{\pi^2}{L^2}\sqrt{\frac{EI}{m_c+m_f}},\ e=\frac{\pi}{L}(e_0a),\ K_e=\frac{L^4}{\pi^4}\frac{1}{EI}k_e,$$
$$K_p=\frac{L^2}{\pi^2}\frac{1}{EI}k_p,\ T=\frac{L^2}{\pi^2}\frac{1}{EI}F,\ U=\frac{L}{\pi}\sqrt{\frac{m_f}{EI}},\ r=\sqrt{\frac{I}{A_c}} \tag{15}$$

Analytical solution

The PPM was first proposed in 1999 in [41]. An expanding parameter is introduced by a linear transformation

$$f=\varepsilon u+a \tag{16}$$

where ε is the perturbation parameter, by substituting Eq. (16) into an original equation to have no secular term in the equation; we can obtain the unknown constant parameters *a* and *b*. Then, the solution is expanded in the form

$$u=\sum_{i=0}^{n}\varepsilon^i u_i=u_0+\varepsilon u_1+\varepsilon^2 u_1+\cdots \tag{17}$$

$$q=A_0\cos(\omega t)-\frac{1}{8}\frac{A_0^3\cos(\omega t)\alpha_2}{\omega^2}+\frac{1}{8}\frac{A_0^3\cos^3(\omega t)\alpha_2}{\omega^2} \tag{18}$$

$$q=0.9771563284\cos(0.04831221268\pi^2 t)+0.02284367164\cos^3(0.04831221268\pi^2 t) \tag{19}$$

To validate the present solution of the problem and find the accuracy, the comparison between our solution and numerical result is done. Good agreement between the present PPM and the numerical solution is observed in Fig. 2.

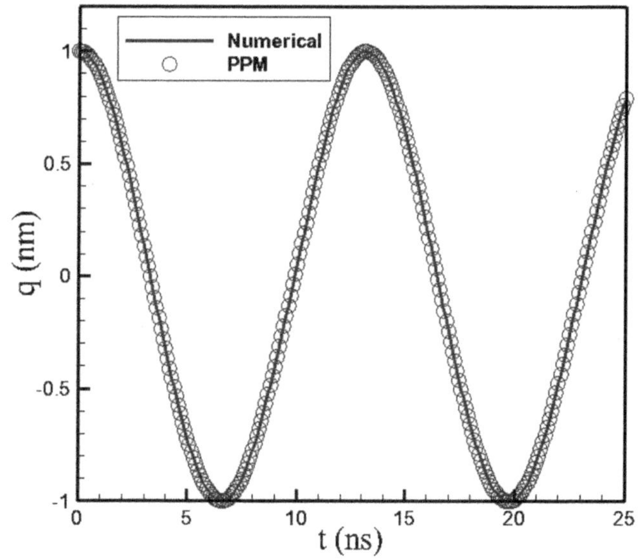

Fig. 2 Comparison between PPM solution and numerical result

Fourth-order Runge–Kutta numerical method

It is obvious that the type of the current problem is the boundary value problem (BVP), and the appropriate numerical method needs to be selected. The numerical solution is performed using the algebra Maple package to solve the present problem. The available methods in this software are a combination of the base scheme (midpoint or trapezoid), and a method enhancement scheme (deferred corrections or Richardson extrapolation). The trapezoid method is generally efficient for typical problems, but the midpoint method can handle harmless end-point singularities that the trapezoid method cannot. The midpoint method, also known as the fourth-order Runge–Kutta method, improves the Euler method by adding a midpoint in the step, which increases the accuracy by one order [42]. Thus, the fourth-order Runge–Kutta method is used as a suitable numerical technique in this study.

Results and discussion

In this section, the effects of the nonlocal parameters (e_0a), axial tensions (F), Winkler constants (K_e), and Pasternak constants (K_p) on nonlinear frequency are studied in the following figures.

Figure 3 shows the behavior of nonlinear frequency for different values of nonlocal parameters. The figure shows that the nonlinear frequency increases with an increase in the nonlocal parameter. The reason is that the nonlocal

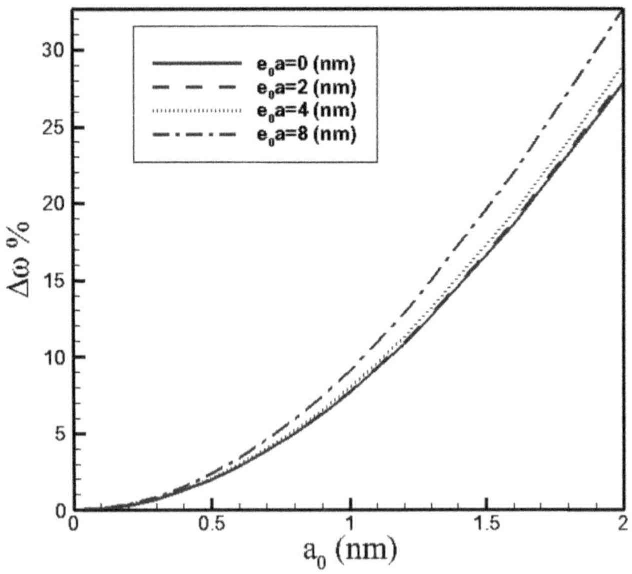

Fig. 3 Variation of nonlinear frequency against the maximum nonlinear amplitude for different nonlocal parameters

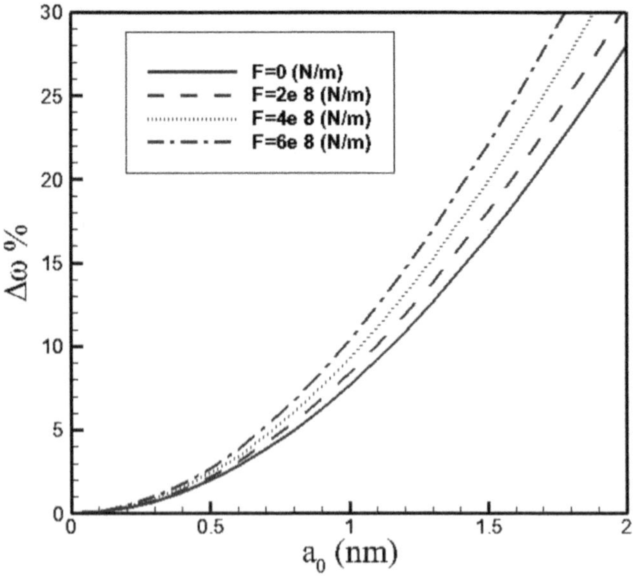

Fig. 4 Variation of nonlinear frequency against the maximum nonlinear amplitude for different axial tensions

theory introduces a more flexible model, and with increasing the flexibility, the effect of the nonlinearity on the model becomes more significant. The Pasternak model expresses the base of the SWCNT. The Pasternak-type foundation, also named the two-parameter foundation model, models the interaction between the medium and the nanotube using two different parameters. These two parameters are: Winkler constant (K_e), which shows normal pressure and Pasternak constant (K_p), which express transverse shear stress due to the interaction of shear deformation of the surrounding elastic medium.

Figure 4 shows the nonlinear frequency variation against the nonlinear amplitude as a function of axial tension. It is shown that the axial tension of the SWCNT can decrease the difference between the nonlinear and the linear resonant frequency, and this effect is profound for high vibration amplitude. It means that increasing the axial tension F can control the nonlinearity.

Figures 5 and 6 show the influences of Winkler constants (K_e) and Pasternak constants (K_p) on nonlinear frequency variation. Figure 5 indicated that by increasing the Winkler constant, the nonlinear frequency decreases, especially for low vibration amplitudes. This means that as the nanotube vibrates in a stiff medium, the nonlinear frequency turns to the linear frequency. It means that for low amplitudes and stiff elastic foundations, the linear simulation of the SWCNT shows a precise theoretical model for transverse flow-induced vibrations.

Figure 6 shows the effects of Pasternak constant on the nonlinear frequency. It can be seen an increase in the shear stiffness of the medium results in the decrease in the nonlinear frequency for the small vibration

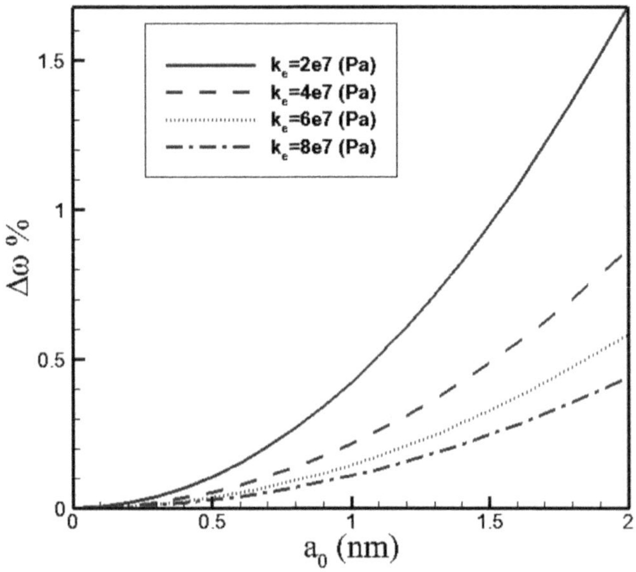

Fig. 5 Variation of nonlinear frequency against the maximum nonlinear amplitude for different Winkler constants

amplitudes, and also the nonlinear flow-induced frequency reduces to the linear. The influence of the effect nonlocal parameter on the nonlinear frequency variation against the flow velocity is shown in Fig. 7. It is resulted that the influence of the nonlocal parameter is greater at higher flow velocities in comparison with lower flow velocities. This effect is more significant when the nonlocal parameter increases.

Figure 8 shows the nonlinear frequency variation against the flow velocity for various axial tensions. The result shows that for low flow velocities, the effect of axial

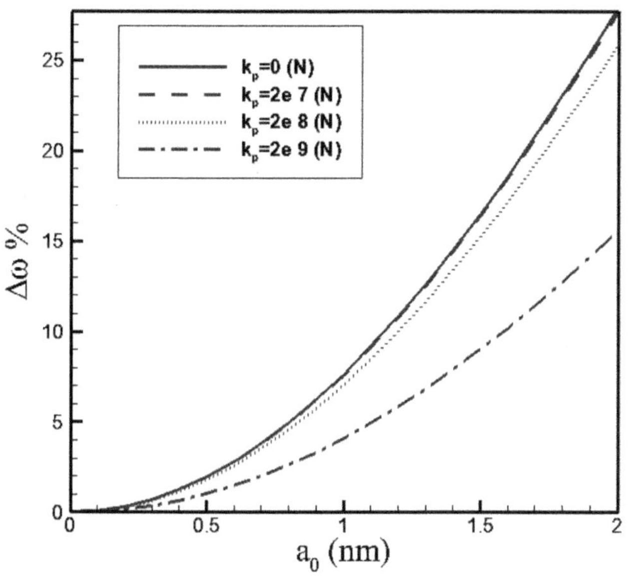

Fig. 6 Variation of nonlinear frequency against the maximum nonlinear amplitude for different Pasternak constants

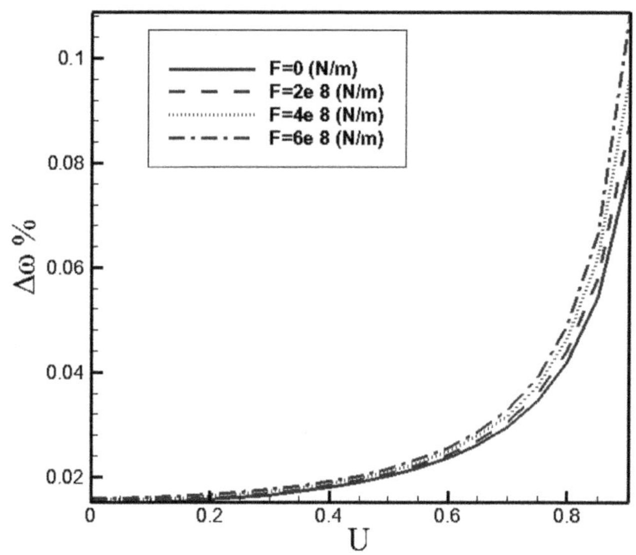

Fig. 8 Variation of nonlinear frequency against the dimensionless fluid velocity for different axial tensions

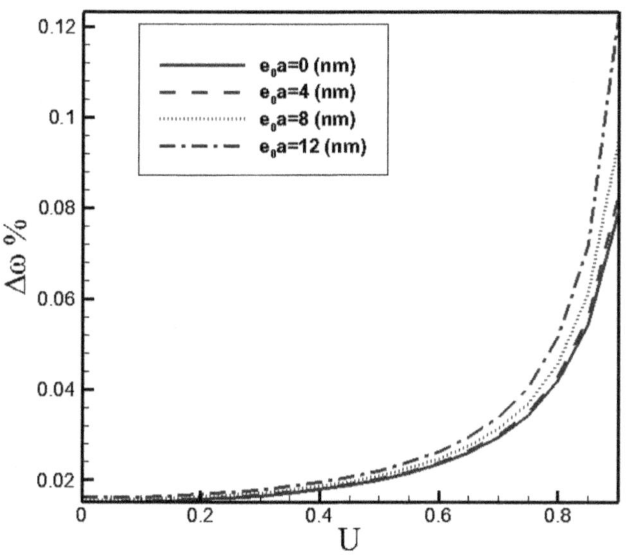

Fig. 7 Variation of nonlinear frequency against the dimensionless fluid velocity for different nonlocal parameters

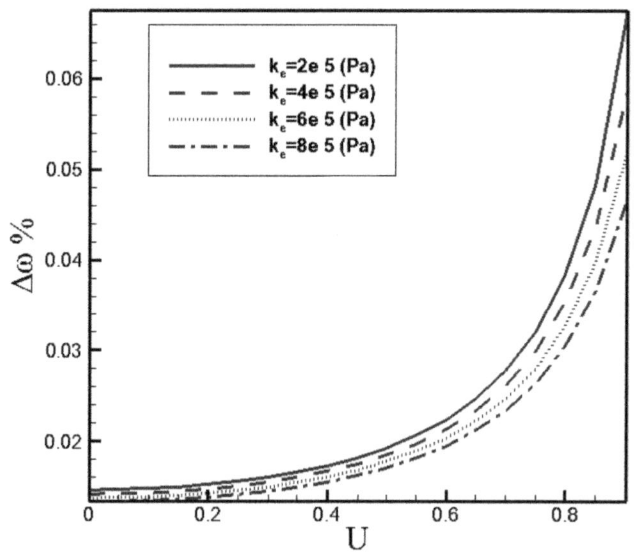

Fig. 9 Variation of nonlinear frequency against the dimensionless fluid velocity for different Winkler constants

tension on the nonlinear frequency variation is little. For high flow velocities, the nonlinear frequency variation decreases with increment in axial tensions.

Figure 9 shows the influence of the Winkler constant on the nonlinear frequency variation against the flow velocities. It shows that the nonlinear frequency variation does not change greatly for low fluid velocities, and the mediums with rigid elastic properties origin the difference between the nonlinear and linear frequencies to remain unchanged with respect to flow velocity. Furthermore, for flexible mediums, the nonlinear frequency variation increases with the flow velocity. The effect of the Pasternak constant on the nonlinear frequency variation with the dimensionless flow velocity is shown in Fig. 10. The result shows that for low fluid velocities ($U < 0.5$) and as the shear stiffness of the elastic medium increases, the nonlinear frequency variation decreases, and for the higher flow velocities, it remains constant. This shows that the nonlinear vibration behavior of the SWCNT is independent of the fluid flow.

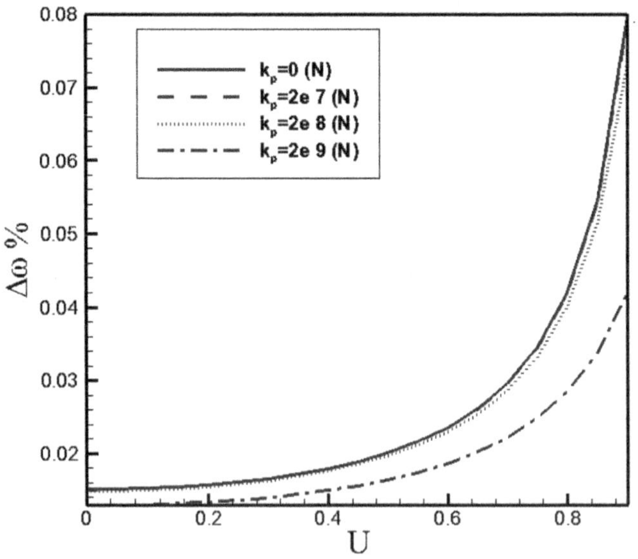

Fig. 10 Variation of nonlinear frequency against the dimensionless fluid velocity for different Pasternak constant

Conclusion

In this study, the PPM is used to solve the nonlinear vibration model of a fluid-conveying SWCNT embedded in a Pasternak foundation. The approximate solution for the nonlinear frequency variation is obtained by the proposed technique. The present results are in excellent agreement with the numerical ones. In addition, this method is powerful and efficient technique for finding science and engineering nonlinear differential equations.

The results show that the nonlinear flow-induced frequency alters from the linear frequency greatly when the amplitude, flow velocity, and nonlocal parameter are high, while for the CNTs embedded in the mediums of high Pasternak parameters, the nonlinearity of the model does not demonstrate a significant effect on the frequency. Moreover, the axial tension restricts the nonlinear effect and limits the flow induced-vibration of the nanotube at high flow velocity and for high vibration amplitudes.

References

1. Iijima, S.: Helical microtubules of graphitic carbon. Nature **354**, 56–58 (1991)
2. Dai, H., Hafner, J.H., Rinzler, A.G., Colbert, D.T., Smalle, R. E.: Nanotubes as nanoprobes in scanning probe microscopy. Nature **384**, 147–150 (1996)
3. Thostenson, E.T., et al.: Advances in the science and technology of carbon nanotubes and their composites: a review. Compos. Sci. Technol. **61**, 1899–1912 (2001)
4. Yazdi, S.H., Mashadi, M.M.: A nano-tuneable pressure switch system design based on single wall carbon nanotubes. J. Appl. Sci. **7**, 1442–1445 (2007)
5. Yoon, J., et al.: Vibration and instability of carbon nanotubes conveying fluid. Compos. Sci. Technol. **65**, 1326–1336 (2005)
6. Natsuki, T., et al.: Wave propagation in single-and double-walled carbon nanotubes filled with fluids. J. Appl. Phys. **101**, 034319 (2007)
7. Reddy, C., et al.: Free vibration analysis of fluid-conveying single-walled carbon nanotubes. Appl. Phys. Lett. **90**, 133122 (2007)
8. Wang, L., Ni, Q.: A reappraisal of the computational modelling of carbon nanotubes conveying viscous fluid. Mech. Res. Commun. **36**, 833–837 (2009)
9. Zhang, Y., et al.: Thermal effect on transverse vibrations of double-walled carbon nanotubes. Nanotechnology **18**, 445701 (2007)
10. Rasekh, M., Khadem, S.: Nonlinear vibration and stability analysis of axially loaded embedded carbon nanotubes conveying fluid. J. Phys. D Appl. Phys. **42**, 135112 (2009)
11. Eringen, A.C.: Nonlocal continuum field theories. Springer, Belrin (2002)
12. Zhang, Y., et al.: Free transverse vibrations of double-walled carbon nanotubes using a theory of nonlocal elasticity. Phys. Rev. B **71**, 195404 (2005)
13. Sudak, L.: Column buckling of multiwalled carbon nanotubes using nonlocal continuum mechanics. J. Appl. Phys. **94**, 7281–7287 (2003)
14. Fu, Y., et al.: Analysis of nonlinear vibration for embedded carbon nanotubes. J. Sound Vib. **296**, 746–756 (2006)
15. Yan, Y., et al.: Dynamical behaviors of fluid-conveyed multi-walled carbon nanotubes. Appl. Math. Model. **33**, 1430–1440 (2009)
16. Nasouri, Komeil, Valipour, Peiman: Fabrication of polyamide 6/carbon nanotubes composite electrospun nanofibers for microwave absorption application. Polym. Sci. Ser. A **57**(3), 359–364 (2015)
17. Zolfagharian, A., Valipour, P., Ghasemi, S.E.: Fuzzy force learning controller of flexible wiper system. Neural Comput. Appl. **26**(2), 1–11 (2015). doi:10.1007/s00521-015-1869-0
18. Zolfagharian, A., Noshadi, A., Ghasemi, S.E., Md. Zain, M.Z.: A nonparametric approach using artificial intelligence in vibration and noise reduction of flexible systems. Proc. Inst. Mech. Eng. Part C J. Mech. Eng. Sci. **228**(8), 1329–1347 (2014)
19. Zolfagharian, A., Noshadi, A., Khosravani, M.R., Md. Zain, M.Z.: Unwanted noise and vibration control using finite element analysis and artificial intelligence in flexible wiper system. Appl. Math. Model. **38**, 2435–2453 (2014)
20. Zolfagharian, A., Ghasemi, S.E., Imani, M.: A multi-objective, active fuzzy force controller in control of flexible wiper system. Latin Am. J. Solids Struct. **11**(9), 1490–1514 (2014)
21. Valipour, P., Ghasemi, S.E.: Numerical investigation of MHD water-based nanofluids flow in porous medium caused by shrinking permeable sheet. J Braz. Soc. Mech. Sci. Eng. **38**, 859–868 (2016)
22. Ghasemi, S.E., Jalili, Palandi S., Hatami, M., Ganji, D.D.: Efficient analytical approaches for motion of a spherical solid particle in plane couette fluid flow using nonlinear methods. J. Math. Comput. Sci. **5**(2), 97–104 (2012)
23. Ahmadi Asoor, A.A., Valipour, P., Ghasemi, S.E.: Investigation on vibration of single-walled carbon nanotubes by variational iteration method. Appl. Nanosci. **6**, 243–249 (2016)
24. Ghasemi, S.E., Hatami, M., Ganji, D.D.: Analytical thermal analysis of air-heating solar collectors. J. Mech. Sci. Technol. **27**(11), 3525–3530 (2013)
25. Mohammadian, E., Ghasemi, S.E., Poorgashti, H., Hosseini, M., Ganji, D.D.: Thermal investigation of Cu-water nanofluid between two vertical planes. Proc. IMechE Part E J Process Mech. Eng. **229**(1), 36–43 (2015)

26. Ghasemi, Seiyed E., Zolfagharian, A., Hatami, M., Ganji, D.D.: Analytical thermal study on nonlinear fundamental heat transfer cases using a novel computational technique. Appl. Therm. Eng. (2015). doi:10.1016/j.applthermaleng.2015.11.120
27. Ghasemi, S.E., Hatami, M., Ganji, D.D.: Thermal analysis of convective fin with temperature-dependent thermal conductivity and heat generation. Case Stud. Therm. Eng. **4**, 1–8 (2014)
28. Ghasemi, S.E., Valipour, P., Hatami, M., Ganji, D.D.: Heat transfer study on solid and porous convective fins with temperature-dependent heat generation using efficient analytical method. J. Cent. South Univ. **21**, 4592–4598 (2014)
29. Ghasemi, S.E., Zolfagharian, A., Ganji, D.D.: Study on motion of rigid rod on a circular surface using MHPM. Propuls. Power Res. **3**(3), 159–164 (2014)
30. Ghasemi, S.E., Hatami, M., Mehdizadeh Ahangar, G.H.R., Ganji, D.D.: Electrohydrodynamic flow analysis in a circular cylindrical conduit using least square method. J. Electrostat. **72**, 47–52 (2014)
31. Ghasemi, S.E., Vatani, M., Ganji, D.D.: Efficient approaches of determining the motion of a spherical particle in a swirling fluid flow using weighted residual methods. Particuology **23**, 68–74 (2015)
32. Darzi, M., Vatani, M., Ghasemi, S.E., Ganji, D.D.: Effect of thermal radiation on velocity and temperature fields of a thin liquid film over a stretching sheet in a porous medium. Eur. Phys. J. Plus **130**, 100 (2015)
33. Ghasemi, S.E., Hatami, M., Kalani, Sarokolaie A., Ganji, D.D.: Study on blood flow containing nanoparticles through porous arteries in presence of magnetic field using analytical methods. Phys. E **70**, 146–156 (2015)
34. Atouei, S.A., Hosseinzadeh, Kh, Hatami, M., Ghasemi, Seiyed E., Sahebi, S.A.R., Ganji, D.D.: Heat transfer study on convectiveeradiative semi-spherical fins with temperature-dependent properties and heat generation using efficient computational methods. Appl. Therm. Eng. **89**, 299–305 (2015)
35. Ghasemi, Seiyed E., Vatani, M., Hatami, M., Ganji, D.D.: Analytical and numerical investigation of nanoparticles effect on peristaltic fluid flow in drug delivery systems. J. Mol. Liq. **215**, 88–97 (2016)
36. Vatani, M., Ghasemi, S.E., Ganji, D.D.: Investigation of micropolar fluid flow between a porous disk and a nonporous disk using efficient computational technique. Proc IMechE Part E J. Process Mech. Eng. (2014). doi:10.1177/0954408914557375
37. Valipour, P., Ghasemi, S.E., Vatani, M.: Theoretical investigation of micropolar fluid flow between two porous disks. J. Cent. South Univ. **22**, 2825–2832 (2015)
38. Ghasemi, Seiyed E., Hatami, M., Hatami, J., Sahebi, S.A.R., Ganji, D.D.: An efficient approach to study the pulsatile blood flow in femoral and coronary arteries by differential quadrature method. Phys. A **443**, 406–414 (2016)
39. Reddy, J., Pang, S.: Nonlocal continuum theories of beams for the analysis of carbon nanotubes. J. Appl. Phys. **103**, 023511 (2008)
40. Reddy, J.: Nonlocal nonlinear formulations for bending of classical and shear deformation theories of beams and plates. Int. J. Eng. Sci. **48**, 1507–1518 (2010)
41. He, J.: Recent developments in asymptotic methods for nonlinear ordinary equations. Int. J. Comput. Num. Anal. Appl. **2**, 127–190 (2002)
42. Aziz, A.: Heat conduction with maple. RT Edwards, Philadelphia (2006)

18

Experimental and theoretical study on the corrosion resistance of Zr–Co–Al–Nb metallic glasses

W. K. Le[1] · Z. Z. Yuan[1] · X. Y. Zhang[1]

Abstract The corrosion resistance of $(Zr_{56}Al_{16}Co_{28})_{100-X}Nb_X(X = 0,\ 2,\ 4)$ metallic glasses in Hanks' solution were investigated by electrochemical polarization measurements. Polarization curves demonstrate that the addition of niobium into Zr–Co–Al amorphous alloys improves the corrosion resistance. Composition and chemical status of the elements in the passive film were characterized by XPS, and first-principles calculations based on density functional theory (DFT) were used to analyse the mechanism theoretically.

Keywords Metallic glasses · Corrosion · DOS · First-principles calculation

Introduction

Zr-based metallic glasses are expected to have medical applications as biomaterials, due to their high glass-forming ability (GFA), high strength and elastic strain limit, relatively low Young's modulus and excellent corrosion resistance [1–4]. Since the Inoue group has developed a nickel- and copper-free glassy Zr–Co–Al ternary alloy with a large supercooled liquid region and good mechanical properties [5], more and more attention has been paid to these metallic glasses. For medical and other applications, the corrosion resistance of the glassy Zr–Co–Al alloy needs to be evaluated and further improved.

It has been reported that addition of alloying element can remarkably enhance the corrosion resistance of some Zr-based metallic glasses, and to analyse how alloying affects the corrosion resistance of metallic glasses, X-ray photoelectron spectroscopy (XPS) was used to judge the composition of the passive film [6–8]. Unfortunately, these complex amorphous structures pose so big challenge to researcher that almost no report about theoretical analysis of how alloying affects the corrosion resistance of master alloy can be seen.

This work addresses two questions. First, which structural model was used to describe the amorphous structure? cluster packing that characterized by the icosahedral order proposed previously for some simpler metallic glasses? Second, how to conduct a theoretical analysis of alloying have an effect on the corrosion resistance of master alloy.

In this paper, Niobium was chosen as the doped element in Zr–Co–Al system. The corrosion resistance of the Nb-modified metallic glasses was tested in artificial body fluid by electrochemical polarization measurement. First-principles calculations based on density functional theory (DFT) were first used to analyse the result [9]. The new metallic glasses are promising for biomedical use; the present results provide a reference for investigating the corrosion resistance based on electron theory.

Experiment

Experiment procedure

The alloy ingots with the nominal compositions of $(Zr_{56}Al_{16}Co_{28})_{100-X}Nb_X(X = 0,\ 2,\ 4)$ were prepared by arc melting mixtures of the pure metals (the purity is above 99.9%) in an argon atmosphere. Each ingot was then

✉ W. K. Le
lwk_lut@126.com

[1] State Key Laboratory of Advanced Processing and Recycling of Non-ferrous Metals, Lanzhou University of Technology, Lanzhou 730050, Gansu, China

reversed and remelted four times to ensure chemical homogeneity. From the alloy ingots, ribbon samples with a cross-section of 0.05×3 mm^2 were fabricated by melt spinning. The structure of the specimens was examined by X-ray diffraction (XRD, Rigaku D/max-2400) using Cu Kα radiation at tube voltage of 40 kV and tube current of 150 mA. The thermal stability associated with glass transition, supercooled liquid, and crystallization for the glassy alloys was investigated by differential scanning calorimetry(DSC, NETZSCH STA 449F3) at a heating rate of 20 K/min in an argon atmosphere.

For electrochemical corrosion tests, electrochemical polarization was conducted in a three-electrode cell composed of a working electrode, a saturated calomel reference electrode (SCE), and a platinum counter electrode. The electrolyte used was Hanks' solution (a simulated body fluid) with pH 7.4 open to air, which was prepared from reagent-grade chemicals and deionized water. The potentiodynamic polarization curves of the specimens were recorded at a potential sweep rate of 1 mV/s when the open-circuit potential became almost steady after immersion in Hanks' solution for at least 20 min, which is 8.0 g/L·NaCl, 0.1 g/L $MgSO_4 \cdot 7H_2O$, 0.4 g/L KCl, 0.1 g/L $MgCl_2 \cdot 6H_2O$, 0.14 g/L·$CaCl_2$, 0.152 g/L $Na_2HPO_4 \cdot 12H_2O$, 0.06 g/L·KH_2PO_4, 1 g/L·$C_6H_{12}O_6$. The fracture surface was examined by scanning electron microscopy (SEM, JEOL JSM-6700F).

Experimental results and discussion

Figure 1 shows DSC curves of the melt-spun $(Zr_{56}Al_{16}Co_{28})_{100-X}Nb_X (X = 0, 2, 4)$ glassy alloy ribbons. All the alloys exhibit the sequent transition of glass transition, followed by supercooled liquid region and then crystallization, accompanying the distinctly separated two exothermic peaks. No appreciable difference in crystallization process is recognized between the melt-spun ribbon alloys, indicating the formation of a similar glassy phase.

The outer surfaces were smooth, and no trace of precipitation of crystalline phase was seen. The absence of a crystalline phase in the ribbon alloys was confirmed by X-ray diffraction [10]. X-ray diffraction patterns of the ribbon alloys exhibit no diffraction peak corresponding to a crystalline state, as shown in Fig. 2.

Figure 3 shows the potentiodynamic polarization curves of $(Zr_{56}Al_{16}Co_{28})_{100-X}Nb_X (X = 0, 2, 4)$ amorphous alloys in Hanks' solution; there is no obvious passivation region but pitting directly. The Nb-free sample shows lowest corrosion potential (E_{corr}) and highest corrosion current density (I_{corr}), but with minor addition of niobium E_{corr} increases and I_{corr} decreases. As we know, the value of E_{corr} indicates the corrosion tendency, while the value of I_{corr} indicates the corrosion rate. Higher E_{corr} indicates a small tendency to corrosion, while a low I_{corr} indicates a slow corrosion rate when a corrosion occurs. The corrosion data including the corrosion potential (E_{corr}) and the corrosion current density (I_{corr}) are summarized in Table 1. It can be concluded that the addition of niobium to Zr–Co–Al amorphous alloys can improve the corrosion resistance [11].

To better understand the corrosion mechanism of the Zr–Co–Al–Nb amorphous alloy, the samples were immersed in 3.5% solutions for 186 days; then, the composition and chemical status of the elements in the passive film were characterized by XPS, as shown in Fig. 4.

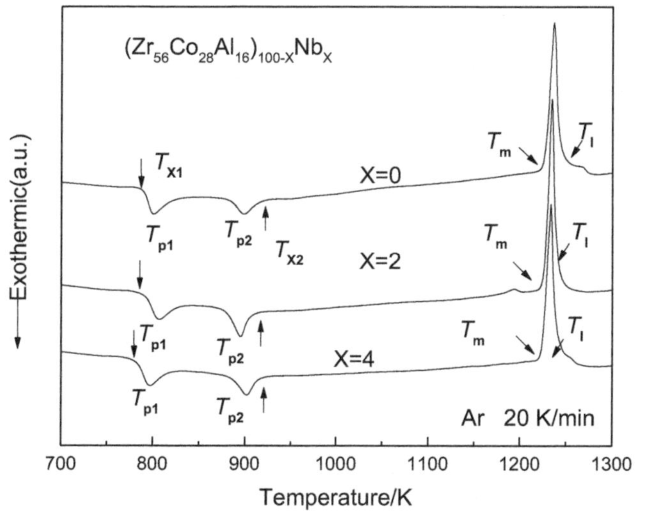

Fig. 1 DSC curves of $(Zr_{56}Al_{16}Co_{28})_{100-X}Nb_X (X = 0, 2, 4)$ alloy ribbons at heating rate of 20 K/min

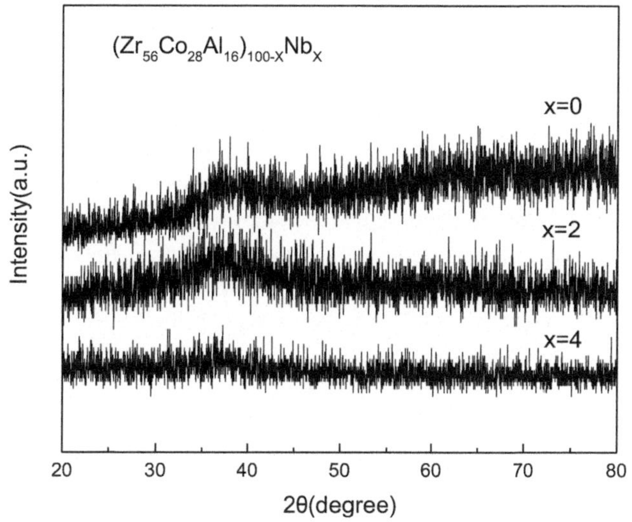

Fig. 2 XRD patterns of $(Zr_{56}Al_{16}Co_{28})_{100-X}Nb_X (X = 0, 2, 4)$ alloy ribbons

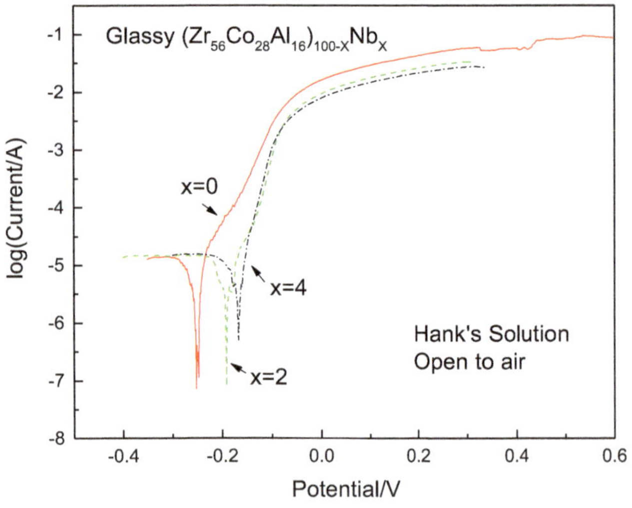

Fig. 3 Polarization curves for $(Zr_{56}Al_{16}Co_{28})_{100-X}Nb_X (X = 0, 2, 4)$ amorphous alloy in Hank's solution

Table 1 Values of E_{corr} and I_{corr} for $(Zr_{56}Al_{16}Co_{28})_{100-X}Nb_X (X = 0, 2, 4)$ amorphous alloy in Hank's solution

Samples	E_{corr}/V	$I_{corr}/\times 10^{-6}$A cm^{-2}
Nb0	−0.248	11.780
Nb2	−0.189	5.418
Nb4	−0.167	3.113

The Zr 3d spectrum consists of two peaks by multiplet splitting approximately at 182.21 and 184.82 eV, which can be assigned to Zr $3d_{5/2}$ and Zr $3d_{3/2}$ electrons originating from Zr^{4+} oxide state (ox). The Al 2p spectrum was decomposed into three spectra peaked at 71.42 eV corresponding to Al $2p_{3/2}$ electrons from metallic state, while, at 73.62 and 74.40 eV, Al $2p_{3/2}$ electrons from Al^{3+} oxide state. The O 1 s spectrum is fitted by two peaks, at 530.53 and at 532.31 eV assigned to metal oxide and Al_2O_3, respectively. Also, very little Co (about 0.3%) was detected in XPS, which is believed not to play a critical role in the outer passive film [7].

It has been reported previously that the high corrosion resistance of the Nb-doped Zr–Co–Al glass alloys is attributed to their single glass phase nature and formation of Zr^{4+}-and Al^{3+}-enriched oxide surface film in the corrosive solutions. XPS analysis revealed that the oxides formed on the sample surface consist mainly of Zr-oxide and Al-oxide for both Nb-free and Nb-bearing metallic glasses. As compared with the Nb-free metallic glasses, the addition of Nb increases significantly the amount of Zr-oxide, but slightly decreases the amount of Al-oxide. The two kinds of oxides are of a dense, compact and stable structure, and have a highly protective ability against corrosion; however, in the chloride ion contained solution,

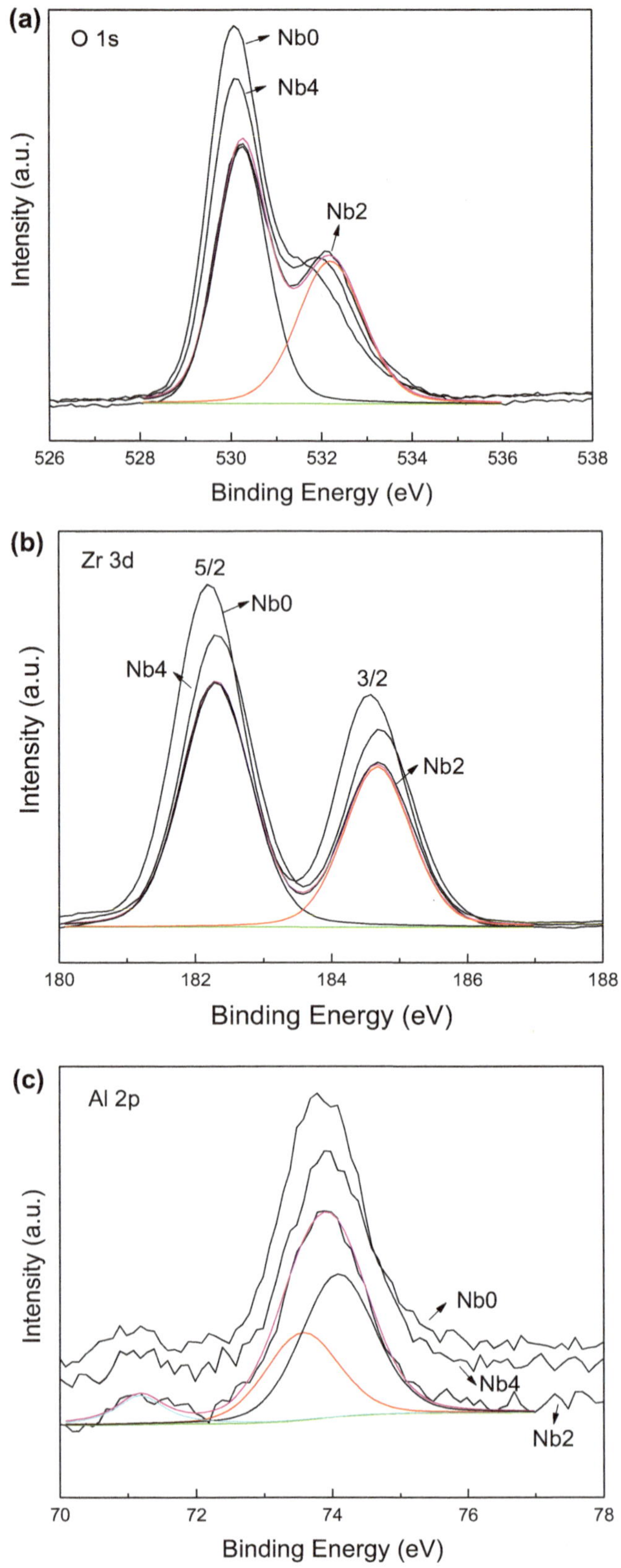

Fig. 4 XPS spectra of Zr 3d, Al 2p, Co 2p, O 1 s electrons of $(Zr_{56}Al_{16}Co_{28})_{100-X}Nb_X (X = 0, 2, 4)$ amorphous alloy: **a** O 1 s, **b** Zr 3d, **c** Al 2p. The *black curves* represent raw XPS spectra, and the *colored one fitted curves*

Zr-oxide is more protectable than Al-oxide; a direct comparison on the corrosion resistance between pure Zr and pure Al in chloride medium revealed higher pitting resistance of Zr than that of Al, suggesting that Zr-oxide was more stable than Al-oxide. Increasing Zr content and decreasing Al content in the surface films could enhance the corrosion resistance of those Zr-based amorphous alloy in chloride medium [12]. Thus, the addition of Nb is favorable for the amorphous alloy to form a protective surface film with higher chemical stability resulting in higher corrosion resistance in chloride medium.

However, the reason why Nb addition suppressed the formation of Al oxides, but promoted the formation of Zr-oxide is still unknown, which needs more comprehensive investigations.

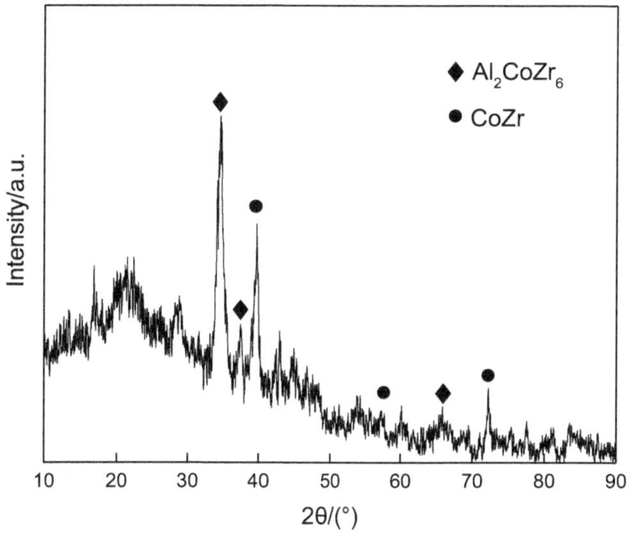

Fig. 5 XRD patterns of $Zr_{56}Al_{16}Co_{28}$ amorphous alloy after heating to 900 K

Calculation

Calculation model

For metallic glasses, to explain their macroscopic properties, it is important to understand their internal structure. Even though scholars proposed that some simpler amorphous structures can be described in terms of cluster packing characterized by the icosahedral order, it is not applicable to the structure of multicomponent metallic glasses. What is more, supposing metallic glasses can be seen as consisting of single cluster packing, when the single cluster packing was considered as a whole, its spatial arrangement must comply with laws, which is not consistent with the fact that amorphous alloy has long-range disorder. Thus, any single structure or cluster is not appropriate to describe amorphous structures, and the description of the atomic structure in an amorphous alloy cannot be exact.

The Inoue group in Sendai has developed the Zr–Al–Co alloy system. Wada et al. [13] found that the differential thermal analysis (DTA) curves of $Zr_{56}Al_{16}Co_{28}$, $Zr_{57}Al_{15}Co_{28}$ and $Zr_{58}Al_{14}Co_{28}$ amorphous alloy approximately the same, indicating the formation of a similar glassy phase. The microstructure of $Zr_{57}Al_{15}Co_{28}$ alloy may consist of ZrCo and Zr_6Al_2Co and small-volume fraction of ZrAlCo phases. With deviating alloy compositions from $Zr_{57}Al_{15}Co_{28}$, the alloys that melted and solidified with the similar phases were confirmed by crystallization process, which is consistent with the reported ternary equilibrium phase diagram [14]. Figure 5 shows the XRD patterns of $Zr_{56}Al_{16}Co_{28}$ amorphous alloy after heating to 900 K with a heating rate of 20 K/min; no Zr–Al–Co phase was found, which was probably because the amount was so small.

The amorphous alloy has long-range disorder, special structure of the short-range order. When amorphous alloys are crystallized, corresponding crystalline phase was produced. The atomic arrangement in short-range of an amorphous alloy tends to be the same as the atomic arrangement of the corresponding crystalline phase; otherwise, atomic arrangement cannot change from disorder to order immediately when crystallized. The amorphous alloy is constituted of quasi-equivalent cluster.

To investigate the structure of amorphous alloy, a simple and accurate method based on quasi-equivalent cluster was proposed. The $Zr_{56}Al_{16}Co_{28}$ amorphous alloy is constituted of α quasi-equivalent cluster, β quasi-equivalent cluster and γ quasi-equivalent cluster, and the atomic arrangement of α quasi-equivalent cluster, β quasi-equivalent cluster, and γ quasi-equivalent cluster is similar to ZrCo phase, Zr_6Al_2Co phase, and ZrAlCo phase, respectively, but these clusters in amorphous alloy are disorganized. Why this method can work well with the challenge we face? This method can well explain at least the following three basic questions. First, when the alloy melted and solidified, the microstructure of $Zr_{57}Al_{15}Co_{28}$ alloy consists of ZrCo and Zr_6Al_2Co and small-volume fraction of Zr–Al–Co phases. Second, amorphous alloy has long-range disorder, special structure of the short-range order; so, amorphous structures described in terms of cluster packing characterized by the icosahedral order is meaningful only in the statistical sense. Third, the description of the atomic structure in amorphous alloy cannot be exact; the atomic-level structure of amorphous alloy remains elusive and intriguing.

Doping position

When alloying element Niobium was added to $Zr_{56}Al_{16}Co_{28}$ amorphous alloy, it is easier to substituted atomic, since $r_{Zr} = 0.160$ nm, $r_{Co} = 0.126$ nm, $r_{Al} = 0.143$ nm, $r_{Nb} = 0.148$ nm, the gap of atomic radius of Nb and Zr, Nb and Al is small, and Nb is next to Zr in the periodic table of chemical elements. If the element Nb is interstitial atom, the lattice distortion can be high; so, cluster becomes unstable.

Table 2 shows the calculation results of Nb-doped α quasi-equivalent cluster supercell systems. From these figures, we may see that binding energy of Nb atom substitutes for the position of Zr atom is lower than Co atom. Table 3 shows the calculation results of Nb-doped β quasi-equivalent cluster supercell systems; from the table, we know that binding energy of Nb atom substitutes for the position of Zr atom is the lowest; therefore, it is easier for Nb atom substitutes for the position of Zr atom.

Calculation results and discussion

Figure 6a is the spatial structure of α quasi-equivalent cluster; Fig. 6b is the spatial structure of β quasi-equivalent cluster, and γ quasi-equivalent cluster was not considered in this study, since it is far few in amorphous alloy. The density of states (DOS) calculations were performed within the framework of the generalized gradient approximation (GGA), as implemented within the CASTEP, a density functional plane-wave pseudopotential program in Materials Studio. PW91 was employed for the exchange correlation functional. An energy cut-off of 300 eV was used for the plane-wave expansion. For geometric optimization, the Brillion zone integration was sampled using a 4 × 4 × 8 Γ-centered k-point grid.

Figure 7 shows the partial density of states (PDOS) of β quasi-equivalent cluster with different doping ratios of niobium which were calculated using LDA. It is shown in Fig. 6a that the majority contribution of PDOS is from the d states of Zr. When niobium doped β quasi-equivalent cluster with 2%, the sharp peak appears at the Fermi level in the spin-up region for Zr-d atoms. When it comes to 4%, the sharp peak appears at low-energy states for Zr-d atoms. The Al-s and Al-p atoms make a small contribution as shown in Fig. 7c. The contribution of Co atoms is very small at the conduction region (Fig. 7b). Figure 8 summarizes the results of the PDOS of α quasi-equivalent cluster with different doping ratios of niobium. Similar to Fig. 7, the majority contribution of PDOS is from the d states of Zr.

When niobium was added to Zr–Co–Al metallic glass, its was improved. Since the amount of niobium is very small, it is impossible to change the crystal structure in the long-range. When niobium was added to Zr–Co–Al metallic glasses, spatial structure of the atoms within short-range was changed, Fermi level is higher, atoms especially Zr–d are more active, which makes it easier for the passivation film to produce and changes the structure of protective surface layers; thus, the corrosion resistance

Table 2 Total energy and difference between binding energy of Nb-doped α quasi-equivalent cluster supercell systems (eV)

Supercell	$E_{tot}(Zr_{n-1}Co_nNb)$	$E_{tot}(Zr)$	E_{tot}
2 × 2 × 2	−18868.71	−2548.18	−21416.89
2 × 2 × 1	−9568.84	−2548.18	−12117.02
1 × 2 × 1	−4919.10	−2548.18	−7467.28
Supercell	$E_{tot}(Zr_nCo_{n-1}Nb)$	$E_{tot}(Co)$	E_{tot}
2 × 2 × 2	−19106.40	−2072.15	−21178.55
2 × 2 × 1	−9806.42	−2072.15	−11878.57
1 × 2 × 1	−5156.50	−2072.15	−7228.65

Table 3 Total energy and difference between binding energy of Nb-doped β quasi-equivalent cluster supercell systems (eV)

Supercell	$E_{tot}(Zr_{6n-1}Co_nAl_{2n}Nb)$	$E_{tot}(Zr)$	E_{tot}
2 × 2 × 2	−71037.02	−2548.18	−73585.20
2 × 2 × 1	−35654.22	−2548.18	−38202.40
1 × 2 × 1	−17962.14	−2548.18	−20510.32
Supercell	$E_{tot}(Zr_{6n}Co_{n-1}Al_{2n}Nb)$	$E_{tot}(Co)$	E_{tot}
2 × 2 × 2	−71273.25	−2072.15	−73345.40
2 × 2 × 1	−35890.03	−2072.15	−37962.18
1 × 2 × 1	−18198.02	−2072.15	−20270.17
Supercell	$E_{tot}(Zr_{6n}Co_nAl_{2n-1}Nb)$	$E_{tot}(Al)$	E_{tot}
2 × 2 × 2	−72260.95	−105.18	−72366.13
2 × 2 × 1	−36877.56	−105.18	−36982.74
1 × 2 × 1	−19185.49	−105.18	−19290.67

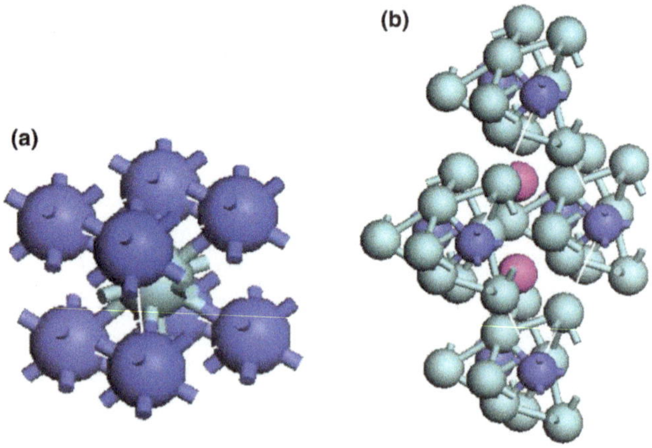

Fig. 6 Spatial structure of quasi-equivalent cluster: **a** α, **b** β

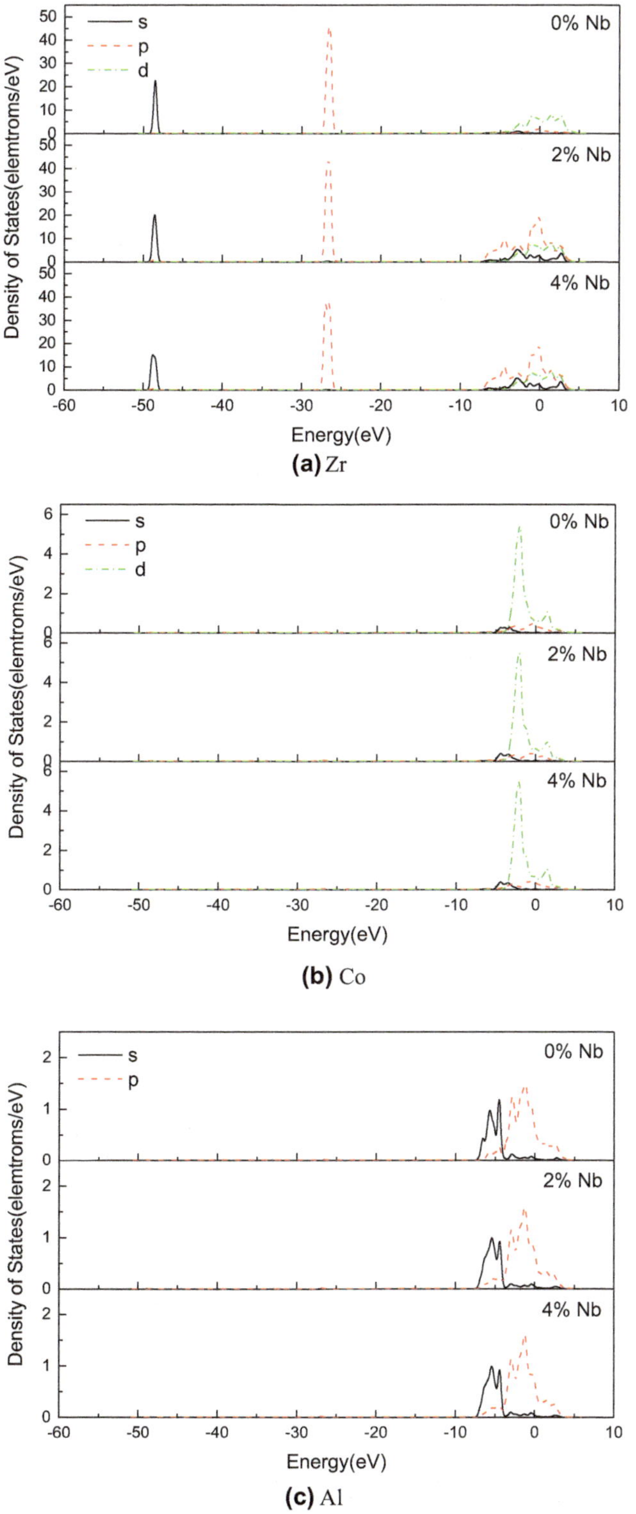

Fig. 7 Partial density of states(PDOS) of β quasi-equivalent cluster with doping ratio of niobium **a** Zr, **b** Co, **c** Al

ability was better. The calculation results are consistent with the experimental results, but how the niobium changes the arrangement of extranuclear electrons of

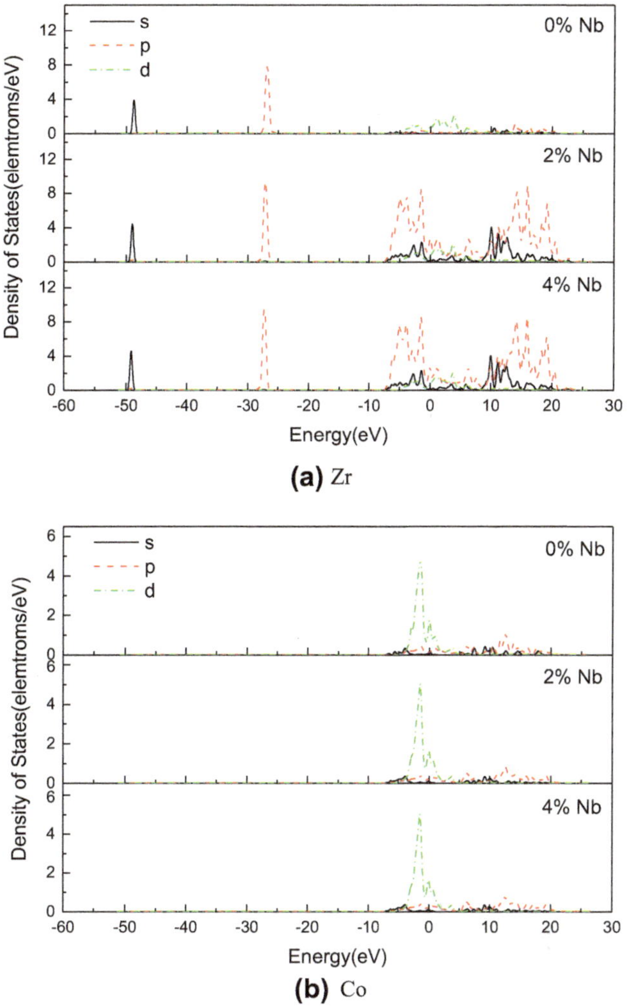

Fig. 8 Partial density of states (PDOS) of α quasi-equivalent cluster with different doping ratios of niobium, **a** Zr, **b** Co

atoms is still unknown, which needs more comprehensive investigations.

Conclusions

The effects of niobium addition on corrosion resistance of Zr–Co–Al ternary metallic glasses were investigated; results show that the addition of niobium into Zr–Co–Al amorphous alloys makes the value of I_{corr} lower, decreases the corrosion rate, and improves the corrosion resistance. XPS analysis revealed that the addition of Nb increases significantly the amount of Zr-oxide, but slightly decreases the amount of Al-oxide in the surface films. First-principles calculations based on density functional theory (DFT) were used to analyse the result. Niobium-doped amorphous alloy makes Zr–d more active, which may change the structure of passive film; thus, corrosion resistance was improved.

References

1. Inoue, A.: Acta Mater. **48**, 279–306 (2000)
2. Li, Y.H., Zhang, W., Dong, C., Qiang, J.B., Xie, G.Q., Fujita, K., Inoue, A.: J. Alloys Compd. **536S**, S117–S121 (2012)
3. Zhang, B., Liao, S., Shu, X., Xie, H.: J. Theor. Appl. Phys. **10**, 129–137 (2016)
4. Li, Y.H., Zhang, W., Dong, C., Qiang, J.B., Fukuhara, M., Makino, A., Inoue, A.: Mater. Sci. Eng. A **528**, 8551–8556 (2011)
5. Zhang, T., Inoue, A.: Mater. Trans. **43**, 267–270 (2002)
6. Kawashima, A., Wada, T.K.: Ohmura, Xie, G.Q., Inoue, A. Mater. Sci. Eng. A **5422**, 140–146 (2012)
7. Zhang, C., Li, N., Pan, J., Guo, S.F., Zhang, M., Liu, L.: J. Alloys Compd. **504S**, S163–S167 (2010)
8. Qin, C.L., Zeng, Y.Q., Louzguine, D.V., Nishiyama, N., Inoue, A.: J. Alloys Compd. **504S**, S172–S175 (2010)
9. Afshar, M., Babaei, M., Kordbacheh, A.H.: J. Theor. Appl. Phys. **8**, 59 (2014)
10. Pang, S.J., Zhang, T., Asami, K., Inoue, A.: J. Mater. Res. **18**, 1652–1658 (2003)
11. Pi, J.H., Pan, Y., Wu, J.L., He, X.C.: Rare Met. Mater. Eng. **43**, 0032–0035 (2014)
12. Hua, N., Huang, L., Chena, W.Z., He, W., Zhang, T.: Mater. Sci. Eng. C **44**, 400–410 (2014)
13. Wada, T., Qin, F.X., Wang, X.M., Yoshimura, M., Inoue, A.: Mater. Res. Soc. **24**, 2941–2948 (2009)
14. ASM Alloy Phase Diagrams Center: http://www.asminternational.org/asmenterprise/apd/. Accessed 16 June 2015

Shifted Tietz–Wei oscillator for simulating the atomic interaction in diatomic molecules

Babatunde J. Falaye[1] · **Sameer M. Ikhdair**[2,4] · **Majid Hamzavi**[3]

Abstract The shifted Tietz–Wei (sTW) oscillator is as good as traditional Morse potential in simulating the atomic interaction in diatomic molecules. By using the Pekeris-type approximation, to deal with the centrifugal term, we obtain the bound-state solutions of the radial Schrödinger equation with this typical molecular model via the exact quantization rule (EQR). The energy spectrum for a set of diatomic molecules ($NO(a^4\Pi_i)$, $NO(B^2\Pi_r)$, $NO(L'^2\phi)$, $NO(b^4\Sigma^-)$, $ICl\left(X^1\Sigma_g^+\right)$, $ICl(A^3\Pi_1)$ and $ICl(A'^3\Pi_2)$ for arbitrary values of n and ℓ quantum numbers are obtained. For the sake of completeness, we study the corresponding wavefunctions using the formula method.

Keywords Exact quantization rule · Formula method · Shifted Tietz-Wei potential

✉ Babatunde J. Falaye
fbjames11@physicist.net; babatunde.falaye@fulafia.edu.ng

Sameer M. Ikhdair
sameer.ikhdair@najah.edu; sikhdair@gmail.com

Majid Hamzavi
majid.hamzavi@gmail.com

1 Applied Theoretical Physics Division, Department of Physics, Federal University Lafia, P. M. B. 146, Lafia, Nigeria

2 Department of Physics, Faculty of Science, An-Najah National University, New campus, P. O. Box 7, Nablus, West Bank, Palestine

3 Department of Physics, University of Zanjan, Zanjan, Iran

4 Department of Electrical Engineering, Near East University, Nicosia, Northern Cyprus, Mersin 10, Turkey

Introduction

By employing the dissociation energy and the equilibrium bond length for a diatomic molecule as explicit parameters, Jia et al. [1] generated improved expressions for some well-known potentials including Rosen–Morse, Manning–Rosen, Tietz and Frost–Musulin potential energy functions. These authors found that the well-known Tietz potential function is conventionally defined in terms of five parameters but it actually has only four independent parameters. Furthermore, the Wei [2] and Tietz potential functions [3] are exactly same solvable empirical functions.

Wang et al. [4] also generated improved expressions for two versions of the Schiöberg potential energy function which are the Rosen–Morse and Manning–Rosen potential functions. By choosing the experimental values of the dissociation energy, equilibrium bond length and equilibrium harmonic vibrational frequency as inputs, the authors obtained the average deviations of the energies calculated with the potential model from the experimental data for five diatomic molecules, and find that no one of six three-parameter empirical potential energy functions is superior to the other potentials in fitting experimental data for all molecules examined.

All these efforts were made in an attempt to find a most suitable molecular potential in its description of diatomic molecules. Following Refs. [1, 4], we suggest sTW as a modification for the TW [2–8]. This potential can be written as

$$V(r) = V_e\left[\frac{2(c_h - 1)e^{-b_h(r-r_e)} - (c_h^2 - 1)e^{-2b_h(r-r_e)}}{(1 - c_h e^{-b_h(r-r_e)})^2}\right], \tag{1}$$

where $b_h = \beta(1 - c_h)$, r_e is the molecular bond length, β is the Morse constant, V_e is the potential well depth and c_h is an optimization parameter obtained from ab initio or

Rydberg–Klein–Rees (RKR) intramolecular potentials. r is the internuclear distance. When the potential constant approaches zero, i.e. $c_h \to 0$, the sTW potential reduces to the Morse potential [9]. This potential is just the TW potential shifted by dissociation energy D_e. The shape of this potential is shown in Fig. 1a for different molecules.

Figure 1b compare between TW diatomic molecular potential, sTW diatomic molecular potential and the Morse potential using the parameters set for $H_2\left(X^1\Sigma_g^+\right)$ diatomic molecule. As it can be seen from this plot, the shifted Tietz–Wei and the Morse potentials are very close to each other for large values of r in the regions $r \approx r_e$ and $r > r_e$, but they are very different at $r \approx 0$. This implies that the shifted Tietz–Wei potential is as good as traditional Morse potential and better than the Tietz–Wei potential in stimulating the atomic interaction for diatomic molecules.

The scheme of our presentation is as follows. In the next section we give basic ingredient of exact quantization rule and all necessary formulas for our calculations. We solve the radial Schrödinger equation for the sTW and also obtain the rotational-vibrational energy spectrum for some diatomic molecules. Finally, results and conclusions are presented.

A brief review of the exact quantization rule

Here we give a brief review on the EQR. The details can be found in Refs. [10–24]. In 2005, Ma and Xu [12, 13] by carefully studying one-dimensional Schrödinger equation, have extended results to three-dimensional case by simply making the replacements $x \to r$ and $V(x) \to V_{\text{eff}}(r)$:

$$\int_{r_a}^{r_b} k(r)dr = N\pi + \int_{r_a}^{r_b} \phi(r)\left[\frac{dk(r)}{dr}\right]\left[\frac{d\phi(r)}{dr}\right]^{-1} dr,$$
$$k(r) = \sqrt{\frac{2\mu}{\hbar^2}[E - V_{\text{eff}}(r)]}, \tag{2}$$

where r_A and r_B are two turning points determined by $E = V_{\text{eff}}(r)$. The $N = n + 1$ is the number of the nodes of $\phi(r)$ in the region $E_{n\ell} = V_{\text{eff}}(r)$ and is larger by one than the number n of the nodes of wave function $\psi(r)$. The first term $N\pi$ is the contribution from the nodes of the logarithmic derivative of wave function, and the second is called the quantum correction.

In this approach, the energy spectrum equation is obtained by solving the two integrals involved in Eq. (2). This quantization rule has been used in many physical systems to obtain the exact solutions of many exactly solvable quantum systems [10–24]. EQR is a very important foundation to proper quantization rule (PQR) [10, 25–27].

The energy spectrum

To study any quantum physical model characterized by the diatomic molecular potential given by Eq. (1), we need to solve the following Schrödinger equation for spherically symmetric potential in any arbitrary dimensional space:

$$\left(\frac{d^2}{dr^2} + \frac{D-1}{r}\frac{d}{dr} - \frac{\ell(\ell + D - 2)}{r^2} + \frac{2\mu}{\hbar^2}(E_{n\ell} - V(r))\right)\psi_{n,\ell,m}(r, \Omega_D) = 0. \tag{3}$$

Now, by defining the wavefunction $\psi_{n,\ell,m}(r, \Omega_D)$ as $r^{(1-D)/2}R_{n\ell}(r)Y_{\ell m}(\theta, \phi)$ and taking $V(r)$ as the sTW

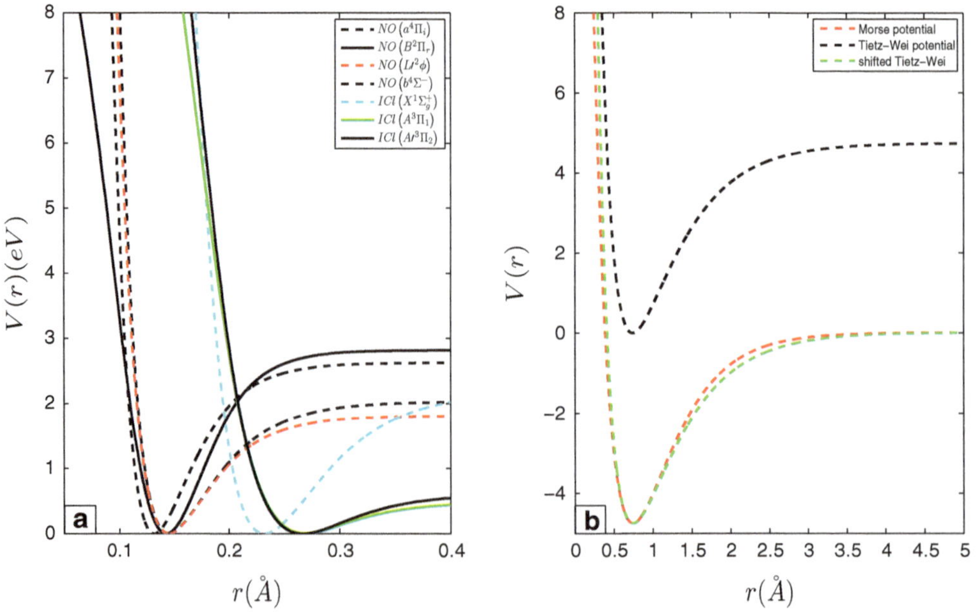

Fig. 1 **a** Shape of sTW diatomic molecular potential for different diatomic molecules. **b** Shape of the sTW, TW and Morse oscillator potentials for $H_2\left(X^1\Sigma_g^+\right)$ diatomic molecule with $c_h = 0.170066$, $b_h = 1.61890\,\text{Å}^{-1}$, $r_e = 0.7416\,\text{Å}$ and $V_e = 38318\,\text{cm}^{-1}$

diatomic molecular potential, the radial part of Eq. (3) can be found as

$$\frac{d^2R_{n\ell}(r)}{dr^2}+\frac{2\mu}{\hbar^2}\times\left[E_{n\ell}-V_e\left[\frac{2(c_h-1)e^{-b_h(r-r_e)}-(c_h^2-1)e^{-2b_h(r-r_e)}}{1-c_he^{-b_h(r-r_e)}}\right]^2-\frac{(\tilde{\eta}^2-\frac{1}{4})\hbar^2}{2\mu r^2}\right]R_{n\ell}(r)=0, \tag{4}$$

where n, ℓ and $E_{n\ell}$ denote the principal quantum number, orbital angular momentum number and the bound state energy eigenvalues of the system under consideration (i.e., $E_{n\ell}<0$), respectively. The parameter $\tilde{\eta}=\ell+\frac{1}{2}(D-2)$ which is a linear combination of the spatial dimensions D and the angular momentum quantum number ℓ. It is well known that for $\ell=0$, problem in the form (4) is exactly solvable. But for $\ell\neq0$, it is not due to the centrifugal barrier. Therefore, in order to solve the above equation for $\ell\neq0$ states, Hamzavi et al. [6] found that the following formula is a good approximation scheme to deal with the centrifugal barrier:

$$\frac{1}{r^2}\approx\frac{1}{r_e^2}\left(D_0+D_1\frac{e^{-b_h(r-r_e)}}{1-c_he^{-b_h(r-r_e)}}+D_2\frac{e^{-2b_h(r-r_e)}}{(1-c_he^{-b_h(r-r_e)})^2}\right), \tag{5}$$

with

$$D_0=1-\frac{1}{\alpha}(1-c_h)(3+c_h)+\frac{3}{\alpha^2}(1-c_h)^2,\quad \lim_{c_h\to0}D_0=1-\frac{3}{\alpha}+\frac{3}{\alpha^2} \tag{6a}$$

$$D_1=\frac{2}{\alpha}(1-c_h)^2(2+c_h)-\frac{6}{\alpha^2}(1-c_h)^3,\quad \lim_{c_h\to0}D_1=\frac{4}{\alpha}-\frac{6}{\alpha^2} \tag{6b}$$

$$D_2=-\frac{1}{\alpha}(1-c_h)^3(1+c_h)+\frac{3}{\alpha^2}(1-c_h)^4,\quad \lim_{c_h\to0}D_2=-\frac{1}{\alpha}+\frac{3}{\alpha^2}, \tag{6c}$$

Constant $\alpha=b_hr_e$ has been introduced for mathematical simplicity. Now, by inserting this approximation into Eq. (4) and then introducing a new transformation of the form $r\to\zeta=\frac{r-r_e}{r_e}$ through the mapping function $\zeta=f(r)$ with r in the domain $[0,\infty)$ or ζ in the domain $[-1,\infty]$, we obtain the following second order differential equation:

$$\frac{1}{r_e^2}\frac{d^2R_{n\ell}(\zeta)}{d\zeta^2}+\frac{2\mu}{\hbar^2}[E_{n\ell}-V_{\text{eff}}(\zeta)]R_{n\ell}(\zeta)=0,\quad\text{with}$$

$$V_{\text{eff}}(\zeta)=\left[\mathcal{A}+\frac{\mathcal{B}}{e^{\alpha\zeta}-c_h}+\frac{\frac{\ell(\ell+1)D_2}{r_e^2}+\frac{2\mu V_e}{\hbar^2}(c_h-1)^2}{(e^{\alpha\zeta}-c_h)}+\mathcal{F}\left(e^{\alpha\zeta}-c_h\right)^2\right],$$

$$\mathcal{A}=\left(\frac{\tilde{\eta}^2-\frac{1}{4}}{r_e^2}\right)\frac{D_0\hbar^2}{2\mu},\ \mathcal{B}=\left(\frac{\tilde{\eta}^2-\frac{1}{4}}{r_e^2}\right)\frac{D_1\hbar^2}{2\mu}+2V_e(c_h-1),$$

$$\mathcal{F}=\left(\frac{\tilde{\eta}^2-\frac{1}{4}}{r_e^2}\right)\frac{D_2\hbar^2}{2\mu}+V_e(c_h-1)^2. \tag{7}$$

If we define $\tilde{\varsigma}=\frac{1}{e^{\alpha\zeta}-c_h}$, then we can obtain the two turning points $\tilde{\varsigma}_a$ and $\tilde{\varsigma}_b$ as well as their sum and product properties by solving $V_{\text{eff}}(\zeta)-E_{n\ell}=0$ or $V_{\text{eff}}(\tilde{\varsigma})-E_{n\ell}=0$ as:

$$\begin{aligned}\tilde{\varsigma}_a&=-\frac{\mathcal{B}}{2\mathcal{F}}-\frac{1}{2\mathcal{F}}\sqrt{\mathcal{B}^2-4\mathcal{F}(\mathcal{A}-E_{n\ell})},\quad\text{and}\\ \tilde{\varsigma}_b&=\frac{\mathcal{B}}{2\mathcal{F}}+\frac{1}{2\mathcal{F}}\sqrt{\mathcal{B}^2-4\mathcal{F}(\mathcal{A}-E_{n\ell})}\\ \tilde{\varsigma}_a+\tilde{\varsigma}_b&=-\frac{\mathcal{B}}{\mathcal{F}},\quad \tilde{\varsigma}_a\tilde{\varsigma}_b=\frac{\mathcal{A}-E_{n\ell}}{\mathcal{F}}\quad\text{and}\\ k(y)&=\sqrt{\frac{2\mu\mathcal{F}}{\hbar^2}}[-(\tilde{\varsigma}-\tilde{\varsigma}_a)(\tilde{\varsigma}-\tilde{\varsigma}_b)]^{1/2}.\end{aligned} \tag{8}$$

Now, we can write the non-linear Riccati equation for the ground state is as

$$-\alpha\frac{\tilde{\varsigma}(1+\tilde{\varsigma})}{r_e}\phi_0'(\tilde{\varsigma})+\phi_0^2(\tilde{\varsigma})+\frac{2\mu}{\hbar^2}\left[E_{0\ell}-V_{eff}(\tilde{\varsigma})\right]=0. \tag{9}$$

Since the logarithmic derivative $\phi_0(\tilde{\varsigma})$ for the ground state has one zero and no pole, it has to take the linear form in $\tilde{\varsigma}$. Thus, we assume the following solution for the ground states

$$\phi_0(\tilde{\varsigma})=A+B\tilde{\varsigma}. \tag{10}$$

By putting Eqs. (10) into (9), we can solve the non-linear Riccati equation. After proper comparison, it is straightforward to obtain the ground state energy and values of A and B as

$$E_{0\ell}=\mathcal{A}-\frac{\hbar^2A^2}{2\mu}\quad\text{with}\quad A=\frac{\mu}{\hbar^2}\frac{\mathcal{B}-\mathcal{F}/c_h^2}{B}+\frac{B}{2}\quad\text{and}$$

$$B=\frac{\alpha c_h}{2r_e}+\frac{\alpha c_h}{2r_e}\sqrt{1+\frac{8\mu\mathcal{F}r_e^2}{\alpha^2\hbar^2c_h^2}}. \tag{11}$$

Since we now have all basic ingredient required to perform our calculations, thus, we proceed to calculating integrals (2)

$$\begin{aligned}\int_{r_a}^{r_b} k(r)\mathrm{d}r &= -\frac{r_\mathrm{e}}{\alpha}\int_{\tilde{\xi}_a}^{\tilde{\xi}_b}\frac{k(\tilde{\xi})}{a\tilde{\xi}(1+\tilde{\xi})}d\tilde{\xi} = -\frac{r_\mathrm{e}}{\alpha}\int_{\tilde{\xi}_{0a}}^{\tilde{\xi}_{0b}}\\ &\quad\times\sqrt{\frac{2\mu\mathcal{F}}{\hbar^2}}\frac{[-(\tilde{\xi}-\tilde{\xi}_a)(\tilde{\xi}-\tilde{\xi}_b)]^{1/2}}{\tilde{\xi}(1+\tilde{\xi}c_\mathrm{h})}\mathrm{d}\tilde{\xi}\\ &= -\frac{\pi r_\mathrm{e}}{\alpha}\sqrt{\frac{2\mu\mathcal{F}}{\hbar^2}}\\ &\quad\times\left[\frac{\sqrt{(1+\tilde{\xi}_a c_\mathrm{h})(1+\tilde{\xi}_b c_\mathrm{h})}}{c_\mathrm{h}}-\frac{1}{c_\mathrm{h}}-\sqrt{\tilde{\xi}_a\tilde{\xi}_b}\right]\\ &= -\frac{\pi r_\mathrm{e}}{\alpha}\sqrt{\frac{2\mu\mathcal{F}}{\hbar^2}}\\ &\quad\times\left[\frac{\sqrt{(\mathcal{A}-E_{n\ell})c_\mathrm{h}^2-\mathcal{B}c_\mathrm{h}+\mathcal{F}}}{\mathcal{F}c_\mathrm{h}}-\frac{1}{c_\mathrm{h}}-\sqrt{\frac{\mathcal{A}-E_{n\ell}}{\mathcal{F}}}\right],\end{aligned} \tag{12}$$

where we have used the following standard integral

$$\begin{aligned}&\int_{x_a}^{x_b}\frac{\sqrt{(x_a-x)(x-x_b)}}{x(1+Qx)}dx\\ &= \pi\left[\frac{\sqrt{(Qx_a+1)(Qx_b+1)}}{Q}-\frac{1}{Q}-\sqrt{x_a x_b}\right].\end{aligned} \tag{13}$$

Furthermore, we can find the integral on the right hand side as

$$\begin{aligned}&\int_{r_{0a}}^{r_{0a}}\phi(r)\left[\frac{\mathrm{d}k_0(r)}{\mathrm{d}r}\right]\left[\frac{\mathrm{d}\phi(r)}{\mathrm{d}r}\right]^{-1}\mathrm{d}r\\ &= -\frac{r_\mathrm{e}}{\alpha}\int_{\tilde{\xi}_{0a}}^{\tilde{\xi}_{0b}}\left[\frac{\mathrm{d}k(r)}{\mathrm{d}r}\right]\left[\frac{\mathrm{d}\phi(\tilde{\xi})}{\mathrm{d}\tilde{\xi}}\right]^{-1}\left(\frac{A}{B}+\tilde{\xi}\right)\mathrm{d}\tilde{\xi}\\ &= \frac{r_\mathrm{e}}{2\alpha}\int_{\tilde{\xi}_{0a}}^{\tilde{\xi}_{0b}}\sqrt{\frac{2\mu\mathcal{F}}{\hbar^2}}\frac{[2\tilde{\xi}-(\tilde{\xi}_a+\tilde{\xi}_b)]\left[\frac{A}{B}+\tilde{\xi}\right]}{\tilde{\xi}(1+\tilde{\xi}c_\mathrm{h})\sqrt{-(\tilde{\xi}-\tilde{\xi}_a)(\tilde{\xi}-\tilde{\xi}_b)}}\mathrm{d}\tilde{\xi}\\ &= \frac{r_\mathrm{e}}{2\alpha}\sqrt{\frac{2\mu\mathcal{F}}{\hbar^2}}\int_{\tilde{\xi}_{0a}}^{\tilde{\xi}_{0b}}\left[\left(\frac{A}{B}-\frac{1}{c_\mathrm{h}}\right)\left(1+\frac{\tilde{\xi}_{0a}+\tilde{\xi}_{0b}}{2}c_\mathrm{h}\right)\frac{1}{1+\tilde{\xi}c_\mathrm{h}}\right.\\ &\quad\left.+\frac{1}{c_\mathrm{h}}\cdot-\frac{A}{B}\left(\frac{\tilde{\xi}_{0a}+\tilde{\xi}_{0b}}{2}\right)\frac{1}{\tilde{\xi}}\right]\times\frac{\mathrm{d}\tilde{\xi}}{\sqrt{-(\tilde{\xi}-\tilde{\xi}_a)(\tilde{\xi}-\tilde{\xi}_b)}}\\ &= \left[\left(\frac{A}{B}-\frac{1}{c_\mathrm{h}}\right)\left(1-\frac{\mathcal{B}c_\mathrm{h}}{2\mathcal{F}}\right)\frac{1}{\sqrt{1-\frac{\mathcal{B}c_\mathrm{h}}{\mathcal{F}}+\frac{A-E_{0\ell}}{\mathcal{F}}c_\mathrm{h}^2}}\right.\\ &\quad\left.+\frac{A}{B}\left(\frac{\mathcal{B}}{2\mathcal{F}}\right)\frac{\sqrt{\mathcal{F}}}{\sqrt{\mathcal{A}-E_{0\ell}\mathcal{F}}}\right]\frac{\pi r_\mathrm{e}}{\alpha}\sqrt{\frac{2\mu\mathcal{F}}{\hbar^2}}\\ &= \frac{\pi r_\mathrm{e}}{\alpha}\sqrt{\frac{2\mu\mathcal{F}}{\hbar^2}}\left[\frac{1}{\mathcal{B}c_\mathrm{h}}\sqrt{\frac{2\mu\mathcal{F}}{\hbar^2}}+\frac{1}{c_\mathrm{h}}\right].\end{aligned} \tag{14}$$

Using the results in Eqs. (12) and (14) together with Eq. (1), we can find the energy eigenvalues equation for the sTW diatomic molecular potential as

$$\begin{aligned}E_{n\ell}^D &= \frac{\hbar^2\left(\tilde{\eta}^2-\frac{1}{4}\right)D_0}{2\mu r_\mathrm{e}^2}-\frac{\alpha^2\hbar^2}{2\mu r_\mathrm{e}^2}\\ &\quad\times\left[\frac{(\delta+n)^2+\frac{\left(\tilde{\eta}^2-\frac{1}{4}\right)}{\alpha^2 c_\mathrm{h}^2}(D_1c_\mathrm{h}-D_2)+\frac{2\mu V_\mathrm{e}r_\mathrm{e}^2}{\alpha^2\hbar^2}\left(1-\frac{1}{c_\mathrm{h}^2}\right)}{2(\delta+n)}\right]^2\\ &\text{with } \delta=\frac{1}{2}+\frac{1}{2}\sqrt{1+\frac{4}{c_\mathrm{h}^2}\left(\frac{D_2\left(\tilde{\eta}^2-\frac{1}{4}\right)}{\alpha^2}+\frac{2\mu V_\mathrm{e}r_\mathrm{e}^2}{\alpha^2\hbar^2}(1-c_\mathrm{h})^2\right)}.\end{aligned} \tag{15}$$

In three-dimensional ($D=3$) space, it can be reduced to the form

$$\begin{aligned}E_{n\ell} &= \frac{\hbar^2\ell(\ell+1)D_0}{2\mu r_\mathrm{e}^2}-\frac{\alpha^2\hbar^2}{2\mu r_\mathrm{e}^2}\\ &\quad\times\left[\frac{(\delta+n)^2+\frac{\ell(\ell+1)}{\alpha^2 c_\mathrm{h}^2}(D_1c_\mathrm{h}-D_2)+\frac{2\mu V_e r_e^2}{\alpha^2\hbar^2}\left(1-\frac{1}{c_h^2}\right)}{2(\delta+n)}\right]^2\\ &\text{with } \delta=\frac{1}{2}+\frac{1}{2}\sqrt{1+\frac{4}{c_h^2}\left(\frac{D_2\ell(\ell+1)}{\alpha^2}+\frac{2\mu V_e r_e^2}{\alpha^2\hbar^2}(1-c_h)^2\right)}.\end{aligned} \tag{16}$$

The Eigenfunctions

For the sake of completeness, we study the corresponding wavefunctions for this potential. For this purpose we introduce a new transformation of the form $t=e^{-b_\mathrm{h}(r-r_\mathrm{e})}\in(e^{\alpha},0)$ in Eq. (7), which maintained the finiteness of the transformed wave functions on the boundary conditions. Thus, we can find

$$\begin{aligned}&\frac{\mathrm{d}^2U_{n\ell}(t)}{\mathrm{d}t^2}+\frac{1}{t}\frac{\mathrm{d}U_{n\ell}(t)}{\mathrm{d}t}+\frac{1}{t^2(1-c_\mathrm{h}t)^2}\\ &\quad\times\left\{\left[\frac{2\mu r_\mathrm{e}^2E_{n\ell}}{\hbar^2\alpha^2}-\frac{\left(\tilde{\eta}^2-\frac{1}{4}\right)}{\alpha^2}D_0\right]\right.\\ &\quad+\left[-2c_\mathrm{h}\left(\frac{2\mu r_\mathrm{e}^2(E_{n\ell}+V_\mathrm{e})}{\alpha^2\hbar^2}-\frac{\left(\tilde{\eta}^2-\frac{1}{4}\right)}{\alpha^2}D_0\right)\right.\\ &\quad\left.+\frac{4\mu r_\mathrm{e}^2V_\mathrm{e}}{\hbar^2\alpha^2}-\frac{\left(\tilde{\eta}^2-\frac{1}{4}\right)}{\alpha^2}D_1\right]t\\ &\quad+\left[c_\mathrm{h}^2\left(\frac{2\mu r_\mathrm{e}^2(E_{n\ell}+V_\mathrm{e})}{\alpha^2\hbar^2}-\frac{\left(\tilde{\eta}^2-\frac{1}{4}\right)}{\alpha^2}D_0\right)\right.\\ &\quad\left.\left.+\frac{\left(\tilde{\eta}^2-\frac{1}{4}\right)}{\alpha^2}(D_1c_\mathrm{h}-D_2)-\frac{2\mu r_\mathrm{e}^2V_\mathrm{e}}{\hbar^2\alpha^2}\right]t^2\right\}U_{n\ell}(t)=0.\end{aligned} \tag{17}$$

Following the procedure described in Ref. [28], we can write the solution $U_{n\ell}(t)$ in terms of hypergeometric polynomials and thus the wave function takes the form

Table 1 Model parameters of the diatomic molecules studied in the present work

Molecules (states)	c_h	$\mu/10^{-23}$ (g)	b_h (Å^{-1})	r_e (Å)	D (cm^{-1})
NO $(a^4\Pi_i)$	0.0082003	1.249	2.408413	1.451	16,361
NO $(B^2\Pi_r)$	−0.482743	1.249	3.42650	1.428	22,722
NO $(L'^2\phi)$	−0.073021	1.249	2.73796	1.451	14,501
NO $(b^4\Sigma^-)$	−0.085078	1.249	3.01538	1.318	21,183
ICl $\left(X^1\Sigma_g^+\right)$	−0.086212	4.55237	2.008578	2.3209	17,557
ICl $(A^3\Pi_1)$	−0.167208	4.55237	2.542557	2.6850	3814.7
ICl $(A'^3\Pi_2)$	−0.157361	4.55237	2.373450	2.6650	4875

Table 2 The bound states energy eigenvalues ($D = 2$ and 3) for set of diatomic molecules for various n and rotational ℓ quantum numbers in sDF diatomic molecular potential

D	n	ℓ	NO $(a^4\Pi_i)$	NO $(B^2\Pi_r)$	NO $(L'^2\phi)$	NO $(b^4\Sigma^-)$	ICl $\left(X^1\Sigma_g^+\right)$	ICl $(A^3\Pi_1)$	ICl $(A'^3\Pi_2)$
	0	0	−1.971298585	−2.88233770	−1.855428285	−2.695026855	−2.200695845	−0.4861418795	−0.618432865
	1	0	−1.859308585	−3.01386965	−1.972859085	−2.834614855	−2.248819245	−0.5129225795	−0.646829015
		1	−1.859148585	−3.01373250	−1.972723585	−2.834451355	−2.248804845	−0.5129118495	−0.646818105
	2	0	−1.750608585	−3.14759050	−2.093631485	−2.977436555	−2.297420945	−0.5403157895	−0.675774025
		1	−1.750468585	−3.14745280	−2.093493685	−2.977270755	−2.297406745	−0.5403049695	−0.675763045
		2	−1.750098585	−3.14703970	−2.093081285	−2.976773655	−2.297363645	−0.5402725095	−0.675730195
2	3	0	−1.645178585	−3.28347662	−2.217724085	−3.123472855	−2.346500045	−0.5683148895	−0.705263165
		1	−1.645068585	−3.28333841	−2.217584185	−3.123304955	−2.346485745	−0.5683039695	−0.705252125
		2	−1.644678585	−3.28292375	−2.217165185	−3.122801255	−2.346442645	−0.5682712195	−0.705219005
		3	−1.644058585	−3.28223274	−2.216466785	−3.121961855	−2.346370345	−0.5682166495	−0.705163785
	4	0	−1.543078585	−3.42150558	−2.345115985	−3.272704955	−2.396055245	−0.5969133695	−0.735291785
		1	−1.542938585	−3.42136686	−2.344973985	−3.272534755	−2.396040945	−0.5969023495	−0.735280655
		2	−1.542598585	−3.42095075	−2.344548485	−3.272024355	−2.395997445	−0.5968693195	−0.735247275
		3	−1.541978585	−3.42025729	−2.343839285	−3.271173455	−2.395924945	−0.5968142695	−0.735191635
		4	−1.541118585	−3.41928647	−2.342846285	−3.269982555	−2.395823745	−0.5967371995	−0.735113745
	0	0	−1.971278585	−2.88230356	−1.855395085	−2.694986655	−2.200692145	−0.4861392095	−0.618430175
	1	0	−1.859278585	−3.01383536	−1.972825385	−2.834574055	−2.248815545	−0.5129198795	−0.646826285
		1	−1.859008585	−3.01356106	−1.972554585	−2.834247155	−2.248787045	−0.5128984495	−0.646804535
	2	0	−1.750568585	−3.14755607	−2.093596885	−2.977395155	−2.297417445	−0.5403130995	−0.675771295
		1	−1.750288585	−3.14728068	−2.093321985	−2.977063955	−2.297388745	−0.5402914395	−0.675749345
		2	−1.749808585	−3.14672991	−2.092771985	−2.976400855	−2.297331445	−0.5402481695	−0.675705525
3	3	0	−1.645158585	−3.28344206	−2.217689085	−3.123430855	−2.346496645	−0.5683121595	−0.705260395
		1	−1.644908585	−3.28316563	−2.217409585	−3.123095155	−2.346467645	−0.5682903295	−0.705238345
		2	−1.644418585	−3.28261279	−2.216850885	−3.122423255	−2.346410245	−0.5682466595	−0.705194145
		3	−1.643678585	−3.28178358	−2.216012785	−3.121415755	−2.346323545	−0.5681811695	−0.705127915
	4	0	−1.543028585	−3.42147091	−2.345080285	−3.272662455	−2.396051945	−0.5969106095	−0.735288995
		1	−1.542798585	−3.42119348	−2.344796885	−3.272322355	−2.396022645	−0.5968885995	−0.735266735
		2	−1.542328585	−3.42063869	−2.344229385	−3.271641355	−2.395964745	−0.5968445495	−0.735222225
		3	−1.541588585	−3.41980654	−2.343378185	−3.270620455	−2.395877945	−0.5967784695	−0.735155455
		4	−1.540608585	−3.41869709	−2.342243285	−3.269259655	−2.395761945	−0.5966903895	−0.735066435

$$U_{n\ell}(\zeta) = N_{n\ell} e^{-p\alpha\zeta}(1 - c_{\rm h}e^{-\alpha\zeta})^q \, {}_2F_1(-n, n + 2(p + q); 2p + 1, c_{\rm h}e^{-\alpha\zeta}), \tag{18}$$

with

$$p = \sqrt{\left[\left(\frac{\tilde{\eta}^2 - \frac{1}{4}}{\alpha^2}\right)D_0 - \frac{2\mu r_{\rm e}^2}{\hbar^2\alpha^2}E_{n\ell}\right]} \quad \text{and}$$
$$q = \frac{1}{2}\left\{1 + \sqrt{1 + \frac{4}{c_{\rm h}^2}\left[\left(\frac{\tilde{\eta}^2 - \frac{1}{4}}{\alpha^2}\right)D_2 + \frac{2\mu r_{\rm e}^2 V_{\rm e}}{\hbar^2\alpha^2}(1 - c_{\rm h})^2\right]}\right\}, \tag{19}$$

where $N_{n\ell}$ is the normalization constant. For a further detail on the calculation of similar potential models solved using formula method, one is advised to refer to other work [28]

Results and conclusions

In this study, in an attempt to find a more suitable potential that simulate the atomic interaction in diatomic molecules, we suggested sTW diatomic molecular potential as a modification for the TW diatomic molecular potential. The bound state solution of this potential has been found in an arbitrary D-dimension via the improved exact quantization rule.

Further, using the spectroscopic parameters presented in Table 1 which are taken from Ref. [8], we computed rotational-vibrational energy spectrum of some diatomic molecules in 2,3,4,5-dimensions. The results are presented in Tables 2 and 3. In our numerical computations, we have used the following conversions: 1 amu = 931.494028

Table 3 The bound states energy eigenvalues (D = 4 and 5) for set of diatomic molecules for various n and rotational ℓ quantum numbers in sDF diatomic molecular potential

D	n	ℓ	NO ($a^4\Pi_i$)	NO ($B^2\Pi_r$)	NO ($L'^2\phi$)	NO ($b^4\Sigma^-$)	ICl ($X^1\Sigma_g^+$)	ICl ($A^3\Pi_1$)	ICl ($A'^3\Pi_2$)
	0	0	−1.971158585	−2.88220112	−1.855295085	−2.694865755	−2.200681645	−0.4861312495	−0.618422105
	1	0	−1.859148585	−3.01373250	−1.972723585	−2.834451355	−2.248804845	−0.5129118495	−0.646818105
		1	−1.858778585	−3.01332103	−1.972317885	−2.833961155	−2.248762245	−0.5128796695	−0.646785525
	2	0	−1.750468585	−3.14745280	−2.093493685	−2.977270755	−2.297406745	−0.5403049695	−0.675763045
		1	−1.750098585	−3.14703970	−2.093081285	−2.976773655	−2.297363645	−0.5402725095	−0.675730195
		2	−1.749438585	−3.14635128	−2.092393885	−2.975945655	−2.297292045	−0.5402183995	−0.675675405
3	3	0	−1.645068585	−3.28333841	−2.217584185	−3.123304955	−2.346485745	−0.5683039695	−0.705252125
		1	−1.644678585	−3.28292375	−2.217165185	−3.122801255	−2.346442645	−0.5682712195	−0.705219005
		2	−1.644058585	−3.28223274	−2.216466785	−3.121961855	−2.346370345	−0.5682166495	−0.705163785
		3	−1.643188585	−3.28126535	−2.215488785	−3.120786155	−2.346269545	−0.5681402395	−0.705086495
	4	0	−1.542938585	−3.42136686	−2.344973985	−3.272534755	−2.396040945	−0.5969023495	−0.735280655
		1	−1.542598585	−3.42095075	−2.344548485	−3.272024355	−2.395997445	−0.5968693195	−0.735247275
		2	−1.541978585	−3.42025729	−2.343839285	−3.271173455	−2.395924945	−0.5968142695	−0.735191635
		3	−1.541118585	−3.41928647	−2.342846285	−3.269982555	−2.395823745	−0.5967371995	−0.735113745
		4	−1.540048585	−3.41803836	−2.341569785	−3.268451355	−2.395693345	−0.5966380895	−0.735013565
	0	0	−1.970998585	−2.88203041	−1.855128785	−2.694664455	−2.200663845	−0.4861179795	−0.618408615
	1	0	−1.859008585	−3.01356106	−1.972554585	−2.834247155	−2.248787045	−0.5128984495	−0.646804535
		1	−1.858508585	−3.01301246	−1.972013385	−2.833593555	−2.248729945	−0.5128555495	−0.646761065
	2	0	−1.750288585	−3.14728068	−2.093321985	−2.977063955	−2.297388745	−0.5402914395	−0.675749345
		1	−1.749808585	−3.14672991	−2.092771985	−2.976400855	−2.297331445	−0.5402481695	−0.675705525
		2	−1.749048585	−3.14590380	−2.091946885	−2.975407055	−2.297245545	−0.5401832395	−0.675639815
4	3	0	−1.644908585	−3.28316563	−2.217409585	−3.123095155	−2.346467645	−0.5682903295	−0.705238345
		1	−1.644418585	−3.28261279	−2.216850885	−3.122423255	−2.346410245	−0.5682466595	−0.705194145
		2	−1.643678585	−3.28178358	−2.216012785	−3.121415755	−2.346323545	−0.5681811695	−0.705127915
		3	−1.642668585	−3.28067802	−2.214895385	−3.120072855	−2.346208245	−0.5680938395	−0.705039565
	4	0	−1.542798585	−3.42119348	−2.344796885	−3.272322355	−2.396022645	−0.5968885995	−0.735266735
		1	−1.542328585	−3.42063869	−2.344229385	−3.271641355	−2.395964745	−0.5968445495	−0.735222225
		2	−1.541588585	−3.41980654	−2.343378185	−3.270620455	−2.395877945	−0.5967784695	−0.735155455
		3	−1.540608585	−3.41869709	−2.342243285	−3.269259655	−2.395761945	−0.5966903895	−0.735066435
		4	−1.539388585	−3.41731030	−2.340824985	−3.267557955	−2.395617345	−0.5965802895	−0.734955135

MeV/c^2, 1 cm^{-1} = 1.239841875 × 10^{-4} eV, and $\hbar c = 1973.29 eV\text{Å}$.

From Eq. (15), it can be seen that two interdimensional states are degenerate whenever $(n, \ell, D) \rightarrow (n, \ell \pm 1, D \mp 2) \Rightarrow E^D_{n,\ell} = E^{(D\mp2)}_{n,\ell\pm1}$. Thus, a knowledge of $E^D_{n,\ell}$ for D = 2 to 5 provides the information necessary to find $E^D_{n,\ell}$ for other higher dimensions. For example, $E^{(2)}_{0,4} = E^{(4)}_{0,3} = E^{(6)}_{0,2} = E^{(8)}_{0,1}$. This is the same transformational invariance described for bound states of free atoms and molecules [29–31] and demonstrates the existence of interdimensional degeneracies among states of the confined Hulthén potential.

The advantage of the approach employed in this study is that it gives the eigenvalues through the calculation of two integral given by Eq. (2) and solving the resulting algebraic equation. Firstly, we can easily obtain the quantum correction by only considering the solution of the ground state of the quantum system since it is independent of the number of nodes of the wave function for exactly solvable quantum system. The general expressions obtained for the energy eigenvalues and wave functions can be easily reduced to the 3D space (D = 3) and for s-wave (i.e. $\ell = 0$ state). The EQR produce as good results as the PQR, however the procedure followed using PQR is more shorter and quick.

Acknowledgments We thank the kind referees for the positive enlightening comments and suggestions, which have greatly helped us in making improvements to this paper. In addition, BJF acknowledges eJDS (ICTP) and Prof. K. J. Oyewumi, for his encouragement.

References

1. Jia, C.S., Diao, Y.F., Liu, X.J., Wang, P.Q., Liu, J.Y., Zhang, G.D.: Equivalence of the Wei potential model and Tietz potential model for diatomic molecules. J. Chem. Phys. **137**, 014101 (2012)
2. Hua, W.: Four-parameter exactly solvable potential for diatomic molecules. Phys. Rev. A. **42**, 2524 (1990)
3. Tietz, T.: Potential energy function for diatomic molecules. J. Chem. Phys. **38**, 3036 (1963)
4. Wang, P.Q., Zhang, L.H., Jia, C.S., Liu, J.Y.: Equivalence of the three empirical potential energy models for diatomic molecules. J. Mol. Spect. **274**, 5 (2012)
5. Gordillo-Vázquez, F.J., Kunc, J.A.: Radial probability distributions of atoms in diatomic molecules represented by the rotating Morse and Tietz-Hua oscillators. J. Mol. Struct. (THEOCHEM) **425**, 263 (2012)
6. Falaye, B.J., Oyewumi, K.J., Ikhdair, S.M., Hamzavi, M.: Eigensolution techniques, their applications and Fisher's information entropy of the Tietz–Wei diatomic molecular model. Phys. Scr. **89**, 115204 (2014)
7. Sun, G.H, Dong, S.H.: Relativistic treatment of spinless particles subject to a Tietz-Wei oscillator. Commun. Theor. Phys. **58**, 195 (2012)
8. Gordillo-Vázquez, F.J., Kunc, J.A.: Statistical–mechanical calculations of thermal properties of diatomic gases. J. Appl. Phys. **84**, 4693 (1998)
9. Ikhdair, S.M.: Effective Schrödinger equation with general ordering ambiguity position-dependent mass Morse potential. Mol. Phys. **110**, 1415 (2012)
10. Serrano, F.A., Gu, X.Y., Dong, S.H.: Qiang-Dong proper quantization rule and its applications to exactly solvable quantum systems. J. Math. Phys. **51**, 082103 (2010)
11. Dong, S.H., Cruz-Irisson, M.: Energy spectrum for a modified Rosen-Morse potential solved by proper quantization rule and its thermodynamic properties. J. Math. Chem. **50**, 881 (2012)
12. Ma, Z.Q., Xu, B.W.: Quantum correction in exact quantization rules. Europhys. Lett. **69**, 685 (2005)
13. Ma, Z.Q., Xu, B.W.: Exact quantization rules for bound states of the Schrödinger equation. Int. J. Mod. Phys. E. **14**, 599 (2005)
14. Yang, C.N.: Monopoles in quantum field theory. In: Craigie, N.S., Goddard, P., Nahm, W. (eds.) Proceedings of the Monopole Meeting, Trieste, Italy, p. 237. World Scientific, Singapore (1982)
15. Gu, X.Y., Dong, S.H.: Energy spectrum of the Manning-Rosen potential including centrifugal term solved by exact and proper quantization rules. J. Math. Chem. **49**, 2053 (2011)
16. Kasri, Y., Chetouani, L.: Application of the exact quantization rule for some noncentral separable potentials. Can. J. Phys. **86**, 1083 (2008)
17. Ikhdair, S.M., Sever, R.: Exact quantization rule to the Kratzer-type potentials: an application to the diatomic molecules. J. Math. Chem. **45**, 1137 (2009)
18. Qiang, W.C., Zhou, R.S., Gao, Y.: Application of the exact quantization rule to the relativistic solution of the rotational Morse potential with pseudospin symmetry. J. Phys. A: Math. Theor. **40**, 1677 (2007)
19. Dong, S.H., Morales, D., Garcia-Ravelo, J.: Exact quantization rule and its applications to physical potentials. Int. J. Mod. Phys. E **16**, 189 (2007)
20. Qiang, W.C., Dong, S.H.: Arbitrary l-state solutions of the rotating Morse potential through the exact quantization rule method. Phys. Lett. A **363**, 169 (2007)
21. Gu, X.Y., Dong, S.H., Ma, Z.Q.: Energy spectra for modified Rosen–Morse potential solved by the exact quantization rule. J. Phys. A: Math. Theor. **42**, 035303 (2009)
22. Ikhdair, S.M., Abu-Hasna, J.: Quantization rule solution to the Hulthén potential in arbitrary dimension with a new approximate scheme for the centrifugal term. Phys. Scr. **83**, 025002 (2011)
23. Falaye, B.J., Ikhdair, S.M., Hamzavi, M.: Energy states of some diatomic molecules: the exact quantization rule approach. ZFN A. **70**, 85 (2015)
24. Ikhdair, S.M., Sever, R.: Any l-state solutions of the Woods–saxon potential in arbitrary dimensions within the new improved quantization rule. Int. J. Mod. Phys. A **25**, 3941 (2010)
25. Qiang, W.C., Dong, S.H.: Proper quantization rule. EPL **89**, 10003 (2010)
26. Serrano, F.A., Cruz-Irisson, M., Dong, S.H.: Proper quantization rule as a good candidate to semiclassical quantization rules. Ann. Phys. (Berlin) **523**, 771 (2011)
27. Falaye, B.J., Ikhdair, S.M., Hamzavi, M.: Spectroscopic study of some diatomic molecules via the proper quantization rule. J. Math. Chem. (2015). doi:10.1007/s10910-015-0491-9
28. Falaye, B.J., Ikhdair, S.M., Hamzavi, M.: Formula method for bound state problems. Few-Body Syst. **56**, 63 (2015)
29. Herrick, D.R.: Degeneracies in energy levels of quantum systems of variable dimensionality. J. Math. Phys. 16, 281 (1975)
30. Herrick, D.R., Stillinger, F.H.: Variable dimensionality in atoms and its effect on the ground state of the helium isoelectronic sequence. Phys. Rev. A **11**, 42 (1975)
31. Frantz, D.D., Herschbach, D.R.: Interdimensional degeneracy and symmetry breaking in *D*-dimensional H_2^+. The J. Chem. Phys. 92 (11), 6668 (1990)

20

Generic scalar potentials in geometric scalar gravity

Nahomi Kan[1] · Kiyoshi Shiraishi[2]

Abstract We discuss a generic form of the scalar potential appearing in the geometric scalar theory of gravity. We find the conditions on the potential by considering weak and strong gravity. The modified black hole solutions are obtained for generic potentials and the inverse problems on a black hole and on a spherical body ('pseudo-gravastar') are investigated.

Keywords Modified gravity · Black holes · Newtonian approximation · Compact objects

Introduction

General relativity (GR) is a unique theory of gravity which passes various experimental tests, at least to our present knowledge. The modifications of GR are, however, studied by many researchers for the reason that they may resolve the cosmological riddles about dark matter and/or dark energy [1–3].

In recent years, Novello and collaborators advocated a novel theory of gravity, called geometric scalar gravity (GSG) [4–8]. In this theory, the dynamic field is a single scalar field, but a normalized derivative of the scalar field expresses a part of the dynamical metric, as well as the scalar field itself. Therefore, it is possible to avoid the difficulty in old scalar gravitation theories not including derivative terms [9–12].[1]

Novello is also insisting [13] that more scalar degrees of freedom are necessary to explain the gravitational field around the spinning source more practically. However, it is a challenging problem for theorists to clarify how GSG, as the simplest model or as a case where many scalar degrees of freedom freeze, can be adjusted to realistic gravity.

As a modified theory of gravity, mimetic gravity [14] has been proposed, where a normalized scalar derivative term is also used for the dynamical metric. If one would construct such a modified theory of gravity from tensor and scalar fields, it is worth examining the behavior of the scalar model in the limit of a few degree of freedom and pursuing how many aspects can be found through the study of the simplest theory. Based on the research, we could advance the study of theories including a vector field, such as TeVeS [15], and more complicated theories.

Novello and collaborators chose the specific form for the potential of the scalar field from which the Schwarzschild spacetime is derived as a spherically symmetric vacuum solution, and further advanced towards the discussion of the (exotic) cosmology based on it [4–8]. However, because the dynamics of gravity in GSG is completely different from GR, so we need not require perfect coincidence of the solution with an exact solution of GR. Actually, we have to check observational validity through description of the parametrized post-Newton treatment of the weak gravitational field (see for example, [16, 17]). Therefore, there may be a finite ambiguity in the form of the permissible potential. At first, we pay attention to this

✉ Kiyoshi Shiraishi
shiraish@yamaguchi-u.ac.jp

Nahomi Kan
kan@gifu-nct.ac.jp

1 National Institute of Technology, Gifu College, Motosu-shi, Gifu 501-0495, Japan

2 Graduate School of Science and Engineering, Yamaguchi University, Yamaguchi-shi, Yamaguchi 753-8512, Japan

[1] For older many works, please see the references therein.

point and expose the necessary condition that the potential should satisfy.

Another important issue of the non-linear theory of gravity is avoidance of the singularity. A singular point is concealed by a horizon of a black hole in GR. There is room of argument for whether GSG can also be used without a correction in the strong gravity, but we wish to know what kind of the potential provides a spherically symmetric black hole solution with a horizon in GSG.

From these two discussions about weak and strong gravity, we consider possible forms of the scalar potential and some restrictions on GSG.

This paper is organized as follows. In "A brief review of GSG", we begin by reviewing GSG to make this paper self-contained. Then, we consider a spherical gravitational field in GSG and clarify the condition on the relation between the scalar potential and parametrized post-Newtonian description in "A static spherical solution and the post-Newtonian parameters". A possible singularity in the spherical vacuum solution in GSG is studied in "A static spherical vacuum solution and singularity". The condition on the scalar potential for emerging a horizon is revealed. In "Inverse problems in GSG", the potential which leads to a slightly modified Schwarzschild solution is obtained, such as an inverse problem. In this section, we also consider the spherical body and investigate behaviors of the energy density and the pressure of the static fluid for a given de Sitter-like spacetime as another inverse problem. Finally, we give summary and prospects in "Summary and prospects".

A brief review of GSG

In this section, we briefly review GSG [4] to make the present paper self-contained. In GSG, the physical metric $q_{\mu\nu}$ is described by a scalar field Φ as

$$q_{\mu\nu} = e^{2\Phi}\left[\eta_{\mu\nu} - \frac{e^{-4\Phi}V(\Phi) - 1}{e^{-4\Phi}V(\Phi)}\frac{\partial_\mu\Phi\partial_\nu\Phi}{w}\right], \tag{2.1}$$

where

$$w \equiv \eta^{\mu\nu}\partial_\mu\Phi\partial_\nu\Phi, \tag{2.2}$$

and $\eta_{\mu\nu}$ is a flat Minkowski metric with the signature $(+\,-\,-\,-)$. The inverse of the metric is then written as

$$q^{\mu\nu} = e^{-2\Phi}\left[\eta^{\mu\nu} + \frac{e^{-4\Phi}V(\Phi) - 1}{w}\eta^{\mu\rho}\eta^{\nu\sigma}\partial_\rho\Phi\partial_\sigma\Phi\right], \tag{2.3}$$

and note that

$$\sqrt{-q} = \sqrt{-\det q_{\mu\nu}} = \frac{e^{6\Phi}}{\sqrt{V(\Phi)}}\sqrt{-\eta}. \tag{2.4}$$

We should also note that $q^{\mu\nu}\partial_\nu\Phi = e^{-6\Phi}V(\Phi)\eta^{\mu\nu}\partial_\nu\Phi$.

We call $V(\Phi)$ the (scalar) potential in the present paper. This function determines the dynamics of the scalar Φ, and only the property of gravity constrains its concrete form, as far as we do not assume any symmetry argument.

The action governing the dynamics of the scalar field Φ with a potential $V(\Phi)$ is given by

$$S_\Phi = \frac{1}{\kappa}\int\sqrt{-q}\sqrt{V(\Phi)}q^{\mu\nu}\partial_\mu\Phi\partial_\nu\Phi \mathrm{d}^4x, \tag{2.5}$$

where κ is a constant. The variation of the action with respect to the scalar Φ is calculated as

$$\delta S_\Phi = -\frac{2}{\kappa}\int\sqrt{-q}\sqrt{V(\Phi)}(\Box\Phi)\delta\Phi \mathrm{d}^4x, \tag{2.6}$$

where

$$\begin{aligned}\Box\Phi &\equiv \frac{1}{\sqrt{-q}}\partial_\mu(\sqrt{-q}q^{\mu\nu}\partial_\nu\Phi)\\ &= e^{-6\Phi}V(\Phi)\left[\frac{1}{\sqrt{-\eta}}\partial_\mu(\sqrt{-\eta}\eta^{\mu\nu}\partial_\nu\Phi) + \frac{w}{2}\frac{\mathrm{d}}{\mathrm{d}\Phi}\ln V(\Phi)\right].\end{aligned} \tag{2.7}$$

The action for matter fields is presumed as

$$S_m = \int\sqrt{-q}\mathcal{L}_m \mathrm{d}^4x, \tag{2.8}$$

and its variation with respect to the metric is written as follows:

$$\delta S_m = -\frac{1}{2}\int\sqrt{-q}\,T^{\mu\nu}\delta q_{\mu\nu}\mathrm{d}^4x, \tag{2.9}$$

where

$$T_{\mu\nu} \equiv \frac{2}{\sqrt{-q}}\frac{\partial\sqrt{-q}\mathcal{L}_m}{\partial q^{\mu\nu}} \tag{2.10}$$

is the energy–momentum tensor of matter fields. Some manipulation leads to the following result on the variation with respect to the scalar Φ:

$$T^{\mu\nu}\delta q_{\mu\nu} = \left[2T + \left(4 - \frac{1}{V}\frac{\mathrm{d}V}{\mathrm{d}\Phi}\right)E\right]\delta\Phi - 2C^\lambda\partial_\lambda\delta\Phi, \tag{2.11}$$

where

$$\begin{aligned}&T = T^{\mu\nu}q_{\mu\nu}, \quad E = \frac{T^{\mu\nu}\partial_\mu\Phi\partial_\nu\Phi}{\Omega},\\ &C^\lambda = \frac{e^{-4\Phi}V - 1}{\Omega}(T^{\lambda\nu} - Eq^{\lambda\nu})\partial_\nu\Phi,\end{aligned} \tag{2.12}$$

with

$$\Omega \equiv q^{\mu\nu}\partial_\mu\Phi\partial_\nu\Phi = e^{-6\Phi}Vw. \tag{2.13}$$

We then obtain the variation of the action for matter fields with respect to Φ as

$$\delta S_m = -\int \left[T + \left(2 - \frac{1}{2V}\frac{\mathrm{d}V}{\mathrm{d}\Phi}\right)E + \nabla_\lambda C^\lambda\right]\delta\Phi\sqrt{-q}\mathrm{d}^4x, \tag{2.14}$$

where

$$\nabla_\lambda C^\lambda \equiv \frac{1}{\sqrt{-q}}(\partial_\lambda\sqrt{-q}\,C^\lambda). \tag{2.15}$$

Now we define the total action of the gravitating system as

$$S = S_\phi + S_m. \tag{2.16}$$

The equation derived from this total action can be written as

$$\sqrt{V(\Phi)}\Box\Phi = \kappa\chi, \tag{2.17}$$

where

$$\chi = -\frac{1}{2}\left[T + \left(2 - \frac{1}{2V}\frac{\mathrm{d}V}{\mathrm{d}\Phi}\right)E + \nabla_\lambda C^\lambda\right]. \tag{2.18}$$

Let us consider the Newtonian limit. The gravitating source can be estimated as

$$T_{00} \approx \rho, \tag{2.19}$$

where ρ is the energy density and the other components of the energy–momentum tensor $T_{\mu\nu}$ can be neglected.

Novello et al. then assumed

$$q_{00} = e^{2\Phi} \approx 1 + 2\Phi_N, \tag{2.20}$$

where Φ_N denotes the static Newtonian potential. The other elements of the metric tensor are taken as those of the Minkowski metric.

We would like to consider the case with the non-zero asymptotic value of the scalar field Φ_∞. The redefinition of the time coordinate yields the metric component $q_{00} = e^{2\Phi_\infty} \to 1$ in the asymptotic region. In this case, since $|\Phi - \Phi_\infty| \approx |\Phi_N| \ll 1$, we find that the dynamical equation (2.17) can be approximated as

$$\sqrt{V(\Phi_\infty)}\nabla^2\Phi = \frac{\kappa}{2}\rho. \tag{2.21}$$

Thus, if we set, with the definition of the Newton constant G

$$\frac{\kappa}{\sqrt{V(\Phi_\infty)}} = 8\pi G, \tag{2.22}$$

we have a correct Newtonian limit.

A static spherical solution and the post-Newtonian parameters

Novello et al. considered a static spherical solution in GSG [4]. We are interested in generic spherically symmetric solutions of this theory. We start with their argument in this section. First, they considered the flat metric in the spherical coordinates,

$$\eta_{\mu\nu}\mathrm{d}x^\mu\mathrm{d}x^\nu = \mathrm{d}T^2 - \mathrm{d}R^2 - R^2\mathrm{d}\Omega^2, \tag{3.1}$$

with $\mathrm{d}\Omega^2$ is the line element on a unit sphere. Because the spherical symmetry enforces that $\Phi = \Phi(R)$, the physical line element is given as

$$\mathrm{d}s^2 = q_{\mu\nu}\mathrm{d}x^\mu\mathrm{d}x^\nu = e^{2\Phi}\mathrm{d}T^2 - \frac{e^{6\Phi}}{V(\Phi)}\mathrm{d}R^2 - e^{2\Phi}R^2\mathrm{d}\Omega^2. \tag{3.2}$$

Now, defining $t \equiv e^{\Phi_\infty}T$ and converting the radial coordinate to $r \equiv e^{\Phi}R$, we find

$$\mathrm{d}s^2 = B(r)\mathrm{d}t^2 - A(r)\mathrm{d}r^2 - r^2\mathrm{d}\Omega^2, \tag{3.3}$$

where

$$B(r) = e^{2(\Phi-\Phi_\infty)}, \tag{3.4}$$

$$A(r) = \frac{e^{4\Phi}}{V(\Phi)}\left(1 - r\frac{\mathrm{d}\Phi}{\mathrm{d}r}\right)^2. \tag{3.5}$$

To obtain an asymptotically flat spacetime, we assume

$$\lim_{r\to\infty}\Phi(r) = \Phi_\infty, \quad \lim_{r\to\infty} r\Phi'(r) = 0, \quad e^{-4\Phi_\infty}V(\Phi_\infty) = 1, \tag{3.6}$$

where $\Phi'(r)$ is the first derivative of $\Phi(r)$.

The equation of motion in vacuum $\Box\Phi = 0$ implies

$$\frac{\mathrm{d}}{\mathrm{d}r}\left[\sqrt{\frac{B(r)}{A(r)}}r^2\frac{\mathrm{d}\Phi}{\mathrm{d}r}\right] = 0. \tag{3.7}$$

Novello et al. [4] insisted that the form for $V(\Phi)$ is chosen so as to realize the exact Schwarzschild solution. So, they chose

$$\Phi_\infty = 0, \quad V(\Phi) = V_{\mathrm{N}}(\Phi) = \frac{1}{4}e^{2\Phi}(1 - 3e^{2\Phi})^2. \tag{3.8}$$

This criterion for the choice of $V(\Phi)$ is too severe for a model of weak gravity, we think. The Newtonian limit has been examined in the previous section. We study here the post-Newtonian limit. The possible asymptotic behavior determines the form of the scalar field as

$$\Phi_N \equiv \Phi - \Phi_\infty = -\frac{GM}{r} - g\frac{G^2M^2}{r^2} + O((GM/r)^3), \tag{3.9}$$

where M is the gravitational mass and g is a constant of dimensionless. The potential $V(\Phi)$ is assumed up to the linear order in Φ_N, corresponding to the perturbative regime and we set

$$e^{-4\Phi}V(\Phi) = 1 + 4k\Phi_N + O(\Phi_N^2). \tag{3.10}$$

Then the metric components $B(r)$ and $A(r)$ are expressed in terms of the parameters as

$$B(r) = 1 - \frac{2GM}{r} + \frac{2(1-g)G^2M^2}{r^2} + O((GM/r)^3), \tag{3.11}$$

$$A(r) = 1 + \frac{2(2k-1)GM}{r} + O((GM/r)^2). \tag{3.12}$$

and we then find that

$$\sqrt{\frac{B(r)}{A(r)}}r^2\frac{d\Phi}{dr} = GM + \frac{2(g-k)G^2M^2}{r} + O(GM(GM/r)^2). \tag{3.13}$$

Therefore up to the order in the present consideration, the combination of the dynamic Eqs. (3.7) and (3.13) tells us $g = k$, so we obtain $B(r) = 1 - \frac{2GM}{r} + \frac{2(1-k)G^2M^2}{r^2} + O((GM/r)^3)$.

According to the metric form with two post-Newtonian parameters β and γ [16, 17]

$$B(r) = 1 - \frac{2GM}{r} + \frac{2(\beta-\gamma)G^2M^2}{r^2} + O((GM/r)^3), \tag{3.14}$$

$$A(r) = 1 + \frac{2\gamma GM}{r} + O((GM/r)^2), \tag{3.15}$$

we can express two of the post-Newtonian parameters β and γ by using the linear coefficient k in $V(\Phi)$ as

$$\beta = k, \quad \gamma = 2k - 1. \tag{3.16}$$

For the experimentally viable values $\beta = \gamma = 1$ [16, 17], we should take $k = 1$. Of course, the expansion of the potential chosen by Novello et al. (3.8) shows that it is in the case.

To summarize, we should select the form of the scalar potential $V(\Phi)$ as

$$e^{-4\Phi}V(\Phi) = 1 + 4\Phi_N + O(\Phi_N^2), \tag{3.17}$$

with $\Phi_N = \Phi - \Phi_\infty$ for viable post-Newtonian parameters β and γ. The equivalence of the gravitational and the inertial mass of the spherical gravitating body then holds. Note that (3.17) is equivalent to

$$\frac{1}{V(\Phi)}\frac{dV(\Phi)}{d\Phi} = 8 \quad \text{at } \Phi = \Phi_\infty. \tag{3.18}$$

Since a condition for the potential is obtained now, we also try to investigate the asymptotic gravitational field around an electric point charge in GSG. We introduce a point charge and consider the spherically symmetric solution. The Lagrangian density for the electromagnetic field is given by

$$\mathcal{L}_{EM} = -\frac{1}{16\pi}F_{\mu\nu}F^{\mu\nu}, \tag{3.19}$$

where $F_{\mu\nu} \equiv \partial_\mu A_\nu - \partial_\nu A_\mu$. The energy–momentum tensor obtained from this Lagrangian is

$$T_{\mu\nu} = -\frac{1}{4\pi}\left(F^2_{\mu\nu} - \frac{1}{4}F^2 q_{\mu\nu}\right), \tag{3.20}$$

where $F^2_{\mu\nu} \equiv F_{\mu\rho}F_{\nu\sigma}q^{\rho\sigma}$ and $F^2 \equiv F^2_{\mu\nu}q^{\mu\nu}$.

A spherically symmetric solution for the equation of motion $\nabla_\mu F^{\mu\nu} = 0$ is given by

$$F_{0r} = \frac{Q}{r^2}\sqrt{B(r)A(r)}, \tag{3.21}$$

where Q is a constant corresponding to the electric charge and we use the standard metric (3.3). Then we find

$$T = 0, \quad E = \frac{1}{8\pi}\frac{Q^2}{r^4}, \quad C^\lambda = 0, \tag{3.22}$$

in χ (2.18) in the right-hand side of the equation for the scalar field (2.17). The equation of motion for the scalar field $\Phi(r)$ becomes

$$\sqrt{\frac{V(\Phi(r))}{A(r)B(r)}}\frac{1}{r^2}\frac{d}{dr}\left[\sqrt{\frac{B(r)}{A(r)}}r^2\frac{d\Phi(r)}{dr}\right] = \frac{\kappa}{16\pi}\left(2 - \frac{1}{2V}\frac{dV}{d\Phi}\right)\frac{Q^2}{r^4}, \tag{3.23}$$

In the asymptotic region of large r, using (3.17), we can approximate the equation as

$$\sqrt{\frac{V(\Phi_\infty)}{A(r)B(r)}}\frac{1}{r^2}\frac{d}{dr}\left[\sqrt{\frac{B(r)}{A(r)}}r^2\frac{d\Phi(r)}{dr}\right] \approx -\frac{\kappa}{8\pi}\frac{Q^2}{r^4}, \tag{3.24}$$

and we can also find an approximate solution

$$B(r) = 1 - \frac{2GM}{r} - \frac{GQ^2}{r^2} + O((GM/r)^3). \tag{3.25}$$

Unfortunately, this does not coincide with the Reissner–Nordström solution in GR, because of the wrong sign in front of the r^{-2} term. We should note that there is no reason to match exactly the solutions to GSG and GR.

Although we do not pursue the charged solution in GSG further in this paper, we suppose that the presence of non-minimal couplings between the vector and scalar fields could change the asymptotic behavior and it will be worth studying this possibility.

A static spherical vacuum solution and singularity

As we have seen in the previous section, the vacuum solution with spherical symmetry satisfies

$$\sqrt{\frac{b(r)}{A(r)}}r^2\frac{\mathrm{d}\Phi}{\mathrm{d}r}=\frac{1}{2}\frac{1}{\sqrt{A(r)b(r)}}r^2\frac{\mathrm{d}b(r)}{\mathrm{d}r}=GMe^{\Phi_\infty}>0, \tag{4.1}$$

where $b(r)\equiv e^{2\Phi(r)}$ and $A(r)$ has been defined as (3.5). In this section, we start with undetermined asymptotic value for the scalar, Φ_∞. The Eq. (4.1) means that $q_{00}=B(r)=b(r)e^{-2\Phi_\infty}$ is a monotonically increasing function of r. Moreover, if the value of b is positive everywhere, the above equation can be rewritten as

$$\frac{\sqrt{V(\Phi(r))/b(r)}}{|2b(r)-rb'(r)|}r^2b'(r)=GMe^{\Phi_\infty}, \tag{4.2}$$

where the prime ($'$) stands for the derivative with respect to r.

If we assume that $b>0$ and $\sqrt{V(\Phi(r))/b(r)}$ is finite in the vicinity of $r=0$, we see that

$$\frac{b'(r)}{b(r)}\to\infty \quad \text{as} \quad r\to 0. \tag{4.3}$$

(and also $\Phi'(r)\to\infty$) from Eq. (4.2). As far as we assume a non-singular potential ($V(\Phi)<\infty$ for $-\infty<\Phi\leq\Phi_\infty$) and non-singular metric $q_{00}\propto b$ for $r>0$, the vacuum solution has a singularity at the origin. This fact might bring about a possibility that a naked singularity emerges after a possible gravitational collapse.[2] The difficulty of obtaining non-singular vacuum solution forces us to examine the solution of which metric becomes singular at finite r, which can be regarded as a horizon radius.[3] If we set $b(r_g)=0$ and $b'(r_g)\neq 0$ at the radius (of infinite-red-shift) $r=r_g>0$, Eq. (4.1) tells us that

$$0<\lim_{r\to r_g}\sqrt{A(r)b(r)}<\infty. \tag{4.4}$$

This condition is a necessary condition that the radius $r=r_g$ can be the horizon radius. Then, because $\lim_{r\to r_g}\Phi(r)=-\infty$, it should be satisfied that

$$0<\lim_{\Phi\to-\infty}\sqrt{V(\Phi)/b(r)}<\infty. \tag{4.5}$$

Further observation leads to that fact that the specific combination $2-\frac{rb'}{b}$ in (4.2), which comes from $A(r)$, becomes negative at $r=r_g$. Since this must be positive for a large r, we conclude that the quantity $2-\frac{rB'}{B}$ vanishes at a certain radius $r=r_0>r_g$. This condition requires

$$V(\Phi(r_0))=0, \tag{4.6}$$

because the both sides of (4.2) should be finite. Therefore the differential equation (4.2) for $b(r)$ is rewritten as

$$\frac{r}{2}\frac{b'(r)}{b(r)}=\left(1-r\frac{\sqrt{V(\Phi(r))/b(r)}}{GMe^{\Phi_\infty}}\right)^{-1} \quad \text{for } r_g<r<r_0, \tag{4.7}$$

$$\frac{r}{2}\frac{b'(r)}{b(r)}=\left(1+r\frac{\sqrt{V(\Phi(r))/b(r)}}{GMe^{\Phi_\infty}}\right)^{-1} \quad \text{for } r>r_0, \tag{4.8}$$

In order to obtain the smooth function (such as $b(r)\in C^\infty$), it should be assumed that

$$V(\Phi)/b=[F(b)]^2, \tag{4.9}$$

and $F(b)$ has a vanishing point, according to Eq. (4.6). Then the horizon radius r_g is determined as

$$\frac{GMe^{\Phi_\infty}}{r_g}=|F(0)|\equiv F_0. \tag{4.10}$$

We will examine the case that the equation has exact analytic solutions. In the simplest case, F is linear in b such that

$$F(b)=F_0(1-f_1b), \tag{4.11}$$

where F_0 and f_1 are positive constants. Adopting a new variable $y\equiv 1-r_g/r$, the differential equation for $b(y)$ becomes

$$\frac{\mathrm{d}b}{\mathrm{d}y}=\frac{2b}{-y+1-F(b)/F_0}=\frac{2b}{-y+f_1b}, \tag{4.12}$$

and the boundary condition is $b(0)=0$. The solution is

$$b=\frac{3}{f_1}y=\frac{3}{f_1}\left(1-\frac{r_g}{r}\right). \tag{4.13}$$

Then, we find $e^{2\Phi_\infty}=b(1)=\frac{3}{f_1}$ and

$$B(r)=e^{2(\Phi-\Phi_\infty)}=1-\frac{r_g}{r}, \tag{4.14}$$

$$A(r)=\frac{3}{4F_0^2f_1}\left(1-\frac{r_g}{r}\right)^{-1}. \tag{4.15}$$

To obtain asymptotic flat spacetime, we must choose

$$F_0^2=\frac{3}{4f_1}. \tag{4.16}$$

Then, remembering $b=e^{2\Phi}$, the scalar potential has the form

$$\begin{aligned}V(\Phi)&=\frac{3}{4f_1}e^{2\Phi}\left(1-f_1e^{2\Phi}\right)^2\\&=\frac{e^{4\Phi_\infty}}{4}e^{2(\Phi-\Phi_\infty)}\left[1-3e^{2(\Phi-\Phi_\infty)}\right]^2,\end{aligned} \tag{4.17}$$

[2] The time-dependent dynamics in GSG has not been studied, however. We should study the gravitational collapse in GSG.

[3] The existence of naked singularity in nature is still controversial, however. For example, see Ref. [18–20].

and $V(\Phi_\infty) = e^{4\Phi_\infty}$. Thus, the post-Newtonian condition (3.17) is satisfied. The spacetime is the Schwarzschild spacetime, and thus $\beta = \gamma = 1$. The Newton limit also tells $r_g = 2GM$, where $G = \kappa/(8\pi\sqrt{V(\Phi_\infty)}) = f_1\kappa/(24\pi)$. The potential (4.17) is essentially the Novello's potential $V_N(\Phi)$, provided that $f_1 = 3$ (which implies $\Phi_\infty = 0$).

Incidentally, choosing the function $F(b)$ as

$$F(b) = F_0(1 - f_p b^p), \tag{4.18}$$

where f_p is a constant, the differential equation for $b(y)$ becomes

$$\frac{db}{dy} = \frac{2b}{-y + f_p b^p}, \tag{4.19}$$

and then the equation is analytically solved as

$$b = \left(\frac{2p+1}{f_p}\right)^{1/p} y^{1/p} = \left(\frac{2p+1}{f_p}\right)^{1/p} \left(1 - \frac{r_g}{r}\right)^{1/p}, \tag{4.20}$$

and $e^{2\Phi_\infty} = \left(\frac{2p+1}{f_p}\right)^{1/p}$. The components of the metric are then

$$B(r) = e^{2(\Phi - \Phi_\infty)} = \left(1 - \frac{r_g}{r}\right)^{1/p}, \tag{4.21}$$

$$A(r) = \frac{1}{4pF_0^2}\left(\frac{2p+1}{f_p}\right)^{1/p}\left(1 - \frac{r_g}{r}\right)^{-2+1/p}. \tag{4.22}$$

If we choose $\frac{1}{4pF_0^2}\left(\frac{2p+1}{f_p}\right)^{1/p} = 1$, the solution describes an asymptotically flat spacetime, i.e.,

$$ds^2 = \left(1 - \frac{r_g}{r}\right)^{1/p} dt^2 - \left(1 - \frac{r_g}{r}\right)^{-2+1/p} - r^2 d\Omega^2. \tag{4.23}$$

The location $r = r_g$ is, however, a singularity of the metric (4.23), which comes from the fact that $db/dr|_{r=r_g}$ does not take a finite value (i.e., $b'(0) = 0$ or $b'(0) = \infty$, where $b'(y)$ is the first derivative of $b(y)$).

The physical condition $\beta = \gamma$ adopted in the previous section, i.e., the absence of $1/r^2$ term in the expansion of $B(r)$ in terms of powers of $1/r$, is transformed as $\lim_{y\to 1} b''(y) = 0$, where $b''(y)$ is the second derivative of $b(y)$. Moreover, we can rewrite the condition (3.18), using the differential equation for $b(y)$, as

$$2b(1)F'(b(1)) = 3F(b(1)), \tag{4.24}$$

where $F'(b)$ is the first derivative of $F(b)$.

We further quest another possible potential, which leads to the existence of a horizon as well as the nice behavior at weak gravity, $\beta = \gamma = 1$. We assume, with constants f_1, f_2 and f_3,

$$F(b) = F_0(1 - f_1 b - f_2 b^2 - f_3 b^3). \tag{4.25}$$

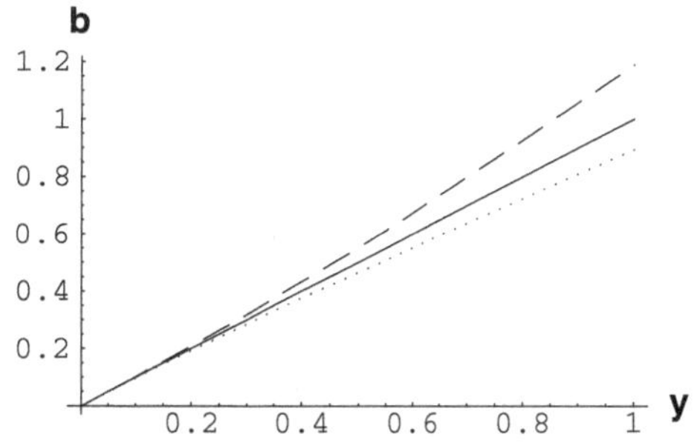

Fig. 1 The solutions for $b(y)$ are plotted against y. The *rigid line* exhibits the case with $f_1 = 3$, $f_2 = 0$ and $f_3 = 0$. The *dotted line* exhibits the case with $f_1 = 3$, $f_2 = 1$ and $f_3 \approx -0.52226$. The *broken line* indicates the case with $f_1 = 3$, $f_2 = -1$ and $f_3 \approx 0.39273$

To satisfy $\lim_{y\to 1} b''(y) = 0$, we must choose a set of parameters, f_1, f_2 and f_3. A set of parameters can be found numerically as $f_1 = 3$, $f_2 = 1$ and $f_3 \approx -0.52226$. Then, $b(1) \approx 0.893544$ and $F_0 \approx 0.448751$. Therefore, the horizon radius is $r_g \approx 2.10645GM$. Another set is $f_1 = 3$, $f_2 = -1$ and $f_3 \approx 0.39273$. Then, $b(1) \approx 1.18827$. and $F_0 \approx 0.601671$. The horizon radius is $r_g \approx 1.81175GM$ in this case. The solutions for $b(y)$ are displayed in Fig. 1. We find that the horizon radius can become larger or smaller than $2GM$ by selecting the parameters, i.e., the form of the scalar potential $V(\Phi)$.

Inverse problems in GSG

An inverse problem on black holes

By now, we have found the conditions for the scalar potential $V(\Phi)$. Vast ambiguity in the form of the potential remains with the conditions. In this section, we consider an 'inverse problem', which require the potential form to obtain a given simple metric component. We assume the simplest 'solution' which is permissible up to the post-Newtonian order,

$$b = \frac{3}{f}\left[y + h\left(y^2 - \frac{1}{3}y^3\right)\right], \tag{5.1}$$

where f (>0) and h (>-1) are constants. Since $y = 1 - r_g/r$, we have

$$b = e^{2\Phi} = e^{2\Phi_\infty}\left[1 - \frac{2GM}{r} + \frac{4}{3}\delta\frac{(2GM)^3}{r^3}\right], \tag{5.2}$$

where

$$e^{2\Phi_\infty} = \frac{3}{f}\left(1 + \frac{2}{3}h\right), \quad \delta = \frac{h}{4}\frac{\left(1 + \frac{2}{3}h\right)^2}{(1+h)^3}, \tag{5.3}$$

$$\frac{r_g}{2GM} = \frac{e^{\Phi_\infty}}{2F_0} = \frac{1 + \frac{2}{3}h}{1+h}, \quad F_0^2 = \frac{3}{4f}\frac{(1+h)^2}{1 + \frac{2}{3}h}. \tag{5.4}$$

If we solve the Eq. (5.2) for $2GM\,/\,r$, we find

$$\frac{2GM}{r}=\frac{1}{\sqrt{\delta}}\sin\left[\frac{1}{3}\arcsin\left\{3\sqrt{\delta}(1-e^{-2\Phi_\infty}b)\right\}\right]\qquad\text{for } \delta>0, \tag{5.5}$$

$$\frac{2GM}{r}=\frac{1}{\sqrt{|\delta|}}\sinh\left[\frac{1}{3}\ \text{arcsinh}\ \left\{3\sqrt{|\delta|}(1-e^{-2\Phi_\infty}b)\right\}\right]\quad\text{for } \delta<0. \tag{5.6}$$

The function $F(b)$ defined in the previous section can be rewritten as

$$F(b)=F_0(1-f\varphi(b)), \tag{5.7}$$

and then we find that the function $\varphi(b)$ satisfies $\varphi'(0)=1$. Substituting the 'solution' into the equation of motion, we get

$$f\varphi(b)=Y(b)+\frac{2fb}{3\left[1+h(2Y(b)-Y(b)^2)\right]}, \tag{5.8}$$

where

$$Y(b)=1-\frac{2\sqrt{1+h}}{\sqrt{h}}\sin\left[\frac{1}{3}\arcsin\left\{\frac{3\sqrt{h}\left(1+\frac{2}{3}h\right)}{2(1+h)^{3/2}}\right.\right.$$
$$\left.\left.\times\, 1-\frac{f}{3+2h}b\Big)\right\}\right]\qquad\text{for } h>0, \tag{5.9}$$

$$Y(b)=1-\frac{2\sqrt{1+h}}{\sqrt{|h|}}\sinh\left[\frac{1}{3}\ \text{arcsinh}\ \left\{\frac{3\sqrt{|h|}\left(1+\frac{2}{3}h\right)}{2(1+h)^{3/2}}\right.\right.$$
$$\left.\left.\times\left(1-\frac{f}{3+2h}b\right)\right\}\right]\qquad\text{for } h<0. \tag{5.10}$$

In the limit of $h\to 0$, we obtain $f\varphi(b)=fb$ as we have already known. On the other hand, another limiting case $h\to\infty$ yields, if we set $f=3+2h$ for further simplicity, $r_g=\frac{4GM}{3}$ and one can find

$$b=1-\frac{2GM}{r}+\frac{4}{27}\frac{(2GM)^3}{r^3}=1-\frac{3r_g}{2r}+\frac{r_g^3}{2r^3}=\frac{3}{2}y^2-\frac{1}{3}y^3. \tag{5.11}$$

The shape of the potential is shown in Fig. 2. Unfortunately the potential varies very slightly with h in the present assumption for parametrization.

An inverse problem on a spherical body: pseudo-gravastar

Next, we consider a spherical configuration of a gravitating perfect fluid. We assume that the outside of the radius of such a 'star' is empty vacuum. The vacuum solutions and

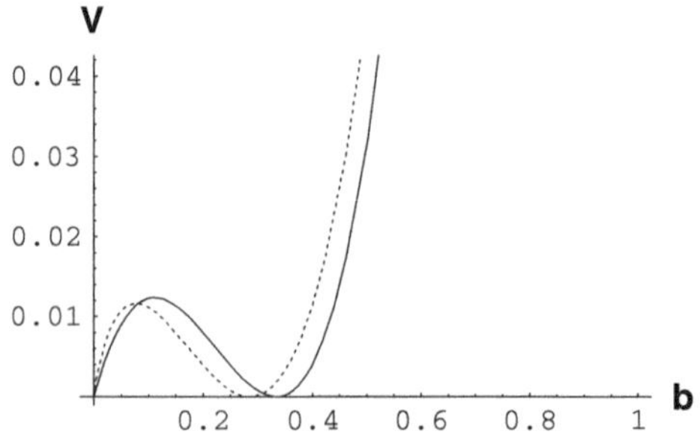

Fig. 2 The shape of the scalar potential is plotted against $b=e^{2\Phi}$ (only shown for small b). The *rigid line* exhibits the case with $h=0$. The *dotted line* exhibits the case with $h=\infty$

the corresponding scalar potential have been assumed and found in the previous subsection.

We assume that the energy–momentum tensor is given in the form of a perfect fluid

$$T_{\mu\nu}=(\rho+p)u_\mu u_\nu-pq_{\mu\nu}, \tag{5.12}$$

where u_μ is the four-velocity which satisfies $q_{\mu\nu}u^\mu u^\nu=1$, ρ is the energy density and p is the pressure of the fluid. In the static case, the four-velocity has only the time-like component. Then, the quantities in (2.12) read

$$T=\rho-3p,\quad E=-p,\quad C^\lambda=0. \tag{5.13}$$

The dynamical Eq. (2.17) is expressed, with the metric (3.3), as

$$\sqrt{\frac{V(\Phi(r))}{A(r)B(r)}}\frac{1}{r^2}\frac{\mathrm{d}}{\mathrm{d}r}\left[\sqrt{\frac{B(r)}{A(r)}}r^2\frac{\mathrm{d}\Phi(r)}{\mathrm{d}r}\right]$$
$$=\frac{V(\Phi)}{r^2e^{3\Phi}|1-r\Phi'|}\left[\frac{\sqrt{V(\Phi)}r^2\Phi'}{e^{\Phi}|1-r\Phi'|}\right]'=\frac{\kappa}{2}\left[\rho(r)-\left(5-\frac{1}{2V}\frac{\mathrm{d}V}{\mathrm{d}\Phi}\right)p(r)\right]. \tag{5.14}$$

where the prime $'$ denotes the derivative with respect to the radial coordinate r. On the other hand, the conservation of the energy–momentum tensor leads to the equation

$$\nabla_\mu T^{\mu\nu}=\frac{1}{\sqrt{-q}}\partial_\mu(\sqrt{-q}\,T^{\mu\nu})+\Gamma^\nu_{\rho\sigma}T^{\rho\sigma}=0, \tag{5.15}$$

where

$$\Gamma^\nu_{\rho\sigma}=\frac{1}{2}q^{\nu\lambda}(\partial_\rho q_{\lambda\sigma}+\partial_\sigma q_{\lambda\rho}-\partial_\lambda q_{\rho\sigma}). \tag{5.16}$$

For the perfect fluid, the equation of conservation becomes

$$p'(r)+[\rho(r)+p(r)]\Phi'(r)=0. \tag{5.17}$$

The Eqs. (5.14) and (5.17), together with a certain equation of state, which gives the relation between ρ and p, completely governs the structure of the spherical body in GSG. This is a direct problem on the structure of a 'star'.

Solving the equations is straightforward but tedious in genaral cases. We leave the study on the generic solutions for various equation of state for another separate work. Here, we assume, the smooth metric inside the 'star' and we investigate the behaviors of the energy density and the pressure as another inverse problem.

In this study, we simply assume that the (00) component of the metric inside a spherical star can be expressed by

$$q_{00} = C_0 - C_2 r^2, \tag{5.18}$$

where C_0 and C_2 are constants. Incidentally, this behavior matches the de Sitter space in GR. Mazur and Mottla proposed 'gravastars', which have the de Sitter spacetime structure inside them, about a decade ago [21–24]. Therefore, the present analysis can be considered as the first step to investigate such simple and interesting exotic objects, say, 'pseudo-gravastars' in GSG (and similar theories mentioned in "Introduction").

Although the gravastars in GR may have a shell of an exotic matter at their surface, we take a simple assumption in this paper, for the 'pseudo-gravastars' in GSG; We define the surface of a star as $p(r_*) = 0$, where r_* is the radius of the star.

We use the result of the previous subsection, so we adopt $V(\Phi) = e^{2\Phi} F(e^{2\Phi})^2$, where F is defined by (5.7)–(5.10) with the choice $f = 3 + 2h$. The metric for the exterior of the star can be described by

$$b(r) = e^{2\Phi(r)} = b_e(r) = 1 - \frac{2GM}{r} + \frac{4\delta}{3}\frac{(2GM)^3}{r^3}, \quad \text{for } r > r_*. \tag{5.19}$$

On the other hand, we assume that the metric is given by the function (5.18) which smoothly connects to $b_e(r)$:

$$b(r) = b_*(r) \equiv b_e(r_*) - \frac{b_e'(r_*)}{2}\left(1 - \frac{r^2}{r_*^2}\right) \equiv e^{2\Phi_*(r)}, \quad \text{for } r < r_*. \tag{5.20}$$

For $r < r_*$, eliminating ρ from Eqs. (5.14, 5.17) yields

$$\left[e^{6\Phi_*(r)} \frac{1}{\sqrt{V(\Phi_*(r))}} \kappa p(r)\right]' = -\frac{2e^{3\Phi_*(r)}\sqrt{V(\Phi_*(r))}\Phi_*'(r)}{r^2(1 - r\Phi_*'(r))}\left[\frac{\sqrt{V(\Phi_*(r))}r^2\Phi_*'(r)}{e^{\Phi_*(r)}(1 - r\Phi_*'(r))}\right]', \tag{5.21}$$

for $r < r_*$. To obtain the solution for $p(r)$, we have only to integrate the equation.

Some numerical examples are shown in Figs. 3, 4, 5 and 6. In each figure, $10\kappa r_g^2 p$ and $\kappa r_g^2 \rho$ are plotted against r/r_g. Figures 3 and 4 are in the case with the parameter $h = 0$ while Figs. 5 and 6 are in the case with $h = \infty$. In Figs. 3

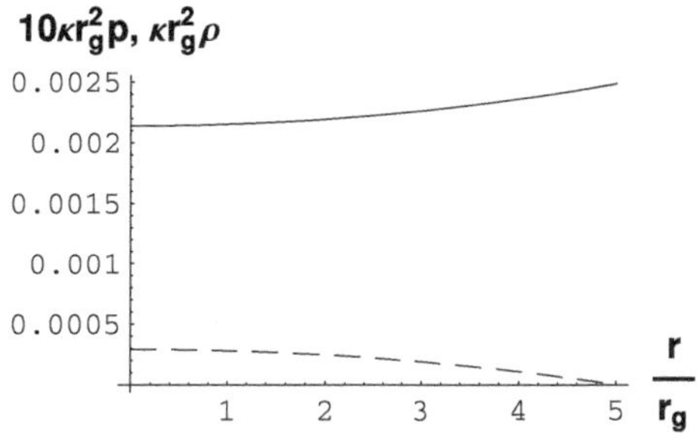

Fig. 3 The energy density (*solid line*) and the pressure (*dashed line*) are plotted against r/r_g in the case with the parameter $h = 0$ and $r_* = 5r_g$

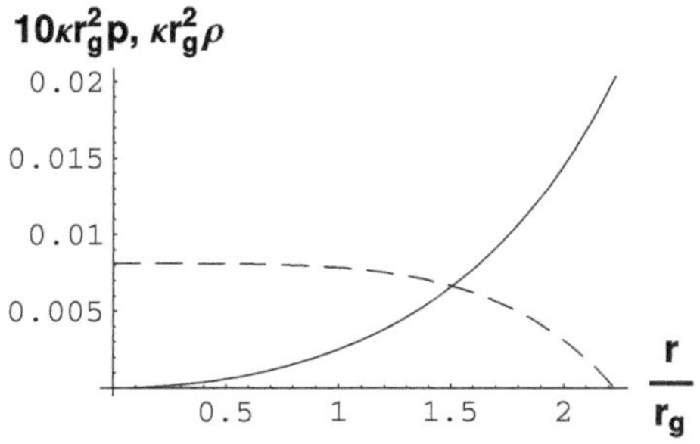

Fig. 4 The energy density (*solid line*) and the pressure (*dashed line*) are plotted against r/r_g in the case with the parameter $h = 0$ and $r_* = 2.228r_g$

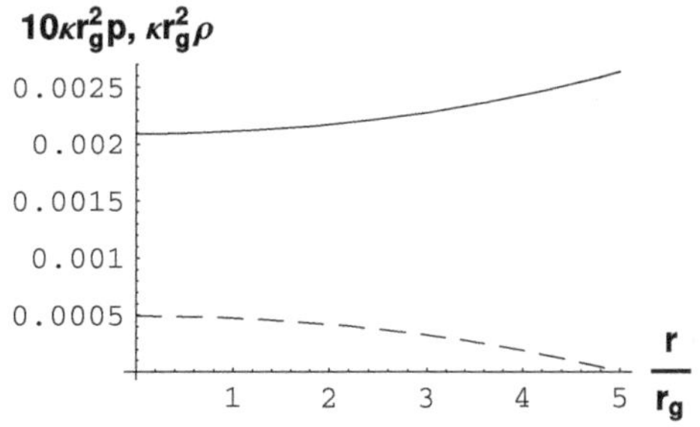

Fig. 5 The energy density (*solid line*) and the pressure (*dashed line*) are plotted against r/r_g in the case with the parameter $h = \infty$ and $r_* = 5r_g$

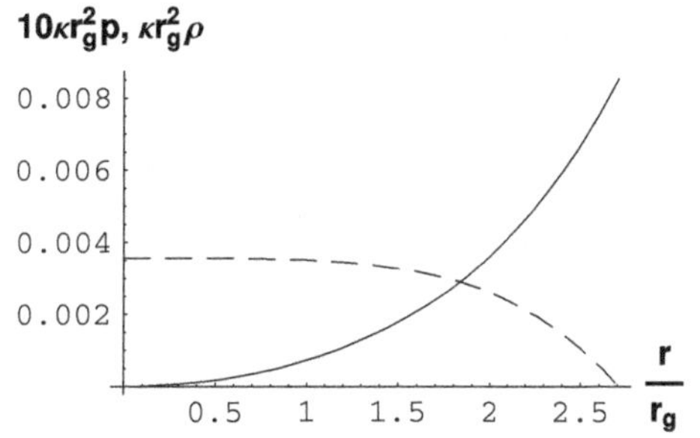

Fig. 6 The energy density (*solid line*) and the pressure (*dashed line*) are plotted against r/r_g in the case with the parameter $h = \infty$ and $r_* = 2.707r_g$

and 5, we choose $r_* = 5r_g$. Figures. 4 and 6 indicate the results with $\rho(0) = 0$.

In both cases of $h = 0$ and $h = \infty$, the behaviors of ρ and p are qualitatively equal, respectively. In Figs. 4 and 6, the values for r_* are considered as the lower bound for the radii of the stars of the present type. Below the critical value for the radius, the energy density at the origin becomes negative.

In contrast to the gravastars in GR, we find that the value of the pressure p can be positive and simply decreasing in the present pseudo-gravastar in GSG. Of course, the geometry of the interior of the star is not an exact de Sitter space, for $q_{11} \neq -(q_{00})^{-1}$ in GSG.

The 'stars' constructed above are, however, not realistic because the energy density ρ also should decrease along with r for a normal equation of state.[4] Nevertheless, we suppose that the characteristic scales of the 'star' in GSG revealed in the simple approach are useful for setting the starting values of numerical calculation.

Summary and prospects

In this paper, we considered the geometric scalar gravity of Novello and collaborators and found conditions of the scalar potential that had a permissible solution in the case of weak and strong gravity. We found that the simple potential of Novello et al. leads to the solution which agrees with the Schwarzschild solution of GR and small corrections to the potential are permitted and the generic potential leads to the 'black hole solution' with a modified mass-horizon radius relation (i.e., $r_g \neq 2GM$), of which asymptotic behavior describes weak gravity up to the post-Newtonian order. We also showed possible solutions for 'pseudo-gravastars' in GSG with generic scalar potentials.

We will try to extend the geometric scalar theory of gravity with additional degrees of freedom. We also wish to study the mass-radius relation (known in GR [25]) which may exist in a solution describing a spherical star in GSG by numerical analyses. In addition, we want to study 'cosmologies' expressed by the generic scalar potentials in geometric scalar gravity. We hope to report soon on these issues.

The quantization of GSG is another very interesting subject. We wish to discuss it in a separate paper.

Note added:

After completion of this manuscript, we have noticed a new preprint arXiv:1508.02665 [gr-qc] by Jardim and Landim [26] on GSG. It discusses the issue of the cosmological constant in GSG and argues an 'inverse problem' to realize the de Sitter spacetime in GSG.

Acknowledgments We thank Prof. Kumar Shwetketu Virbhadra for valuable information on naked singularities. We also thank Prof. Sunny Vagnozzi for providing information on their work on modified gravity including scalar derivative terms [27, 28].

References

1. Nojiri, S., Odintsov, S.D.: Phys. Rep. **505**, 59 (2011)
2. Capozziello, S., De Laurentis, M.: Phys. Rep. **509**, 167 (2011)
3. Clifton, T., Ferreira, P.G., Padilla, A., Skordis, C.: Phys. Rep. **513**, 1 (2012)
4. Novello, M., Bittencourt, E., Moschella, U., Goulart, E., Salim, J.M., Toniato, J.D.: Geometric scalar theory of gravity, JCAP 1306 (2013) 014; JCAP 1401 (2014) 01, E01
5. Novello, M., Bittencourt, E.: The gravitational energy. arXiv: 1311.6984 [gr-qc] (2013)
6. Bittencourt, E., Moschella, U., Novello, M., Toniato, J.D.: Cosmology in geometric scalar gravity. Phys. Rev. **D90**, 123540 (2014)
7. Bittencourt, E., Novello, M., Moschella, U., Goulart, E., Salim, J.M., Toniato, J.D.: Geometric scalar gravity. Nonlinear Phenom. Complex Syst. **17**, 349 (2014)
8. Toniato, J.D.: A teoria geométrica-escalar da gravitação e sua aplicação à cosmologia, Tese de Doutorado, Rio de Janeiro (2014)
9. Kraichman, R.: Phys. Rev. **98**, 1118 (1955)
10. Freund, P.G.O., Nambu, Y.: Phys. Rev. **174**, 1741 (1968)
11. Deser, S., Halpern, L.: Gen. Rel Grav. **1**, 131 (1970)
12. Shapiro, S.L., Teukolsky, S.A.: Phys. Rev. **D47**, 1529 (1993)
13. Novello, M.: Talk in hot topics in modern cosmology, spontaneous workshop IX, 27 April–1 May 2015, Cargèse
14. Chamseddine, A.H., Mukhanov, V.: JHEP **1311**, 135 (2013)
15. Bekenstein, J.D.: Phys. Rev. **D70**, 083509 (2004)
16. Weinberg, S.: Gravitation and Cosmology, pp. 184. Wiley, New York (1972)
17. Hartle, J.B.: Gravity, An Introduction to Einstein's General Relativity, pp. 222. Addison Wesley, New York (2003)
18. Claudel, C.-M., Virbhadra, K.S., Ellis, G.F.R.: J. Math. Phys. **42**, 818 (2001)
19. Virbhadra, K.S., Ellis, G.F.R.: Phys. Rev. **D65**, 103004 (2002)
20. Virbhadra, K.S., Keeton, C.R.: Phys. Rev. **D77**, 124014 (2008)
21. Mazur, P.O., Mottola, E.: Proc. Natl. Acad. Sci. USA (PNAS) **101**, 9545 (2004). arXiv: gr-qc/0109035
22. Cattoën, C., Faber, T., Visser, M.: Class. Q. Grav. **22**, 4189 (2005)
23. Faber, T.: arXiv: gr-qc/0607029
24. DeBenedictis, A., Horvat, D., Ililić, S., Kloster, S., Viswanathan, K.S.: Class. Q. Grav. **23**, 2303 (2006)
25. Buchdahl, H.A.: Phys. Rev. **116**, 1027 (1959)
26. Jardim, I.C., Landim, R.R.: arXiv:1508.02665 [gr-qc]
27. Myrzakulov, R., Sebastiani, L., Vagnozzi, S.: Eur. Phys. J. **C75**, 444 (2015)
28. Myrzakulov, R., Sebastiani, L., Vagnozzi, S., Zerbini, S.: Fund. J. Mod. Phys. **8**, 119 (2015)

[4] As for a possible improvement, we should take shell boundaries and anisotropic pressures for exotic stars into account.

21

Effect of substrate temperature on ZnS films prepared by thermal evaporation technique

Rahul Vishwakarma[1]

Abstract The nanocrystalline ZnS semiconducting thin films of 500 nm thickness have been deposited on glass substrate at different substrate temperatures (T_s) by thermal evaporation technique. The structural property of deposited thin films has been measured by X-ray diffraction, scanning electron microscopy, and Energy dispersive analysis of X-ray. The electrical and optical properties of thin films have been determined by D.C. two point probe and ultraviolet visible spectroscopy measurements. The X-ray diffraction patterns show that thin films have cubic structure. The electrical resistivity of thin films has decreased from 0.36×10^6 to 0.15×10^6 Ω cm as substrate temperature increases from 300 to 400 K. It shows that films have semiconducting in nature. The grain size and electrical conductivity of the thin films have increased as the deposition temperature increased while dislocation density, activation energy, and band gap decreased. The minimum band gap 3.43 eV has been found.

Keywords ZnS films · Grain size · Dislocation density · Electrical resistivity · Band gap · Activation energy

Introduction

The nanocrystalline zinc sulfide (ZnS) semiconductor has attracted much attention from the viewpoint of fabrication of many optoelectronics devices because of their unique properties such as quantum size effect [1, 2] and abnormal luminescence phenomenon [3–5]. The non-stoichiometry is responsible for conductivity in thin film which is created during deposition of thin films. In recent years, advances in formation of nanometer-sized materials continuously unabated and it show that more exciting opportunities are laying ahead if size of crystal structure can be scaled down to nanometers. The semiconductor zinc sulfide belongs to II–VI group in periodic table which has large energy band gap (3.72–3.77 eV) and cubic structure in the ultra-violet visible (UV)-region. Due to these wide band gap materials, it is suitable for use in the blue light-emitting diodes [6], optoelectronics devices such as electron luminescent display [7], cathode luminescent display [8], and multilayer dielectrics filters [9]. The ZnS semiconductor exists in two phases, i.e., cubic phase (sphalerite) and hexagonal phase (wurtzite). The exciting binding energy (38 MeV) of ZnS is higher than thermal energy (25 MeV) at room temperature which can show excitonic emission. For the fabrication of optoelectronics devices to need high conductivity, high mobility, and low band gap. Many growth techniques have been adopted to prepare ZnS thin films, such as sputtering [10], pulsed-laser deposition [11], metal organic chemical vapor deposition [12–15], electron beam evaporation [16], photochemical deposition [17], chemical bath deposition [18], and thermal evaporation technique. Among these methods, the thermal evaporation technique is the most interesting technique because the advantages of this technique have high stability, high reproducibility, high deposition rate, large area deposition, and non-expansive. This technique provides economical and efficient use of evaporate material enabling constant rate of deposition. The aim of the present study is to investigate the effect of substrate temperature on the structural, electrical, and optical properties of ZnS thin films deposited on glass substrate by thermal evaporation

✉ Rahul Vishwakarma
rhl.jaunpur@gmail.com

1 Department of Physics and Electronics, Dr. R M L Avadh University, Faizabad, UP, India

technique. We also studied its variation with substrate temperature.

Experimental

Chemicals

For the preparation of non-stoichiometric semiconducting ZnS thin films, the material used as source is zinc sulfide. All chemicals were purchased from Alfa Aesar, Ltd. USA, which have a high purity of 99.99 %.

Substrate cleaning

The substrate cleaning plays an important role in the deposition of thin films, so commercially available glass substrates with the size of (75 mm × 25 mm × 1 mm) were washed in detergent then in chromic acid and finally washed with double distilled water in ultrasonic cleaner and dried at 400 K in an oven.

Preparation of ZnS thin films

The nanocrystalline ZnS thin films of 500 nm thickness have been deposited on well-cleaned glass substrates at different substrate temperatures (T_s = 300–400 K) by the thermal evaporation technique. The films were fabricated under vacuum of 10^{-4}–10^{-5} Torr. The zinc sulfide is used as the source material. The stoichiometric starting/source materials (ZnS) were taken in molybdenum boat and evaporated in vacuum ($\sim 10^{-5}$ Torr) in the vacuum system equipped with liquid nitrogen trap. The source materials were kept at the distance of 120 mm from the substrate holder in vacuum chamber. The film deposition rate (2.0–15 nm/s) was adjusted by changing the electrical current. The deposition rate was measured by digital film thickness monitor using a quartz crystal sensor set at 6 MHz (DTM-10). Initially, substrates were not exposed to vapor stream by using shutter, and after obtaining constant rate of evaporation, the shutter was opened. During deposition of thin film, substrate was placed normal to the line of sight from the evaporation surface at different polar angles to obtain uniform deposition. The source material about 30 mg was used for the deposition of each thin film. In each cycle of deposition, fresh source material was kept in molybdenum boat. The film thickness and deposition rate were measured by film thickness monitor (FTM, VICO, DTM-10) with quartz crystal sensor vibrating at a frequency 6 MHz. The sensor of FTM was attached parallel to the substrate. The substrate temperature was measured using digital temperature meter (PTS-9601: GELCO) that was attached with thermocouple sensor. The tip of the thermocouple sensor was in contact with the surface of substrate. The different substrate temperatures were obtained by changing the current of substrate heater.

Characterization of thin films

Structural characterization

The X-ray diffraction (XRD) patterns of deposited thin films have been recorded by Rigaku diffractometer using graphite-filtered CuKα_1 radiation (λ = 1.54 Å) at 40 kV, 100 mA with a scanning rate of 3°/min (2θ = 20°–60°). The composition of deposited films was analyzed using scanning electron microscope, with attached energy dispersive analysis of X-rays (EDAX). The surface morphology of thin films was studied by scanning electron microscopy measurement. The crystallite size and dislocation density of deposited thin films were calculated by using Debye–Scherrer's formula [19].

$$D = \frac{0.94\lambda}{\beta \cos\theta} \tag{1}$$

$$\text{Dislocation density} = \frac{1}{D^2}, \tag{2}$$

where λ is wavelength of radiation used, θ is diffraction angle of the concern diffraction peak, β is the full width at half maximum (FWHM) of the diffraction peak corresponding to a particular crystal plane.

Electrical characterization

The electrical resistivity and conductivity of semiconducting ZnS thin filmswere measured by using the D.C. two point probe method. The electrical resistivity and conductivity of thin films were determined by the following equation [20].

$$\rho = \rho_0 \exp\left(\frac{E_0}{KT}\right) \tag{3}$$

where ρ is resistivity, K is Boltzmann constant, and T is absolute temperature.

Optical characterization

The optical absorption spectra were recorded by a shimadzu double beam double monochromator spectrophotometer (UV-2550) in the wavelength range of 300–1500 nm at room temperature with unpolarized radiation on thin films which were deposited on the glass substrate at different substrate temperatures. The absorption coefficient (α) was calculated for deposited thin film in the region of strong absorption using the relation [21].

$$\alpha = \frac{1}{d\{\ln(\frac{1}{T})\}}, \quad (4)$$

where α is absorption coefficient at particular wavelength, T the transmittance at same wavelength, and d is film thickness.

The direct band gap of thin film has calculated by using Tauc relation [20].

$$\alpha h\nu = A(h\nu - E_g)^n \quad (5)$$

where $h\nu$ is photon energy, E_g is band gap, A is constant, and $n = ½$ for direct band gap material.

Results and discussion

Structural analysis of thin films

The X-ray diffraction patterns of the deposited semiconducting ZnS thin films are shown in Figs. 1, 2, 3, 4 and 5. The XRD measurement reveals that the nanocrystalline of ZnS thin films is in cubic structure. The cubic phase of the crystals has identified from the agreement of peak position with standard JCPDS data card no. 65–1691. The nanocrystals of ZnS have different planes (111), (220), and (311) with preferential orientation along (111) plane. Figures 1, 2, 3, 4, and 5, show that intensity of XRD peaks was increased with the increase of substrate temperature 300–400 K. This indicates that a full width half maximum (FWHM) of diffraction peaks was decreased. For this, particle size of deposited film was increased with the increase of substrate temperature. Figure 1 shows that low grain size and high dislocation density while Fig. 5 has

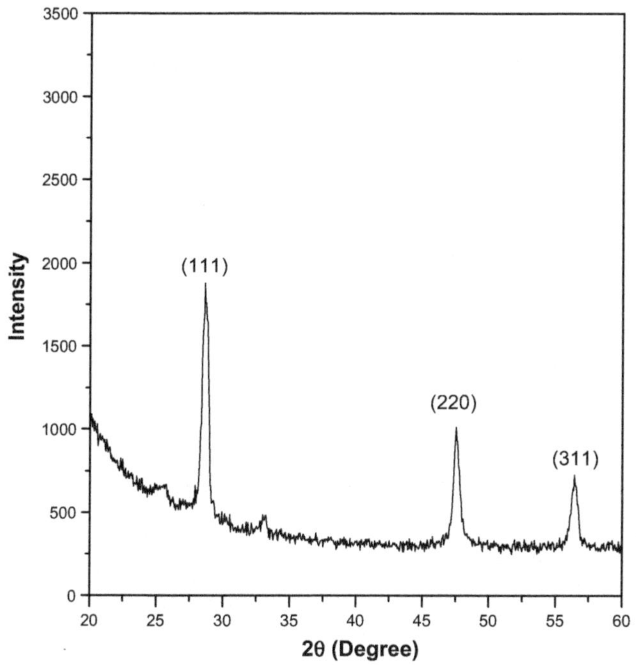

Fig. 1 X-ray diffraction pattern of ZnS thin film at $T_s = 300$ K

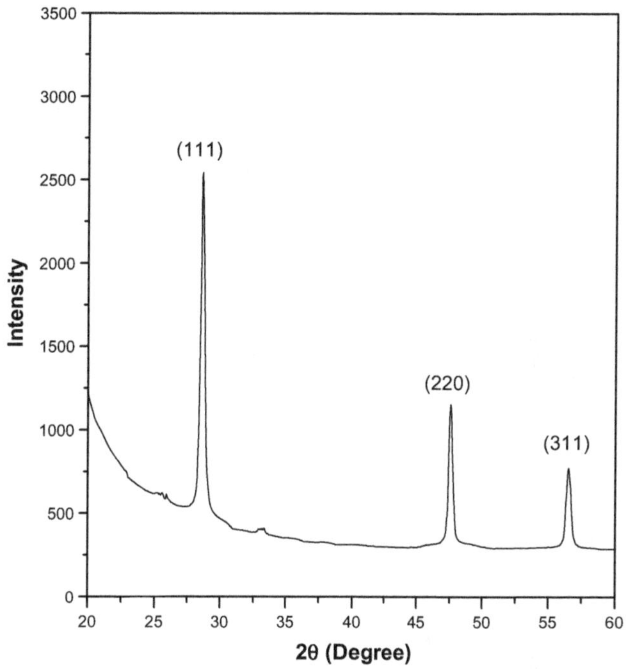

Fig. 2 X-ray diffraction pattern of ZnS thin film at $T_s = 323$ K

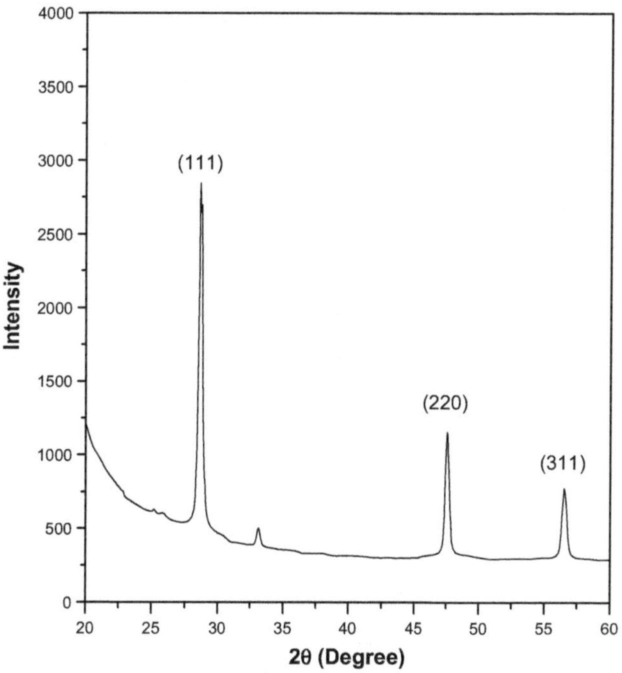

Fig. 3 X-ray diffraction pattern of ZnS thin film at $T_s = 350$ K

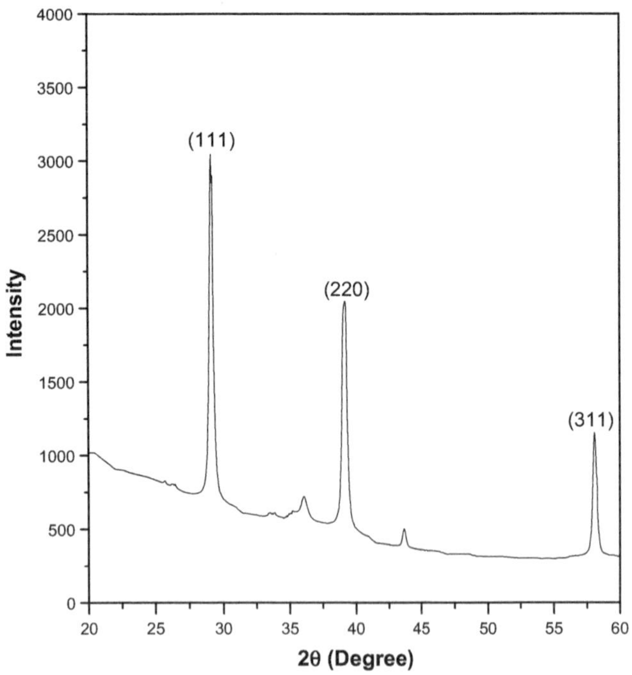

Fig. 4 X-ray diffraction pattern of ZnS thin film at $T_s = 373$ K

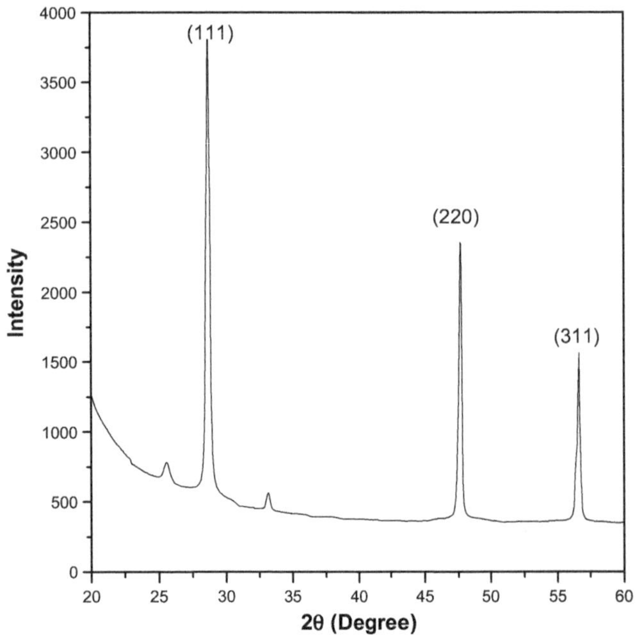

Fig. 5 X-ray diffraction pattern of ZnS thin film at $T_s = 400$ K

highest grain size and lowest dislocation density. Thus, crystallinity is improved with the increase of substrate temperature. The experimental and standard d values of ZnS materials are given in Table 1. The calculated grain size and dislocation density of thin films are given in Table 2.

Table 1 Experimental and Standard d values

S. no	(h k l) plane	Experimental *d* value (Å)	Standard *d* value (Å)
1	111	3.117	3.123
2	220	1.912	1.912
3	311	1.629	1.633

From the EDAX, films were non-stoichiometric with the composition Zn = 52.46 % and S = 47.54 %. There were small differences between composition of Zn and S in films and source material nearly equal to 1 %.

The variations of grain size or particle size (nm) and dislocation density of semiconducting thin films with different substrate temperatures are shown in Fig. 6. It was cleared from Fig. 6; the grain size of the films have been increased from 36.12 to 43.82 nm as the substrate temperature increased from 300 to 400 K while dislocation density decreased from 0.76×10^{15} to 0.52×10^{15} line/m^2. The maximum grain size 43.82 nm and minimum dislocation density 0.52×10^{15} line/m^2 have been obtained and its value nearly saturated above at 400 K. This indicates that the growth in grain size was constant above the temperature 400 K. The highest grain size of thin film indicates the highest conductivity and lowest resistivity of film. Thus, crystallinity of thin films has improved with the increase of substrate temperature. This may be due to collapse of nano-particles with each other and decrease crystal defects in thin films at higher substrate temperature.

The maximum grain size 43.82 nm and minimum dislocation density 0.52×10^{15} line/m^2 have been observed for the ZnS thin film deposited at temperature 400 K whereas Haque et al. [22] have reported the grain size of 34.08 nm and dislocation density 0.86×10^{15} line/m^2 for ZnS thin films by R F Magnetron sputtering. Borah et al. [23] have reported the grain size 7.0 nm for ZnS crystal by a chemical route method. Chauhan et al. [24] and Lu et al. [25] have reported the grain size 4.0 and 3.0 nm, respectively, for ZnS nanoparticles by chemical precipitation method.

The surface morphology of ZnS thin films is shown in Figs. 7, 8, 9, 10, and 11. It was cleared from Figs. 7, 8, 9, 10, and 11 that the fabricated thin films were homogenous, without crack and pin holes. In scanning electron micrograph as shown in Figs. 7, 8, 9, 10 and 11, crystal or grain size has been increased with the increases of substrate temperature from 300 to 400 K. The grain size was lowest while dislocation density and strain were highest in Fig. 7. The grain size in Fig. 11 was highest while dislocation density and strain were lowest at 400 K than other surface micrographs. This is a good agreement with X-ray diffraction of thin films.

Table 2 Structural, electrical, and optical parameters

Substrate temperature (T_s) (K)	Structural parameters		Electrical parameters			Optical parameters
	Grain size (nm)	Dislocation density × 10^{15} line/m^2	Resistivity × 10^6 (Ω cm)	Conductivity × 10^{-6} (1/Ω cm)	Activation energy (eV)	Band gap (eV)
300	36.12	0.76	0.36	2.77	1.24	3.52
323	39.64	0.63	0.28	3.57	1.06	3.47
350	42.17	0.56	0.19	5.26	0.95	3.44
373	43.56	0.52	0.16	6.25	0.88	3.43
400	43.82	0.52	0.15	6.66	0.86	3.43

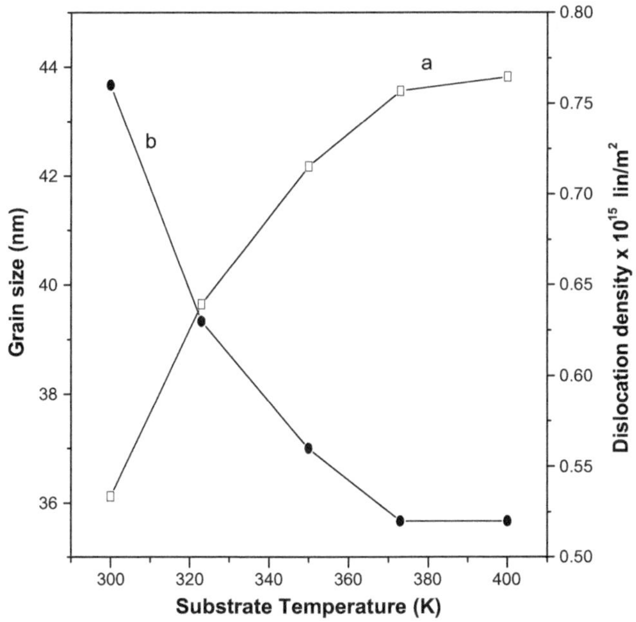

Fig. 6 Variation of **a** grain size and **b** dislocation density with different substrate temperatures

Electrical analysis of thin films

The calculated value of electrical resistivity, electrical conductivity, and activation energy of thin films is given in Table 2. The electrical resistivity of thin films decrease with the increase of substrate temperature. This shows that the film was semiconductor in nature. The decrease in electrical resistivity is due to the improvement in crystallinity of film as the substrate temperature was increased from 300 to 400 K. The activation energy of semiconducting ZnS thin films has been determined from slope of log of resistivity (log ρ) with reciprocal of temperature $(1/T) \times 10^3$ K. The variation of electrical resistivity and conductivity of thin films with different substrate temperatures is shown in Fig. 12.

Fig. 7 Scanning electron micrograph of ZnS film at room temperature

Fig. 8 Scanning electron micrograph of ZnS film at $T_s = 323$ K

It is clear from Fig. 12 that electrical resistivity of film decreased from 0.36×10^6 to 0.15×10^6 Ω cm as substrate temperature increased from 300 K to 400 K. The electrical conductivity increased from 2.77×10^{-6} to 6.66×10^{-6} 1/Ω cm with the increase of substrate

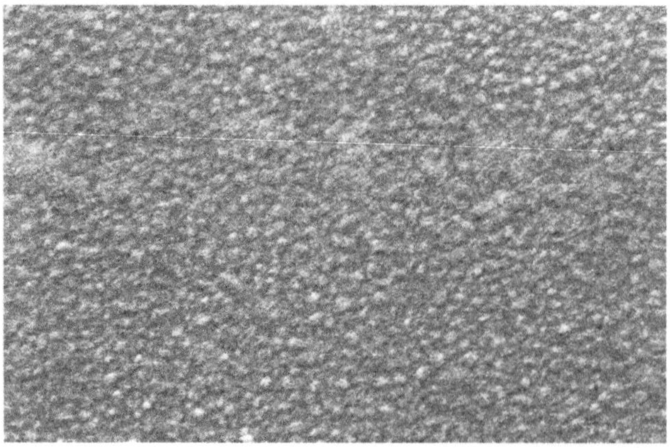

Fig. 9 Scanning electron micrograph of ZnS film at $T_s = 350$ K

Fig. 10 Scanning electron micrograph of ZnS film at $T_s = 373$ K

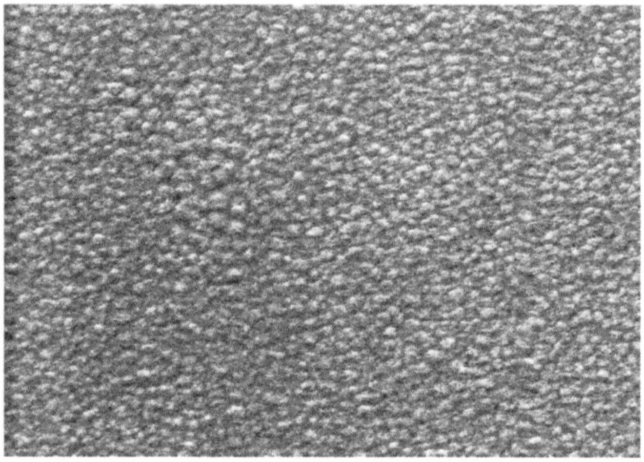

Fig. 11 Scanning electron micrograph of ZnS film at $T_s = 400$ K

temperature from 300 K to 400 K. This indicates that films are semiconducting in nature. The decrease in electrical resistivity is due to the improvement in crystallinity of thin

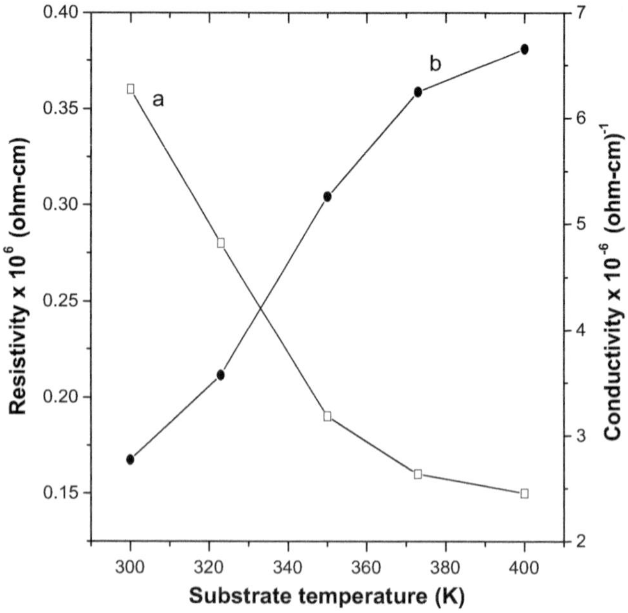

Fig. 12 Variation of **a** electrical resistivity and **b** conductivity with different substrate temperatures

films, mobility of charge carrier, and concentration of charge carrier with substrate temperature. The minimum electrical resistivity 0.15×10^6 Ω cm and high electrical conductivity 6.66×10^{-6} 1/Ω cm have been obtained at $T_s = 400$ K. The value of electrical resistivity and conductivity of thin films were nearly saturated above 400 K. This may be due to increment in grain size has constant with increase of substrate temperature.

The variation of activation energy of thin films with different substrate temperatures is shown in Fig. 13. It is cleared from Fig. 13 that the activation energy has been decreased from 1.24 to 0.86 eV as the substrate temperature increased from 300 to 400 K and nearly saturated above at 400 K. This may be due to distance between valence band and conductance has decreased with increase of substrate temperature from 300 to 400 K and above 400 K, it has constant. The minimum activation energy obtained at $T_s = 400$ K was 0.86 eV.

The lowest resistivity 0.15×10^6 Ω cm and high electrical conductivity 6.66×10^{-6} 1/Ω cm have been found for the ZnS thin films deposited at temperature of 400 K, whereas Ozutok et al. [26] have reported the resistivity 0.49×10^6 Ω cm and conductivity 2.05×10^{-6} 1/Ω cm for ZnS thin film deposited by Spray pyrolysis technique. Ubale et al. [20] have reported the resistivity 0.18×10^6 Ω cm and activation energy 1.29 eV for ZnS thin films deposited by Chemical bath technique. Shinde et al [27] have reported the activation energy 0.89 eV for ZnS film deposited by chemical bath deposition.

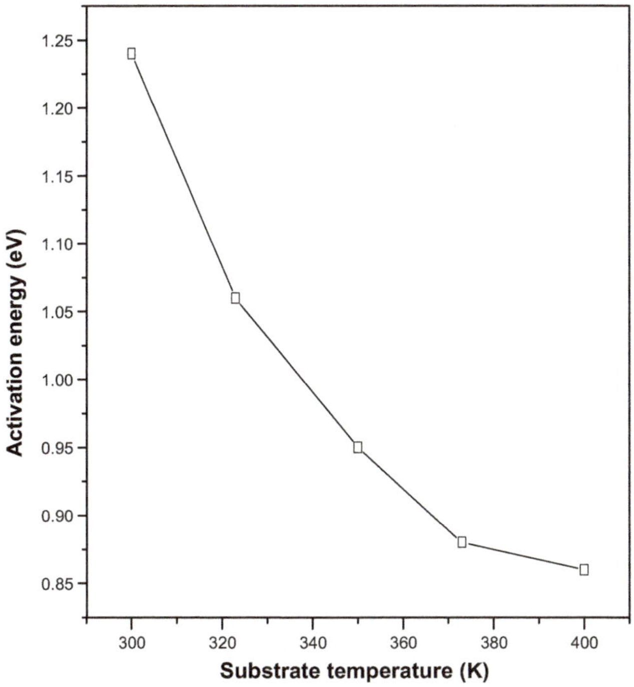

Fig. 13 Variation of activation energy with substrate temperature

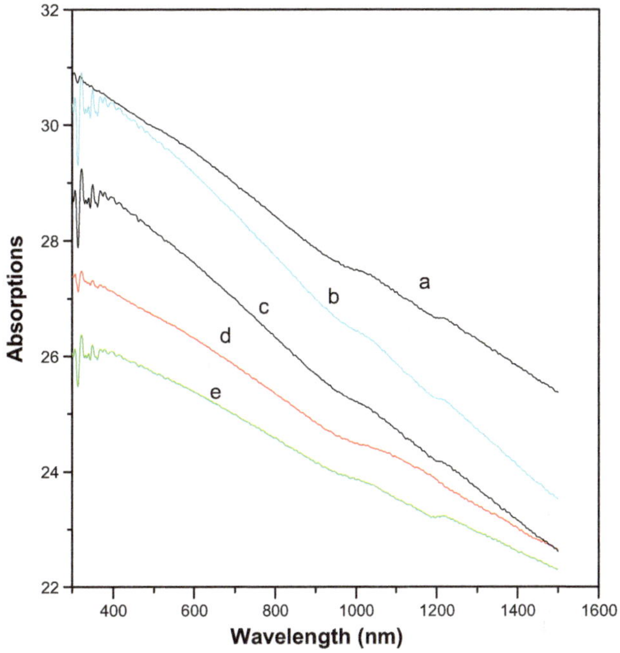

Fig. 14 Variation of absorption with wavelength at **a** 300 K, **b** 323 K, **c** 350 K, **d** 373 K, **e** 400 K

Optical analysis of thin films

The optical absorption spectra of ZnS thin films at different substrate temperatures are shown in Fig. 14. It is clear from Fig. 14 that the optical absorption decreases with the increase of wavelength and minimum absorption obtained at 1500 nm. The optical absorption also decreased with the increase of substrate temperature because grain boundary scattering was reduced with the increase of substrate temperature.

This indicates that grain size of crystal increased with increase of substrate temperature. This may be increase in transmittance of thin films with substrate temperature. The variation of $(\alpha h\nu)^2$ with photon energy (eV) is shown in Fig. 15. The extrapolation of straight-line portion of plot gives the value of direct band gap. Using these spectra, we determined the band gap of ZnS thin films. The band gap obtained was 3.49–3.43 eV at temperature range 350–400 K. The minimum band gap obtained was 3.43 eV at 400 K. This shows that band gap of thin film decreased with the increase of substrate temperature.

The minimum direct band gap (3.43 eV) has been observed for the ZnS thin film deposited at substrate temperature 400 K. Shinde et al [27] have reported the band gap 3.51 eV for ZnS thin films deposited by chemical bath deposition. Kumar et al. [28] have reported the band gap 3.50 eV for ZnS thin film by vacuum evaporation technique.

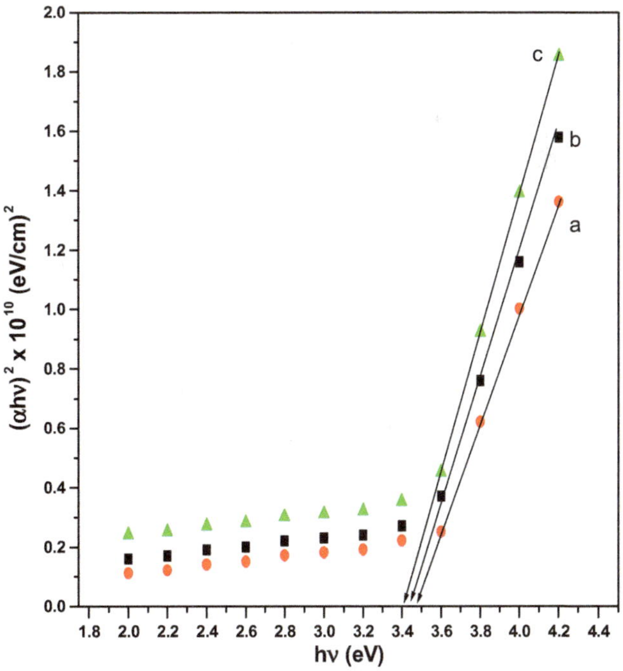

Fig. 15 Variation of $(\alpha h\nu)^2$ with photon energy at **a** 350 K, **b** 373 K, **c** 400 K

The structural, electrical, and optical properties of deposited film studies support this decrease in this resistivity due to improvement in crystallinity of the film which would increase the charge carrier, mobility, and decrease in defect levels with increase in the temperature.

Conclusion

The non-stoichiometric semiconducting ZnS thin films have been deposited on glass substrate by thermal evaporation technique at different substrate temperatures. The maximum grain size 43.82 nm has been obtained. The SEM shows the agglomeration of nano-crystals. The XRD analysis shows that films have cubic structure. The minimum resistivity $0.15 \times 10^6\ \Omega$ cm and maximum conductivity $6.66 \times 10^{-6}\ 1/\Omega$ cm of film have been obtained. The measurement of optical absorption spectra indicates that band gap decreased with the increase of substrate temperature. The minimum band gap 3.43 eV has been observed.

Acknowledgments I am thankful to our institute for providing XRD and SEM facilities. I am also thankful to our friend for help in preparation of thin films.

References

1. Li, L.-P., Qiu, X.-Q., Li, G.-S.: Correlation between size-induced lattice variations and yellow emission shift in ZnO nanostructures. Appl. Phys. Lett. **87**(12), 124101–124103 (2005)
2. Schmidt, H.M., Weller, H.: Synthesis of ZnS nano particles by solid liquid chemicals method. Chem. Phys. Lett. **129**(6), 615–618 (1986)
3. Xin, X.-S., Zhou, B.-B., Du, H., Xu, X.-Q., Su, W.-H.: Preparation of ZnS nano particles and study its optical property. Chem. Res. Appl. **17**(3), 303–306 (2005)
4. Cheng, X.W., Li, X., Zhai, F.-F., Yu, X., Long, X., Lin, P.-T.: Fabrication and Photoluminescence study on ZnO nanostructured particles. J. Aeronaut. Mater. **29**(1), 92–95 (2009)
5. Nasuno, Y., Kohama, N., Nishimura, K., Hayakawa, T., Taniguchi, H., Shimizu, M.: Size-dependent melting behavior of Zn nanowire arrays. Appl. Phys. Lett. **88**(7), 1909–1911 (2006)
6. Katyama, H., Oda, S., Kulimato, H.: Study of ZnS nano particals. Appl. Lett. **27**, 657–661 (1975)
7. Mach, R., Muller, G.O.: Physical concept of high filed, thin film of ZnS. Phys. Status Solidi A **69**, 11–17 (1982)
8. Ledger, A.M.: Inhomogeneous interface laser mirror coatings. Appl. Opt. **18**, 2979–2983 (1979)
9. Jones, P.L., Moore, D.M., Smith, S.C.: A new method for melting and recrystallization of lanthanum ZnS nanoparticle. J Phys E **9**, 312–316 (1976)
10. Nitta, A., Tanakab, K., Maekawab, Y., Kusabirakib, M., Aozasa, M.: Effects of gas impurities in the sputtering environment on the stoichiometry and crystallinity of ZnS: Mn electroluminescent-device active layers. Thin Solid Films **384**, 261–268 (2001)
11. Yano, S., Schroeder, R., Ullrich, B., Sakai, H.: Absorption and photocurrent properties of thin ZnS films formed by pulsed-laser deposition on quartz. Thin Solid Films **423**, 273–276 (2003)
12. Feng, Q.J., Shen, D.Z., Zhang, J.Y., Liang, H.W., Zhao, D.X., Lua, Y.M., Fan, X.W.: Highly aligned ZnS nanorods grown by plasma-assisted metalorganic chemical vapor deposition. J. Crystal Growth **285**, 561–565 (2005)
13. Zhang, Z., Shen, D.Z., Zhang, J.Y., Shan, C.X., Lu, Y.M., Liu, Y.C., Li, B.H., Zhao, D.X., Yao, B., Fan, X.W.: The growth of single cubic phase ZnS thin films on silica glass by plasma-assisted metalorganic chemical vapor deposition. Thin Solid Films **513**, 114–117 (2006)
14. Gunasekaran, M., Gopalakrishnan, R., Ramasamy, P.: Deposition of ZnS thin films by photochemical deposition technique. Mater. Lett. **58**, 67–70 (2003)
15. Zhang, X.T., Liu, Y.C., Zhang, L.G., Zhang, J.Y., Lu, Y.M., Shen, D.Z., Xu, W., Zhong, G.Z., Fan, X.W., Kong, Z.G.: Photoluminescence and optically pumped ultraviolet lasing from nanocrystalline ZnO thin films prepared by thermal oxidation of high-quality ZnS thin films. Chin. Phys. Lett. **19**, 127–130 (2002)
16. Wang, S., Fu, X., Xia, G., Wang, J., Shao, J., Fan, Z.: Structure and optical properties of ZnS thin films grown by glancing angle deposition. Appl. Surf. Sci. **252**, 8734–8737 (2006)
17. Fathy, N., Kobayashi, R., Ichimura, M.: Preparation of ZnS thin films by the pulsed electrochemical deposition. Mater. Sci. Eng., B **107**, 271–276 (2004)
18. Roy, P., Ota, J.R., Srivastava, S.K.: Crystalline ZnS thin films by chemical bath deposition method and its characterization. Thin Solid Films **515**, 1912–1917 (2006)
19. Ramasamy, V., Praba, K., Murugadoss, G.: Study of optical and thermal properties in nical doped ZnS nanoparticles using Surfactants. Superlattices Microstruct. **51**, 699–714 (2012)
20. Ubale, A.U., Kulkarni, D.K.: Preparation and study of thickness dependent electrical characteristics of zinc sulfide thin films. Bull. Mater. Sci. **28**(1), 43–47 (2005)
21. Cottrell, A.: An introduction to metallurgy, 2nd edn, pp. 173–178. Edward Arnold Ltd, London (1975)
22. Haque, F., Rahman, K.S., Islam, M.A., Rasid, M.J., Alam, M.M., Alothman, Z.A., Soplan, K., Amin, N.: Growth optimization of ZnS thin films by R F Magnetran sputtering as prospective buffer layer in thin film solar cell. Chalcogenide Lett. **11**, 189–197 (2014)
23. Borah, J.P., Barman, J., Sarma, K.C.: Structural and optical properties of ZnS nanoparticles. Chalcogenide Lett. **5**, 201–208 (2008)
24. Chauhan, R., Kumar, A., Chaudhary, R.P.: Characterization of chemically synthesized Mn doped ZnS nanoparticals. Chalcogenide Lett. **9**, 151–156 (2012)
25. Lu, S.W., Lee, B.I., Wang, Z.L., Tong, W., Wagner, B.K., Park, W., Summers, C.J.: Synthesis and photoluminescence enhancement of Mn doped ZnS nanocrystals. J. Luminensence **92**, 73–78 (2001)
26. Ozutok, F., Erturk, K., Bilgin, V.: Growth, electrical and optical properties of ZnS: Mn thin films. Acta Phys. Pol., A **121**, 221–223 (2012)
27. Shinde, M.S., Ahirrao, P.B., Patil, I.J., Patil, R.S.: Studies on nanocrystalline ZnS thin films prepared by modified chemical bath deposition method. Indian J. Pure Appl. Phys. **49**, 765–768 (2011)
28. Kumar, P., Kumar, A., Dixit, P.N., Sharma, T.P.: Optical, structural and electrical properties of zinc sulphide vacuum evaporated thin films. Indian J. Pure Appl. Phys. **44**, 690–693 (2006)

22

High-pressure and temperature-induced structural, elastic, and thermodynamical properties of strontium chalcogenides

Dinesh Varshney[1] · S. Jain[1,2] · S. Shriya[1] · R. Khenata[3]

Abstract Pressure- and temperature-dependent mechanical, elastic, and thermodynamical properties of rock salt to CsCl structures in semiconducting SrX (X = O, S, Se, and Te) chalcogenides are presented based on model interatomic interaction potential with emphasis on charge transfer interactions, covalency effect, and zero point energy effects apart from long-range Coulomb, short-range overlap repulsion extended and van der Waals interactions. The developed potential with non-central forces validates the Cauchy discrepancy among elastic constants. The volume collapse (V_P/V_0) in terms of compressions in SrX at higher pressure indicates the mechanical stiffening of lattice. The expansion of SrX lattice is inferred from steep increase in V_T/V_0 and is attributed to thermal softening of SrX lattice. We also present the results for the temperature-dependent behaviors of hardness, heat capacity, and thermal expansion coefficient. From the Pugh's ratio ($\phi = B_T/G_H$), the Poisson's ratio (ν) and the Cauchy's pressure (C_{12}–C_{44}), we classify SrO as ductile but SrS, SrSe, and SrTe are brittle material. To our knowledge these are the first quantitative theoretical prediction of the pressure and temperature dependence of mechanical stiffening, thermally softening, and brittle nature of SrX (X = O, S, Se, and Te) and still await experimental confirmations.

✉ Dinesh Varshney
vdinesh33@rediffmail.com

1 Materials Science Laboratory, School of Physics, Vigyan Bhavan, Devi Ahilya University, Khandwa Road Campus, Indore 452001, India

2 Department of Physics, S. D. Bansal College of Engineering, Indore 453331, India

3 Laboratoire de Physique Quantique et de Modélisation Mathématique (LPQ3 M), Département de Technologie, Université de Mascara, 29000 Mascara, Algeria

Keywords Chalcogenides · High pressure · Elastic properties · Mechanical properties · Thermal expansion

Introduction

Strontium chalcogenides SrX (X = O, S, Se, and Te) witnesses a pressure-induced phase transition from the six folds coordinated rock salt structure ($B1$) to the eightfold coordinated CsCl structure ($B2$) [1]. Strontium chalcogenides SrX (X = O, S, Se, and Te) have wide range of scientific and technological application in television picture tube glass, catalysis, microelectronics, luminescent devices, radiation dosimetry, fast high resolution optically stimulated luminescent devices, and infrared sensitive device [2].

Many theoretical investigations concerning transition pressure, volume collapse, higher order elastic constants, bulk modulus have been investigated using quantum mechanical calculations. The full-potential linear augmented plane-wave method (FP-LAPW) has been exercised to probe the ground state properties such as lattice parameter, bulk modulus, and its pressure derivative, elastic constant and structural stability of the compound SrS, SrSe, and SrTe, respectively [1–12].

The Burger mechanism was used to study lattice dynamics of SrX (X = S, Se and Te) and their pressure dependence in both $B1$ and $B2$ phases to predict the temperature and pressure dependence of thermal expansion coefficient, bulk modulus, and the heat capacity [2]. The Energy-dispersive X-ray diffraction (EDXD) observes the transition pressure (P_T) in SrO = 36 GPa [3, 4], SrS = 18

GPa [3, 7], SrSe = 14 GPa [3], and SrTe = 12 GPa, respectively [3, 5]. A percentage volume collapse at P_T in going from ($B1$ to $B2$) of about 13.0 in SrO [3, 4], 11.4 in SrS [3, 7], 10.7 in SrSe [3], and 11.1 in SrTe [3, 5] is reported using high-pressure X-rays diffraction studies. The elastic stiffness constant is obtained using ultrasonic pulse-echo technique [6].

Structural properties of the $B1$ and $B2$ phases of alkaline earth oxides and their phase transition have been investigated using ab initio linear muffin-tin-orbital (LMTO) method [9]. The phase transition of SrS from NaCl ($B1$) structure to CsCl ($B2$) structure is investigated by means of ab initio plane-wave pseudo potential density function theory [10]. The Born–Mayer potential-based calculations also predict the phase-transition pressure, bulk modulus, first-order pressure derivatives of bulk modulus in SrX (X = S, S, Se, and Te) [11]. The elastic properties and their pressure dependence of $B1$-type alkaline earth oxides are calculated using the ab initio full-potential linear muffin-tin-orbital (FP-LMTO) [12].

In view of the reported literature on technological important monochalcogenides as SrX (X = O, S, Se, and Te) [1–12], we notice that both the quantum calculations and the lattice model calculations predict the pressure-induced ground state properties and no efforts have been made to explore the temperature-dependent elastic, thermal, and thermodynamical properties of these chalcogenides. In view of these, we made systematic efforts to investigate (a) pressure-induced, and (b) temperature-induced properties. We further made efforts to compare them by means of thermodynamical variables as pressure and temperature. We further note that we properly incorporate the (a) the charge transfer interactions, (b) the covalency effect, and (c) the zero point energy effects in the model Hamiltonian.

For SrX (X = O, S, Se, and Te), although structural transitions are known to some extent, but to the best of our knowledge, no systematic efforts have been made to explore the high-pressure and high-temperature dependence of the aggregate elastic constants, their anisotropy, Melting temperature, Poisson's ratio, elastic wave velocity, Grüneisen parameter, isothermal (adiabatic) compressibility, Shear (Young's) modulus, Hardness, Lamé's constant, Kleinman parameter, and thermodynamic properties as Debye temperature, heat capacity and thermal expansion coefficient either in rock salt or CsCl structures.

The method

The understanding of pressure-dependent structural properties as first-order structural phase transition, associated volume collapse; elastic properties as ductility, mechanical stiffening, thermal softening, anisotropy in elastic constants, Shear (Young's) modulus, hardness, Lame's constant, Kleinman's parameter, shear and longitudinal elastic wave velocity as well thermodynamics properties viz Debye temperature, melting temperature, heat capacity and thermal expansion coefficient of SrX needs the formulation of an effective interatomic potential. The idea we have in mind follows: the change in force constants is small, the short-range interactions are effective up to the second-neighbor ions, and the atoms are held together with harmonic elastic forces without any internal strains within the crystal. Usually, the applications of pressures cause an increase in the overlap of adjacent ions in a crystal and hence, charge transfer takes place between the overlapping electron shells. The transferred charges interact with all others of the lattice via Coulomb's law and give rise to charge transfer interactions under pressure. We have also incorporated zero point energy effect, although it has a small effect in Gibbs free energy.

An isolated phase is stable only when it's free energy is minimized for the specified thermodynamic conditions. As the temperature or pressure or any other variable acting on the systems is altered, the free energy changes smoothly and continuously. A phase transition is said to occur when the changes in structural details of the phase are caused by such variations of free energy. The SrX ceramic transforms from their initial $B1$ to $B2$ structure under pressure. The stability of a particular structure is decided by the minima of Gibbs's free energy, $G = U + PV - TS$, U is internal energy, which at 0 K corresponds to the cohesive energy, S is the vibrational entropy at absolute temperature T, pressure P, and volume V.

We must mention that the calculations presented here assume zero temperature, i.e., the frozen ionic degrees of freedom. Although, the experimental results are obtained at ambient temperature inferring a certain small temperature dependence of the transition pressures in the range of low temperatures as well on polycrystalline or bulk samples. It is worth considering the lattice calculation results as representative of the results that would be obtained under the actual experimental conditions. At zero temperature, the thermodynamically stable phase at a given pressure P is the one with lowest entropy, and the thermo dynamical potential is the Helmholtz free energy (H).

The Gibbs's free energies for NaCl ($B1$) and CsCl ($B2$) phase is given by Born equation $G_{B1}(r) = U_{B1}(r) + 2Pr^3$ for NaCl ($B1$) and $G_{B2}(r') = U_{B2}(r') + 1.54Pr'^3$ for CsCl ($B2$) phase become equal at the phase-transition pressure P and at zero temperature, i.e., $G_{B1} = G_{B2}$ [13]. Here, the abbreviation U_{B1} (r) to the Rock Salt ($B1$) and U_{B2} (r') stands for the CsCl ($B2$) phase. Here, V_{B1} ($=2r^3$) and V_{B2} ($=1.54r'^3$) as the unit cell volume and are being the nearest neighbor distance for $B1$ (NaCl) and $B2$ (CsCl) phase, and its relevant expressions are

$$U_{B1} = (-\alpha_M Ze^2/r)[Z + 2nf(r)] - Cr^{-6} - Dr^{-8} + nb\beta_{ij}\exp\left[(r_i + r_j - r_{ij})/\rho\right] + (n'b/2)\left[\beta_{ii}\exp((2r_i - kr_{ij})/\rho) + \beta_{jj}\exp((2r_j - kr_{ij})/\rho)\right] + \left[\{\hbar < \omega^2 > ^{1/2}/2\right] \quad (1)$$

$$U_{B2} = (-\alpha_M' Ze^2/r')[Z + 2mf(r')] - Cr'^{-6} - Dr'^{-8} + mb\beta_{ij}\exp\left[\left(r_i + r_j - r_{ij}'\right)/\rho\right] + (m'b/2)\left[\beta_{ii}\exp\left(\left(2r_i - k'r_{ii}'\right)/\rho\right) + \beta_{jj}\exp\left(\left(2r_j - k'r_{jj}'\right)/\rho\right)\right] + \left[\{\hbar < \omega^2 > ^{1/2}/2\}\right] \quad (2)$$

Due to complex nature of bonds in SrX (X = O, S, Se, and Te,) the ionic charge for Sr and X atom cannot be determined uniquely. The calculations of the Madelung energy are thus modified by incorporating the covalency effects [14–17]. The charge in above equations is thus written incorporating the polarization of a spherical shaped dielectric in displacing the constituent positive ions. The charge transfer interactions caused by the deformation of the electron shells of the overlapping ions and the covalency effects are the major attributes of long-range Coulomb effects. The IIA–VIA semiconducting compound contains covalent bonds so that some electrons are distributed over the region between neighboring atoms; in such situation energies due to charge dipole–dipole and charge dipole–quadruple interaction. The induced charge dipole–dipole and charge dipole–quadruple (van der Waals) interaction are the third and fourth terms, which are the short-range vdW attractive potential energies.

The fifth and sixth terms are short-range (SR) repulsive energy due to the overlap repulsion between ij, ii and jj ions. α_m (α_m') are the Madelung constants for $B1$ ($B2$) phases. β_{ij} are the Pauling coefficient defined as $\beta_{ij} = 1 + (Z_i/n_i) + (Z_j/n_j)$ with $Z_i(Z_j)$ and $n_i(n_j)$ as the valence and the number of electrons in the outermost orbit. The number of the nearest neighbors for unlike are n (=6) and for like n' (=6) in case of $B1$ (NaCl). Similarly, numbers of the nearest neighbors for unlike m (=8) and m' (=3) are for $B2$ (CsCl). The Ze is the ionic charge, k (k') is being the structure factor for $B1$ ($B2$) structures, and b (ρ) is the hardness (range) parameters. The nearest neighbor ion separations for $B1$ ($B2$) structures are being r (r'), respectively.

The last term is the lowest possible energy of the system and is due to the zero point energy. Here, $\langle\omega^2\rangle^{1/2}$ ($=k_B\theta_D/\hbar$) is the mean square frequency related to the Debye temperature θ_D. The Debye temperature can be known either from Heat capacity measurements or from the Bulk modulus value using $\theta_D = (\hbar/k_B)\sqrt{(5r_0B_T/\mu)}$. Herein, r_0, B, and μ are the equilibrium distance, Bulk modulus, and reduced mass of the compounds. Henceforth, model potential for ground state incorporates the attractive, repulsive, and zero point energy.

We use the variational approach to deduce the overall vdW coefficients C (charge dipole–dipole) and D (charge dipole–quadruple) [18]. The short-range vdW coefficients due to induced charge dipole–dipole and charge dipole–quadruple interactions caused by Sr atom and X atom are

$$c_{ij} = \frac{3}{2}\frac{e\hbar}{\sqrt{m_e}}\alpha_i\alpha_j\left[\left(\frac{\alpha_i}{N_i}\right)^{1/2} + \left(\frac{\alpha_j}{N_j}\right)^{1/2}\right]^{-1}, \quad (3)$$

$$d_{ij} = \frac{27}{8}\frac{\hbar^2}{m_e}\alpha_i\alpha_j\left[\left(\frac{\alpha_i}{N_i}\right)^{1/2} + \left(\frac{\alpha_j}{N_j}\right)^{1/2}\right]^2 \times\left[\left(\frac{\alpha_i}{N_i}\right) + \frac{20}{3}\left(\frac{\alpha_i\alpha_j}{N_iN_j}\right)^{1/2} + \left(\frac{\alpha_j}{N_j}\right)\right]^{-1}. \quad (4)$$

$$c_{ii} = \frac{3}{2}\frac{e\hbar}{\sqrt{m_e}}\alpha_i\alpha_i\left[\left(\frac{\alpha_i}{N_i}\right)^{1/2} + \left(\frac{\alpha_i}{N_i}\right)^{1/2}\right]^{-1}, \quad (5)$$

$$d_{ii} = \frac{27}{8}\frac{\hbar^2}{m_e}\alpha_i\alpha_i\left[\left(\frac{\alpha_i}{N_i}\right)^{1/2} + \left(\frac{\alpha_i}{N_i}\right)^{1/2}\right]^2 \times\left[\left(\frac{\alpha_i}{N_i}\right) + \frac{20}{3}\left(\frac{\alpha_i\alpha_i}{N_iN_i}\right)^{1/2} + \left(\frac{\alpha_i}{N_i}\right)\right]^{-1} \quad (6)$$

$$c_{jj} = \frac{3}{2}\frac{e\hbar}{\sqrt{m_e}}\alpha_j\alpha_j\left[\left(\frac{\alpha_j}{N_j}\right)^{1/2} + \left(\frac{\alpha_j}{N_j}\right)^{1/2}\right]^{-1}, \quad (7)$$

$$d_{jj} = \frac{27}{8}\frac{\hbar^2}{m_e}\alpha_j\alpha_j\left[\left(\frac{\alpha_j}{N_j}\right)^{1/2} + \left(\frac{\alpha_j}{N_j}\right)^{1/2}\right]^2 \times\left[\left(\frac{\alpha_j}{N_j}\right) + \frac{20}{3}\left(\frac{\alpha_j\alpha_j}{N_jN_j}\right)^{1/2} + \left(\frac{\alpha_j}{N_j}\right)\right]^{-1}. \quad (8)$$

In the above equations, the notations: m_e, e, and Z are mass of the electron, charge, and valence of the constituent metallic element, respectively. The symbols α_i and α_j represents the polarizabilities of ith and jth ion, respectively. The effective numbers of electrons responsible for polarization are symbolized by N_i and N_j. The lattice sums S_{ij}, and T_{ij} enable one to compute the overall vdW coefficients C and D in terms of c and d values determined from Eqs. 1 to 8. The lattice sums S_{ij}, and T_{ij} are expressed as [19]:

$$C = c_{ij}S_{ij} + c_{ii}S_{ii} + c_{jj}S_{jj} \quad (9)$$

$$D = d_{ij}T_{ij} + d_{ii}T_{ii} + d_{jj}T_{jj}. \quad (10)$$

The binary alloy as SrX is a tetrahedrally coordinated covalent material and the complex chemical bonding corroborate both ionic and covalent nature. The Coulomb interaction between ions of Sr and *X* atoms leads to charge transfer interactions. Apart this, the covalent character of bond bending and stretching also needs to be incorporated in the potential. Thus, the second term in Eqs. (1, 2) is an algebraic sum of non-central many-body forces as the charge transfer force parameter and the force parameter arises due to covalent nature, i.e., $f(r) = f_{cti} + f_{cov}$. The charge transfer between ions of Si and C atoms is denoted in terms of a force parameter f_{cti} and is expressed as [20–22]:

$$f_{cti} = f_0 \exp(-r/\rho) \tag{11}$$

Here, r_i (r_j) is the ionic radii of $i(j)$ ion.

The complex chemical bonding in II–VI semiconducting compounds infers SrX as partially ionic and partially covalent in bonding. The attractive forces due to covalent nature thus modifies the charge and is now the effective charge. The polarization effects originate from changes in covalency due to Sr–Sr, Sr–*X*, and *X*–*X* interacting electric fields. The covalency term in the interaction potential is thus expressed as [16, 17]:

$$f_{cov}(r) = \frac{4e^2 V_{sp\sigma}^2}{r_0 E_g^3} \tag{12}$$

Here, $V_{sp\sigma}$ represents the transfer matrix element between the outermost *p* orbital and the lowest excited of *s* state. The transfer energy of electron from *p* to the *s* orbital is denoted as E_g. The effective charge e_S^* of SrX is related with the number of electrons transferred to the unoccupied orbital's from its surrounding nearest neighbor. The electron density is thus $n_c = 1 - e_S^*/e$. Thus in Sr*X* for overlap distortion effect $e_S^* \neq e$. The transfer matrix element $V_{sp\sigma}$ and the transfer energy E_g is related to electron density as $n_c/12 \cong V_{sp\sigma}^2/E_g^2$. The effective charge e_S^* is thus

$$\frac{V_{sp\sigma}^2}{E_g^2} = \frac{1 - e_s^*}{12} \tag{13}$$

The transfer energy E_g is

$$E_g = E - I + (2\alpha - 1)\frac{e^2}{r} \tag{14}$$

Here, *E* is the electron affinity for C and *I* is the ionization potential of constituent atom.

The Szigeti effective charge e_S^* $(=Ze)^*$ is written in terms of the optical static dielectric constant ε_0 and the high frequency dielectric constant ε_∞ as [23–31]:

$$e_s^{*2} = \frac{9\mu\omega_{TO}^2(\varepsilon_0 - \varepsilon_\infty)}{4\pi N_k(\varepsilon_\infty + 2)^2} \tag{15}$$

and

$$\frac{e_s^{*2}}{e^2} = \frac{9V\mu\omega_{TO}^2(\varepsilon_0 - \varepsilon_\infty)}{4\pi e^2(\varepsilon_\infty + 2)^2} \tag{16}$$

The symbol μ is the reduced mass, N_k is the number of atoms present per unit cell volume, i.e., $N_k = 1/V$, ω_{TO} is the long wavelength transverse optical phonon frequency. Thus, for Sr*X* e_S^* deviates from *e* and is attributed to covalent nature of Sr–Sr, Sr–*X*, and *X*–*X* bonds.

Usually material documents a transition when the thermo dynamical potential relevant to the given ensemble of the lower pressure phase equals that for some other structure in the absence of any barrier. The low-pressure phase becomes the stable phase above this coexistence pressure. After determining the stable phase we also compute the higher order elastic constants, their pressure derivatives and anisotropy. With these understanding of interatomic potential in Sr*X*, we have four material parameters, namely, modified ionic charge; hardness, range, force parameter [Z_m, *b*, ρ, $f(r)$]. Values of them can be deduced from equilibrium conditions [32–37].

Results and discussion

The application of pressure, temperature and magnetic field probably transform materials from one structure to another. The relative stability of two crystal structures requires an extremely accurate prediction. The interatomic interaction potential with charge transfer interactions caused by ions of Sr and X atom and covalent nature of Sr–Sr, Sr–*X*, and *X*–*X* bonds are effective in studying the structural phase transitions and elastic properties of tetrahedrally coordinated semiconductors Sr*X*. We evaluate the phase-transition pressure by computing the Gibbs free energy $G = U + PV - TS$ for the rock salt and cesium chloride phases. We note that the Gibbs free energy is thus the enthalpy H $(=U + PV)$ at $T = 0$ K.

While doing high-pressure experiments, the huge applied pressure causes a reduction of the material volume. The temperature variations during the experiments will normally produce much smaller changes in the relative stabilities of different phases. Thus, the Gibbs free energy at zero temperature, which is the enthalpy *H* is measured. The thermodynamically stable phase of crystal at 0 K and at ambient pressure *P* is the one with the lowest enthalpy. Thus, the zero temperature theoretical calculations are valid with experiment. In a situation when temperature variations are large during experiment for certain materials the effects of finite temperature may be significant. With these assumptions, we investigate structural and elastic properties of SrX in an ordered way.

The thermo dynamical potential G or H in SrX is computed involving modified ionic charge, hardness, range, and charge transfer parameters [Z_m, b, ρ, $f(r)$] as [23–31]:

$$\left|\frac{dU(r)}{dr}\right|_{r=r_0} = 0 \tag{17}$$

Also, the bulk modulus (B_T):

$$\left|\frac{d^2U(r)}{dr^2}\right|_{r=r_0} = (9kr_0)^{-1}B_T \tag{18}$$

We first deduce vdW coefficients C and D involved in expressions (1) and (2) from the Slater-Kirkwood variational method [18], for SrX material parameters, and are enlisted in Table 1. The charge dipole–dipole and charge dipole–quadruple vdW coefficients are influenced by electronic polarizabilities. The polarizability values have been obtained from least-squares fit of experimental refraction data using additive rule and a Lorentz factor of $4\pi/3$ [38, 39]. We consider that the SrX to be partially ionic and covalent to discuss their structural, mechanical, elastic, and thermo dynamical properties in a systematic manner.

As a next step, we use the experimental data on lattice constant (a) [3, 5–7, 40] the bulk modulus (B_T) [3], ionic (Ze), effective charge (e_s^*), and the second-order aggregate elastic constant C_{12} (C_{44}) [6, 8] for determining the material parameters. The computed strontium SrX (X = O, S, Se, and Te) chalcogenides, material parameter of hardness (b), range (ρ), and non-central many-body forces arose due to charge transfer (f_{cti}) and covalency (f_{cov}) for SrX is illustrated in Table 1. The effective charge e_S^* depends on the values of optical dielectric constant ε_s and the high-frequency dielectric constant ε_∞. The value of long wave length transverse optical phonon frequency ω_{TO} is taken from [2] to have the effective charge e_S^* and hence the covalency contribution.

We then minimize the Gibbs's free energies $G_{B1}(r)$ and $G_{B2}(r')$ for the equilibrium interatomic spacing (r) and (r') to determine structural phase transition of SrX (X = O, S, Se, and Te). Table 1 shows the optimized values of equilibrium interatomic spacing in $B1$ and $B2$ phases. The Gibbs's free energy $G_{B1}(r)$ [$G_{B2}(r')$] as functions of pressure (P) for SrX is discerned in Fig. 1a. At zero pressure, the Gibb's free energy for $B1$ crystal phase is more negative therefore; it is thermodynamically and mechanically

Table 1 Estimated and input crystal data: vdW coefficients [c_{ii}, c_{ij}, c_{jj}, C, d_{ii}, d_{ij}, d_{jj}, D], lattice constant (a_0), bulk modulus (B_T), second-order elastic constant C_{12} (C_{44}), optimized value of ionic radii r_i (r_j), hardness (b), range (ρ), charge transfer parameter $f(r)$, equilibrium distance (r_0), Gibb's free energy $G_{B1}(r)$; $G_{B2}(r')$, and cohesive energy for SrX (X = O, S, Se, and Te) in $B1$ phase at zero pressure

S. no.	Input parameters	SrO	SrS	SrSe	SrTe
1	c_{ii} (10^{-60} erg cm^6)	28.266	28.266	28.266	28.266
2	c_{ij} (10^{-60} erg cm^6)	52.249	95.1	108.314	132.858
3	c_{jj} (10^{-60} erg cm^6)	100.997	384.424	520.886	847.487
4	C (10^{-60} erg cm^6)	461.363	1000	1210	1667
5	d_{ii} (10^{-76} erg cm^8)	13.958	13.958	13.958	13.958
6	d_{ij} (10^{-76} erg cm^8)	33.409	87.421	109.342	156.948
7	d_{jj} (10^{-79} erg cm^8)	76.245	453.118	679.393	1300
8	D (10^{-76} erg cm^8)	241.406	724.114	949.36	1490
9	a_0 (Å)	5.160 [6]	6.024 [7]	6.234 [3]	6.659 [5]
10	B_T (GPa)	91 [3]	58 [3]	45 [3]	40 [3]
11	C_{12} (GPa)	45 [6]	17.5 [8]	13.4 [8]	7.8 [8]
12	C_{44} (GPa)	56 [6]	62.5 [8]	53.7 [8]	44.8 [8]
13	r_i (Å)	1.01	1.01	1.01	1.01
14	r_j (Å)	1.37	1.37	1.40	1.44.2.11
15	b (10^{-12} erg)	1.64	5.469	6.227	9.063
16	ρ (10^{-9} cm)	2.66	3.12	3.54	3.71
17	$f(r)$ (10^{-3})	0.6261	1.303	5.026	5.495
18	Equilibrium distance (Å) r_0 ($B1$)	2.67	3.11	3.26	3.46
19	Equilibrium distance (Å) r_0' ($B2$)	2.8	3.25	3.43	3.64
20	Gibb's free energy (kJ/mol) G_{B1} (r)	−1122	−1196	−1759	−1305
21	Gibb's free energy (kJ/mol) G_{B2} (r')	−1123	−1197	−1716	−1035
22	Cohesive energy (eV)	−33.93, −32.91 [11]	−60.70, −28.79 [11]	−61.43, −27.42 [11]	−67.26 −26.10 [11]

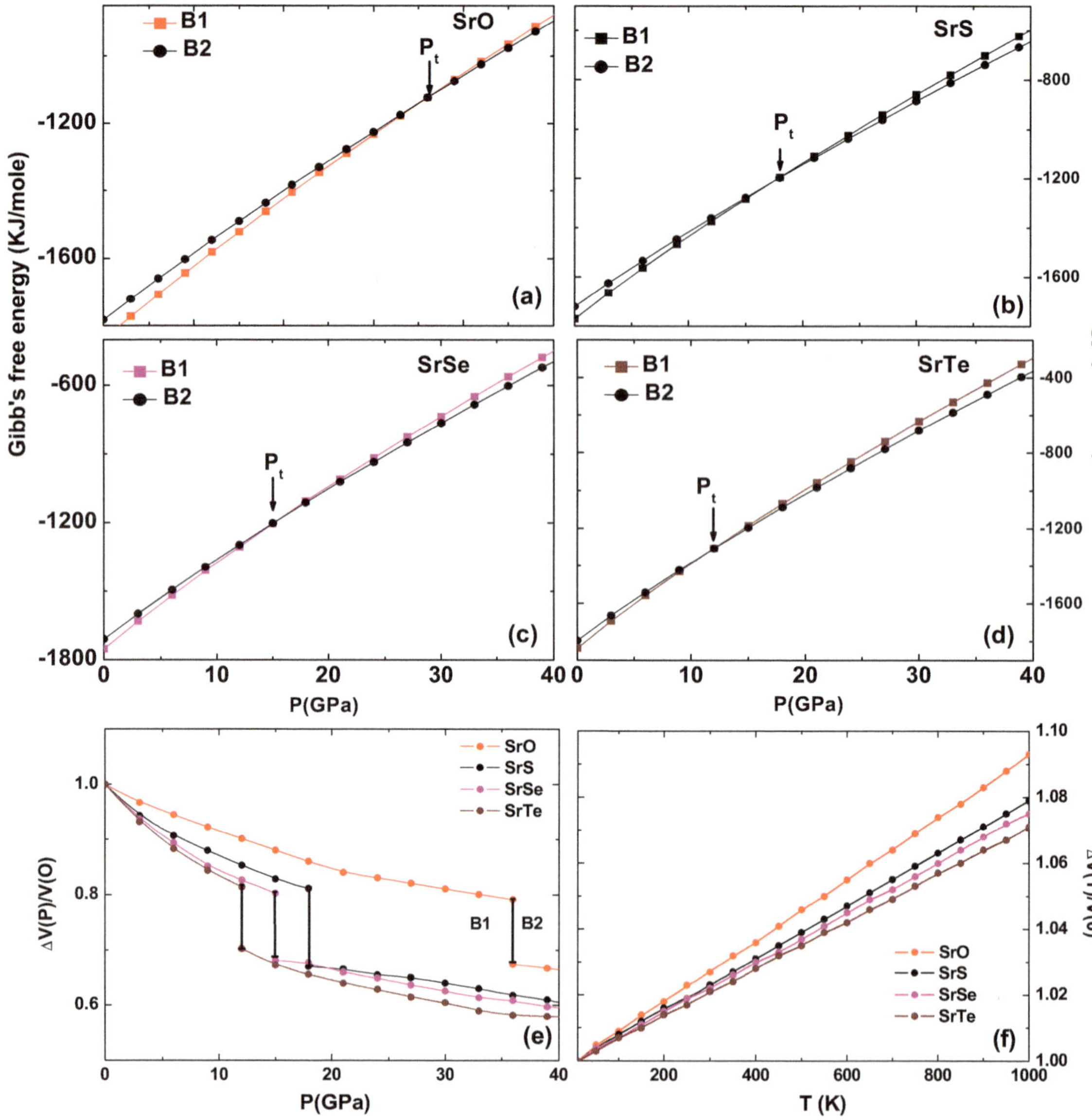

Fig. 1 Variation of Gibb's free energy for *B1* and *B2* phases with pressure and normalized volume with pressure and temperature

stable, while the *B*2 is not. On the other hand, above the phase-transition pressure (P_T), the Gibb's free energy for *B*1 system becomes more negative than *B*2 phase, implying *B*1 phase will be more stable. With the above-deduced material parameters, we have computed the cohesive energy for Sr*X* (*X* = O, S, Se, and Te), which is consistent with earlier deduced value [11].

Sr*X* shows a crystallographic transition from *B*1 to *B*2 in certain pressure range. Table 2 illustrates the computed phase-transition pressure (P_T) and compared with available experimental data [3–5] and theoretical results [1, 2, 11]. P_T for Sr*X* (*X* = O, S, Se, and Te) is consistent with the experimental and other reported values and is attributed to proper formulation of interatomic potential, which considers the various interactions explicitly the non-central many-body forces as charge transfer interactions and covalency effects, as well as use of materials parameter based on experimental data. We may comment that any computational technique has its own limitations related to the chosen materials basic parameters, basic sets, as well the accuracy and precisions used apart from the approximations laid in the method. Needless to suggest that there is always a variation in estimated parameters by each technique.

The values of relative volumes associated with various compressions are estimated from [41]

$$\frac{V_P}{V_0} = \left(1 + \frac{B'}{B_0}P\right)^{-1/B'}, \tag{19}$$

Here, V_0 (B_0) is the cell volume (bulk modulus) of Sr*X* at ambient conditions and V_P is at finite pressure. The symbol

Table 2 Calculated transition pressure (P_T), volume collapse (%), aggregate second-order elastic constants (C_{11}, C_{12} and C_{44}), aggregate bulk modulus (B_T), tetragonal moduli (C_S), pressure derivatives of SOECs (dB_T/dP, dC_{44}/dP and dC_S/dP), anisotropy parameter (γ_i^2), Cauchy discrepancy (Δ_1^2), isotropic shear modulus (G_H), Voigt's shear modulus (G_V), Reuss's shear modulus (G_R), Young's modulus (E), Poisson ratio (ν), Grüneisen parameter (γ_G) for SrX (X = O, S, Se, and Te) in *B1* phase at zero pressure

S. no.	Property		Present	Expt.	Others
1	Transition pressure P_T (GPa)	SrO	35.5	36.0 [3, 4]	30.9–35.3 [1, 11]
		SrS	17.6	18.0 [3]	16.7–17.7 [1, 2]
		SrSe	14.9	14.0 [3]	13.6–16.2 [1, 2]
		SrTe	11.92	12.0 [3, 5]	12.3–14.1 [1, 2]
2	Volume collapse (%)	SrO	11.7	13.0 [3, 4]	11.8 [9]
		SrS	14.1	11.4 [3]	11.5 [2]
		SrSe	12.1	10.7 [3]	10.5 [2]
		SrTe	11.1	11.1 [3, 5]	9.50 [2]
3	C_{11} (GPa)	SrO	133.6		170 [12]
		SrS	79.7		113.9 [10]
		SrSe	57.3		
		SrTe	54.8		
4	C_{12} (GPa)	SrO	34.05		48 [12]
		SrS	17.81		19.4 [10]
		SrSe	14.74		
		SrTe	13.0		
5	C_{44} (GPa)	SrO	46.8		59 [12]
		SrS	26.6		30.3 [10]
		SrSe	23.9		
		SrTe	21.0		
6	B_T (GPa)	SrO	67.3	88 [6]	86–106 [1, 12]
		SrS	38.4		53.9–58.6 [1, 10]
		SrSe	28.9		50.3 [1]
		SrTe	26.9		39.7 [1]
7	C_S (GPa)	SrO	49.8		
		SrS	30.9		
		SrSe	21.3		
		SrTe	20.9		
8	dB_T/dP	SrO	5.40	6.0 [6]	4.08–4.5 [9, 12]
		SrS	5.35		4.19–4.66 [8, 10]
		SrSe	4.92		3.76 [8]
		SrTe	4.94	5.00 [5]	3.23 [8]
9	dC_{44}/dP	SrO	0.266	−0.2 [6]	
		SrS	0.111		
		SrSe	0.104		
		SrTe	4.53		
10	dC_S/dP	SrO	4.87	4.0 [6]	
		SrS	5.05		
		SrSe	4.40		
		SrTe	4.53		
11	$\gamma_1^2 (10^{10}\ Nm^{-2})$	SrO	0.063		
		SrS	0.163		
		SrSe	−0.108		
		SrTe	0.005		
12	$\Delta_1^2\ (10^{10}\ Nm^{-2})$	SrO	−1.276		
		SrS	−0.088		
		SrSe	−0.091		
		SrTe	−0.781		

Table 2 continued

S. no.	Property		Present	Expt.	Others
13	G_H (10^{10} Nm^{-2})	SrO	4.8		
		SrS	2.82		
		SrSe	2.28		
		SrTe	2.08		
14	G_V (10^{10} Nm^{-2})	SrO	4.8		
		SrS	2.83		
		SrSe	2.28		
		SrTe	2.08		
15	G_R (10^{10} Nm^{-2})	SrO	4.8		
		SrS	2.82		
		SrSe	2.28		
		SrTe	2.08		
16	E ($10^{10}Nm^{-2}$)	SrO	11.63		
		SrS	6.81		
		SrSe	5.42		
		SrTe	4.97		
17	ν	SrO	0.212		
		SrS	0.205		
		SrSe	0.188		
		SrTe	0.192		
18	γ_G	SrO	1.523		
		SrS	1.508		
		SrSe	1.403		
		SrTe	1.425		

B' is the pressure derivative of the bulk modulus. Figure 1e documents the estimated value of pressure-dependent radius for both $B1$ and $B2$ structures, the curve of volume collapse (V_P/V_0) with pressure to show phase diagram for Sr*X*. The phase diagram will let us estimate the magnitude of the discontinuity in volume at the transition pressure. The value of relative volumes is shown in Table 2. It is also compared with various experimental [3–5] and other theoretical works [2, 9]. Compressions in Sr*X* at higher pressure indicate the mechanical stiffenening of lattice.

Figure 1f discerns the variation of V_T/V_0 as functions of temperature in $B1$ phase. Here, V_T symbolizes the volume at various temperatures and V_0 at zero temperature and zero pressure volumes, respectively. A steep increase in the ratio V_T/V_0 with increasing temperature infers expansion of Sr*X* lattice and is susceptible to temperature. On the other hand, Sr*X* is compressed at higher pressures as showed in Fig. 1e. Sr*X* lattice is thus thermally softened and mechanical stiffened. The normalized volume V_T/V_0 dependences on temperature is not known for Sr*X*, but the present behavior is consistent with available experimental [42] and theoretical [43] data on Li_2O.

The response of any material that undergoes stress, deforms and then recovers and returns to its original shape after stress ceases leads to the determination of elastic properties. The elastic properties are vital in generating information about the binding characteristic between adjacent atomic planes, anisotropic character of binding and structural stability. Apart from the structural stability of Sr*X* in NaCl ($B1$) and CsCl ($B2$) structures, we now compute the aggregate elastic constants at normal and under hydrostatic pressure. Deduced values are documented in Table 2.

Using the stress–strain coefficients, one determines the second-order aggregate elastic constants C_{ij} under hydrostatic pressure with respect to finite strain. Also, proper parameterization of long-range as Coulomb and non-central many-body forces as charge transfer interactions and covalency effect, short-range as overlap repulsion extended up to the second-neighbor ions, van der Waals interactions and zero point energy effects are differentially needed.

The cubical symmetry of Sr*X* dealt with three independent elastic constants C_{ij}. C_{11} is a response of resistance to deformation by a stress applied on (1,0,0) plane with polarization in the direction $\langle 100 \rangle$. C_{11} probes elasticity in length and a longitudinal strain produces a change in C_{11}. C_{44} refers to the measurement of resistance to deformation with respect to a shearing stress applied across the (100) plane with polarization in the $\langle 010 \rangle$ direction. C_{12} and C_{44} are related to the elasticity in shape, which is a shear

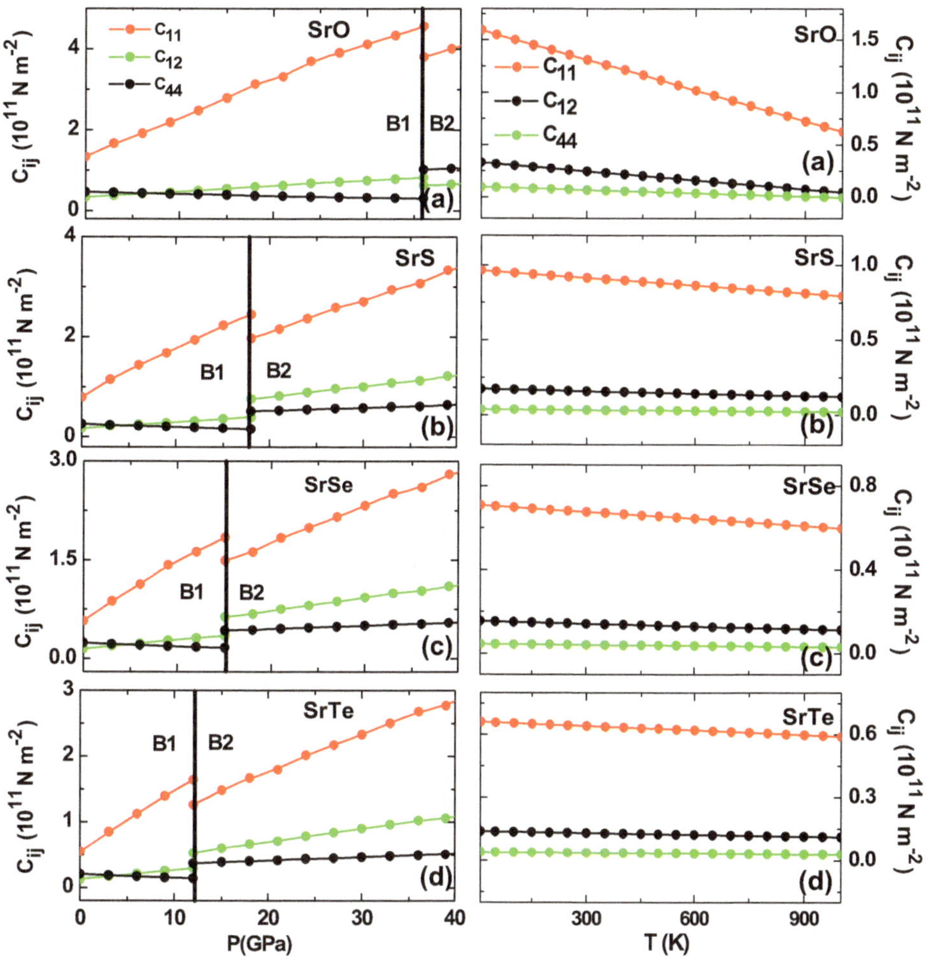

Fig. 2 Variation of aggregate second-order elastic constants (C_{ij}) with pressure and temperature

constant. A transverse strain causes a change in shape without a change in volume and hence C_{12} and C_{44} are less sensitive of pressure as compared to C_{11}.

The variation of three independent second-order aggregate elastic constants (SOECs): C_{11}, C_{12}, and C_{44} with external pressure for Sr*X* in *B*1 and *B*2 phase is first discussed. It is as seen from Fig. 2a that C_{11} and C_{12} increases with increase in pressure in both *B*1 and *B*2 phases. Also, C_{44} decreases with the increase of pressure away from zero till the phase-transition pressures and then increases in *B*1 phase. Similar observations have earlier been reported in Sr*X* [10, 12]. A crossover of C_{12} and C_{44} in CaS has also been reported [44]. At phase-transition pressures, Sr*X* has witnessed a discontinuity in aggregate second-order elastic constants C_{ij}, which identifies the first-order phase transition. Thus, the proposed interaction potential incorporating charge transfer interactions ions of Sr and X atom and covalency effects caused by Sr–Sr, Sr–*X*, and *X*–*X* bonds consistently explains the high-pressure elastic behavior.

The variations in C_{ij} with temperatures (*T*) for Sr*X* are plotted in Fig. 2b. It can be seen that the aggregate elastic constants C_{ij} (*T*) decreases linearly with the temperature in rock salt phase. We note that the pressure dependence of aggregate elastic constants C_{ij} (*P*) documents an increasing trend (please see Fig. 2a). The physical interpretation of temperature dependence of C_{ij} showed that (a) values of C_{11} decreases more steeply with enhancing temperature, (b) C_{12}, and C_{44}, are less sensitive to temperature for SrX, (c) C_{11} is remarkably larger than C_{12}, and C_{44}, and (d) values of all aggregate elastic constants C_{ij} are influenced by temperature dependence indicating that anharmonicity is substantial. Deduced values of C_{ij} with temperatures (*T*) are documented in Table 4. As far as we know, there are no experimental results available for comparison. From the second-order aggregate elastic constants pressure and temperature-dependent behavior, we comment that Sr*X* lattice is mechanical stiffened and thermally softened.

Born criterion for a lattice to be mechanically stable infers that the elastic energy density must be a positive definite quadratic function of strain. The principal minors (alternatively the Eigen values) of the elastic constant matrix should all be positive at ambient conditions. The mechanical stability conditions for a crystal suggest that elastic constants of a cubic crystal follow [13],

$$B_T = (C_{11} + 2C_{12})/3 > 0, \quad (20)$$

$$C_{11}, C_{44} > 0, \quad (21)$$

and

$$C_S = (C_{11} - C_{12})/2 > 0. \quad (22)$$

Here, C_{ij} are conventional aggregate elastic constants and B_T is bulk modulus. We represent, C_{44} and C_S as the shear and tetragonal moduli of a cubic crystal.

Table 2 illustrates the computed values of bulk modulus (B_T), shear moduli (C_{44}), and tetragonal moduli (C_S) which validates the elastic stability criteria for SrX in $B1$ phase. The second-order elastic constants critically depend upon pressure leading to $C_{12}-C_{44} \neq 0$. The mechanically stable phases for cubic crystal satisfy the Born criteria: $C_{12}-C_{44} > 0$. The equilibrium condition leads to $B_1 + B_2 = -1.261Z_m^2$ with emphasis on charge transfer interactions as well covalency effects. For optimized values of r_i (r_j) the Cauchy discrepancy: $C_{12}-C_{44}$ is nonzero at zero pressure and at zero temperature. It is also valid when the many-body non-central forces are not involved in long-range forces. The short-range and long-range effects are naturally of similar order of magnitude. This is due to the fact C_{ij} are calculated at optimized values of equilibrium distances rather than at experimental values.

Table 2 illustrates the calculated values of pressure derivatives of aggregate second-order elastic constants (dB_T/dP, dC_{44}/dP, and dC_S/dP). These are compared with available experimental [5, 6] and theoretical studies [8–10, 12]. For mechanical stability the shear elastic constant C_{44} is nonzero and is known by combining mechanical stability with minimum energy conditions. The high-pressure stability also suggests that the stable phase of the crystal posses the lowest potential energy among the mechanically stable lattices [45].

The elasticity in SrX is thus probed by a non-central many-body force potential, which assumes that the interatomic forces have a certain shape and directionality. The Cauchy discrepancy is defined as $\Delta_1^2 = C_{12}-C_{44}-2P$. Here, Δ_1^2 is a measure of the contribution from the non-central many-body force. However, for pure central interatomic potentials, Cauchy relation is $C_{12} = C_{44} + 2P$.

At zero pressure, the Cauchy discrepancy (Δ_1^2) in Strontium chalcogenides is about −1.276 (SrO), −0.088 (SrS), −0.091 (SrSe), -0.781×10^{10} (SrTe) Nm^{-2}. The Δ_1^2 further enhances on increasing the pressure in both phases as depicted from Fig. 3a. In SrX, larger deviation of Δ_1^2 essentially points to the importance of the many-body non-central (charge transfer and covalency) interaction in the interatomic potential and anharmonic effects are substantial at high pressures. The strength of non-central many-body forces incorporating charge transfer interactions and covalency effects is witnessed by significant deviation in Δ_1^2 at different pressures not only in Rock salt but also in cesium chloride structure, although weak. The importance of many-body non-central forces and anharmonic effects are further explored by analyzing the higher order elastic constants explicitly the third-order elastic constants. Usually, the anharmonic effects are noticeable at high pressure as reflected from elastic constants behavior.

The anisotropy in second-order elastic constants is reflected from geophysical activities of various materials and alloys. The anisotropic parameter γ is unity for isotropic elasticity. As far as cubic crystal is concerned although it is isotropic in structure, but has elastic anisotropy other than unity. This is a consequence of fourth rank tensor property of elasticity.

We define elastic anisotropic parameter γ_1^2 in terms of aggregate C_{ij} as [46]:

$$\gamma_1^2 = \frac{C_{11} - C_{12} - 2C_{44}}{2C_{44}} \quad (23)$$

The pressure dependence of the elastic anisotropic parameter γ_1^2 in SrX is shown in Fig. 3b. It is evident that γ_1^2 increases with advancement of pressure below transition pressure in $B1$ phase of SrX. It is insensitive to pressure in $B2$ phase. A jump has been noted at P_T inferring first-order structural phase transition. The value of anisotropic parameter γ_1^2 for SrX is given in Table 2 at $T = 0$ K and $P = 0$ GPa.

During mechanical processing, explicitly in fabrication the melting ranges of materials and alloys are substantial. The usage of an alloy in the applications as the success of the melting and casting operations depends on the correct selection of temperature. Once solidified and primary processed (rolling or forging), the melting temperature has little significance to designers, engineers and users. The melting temperature influence elevated temperature properties, such as creep strength, the researchers have limited interest. The pressure dependence of the melting temperature: $T_M = 553K + 5.91C_{11}K/\text{GPa}$ for SrX in $B1$ and $B2$ phase is discerned in Fig. 4a. It is noticed that T_M enhances with increased pressure or in other words the resistance to deformation by a stress increases. The value of melting temperature T_M for SrX is given in Table 3 at $T = 0$ K and $P = 0$ GPa An increase in T_M with variations in pressure infers the hardening or stiffening of the lattice. Higher melting temperature symbolizes higher shear modulus (G), and Young's modulus (E) values that we shall see later on. The data on its melting under high pressure are very limited and extremely contradictory, which does not allow one to make any conclusions about congruent or incongruent

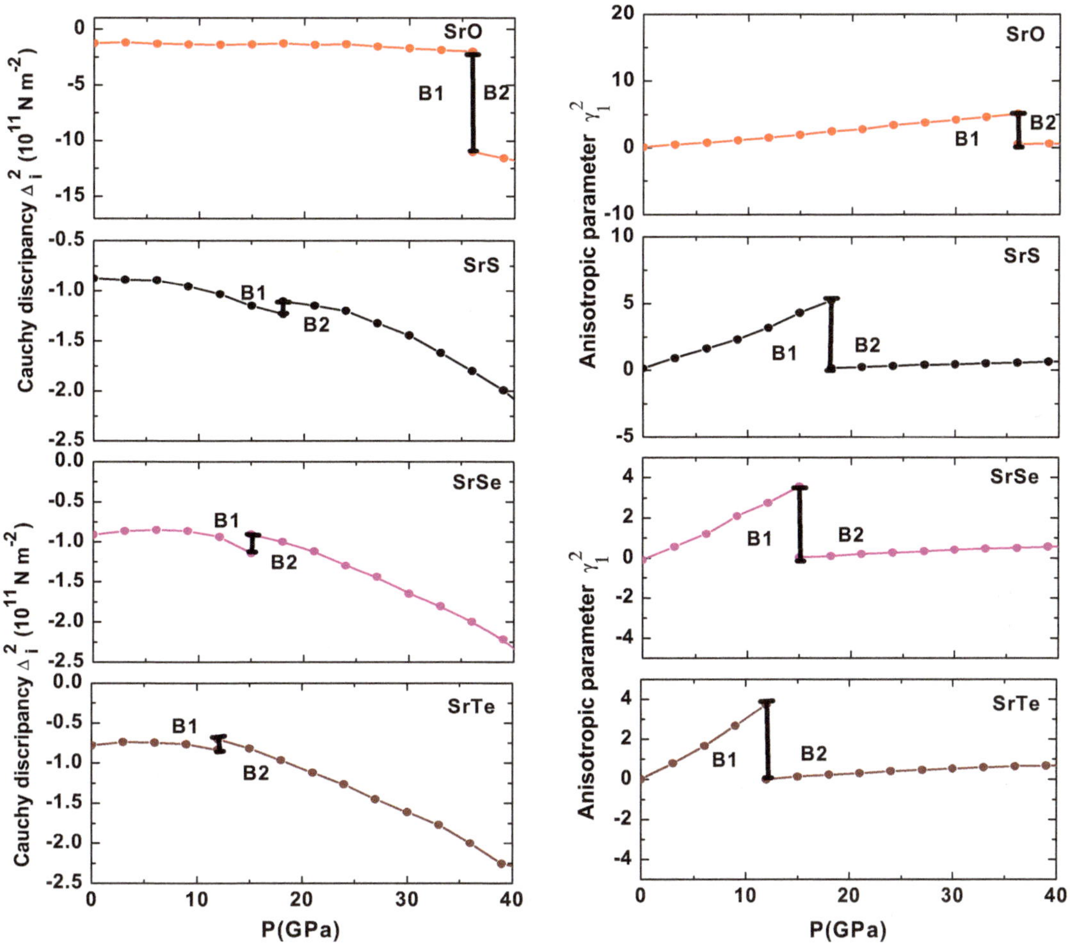

Fig. 3 Variation of Cauchy discrepancy (Δ_1^2) and elastic anisotropy (γ_1^2) in second-order elastic constant (C_{ij}) with pressure and temperature

melting behavior as well the slope of the melting curve of Sr*X*.

Figure 4b shows the temperature dependence of the melting temperature for Sr*X* estimated from the C_{11} elastic constant as discussed previously. At room temperature its value is about 1327 K (SrO), 1094 K (SrS), 952 K (SrSe), 931 K (SrTe) in *B*1 phase. The suppressed T_M with increased temperature indicates that there is a decrease in the resistance to deformation by a stress induced due to temperature. The suppressed T_M infers the weakening of the lattice as a result of thermal softening.

For cubic lattice, three second-order elastic constants and the six non-vanishing third-order elastic constants are obtained from crystal geometry. The anharmonicity of a crystal lattice is successfully probed in terms of higher order elastic constants. The third-order terms in the strain variables are deduced from derivatives of elastic energy (please see Appendix for both rock salt *B*1 and cesium chloride *B*2 phases). For Sr*X* the third-order aggregate elastic constants: C_{111}, C_{112}, C_{166} are negative and C_{123}, C_{144}, C_{456} are positive at $P = 0$ GPa, i.e., in *B*1 phase. We note that no such efforts have been made in the past for third-order elastic constants of Sr*X*. Thus, deduced information on pressure-dependent C_{ijk} will serve as a guideline for future research and its application in materials technology. This information is useful, as the structural strengths of a material are successfully known from the knowledge of elastic constants. The microstructures developed on the applied pressure and temperature can be known from third-order elastic constants variations.

The variation of aggregate third-order elastic constants (TOECs) with pressure for Sr*X* in *B*1 and *B*2 phase is sketched in Fig. 5. For Sr*X* only C_{456} increase with enhancing pressure. Other TOEC's as C_{111}, C_{112}, C_{166} infers a decreasing trend and C_{123}, C_{144} are insensitive with enhancing pressure. Furthermore, aggregate elastic constants C_{ijk} inferring the anharmonic effects are influenced by application of pressure in Sr*X*. In the present interatomic potential, C_{ijk} are affected by the inclusion of second-nearest-neighbor interaction and are influenced by the short-range interactions. Also, many-body non-central forces as long-range Coulomb, charge transfer interactions

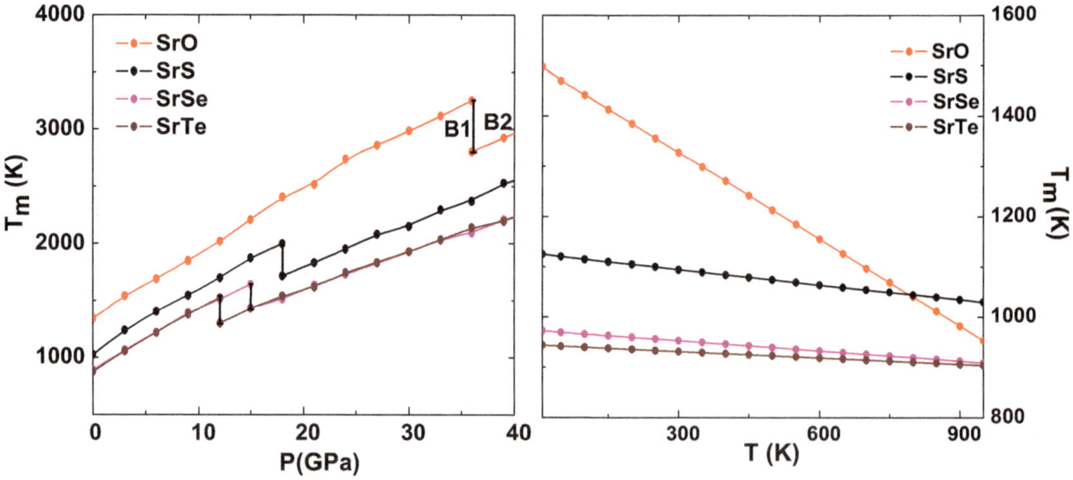

Fig. 4 Variation of melting temperature (T_M) with pressure and temperature

Table 3 Calculated aggregate third-order elastic constant (C_{ijk}), third-order elastic constant anisotropy parameter (γ_i^3), First and second Lame's constant (λ, μ), longitudinal (v_l), shear (v_s) and average elastic wave velocity (v_m), Kleinman parameter (ξ) thermodynamical properties: force constant (f), Reststrahlen frequency (v_0), compressibility (β), average elastic constant (C_s), Debye temperature (θ_D), melting temperature (T_M), force constant (f), hardness (H_v), of SrX (X = O, S, Se, and Te) in *B1* phase at zero pressure

S. no.	Elastic properties	SrO	SrS	SrSe	SrTe
1	C_{111} (10^{10} Nm^{-2})	−252.70	−150.40	−100.70	−95.64
2	C_{112} (10^{10} Nm^{-2})	−17.85	−9.20	−6.99	−6.06
3	C_{123} (10^{10} Nm^{-2})	67.9	40.1	30.5	2.69
4	C_{144} (10^{10} Nm^{-2})	69.0	41.5	34.7	3.08
5	C_{166} (10^{10} Nm^{-2})	−19.78	−10.79	−10.07	−8.86
6	C_{456} (10^{10} Nm^{-2})	6.38	3.84	3.40	3.05
7	γ_1^3	−17.445	−15.711	−23.479	−22.286
8	γ_2^3	−2.784	−2.593	−2.447	−2.42
9	γ_3^3	3.093	−2.947	−2.991	−2.957
10	λ (10^{10} Nm^{-2})	3.53	1.96	1.37	1.3
11	μ (10^{10} Nm^{-2})	4.8	2.82	2.28	2.08
12	v_l (m/s)	5389	4801	3856	3560
13	v_s (m/s)	3258	2925	2390	2197
14	v_m (m/s)	769.822	691.289	565.081	520.064
15	ξ	0.405	0.374	0.407	0.388
16	θ_D (K)	245.578	189.156	145.81	126.839
17	T_M (K)	1343	1024	891.821	876.956
18	f (10^5 dyne/cm)	1.998	1.382	1.188	1.138
19	v_0 (10^{12} Hz)	2.855	1.804	1.257	1.1
20	β (10^{-11} Pa^{-1})	8.231	4.883	3.985	3.577
21	C_s (10^{10}N/m^2)	4.98	3.09	2.13	2.09
22	H_v (10^9 Nm^{-2})	9.970	6.847	6.429	5.760

and covalency are effective in Sr*X*. We comment that pressure-dependent C_{ijk} behavior can have a direct means to describe the interatomic forces at high pressure. It successfully cares the short-range forces, and a balance between long-range and short-range forces.

As a next step, we intend to study the anisotropy among C_{ijk}. The equilibrium condition in the interatomic potential infers $B_1 + B_2 = -1.261\ Z_m^2$. The Cauchy discrepancy Δ_i^3 among third-order elastic constants are: (a) $\Delta_1^3 = C_{112} - C_{166} - 2P$, (b) $\Delta_2^3 = C_{123} - C_{456} - 2P$, (c) $\Delta_3^3 = -C_{144} - C_{456} - 2P$, and (d) $\Delta_4^3 = C_{123} - C_{144} - 2P$. The C_{ijk} at zero pressure, i.e., in *B*1 phase are influenced by contributions from both long-range and short-range interactions. Henceforth, Δ_i^3 is an indicator of the contribution from the non-central many-body force. Figure 6a shows the variation of Δ_i^3 as functions of pressure. The significant deviation in Δ_i^3 is a natural consequence of the non-central many-body forces as charge transfer interactions ions of Sr

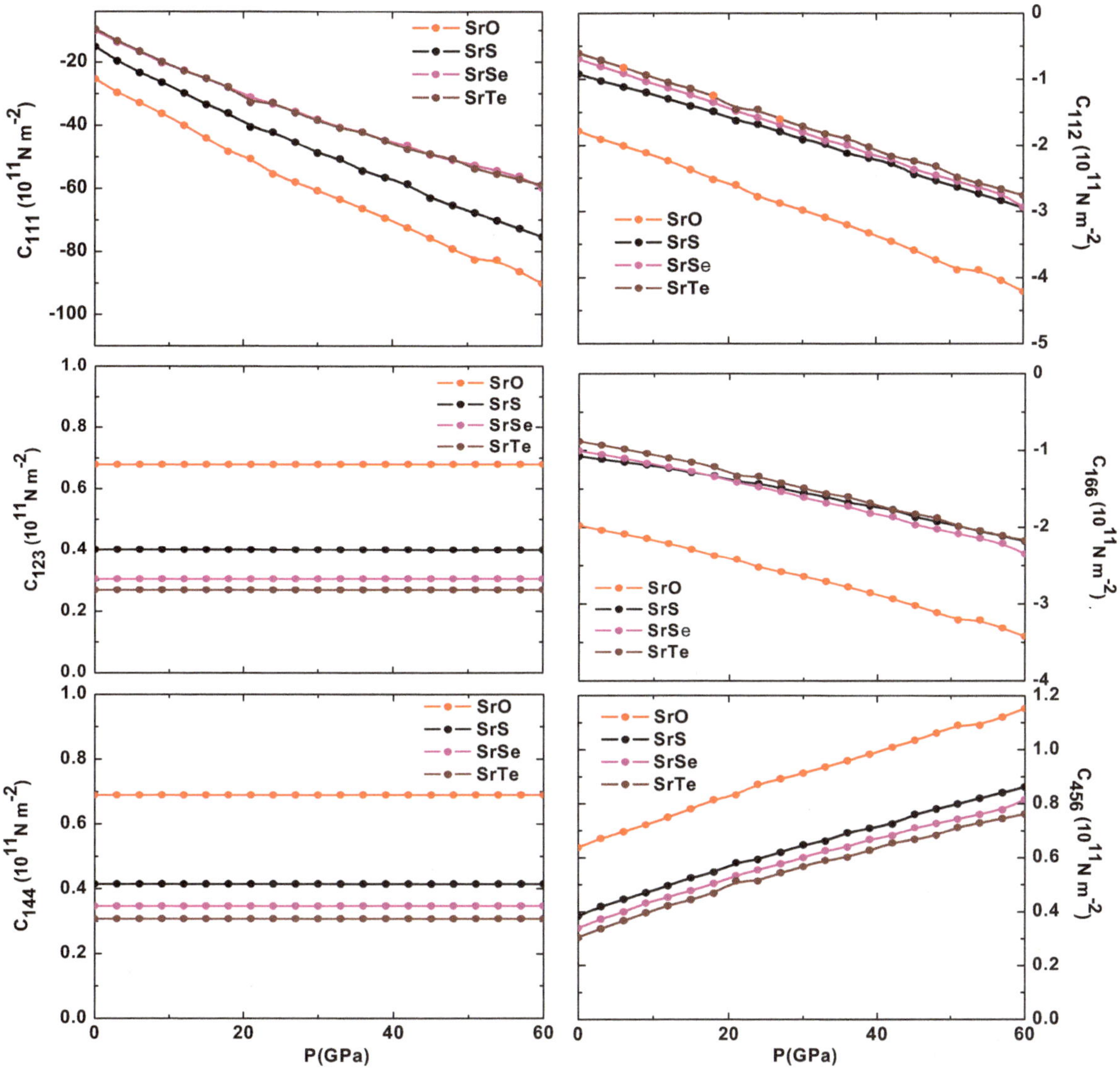

Fig. 5 Variation of aggregate third-order elastic constants (C_{ijk}) with pressure

and X atom and covalency effects caused by Sr–Sr, Sr–X, and X–X bonds apart from short-range interactions as the induced charge dipole–dipole and charge dipole–quadruple (van der Waals) interaction and the overlap repulsion. At zero pressure, Δ_2^3 and Δ_3^3 are positive, while to that Δ_1^3 and Δ_4^3 are negative in $B1$ phase. A growth in Δ_2^3 and Δ_3^3 and decay in with enhanced pressure is witnessed with enhanced pressure till phase-transition pressure. However, all Δ_i^3 is negative in $B1$ phase indicating the importance of non-central many-body forces and anharmonic effects in SrX. As far as we know, there are no experimental results available for Cauchy discrepancy Δ_i^3 in SrX.

For cubic crystal structures, the aggregate third-order elastic constants C_{ijk} discern three anisotropy coefficients and three isotropic coefficients. It is useful to express linear combinations of the anisotropy coefficients and dividing them by the isotropic coefficients. The anisotropy coefficients γ_i^3 follows:

$$\gamma_1^3 = \frac{3C_{112} - C_{111} - 12C_{144} + 12C_{166} - 16C_{456} - 2C_{123}}{2C_{123}} \tag{24}$$

$$\gamma_2^3 = \frac{C_{111} - C_{123} - 2C_{144}}{2C_{144}} \tag{25}$$

$$\gamma_3^3 = \frac{C_{166} - C_{144} - 2C_{456}}{2C_{456}} \tag{26}$$

The pressure dependence of the elastic anisotropic parameter γ_i^3 for both phases is sketched in Fig. 6b. The elastic anisotropy (γ_2^3 and γ_3^3) in $B1$ phase are less sensitive. On the other hand γ_1^3 show variation with increase in pressure in $B1$ phase. The anisotropy factor γ_1^3 is negative

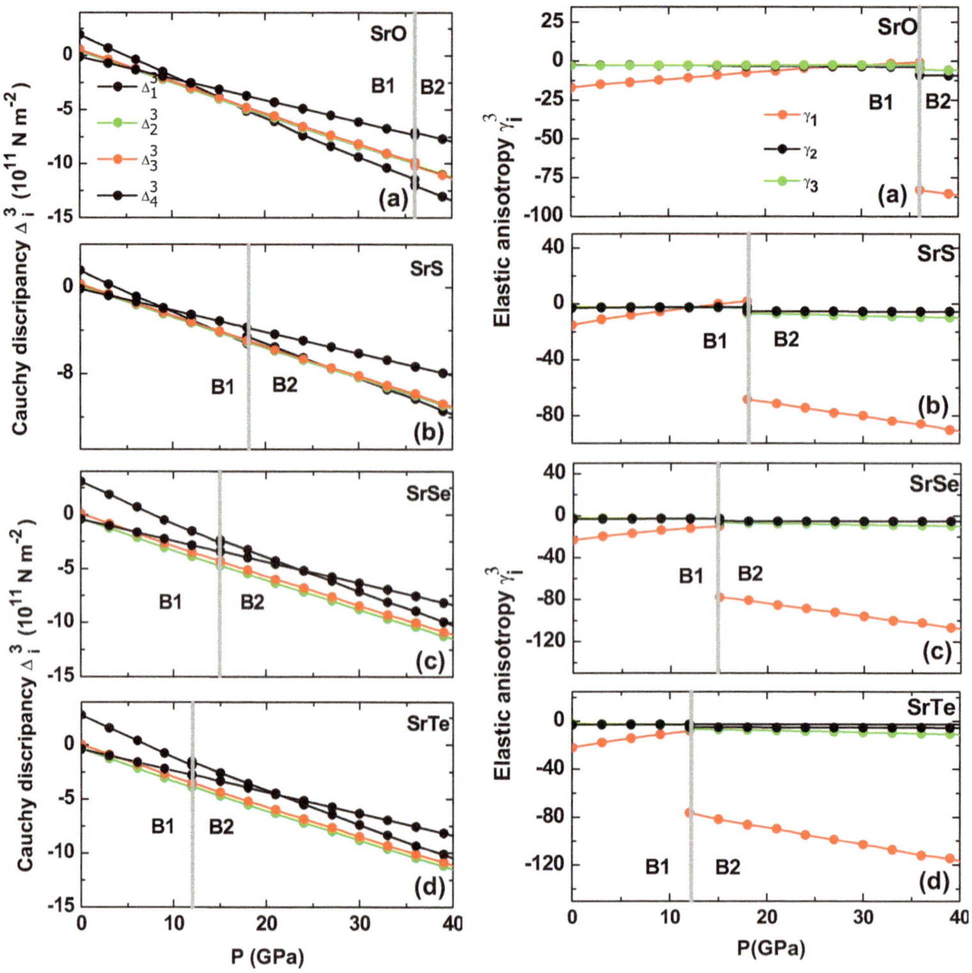

Fig. 6 Variation of Cauchy discrepancy (Δ_i^3) and elastic anisotropy (γ_i^3) in third-order elastic constant with pressure

and shows a decreasing trend with pressure in $B1$ phase at higher pressures. Value of γ_i^3 is given in Table 3 at zero temperature and pressure. As far as we know, there are no experimental and theoretical results available for comparison.

The strength and hardness are key issues for materials useful in technological applications. The mechanical strength and hardness of SrX can be known once elastic constants are known. We now determine elastic moduli at various pressures of SrX which are of substantial importance in engineering, geophysical and seismological application. The elastic properties are important in probing the bonding characteristic between adjacent atomic planes and the anisotropic character of the solid. Crystals leading to elasticity under application of pressure identify the response of a crystal under external strain. This valuable information is characterized by isotropic shear modulus (G_H), and Young's modulus (E). We note that hardness of polycrystalline materials is inhibited in elastic response as G_H and E (Fig. 7).

The resistance to volume change and resistance to reversible deformations upon applied pressures is accounted by Voigt–Reuss–Hill approximation using G_H and E following [47–49]:

$$G_H = \frac{G_V + G_R}{2} \tag{27}$$

Herein, G_V (G_R) is Voigt's (Reuss's) shear modulus corresponding to the upper (lower) bound of G_H values. The G_V (G_R) is obtained from C_{ij} assuming uniform strain throughout the sample as

$$G_V = \frac{C_{11} - C_{12} + 3C_{44}}{5} \tag{28}$$

and

$$\frac{5}{G_R} = \frac{4}{(C_{11} - C_{12})} + \frac{3}{C_{44}} \tag{29}$$

Hardness of materials is usually measured in terms of isotropic shear modulus (G_H), and Young's modulus (E) or bulk modulus (B_T). The bulk modulus infers resistance to

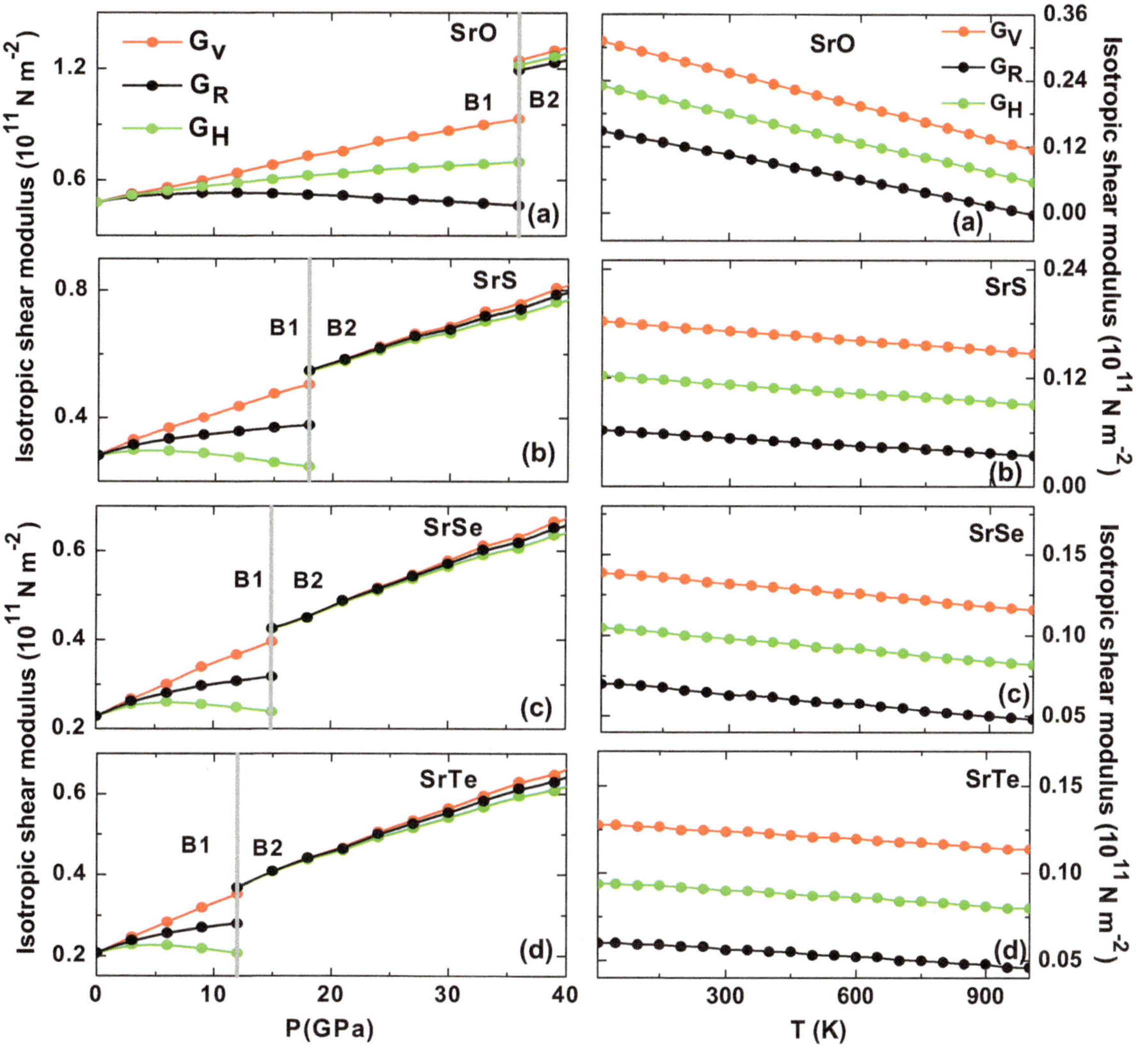

Fig. 7 Variation of isotropic shear modulus (G_H, G_V, and G_R) with pressure and temperature

volume change by applied pressure. The isotropic shear modulus (G_H) measure the resistance to reversible deformations upon shear stress. In such a situation, G_H is a better probe to measure hardness than the bulk modulus (B_T). Table 2 illustrates the calculated isotropic shear modulus (G_H), Voigt's shear modulus (G_V) and Reuss's shear modulus (G_R) for SrX at zero temperature and pressure. SrX (X = O, S Se, Te). A decreasing and then increasing nature of the isotropic shear modulus G_H, G_V and G_R is witnessed in $B1$ phase of SrX. On the other hand, a steep increase in G_H, G_V and G_R is seen with enhanced pressure in $B2$ phase at higher pressures as plotted in Fig. 7a. The explanation of the above characteristics lies in a fact that both G_V and G_R are influenced by (a) combination of C_{11}–C_{12} and (b) C_{44}. It is noticed that G_R initially decreases due to competition of these elastic constants and then start increasing above P_T in all SrX (X = O, S, Se, Te) Chalcogenides. In later phase, increasing trend in shear modulus G_H, G_V and G_R is attributed to a steep increase in $C44$ with pressure. We end up by stating that above transition pressure an increase in G_H, G_V and G_R support our earlier prediction about mechanical stiffening of lattice.

The high temperature studies of materials at ambient pressure leads to the performance of a material in terms of (a) the understanding of vibrational anharmonicity that is associated with the relative interplay of long-range and short-range potential energy function, (b) thermal response in terms of softening or hardening and (c) the elastic behavior of materials as the elasticity, extensibility, acoustic transmission velocity, Debye temperature, specific heat, and thermal conductivity.

The high temperature investigations cause laboratory difficulties as well structural changes make the phenomenon more amenable to interpretation. Figure 7b displays the G_H, G_V and G_R behavior of SrX ceramics as functions of temperature (at zero pressure). We note that

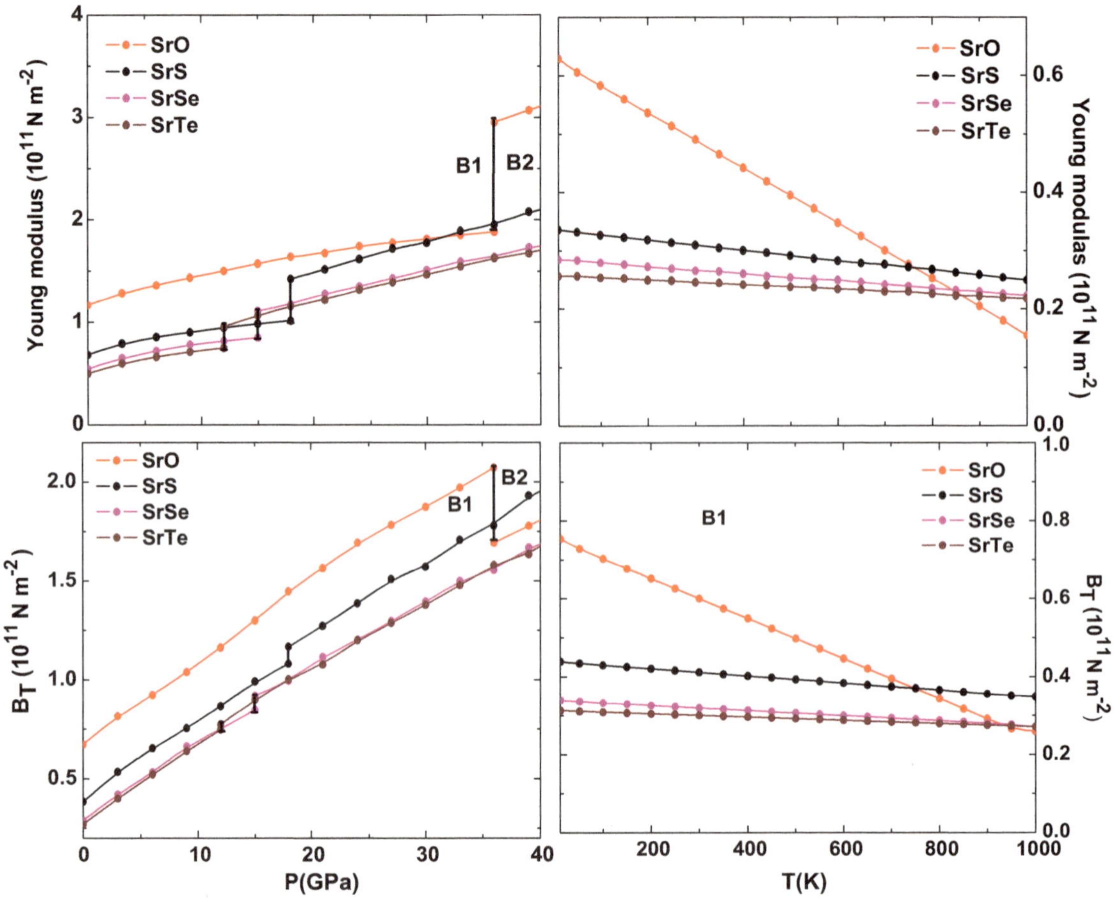

Fig. 8 Variation of Young's modulus (E) and Bulk modulus (B_T) with pressure and temperature

the isotropic shear modulus of SrX is decreasing with enhanced temperature. The steep decrease of G_H, G_V, and G_R is in accordance with aggregate second-order elastic constant C_{ij} behavior with temperature. Suppressed G_H, G_V and G_R as functions of temperature infer the weakening of the lattice as a result of thermal softening.

The tensile modulus as Young's modulus (E) further elaborates the stiffness property. E is also defined in terms of Reuss's shear modulus (G_H), and bulk modulus (B_T):

$$E = \frac{9G_H B_T}{G_H + 3B_T} \tag{30}$$

The tensile strength as Young's modulus (E), is thus computed for SrX (X = O, S, Se, Te) is illustrated in Table 2 for NaCl ($B1$) phase. It is known that the material is stiffer if its Young's modulus is high (E of steel, graphene and diamond is ~200, 1000 and 1220 GPa). Compare to these materials Sr chalcogens are less stiff. The pressure dependence of the Young's modulus (E), of SrX (X = O, S, Se, Te) is sketched in Fig. 8a. It is inferred that E enhances in $B2$ phase as compared to $B1$ phase. The calculated Young's modulus decreases as going from SrO to SrTe, and therefore, Sr chalcogenides are not stiff materials.

The temperature dependence of the tensile strength as Young's modulus (E) of SrX (X = O, S, Se, and Te) is illustrated in Fig. 8b. A decreasing trend is inferred with increase in temperature. Essentially inferring the bond expansion as well bond weakening or in other words thermal induced softening

We now make a comparison of tensile strength E (P) and E (T) of SrX (X = O, S, Se, Te). It is noted that E (P) values are smaller than E (T) for SrX at low pressures and temperatures. It implies that E (P) and E (T) are although susceptible to external variables as pressure and temperature, but it is their Sr and X (X = O, S, Se, and Te) ions in SrX (X = O, S, Se, and Te) that makes SrX (X = O, S, Se, and Te) lattice to be mechanical hard due to bond strengthening and thermal soft due to bond weakening. It is worth commenting that while tailoring the composites based on SrX, one should seek its performance in terms of tensile strength.

Another interest in SrX (X = O, S, Se, and Te).is to probe the ductile and brittle nature of Strontium and

X (*X* = O, S, Se, and Te) based alloys are important and predicted from the knowledge of second-order elastic constants. Ductile materials as Steel and aluminum sustains large strains before rupture, while to that brittle materials as glass and cast iron fractured at lower strains. For materials design and advances in metallurgy as well composite technology, the materials response for applied pressures and temperatures are often gauzed by stress–strain characteristics. Composites are predictive as depending on the external variable (pressure and temperature), the response of the constituent element is tailored in terms of ductile or brittle nature. Also, ductile element in composites may become brittle as the pressure or temperature is increased or decreased. Once the pressure dependence of shear modulus (G_H), and Young's modulus (E) or bulk modulus (B_T) is known, it is worth investigating ductile and brittle nature of Sr*X* (*X* = O, S, Se, Te).

The shear modulus (G_H), measures the resistance to plastic deformation, while the bulk modulus (B_T) probes the resistance to fracture. An empirical relation in terms of the ratio of these moduli is known to differentiate ductile and brittle nature. In accordance, the Pugh's ratio ϕ ($=B_T/G_H$) >1.75, the material sustains large strains before rupture, i.e., the ductile response. On the other hand, for $\phi \leq 1.75$ the material is brittle. Empirically, 1.75 is a critical value that separates ductile and brittle response of materials [50].

From Fig. 9a, the ratio ϕ ($=B_T/G_H$) shows that (a) at zero pressures, semiconducting chalcogenides Sr*X* (*X* = O, S, Se, Te) are ductile in NaCl phase, and (b) with increased pressure brittle nature ϕ ($=B_T/G_H$) >1.75 is documented but SrO shows exceptionally ductile nature in CsCl phase. Hence, semiconducting chalcogenides Sr*X* (*X* = O, S, Se, Te) is ductile only at zero pressure and become brittle after structural phase transformations. To our knowledge, perovskite are ductile that is a test for its deformation before fracture. In materials performance ductility is a powerful probe for quality control. It provides a means to assess the level of impurities and proper processing of a material.

The Poisson's ratio is another measure to differentiate the ductility and brittleness of materials [51]. The critical value of Poisson's ratio (in terms of B_T and G_H) ν is 0.33 that separates ductile and brittle nature of any material. If ν >0.33, the material is ductile, and for ν <0.33, the brittle response of material is observed such as ceramics However, Poisson's ratio lies in between −1.0 and 0.5 which are the lower and upper bounds. The lower bound is a signature of the response of the materials that do not change its shape and the upper bound indicates that the volume is unchanged. The Poisson's ratio ν in terms of bulk modulus B_T and the shear modulus G_H is expressed as [47–49],

$$\nu = \frac{1}{2}\left[3\frac{B_T}{G_H} - 2\right]\left[3\frac{B_T}{G_H} + 1\right]^{-1} \tag{31}$$

From Fig. 9a the Poisson's ratio ν shows that (a) at zero pressure semiconducting SrX (X = O, S, Se, and Te) chalcogenides are brittle in nature, (b) in the NaCl phase, SrX (X = O, S Se, and Te) chalcogenides shows the brittle nature with increasing pressure, (c) at P_T, the value of ν is about 0.348, 0.344, 0.333, and 0.333, for SrX (X = O, S Se, and Te) close to the brittle limit of 0.33 and (d) at high pressures, i.e., in CsCl phase again it documents the brittle nature (ν <0.33). Thus, Frantsevich rule confirms the brittle nature of semiconducting Sr*X* (*X* = O, S Se, Te) chalcogenides. Deduced values of ν at zero pressures are about 0.212, 0.205, 0.188, and 0.192 for semiconducting Sr*X* (*X* = O, S, Se, and Te) chalcogenides and As far as we know, there are no experimental and theoretical results available for comparison. It is clear that Frantsevich empirical rule classify semiconducting Sr*X* (*X* = O, S, Se, and Te) chalcogenides as brittle materials. High Pugh ratio indicates the ductile nature while to that a low value infers more brittleness. There is seemingly a contradiction between Pugh's and Frantsevich empirical rules. However, these empirical rules only differ on the exact border between the bulk modulus B_T and the shear modulus G_H behavior. In view of this, semiconducting Sr*X* (*X* = O, S, Se, and Te) chalcogenides can be considered a borderline case between the classes of ductile and brittle materials.

The Poisson's ratio ν and Pugh's ratio ϕ behavior as a function of temperature are sketched in Fig. 9b. The Poisson's ratio ν and Pugh's ratio ϕ is independent of temperature. Beginning from zero temperature and at high temperatures in NaCl phase, value of ν about 0.361 (SrO), 0.372 (SrS), 0.360(SrSe), 0.363 (SrTe) and the value of ϕ about 3.268 (SrO), 3.585 (SrS), 3.244 (SrSe), 3.314 (SrTe) obtained. The temperature-dependent Poisson's ratio reflects brittle nature of Sr*X* (*X* = O, S Se, Te), while to that a borderline case between the classes of ductile and brittle materials is known from pressure dependence. Poisson's ratio is thus an effective indicator to control the level of impurities and processing of Sr*X* (*X* = O, S, Se, and Te) ceramics. For covalent materials, ν is small ($\nu \sim 0.1$), whereas for metallic materials ν is typically 0.33. As far as we know, there are no experimental and theoretical results available for comparison.

The brittle behavior of Sr*X* (*X* = O, S, Se, and Te) ceramics can also be checked by Cauchy pressures. Based on the calculated C_{12}–C_{44} for Sr*X* (*X* = O, S, Se, and Te) as shown in Tables 1 and 2, it can be seen that the Cauchy pressures are negative. These results again indicate that SiC is brittle in nature. Furthermore, the Poisson's ratio ν is used as a probe to identify the bonding characteristic of the forces than any of the other elastic constants. Estimated

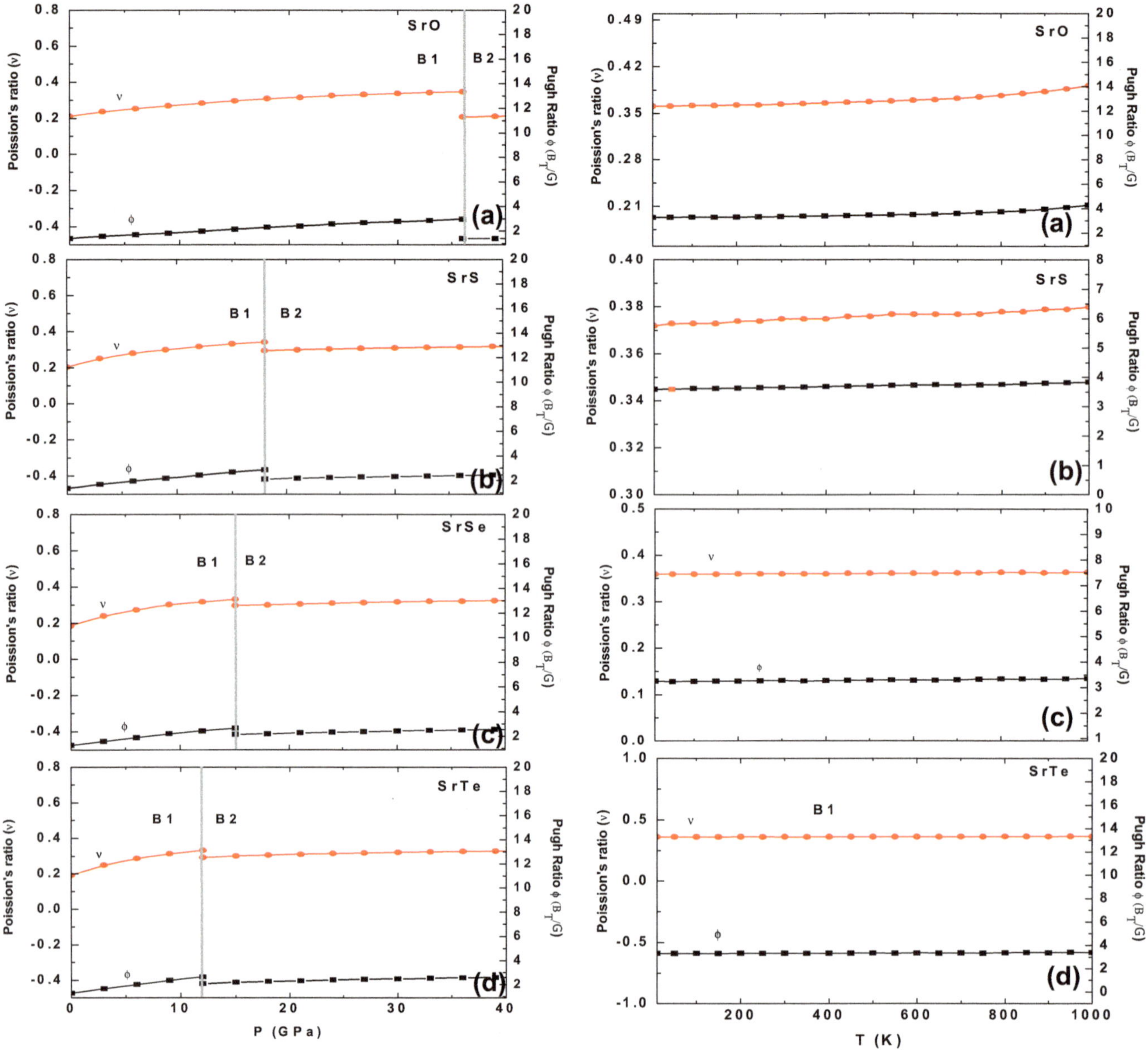

Fig. 9 Variation of Poisson's ratio (ν) and Pugh's modulus ratio (Φ) with pressure and temperature

Poisson's ratio is close to 0.25, which means that Sr*X* (*X* = O, S, Se, and Te) are with predominantly central inter-atomic forces. The larger the Poisson's ratio is, the better the plasticity is henceforth, Sr*X* (*X* = O, S, Se, and Te) is the one which shows a better plasticity.

Usually, the elastic moduli describe only reversible response of a material to small strain near equilibrium. The intrinsic strength of a material reflects permanent plastic deformation at large shear strain. Consequently, to further understand the behavior of Sr*X* (*X* = O, S, Se, and Te) in terms of Vickers hardness: $H_V = 2\ (\phi^2 G_H)^{0.585} - 3$. Here, $\phi = B_T/G_H$. Figure 10a illustrates the theoretical Vickers hardness as a function of pressure. It is clear from the plot that the Vickers hardness H_V decreases in *B*1 phase and then increases in *B*2 phase with increase in pressure, which indicates that Sr*X* (*X* = O, S Se, and Te) becomes hard under pressure implying its good mechanical properties. Figure 10b showed Vickers hardness as a function of temperature for Sr*X* (*X* = O, S, Se, and Te). It is clear from the plot that the Vickers hardness H_V decreases in *B*1 phase, which indicates that Sr*X* (*X* = O, S, Se, and Te) gets softened with enhanced temperature. Apart from the elastic anisotropy of crystals, the hardness is important to discuss their properties because it is highly correlated with the possibility of inducing micro cracks in materials.

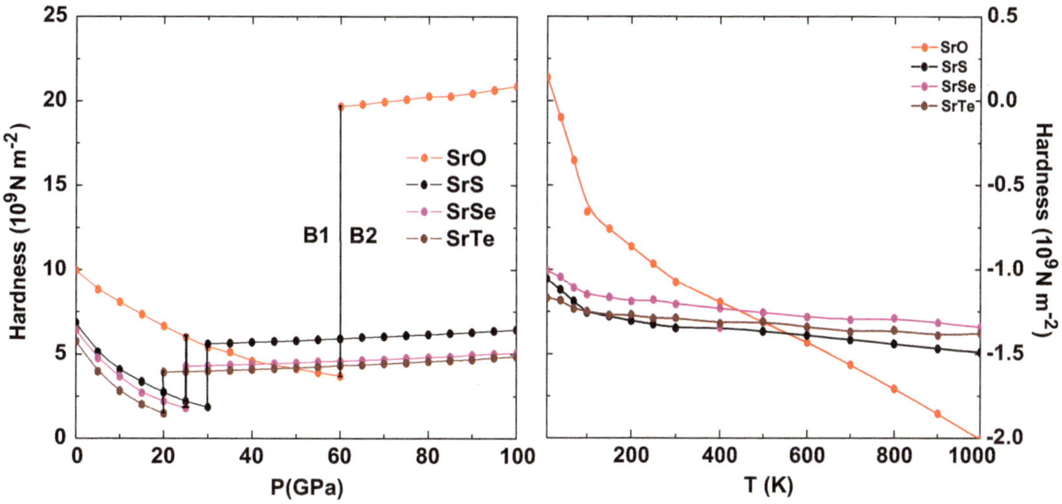

Fig. 10 Variation of hardness (H_V) as a function of pressure and temperature

Sr*X* (X = O, S, Se, and Te) are promising materials with wide range of applicability's with effective mechanical properties. Usually, materials elastic properties are a source of valuable information where materials mechanics is significant as the knowledge of deformational characteristics of materials is essential in engineering design and construction of effective structures. Having, understood the materials elastic behavior using Bulk modulus (B_T), shear modulus (G_H, G_V and G_R), and Young modulus (E), in the following, we will use our calculated elastic constants to discuss the compressional and shear wave velocity in semiconducting Sr*X* (X = O, S, Se, and Te). The velocities of the longitudinal v_l and shear waves v_s are known from the Lamé's constants, λ and μ. The compressional wave with velocity v_l propagates back and forth in a crystal. The shear wave with velocity v_s goes up and down.

The Lamé's constants are related to Young's modulus (E) and Poisson's ratio (ν) (correlating the bulk modulus B_T and the shear modulus G_H). The first Lamé's constant (λ) measures the compressibility of the material. The second Lamé's constant (μ) infers its shear stiffness [47–49]. The Lamé's constants (λ and μ) are expressed as:

$$\lambda = \frac{vE}{(1+v)(1-2v)} \tag{32}$$

$$\mu = \frac{E}{2(1+v)} \tag{33}$$

Figure 11a displays the pressure dependence of the first and second Lamé's constants (λ, μ). Starting from zero pressure and at high pressures, both Lamé's constants (λ, μ) are positive. Note that the Lamé's constant (λ) can be negative, however, for most materials it is also positive. The second Lamé's constant (μ) is positive. An increasing trend of compressibility (λ) of Sr*X* (X = O, S, Se, Te) is witnessed in terms of λ as seen in both rock salt and CsCl structures and is attributed to mechanical hardening of lattice. The decreasing trend in *B*1 phase and increasing trend in *B*2 phase of shear modulus (G_H, G_V and G_R) results in a decreasing behavior of shear stiffness (μ) in *B*1 phase and then enhanced shear stiffness in *B*2 phase. Note that λ and μ together constitute a parameterization of the elastic moduli for homogeneous isotropic media.

The value of pressure-dependent Lamé's constants (λ, μ) is documented in Table 3. Deduced values of Lamé's constants (λ, μ) could not be compared due to lack of data on Sr*X* (X = O, S, Se, and Te). Figure 11b summarizes the temperature-dependent behavior of Lamé's constants (λ, μ), for Sr*X* (X = O, S, Se, and Te). It is noticed that both Lamé's constants (λ, μ) are decreasing with increasing temperature. Thus, both compressibility and shear stiffness showed decreasing trend with temperature variations and identifies once again the thermal softening of Sr*X* (X = O, S, Se, and Te).

The Lamé's constants (λ, μ) determines the longitudinal (shear) wave velocity as

$$v_l = \left[\frac{\lambda + 2\mu}{\rho}\right]^{\frac{1}{2}} \tag{34}$$

$$v_s = \left[\frac{\mu}{\rho}\right]^{\frac{1}{2}} \tag{35}$$

Here, ρ is the density. The average wave velocity v_m has been approximately given by

$$v_m = \left[\frac{1}{3}\left(\frac{2}{v_t^3} + \frac{1}{v_s^3}\right)\right]^{-\frac{1}{3}} \tag{36}$$

The compressional longitudinal (shear) wave velocity thus depends on density of the material as well on Bulk

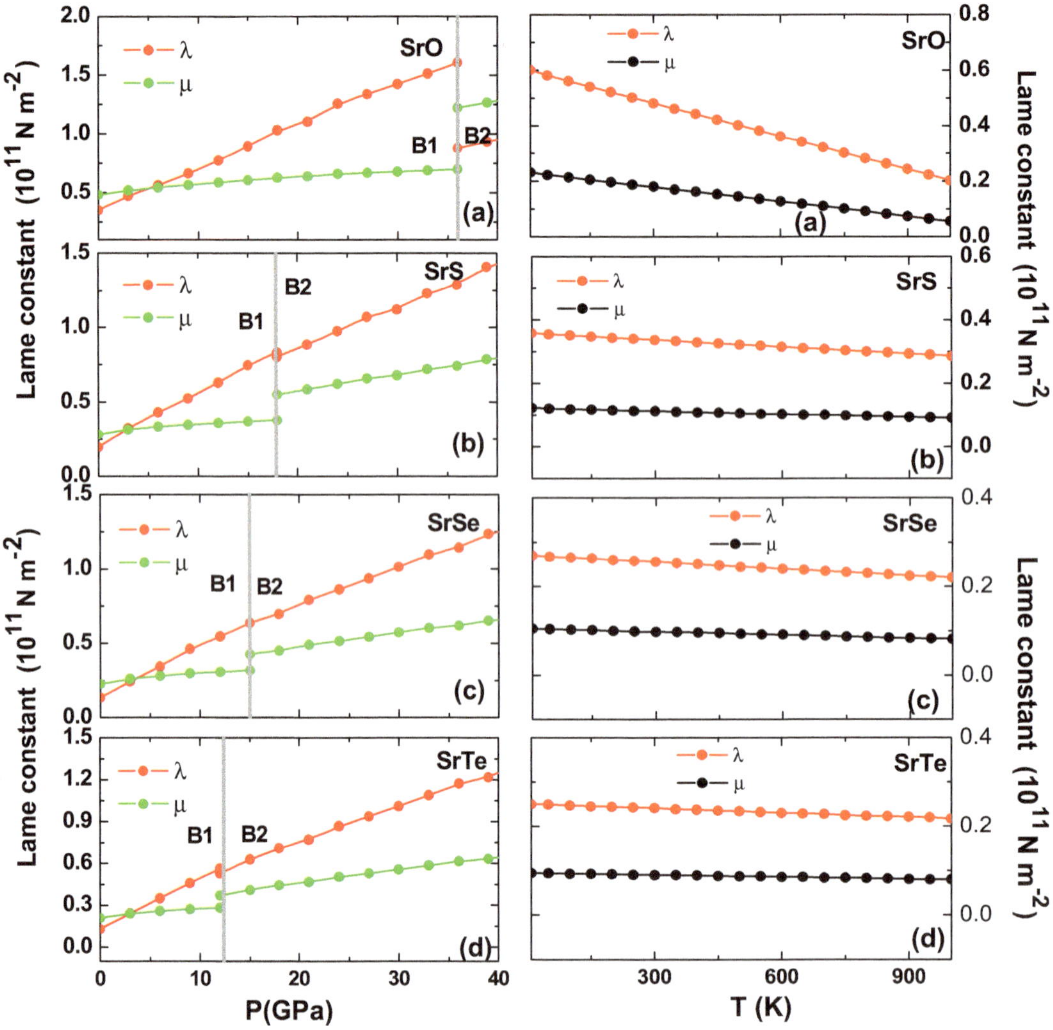

Fig. 11 Variation of Lamé's constant (λ, μ) with pressure and temperature

modulus (B_T), shear modulus (G_H, G_V and G_R), and Young modulus (E). We note that Bulk modulus (B_T) has a strong dependence on the density of the material; Young modulus (E) displays a weak dependence, while the shear modulus (G_H) is independent of density.

Figure 12a displays the pressure dependence of the longitudinal (shear) wave velocity in SrX (X = O, S, Se, Te). In rock salt phase, beginning from zero pressures and at high pressures, compressional wave velocity v_l increases while to that shear wave velocity v_s decreases. Both compression and shear waves increases with enhanced pressure in $B2$ structure. Deduced values of longitudinal, shear and average elastic wave velocities propagating in SrX (X = O, S, Se, and Te) are illustrated in Table 3 at zero temperature and pressure. Figure 12b represents the temperature dependence of the longitudinal (shear) velocity, respectively. It is noticed that both longitudinal (shear) wave velocity decreases in $B1$ phase with enhanced temperature. The values of the longitudinal, shear and average elastic wave velocities propagating in SrX (X = O, S, Se, Te) are documented in Table 4 at room temperature. The high temperature behavior of longitudinal (shear) wave velocity for SrX (X = O, S, Se, and Te) can be considered as predictive studies as they cannot be compared due to unavailability of high temperature data.

The Navier's equation is also used to determine the longitudinal and the shear wave velocity [52, 53]. These are written in terms of Reuss's shear modulus (G_H), and bulk modulus (B_T) as

$$v_l = \left[\frac{3B_T + 4G_H}{3\rho}\right]^{\frac{1}{2}} \tag{37}$$

$$v_s = \left[\frac{G_H}{\rho}\right]^{\frac{1}{2}} \tag{38}$$

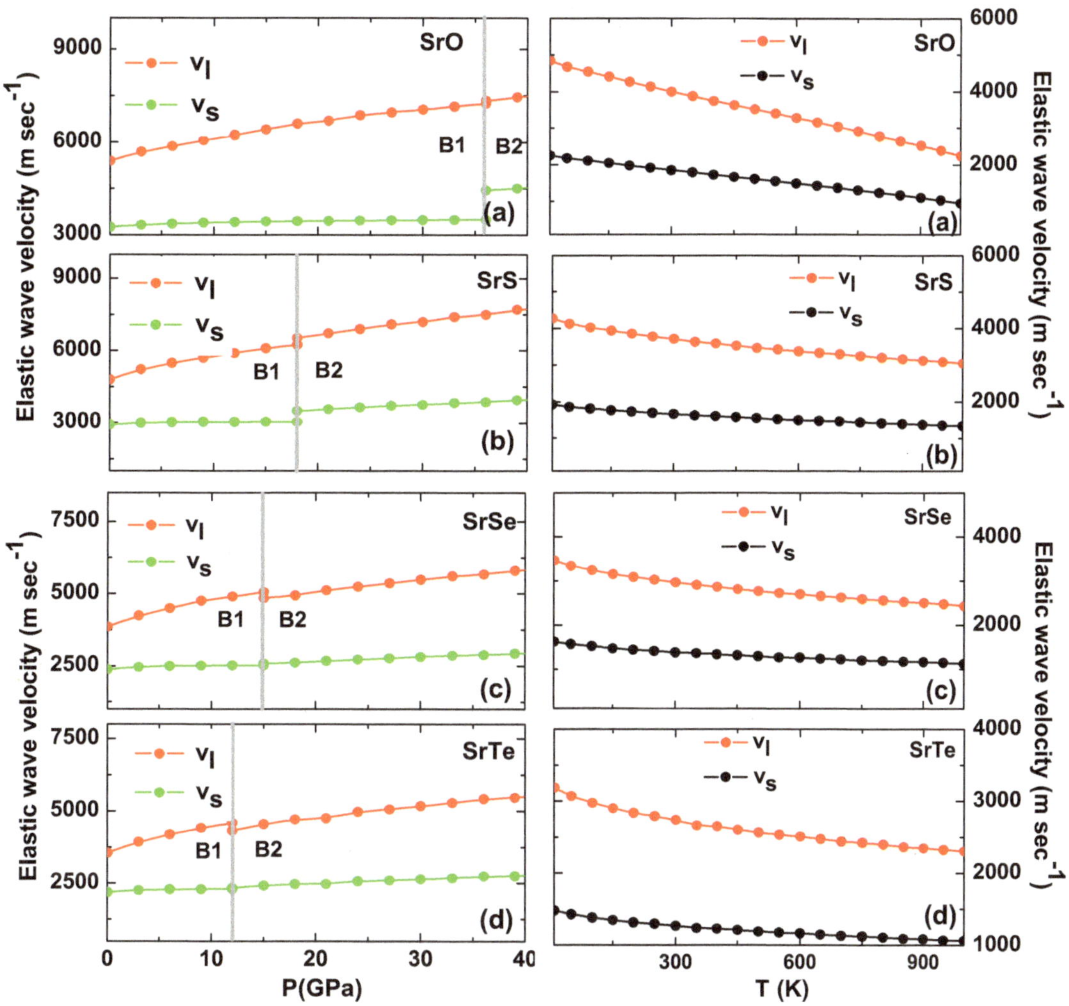

Fig. 12 Variation of elastic wave velocity v_l and v_s with pressure and temperature

Elastic, plastic and molten state properties with pressure as controlling variable are useful for tailoring composites. The Lamé's constants (λ, μ) are of substantial interest for plastic materials. Referring to Eqs. 32 and 33, we note that as the Poisson's ratio (ν) increases, the Lamé's constant (λ, μ) numerically approach the bulk modulus (E). For fluids, the Reuss's shear modulus (G_H) vanishes as the viscosity of the fluid approaches zero. The above is important in context of the present computation and can also be cross-checked from the relation between bulk modulus and Reuss's shear modulus: $B_T = \lambda + 2\ G_H/3$. Furthermore, G_H approaches zero for fluids and hence the Poisson's ratio is $\sim$0.5.

The response to deformations against bond bending or bond-angle distortion is relevant for materials with promising technological applications and also a test to validate the many-body non-central potential that we dealt with. The aggregate elastic constants infers the elastic properties of material that undergo stress, deform and then recover after returns to its original shape after stress ceases. The nature of elastic constants in solids holds a great importance to elucidate the microscopic nature interatomic bonding, equations of state, and vibrational density of states. The above can be understood in terms of Kleinman parameter ξ which describes the relative positions of the cation and anion sub lattices under volume-conserving strain distortions for which positions are not fixed by symmetry [54, 55]. We have explored its applicability to ceramics SiC by using [56]

$$\xi = \frac{C_{11} + 8C_{12}}{7C_{11} + 2C_{12}} \tag{39}$$

The pressure dependence of Kleinman parameter ξ is sketched in Fig. 13c for SrX (X = O, S, Se, and Te) in rock salt ($B1$) and CsCl ($B2$) phases Deduced value of ξ about 0.405 (SrO), 0.374 (SrS), 0.407 (SrSe), 0.388 (SrTe) is documented in Table 3 for SrX (X = O, S, Se, and Te) at zero temperature and pressure. The value of ξ about 0.339

Table 4 Calculated aggregate second-order elastic constants (C_{11}, C_{12} and C_{44}) and aggregate bulk modulus (B_T), average elastic constant (C_S) second-order elastic constant anisotropy parameter (γ_1^2), isotropic shear modulus (G_H), Voigt's shear modulus (G_V), Reuss's shear modulus (G_R), Young's modulus (E), Poisson ratio (ν), first and second-order Lame's constant (λ, μ), longitudinal (v_l), shear (v_s) and average elastic wave velocity (v_m), Kleinman parameter (ξ), Grüneisen parameter (γ_G), and hardness (H_v), thermo dynamical properties: Debye temperature (θ_D), melting temperature of SrX (X = O, S, Se, and Te) in *B1* phase. at temperature of about 300 K

S. no.	Elastic properties	SrO	SrS	SrSe	SrTe
1	C_{11} (10^{10} Nm^{-2})	13.09	9.16	6.76	6.40
2	C_{12} (10^{10} Nm^{-2})	2.46	1.59	1.42	1.31
3	C_{44} (10^{10} Nm^{-2})	0.69	0.34	0.42	0.37
4	B_T (10^{10} Nm^{-2})	6.01	4.11	3.20	3.00
5	C_S (10^{10} Nm^{-2})	5.32	3.79	2.67	2.55
6	γ_1^2	6.703	10.132	5.357	5.878
7	G_H (10^{10} Nm^{-2})	1.8	1.13	0.98	0.9
8	G_V (10^{10} Nm^{-2})	2.54	0.53	1.32	1.24
9	G_R (10^{10} Nm^{-2})	1.06	1.72	0.63	0.56
10	E (10^{10} Nm^{-2})	4.91	3.1	2.66	2.46
11	ν	3.64	0.375	0.361	3.64
12	λ (10^{10} Nm^{-2})	4.8	3.36	2.55	2.41
13	μ (10^{10} Nm^{-2})	2.31	1.13	0.98	0.9
14	v_l (m/s)	4001	3714	2965	2739
15	v_s (m/s)	1852	1663	1381	1268
16	v_m (m/s)	448.2	403.7	335.2	308.0
17	ξ	0.339	0.325	0.361	0.356
18	θ_D (K)	163.514	128.1, 353.6 [10]	103.9	90.209
19	T_M (K)	1327	1094	952.603	931.202
20	H_v (10^9 Nm^{-2})	0.140	−1.188	−1.107	−1.233
21	γ_G	1.54	1.51, 1.64 [10]	1.41	1.43

(SrO), 0.325 (SrS), 0.361 (SrSe), 0.356 (SrTe) is obtained at room temperature and is reported in Table 4 for SrX (X = O, S, Se, and Te). As far as we know, there is no experimental result available for Kleinman parameter and is considered as a prediction study. For both rock salt and CsCl structures, the knowledge of elastic constants at variable pressure is worth investigating aimed for practical applications related to the mechanical properties of a solid: load deflection, thermo elastic stress, internal strain, elastic wave velocities, and fracture toughness.

Physical properties as thermal expansion, heat capacity and Grüneisen parameter can be explained with higher order terms of the interaction potential. The thermal process in terms of heat conduction is hindered if one considers a solid to be perfect harmonic and thermal conductivity will be infinitely large. Thus, the role of anharmonic effects or phonon decay is important as enhanced pressures as well temperatures allows a change in volume or dimensions. The pressure-dependent calculations of elastic constants will provide measures of the accuracy of the calculation of forces in SiC as well the mechanical stiffening or hardening attributed to Sr–Sr, X–X, and Sr–X bond compression and bond strengthening due to lattice vibration.

We express the molecular force in the absence of the Lorentz effective field [23–37]

$$f = \frac{1}{3}\left[\frac{d^2}{dr^2}U_{SR}(r) - \frac{2}{r_0}\frac{d}{dr}U_{SR}(r)\right]_{r=r_0}, \tag{40}$$

The force constant is a function of the second-order derivatives of the short-range (SR) overlap repulsive potential and as well the charge dipole–dipole and charge dipole–quadruple van der Waals potential. The Reststrahlen frequency is obtained from the force constant using

$$\upsilon_0 = \frac{1}{2\pi}\left[\frac{f}{\mu}\right]^{1/2}, \tag{41}$$

Here, μ is the reduced mass of SrX (X = O, S, Se, Te).

The Grüneisen parameter γ_G is a ratio of second and first-order derivatives of the potentials enable us to discuss the anharmonic effects in a crystal. We express γ_G as

$$\gamma_G = -\frac{r_0}{6}\left[\frac{U'''(r_0)}{U''(r_0)}\right] \tag{42}$$

The pressure dependence of Grüneisen parameter is sketched in Fig. 13a for SrX (X = O, S, Se, Te) in rock salt ($B1$) and CsCl ($B2$) phases. A linear decrease of γ_G with pressure in both phases is observed. Deduced γ_G value of about 1.523(SrO), 1.508(SrS), 1.403(SrSe), 1.425(SrTe) in NaCl ($B1$) phase. On the other hand, Fig. 13b discerns temperature dependence of Grüneisen parameter with a

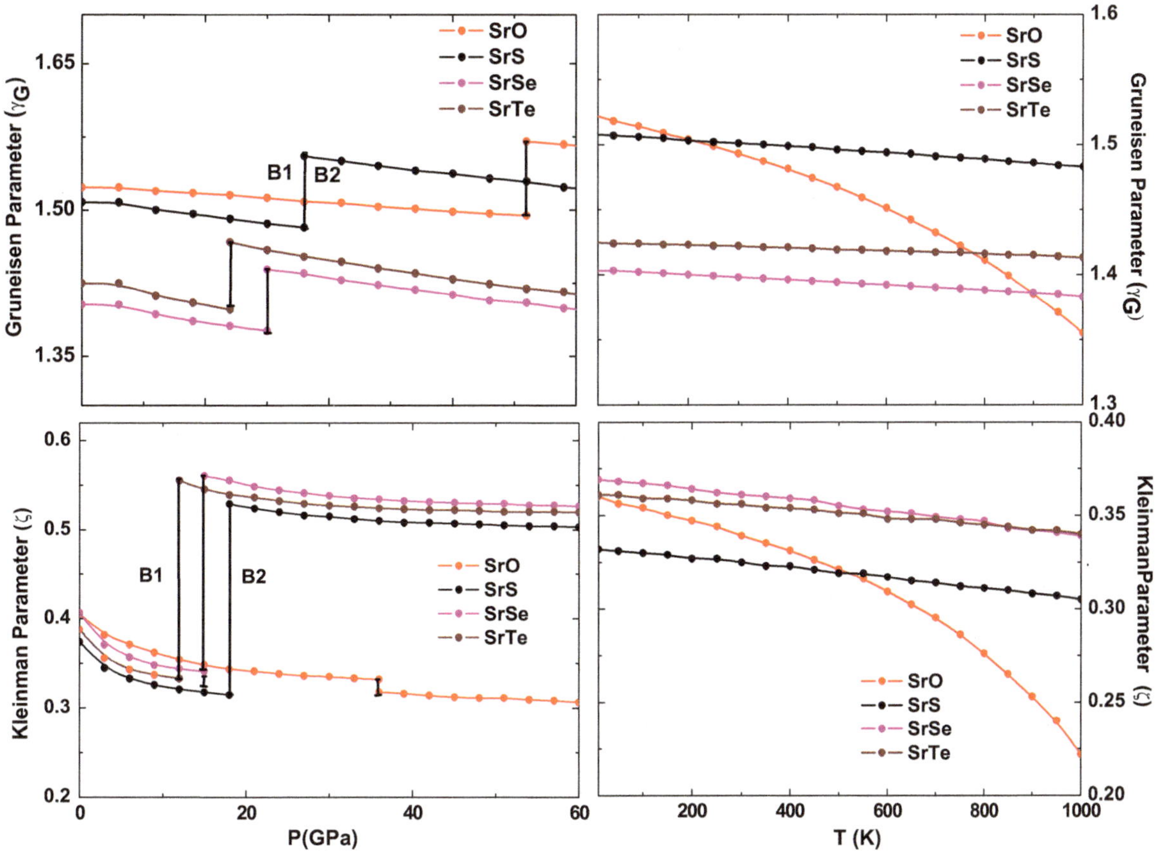

Fig. 13 Variation of Grüneisen parameter (γ_G) and Kleinman parameter (ξ) with pressure and temperature

value of about 1.54 (SrO), 1.51 (SrS), 1.41 (SrSe), 1.43 (SrTe) at 300 K which is comparable with available theoretical result of 1.64 (SrS) [10] As far as we know, there is no experimental result available for Grüneisen parameter and is considered as a prediction study. For most of the solids, Grüneisen parameter ranges from 1.5 to 2.5. As Grüneisen constant γ_G behavior is influenced by ratio of second and first-order derivatives of the potentials, we may thus comment that anharmonicity is significant in SrX (X = O, S, Se, Te). Note that these results are further validated by inelastic neutron scattering measurements to probe phonon frequencies as a function of the crystal volume.

We further compute Debye temperature (θ_D) to shed further light on mechanical stiffened and thermal softened characteristics of SrX following [57–61]:

$$\theta_D^3 = \frac{3.15}{8\pi}\left(\frac{h}{k_B}\right)^3\left(\frac{r}{M}\right)^{\frac{3}{2}}\left(\frac{C_{11}-C_{12}}{2}\right)^{\frac{1}{2}}\left(\frac{C_{11}+C_{12}+2C_{44}}{2}\right)^{\frac{1}{2}}C_{44}^{\frac{1}{2}}, \tag{43}$$

where M is the acoustic mass of SrX. The notations h and k_B are the Planck and Boltzmann constants.

The Debye temperature as functions of pressure is plotted in Fig. 14a for SrO, SrS, SrSe, and SrTe. At zero pressure, θ_D is about 246 (SrO), 189 (SrS), 146 (SrSe), 127 (SrTe) K. It is noticed that with enhanced pressure, θ_D increases in $B1$ phase for SrX (X = O, S, Se, Te). The Debye temperature from $B1$ to $B2$ phase is jumped by 45 (SrO), 40 (SrS), 23 (SrSe), 20 (SrTe) K at P_T. On the other hand, θ_D also showed an increase in $B2$ phase. Deduced value of θ_D at zero pressure is listed in Table 3. Enhanced θ_D in $B1$ phase at higher pressure indicates the mechanical stiffening of lattice.

Deduced value of θ_D at 300 K is listed in Table 4. Debye temperature is a function of temperature and its value varies from technique to technique as well depends on the sample quality. Usually, a standard deviation in θ_D of about 25 K is agreeable. The change in the force constants induced by pressure increases θ_D in $B1$ phase and after transition pressure it continue to increasing which drives the system effectively towards the stiffening of lattice with increasing pressure.

Figure 14b displays the variation of the Debye temperature as functions of temperature at zero pressure. Starting from zero temperature, θ_D decreases rapidly. On the other hand, θ_D enhances with applied pressures at zero temperature. It is worth commenting from high-pressure and high-temperature Debye temperature behavior that (a) the

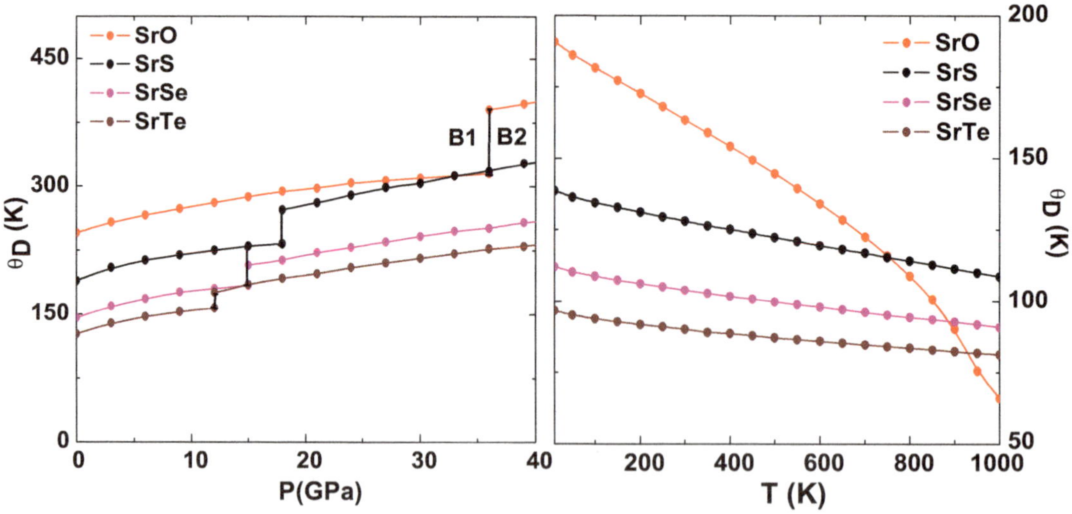

Fig. 14 Variation of Debye temperature (θ_D) as a function of pressure and temperature

pressure-dependent Debye temperature infers the mechanical stiffened bulk modulus due to Sr–Sr, *X*–*X*, and Sr–*X* bond compression and bond strengthening due to lattice vibration and (b) the thermal softening of bulk modulus results from bond expansion and bond weakening due to thermal stress in Sr*X* (*X* = O, S, Se, Te) in rock salt structure.

A comparison of values of $\theta_D(P)$ and $\theta_D(T)$ at zero pressure and zero temperature results: $\theta_D(P) > \theta_D(T)$ implying the susceptibility of θ_D with pressure. It is worth to comment that $\theta_D(P)$ and $\theta_D(T)$ behave differently with applied pressure and temperatures in Sr*X* (*X* = O, S, Se, and Te). This is attributed to the fact vibrational spectrum of Sr*X* (*X* = O, S, Se, and Te) lattice is controlled by aggregate elastic constants C_{ij} behavior. Thus, the understanding of Debye temperature behavior of a material not only provides essential features of the vibrational spectrum but is also mandatory for technological and engineering applications. This quantity is useful as a reference for future experimental studies.

The cumulative effect of both pressure and temperature led us to define aggregate elastic constants C_{av} as

$$C_{av} = \left(\frac{8\pi}{3.15}\right)^{\frac{2}{3}} \left(\frac{k_B}{h}\right)^2 \left(\frac{M}{r}\right) \theta_D^2, \tag{44}$$

This is useful once Debye temperature at zero pressure is known from experiments. Despite various investigations on mechanical and thermodynamical properties, basic properties controlled by elastic constants related with high temperature behaviors are lacking.

To further explore the role of anharmonic effects in terms of heat capacity at constant volume C_v and thermal expansion coefficient α, we express the vibration term A_{vib} [62, 63]:

$$A_{vib}(\theta_D, T) = nk_B T\left[\frac{9\theta_D}{8T} + 3\ln\left\{1 - \exp\left(-\frac{\theta_D}{T}\right)\right\} - D(\theta_D/T)\right] \tag{45}$$

The non equilibrium Gibbs function, $G^*(V; P, T)$, is minimized with respect to volume *V* as

$$\left[\frac{\partial G^*(V; P, T)}{\partial V}\right]_{P,T} = 0 \tag{46}$$

From Eq. (47), we determine heat capacity at constant volume C_v as

$$C_v = 3nk_B\left[4D\left(\frac{\theta_D}{T}\right) - \frac{3\theta_D/T}{e^{\theta_D/T} - 1}\right] \tag{47}$$

Figure 15a documents the variations of heat capacity at constant volume, C_v, with pressure *P* for both *B*1 and *B*2 phase of Sr*X* (*X* = O, S, Se, and Te) at *T* = 50, 100, 300 and 500 K. The normalized heat capacity is $[C_v(P) - C_v(0)]/C_v(0)$, where $C_v(P)$ and $C_v(0)$ are heat capacity at any pressure *P* and at zero pressure. Starting from zero pressure, normalized heat capacity decreases gradually with enhanced pressure in *B*1 phase. This behavior essentially point to the fact that the Sr*X* (*X* = O, S Se, and Te) lattice vibrations energy are controlled by both pressure as well temperature. Note that for higher temperatures $T \rightarrow \theta_D$, the variation in heat capacity with pressure is weak in *B*1 phase. Also, at P_T, reduced jump in between rock salt and CsCl structures can be seen as compared to low temperatures. Similar observations are previously reported on SrO following quasi-harmonic approximation in explaining the thermal expansion coefficient, specific heat, entropy, bulk modulus and the thermal Grüneisen parameter with temperature and pressure in both phases [64].

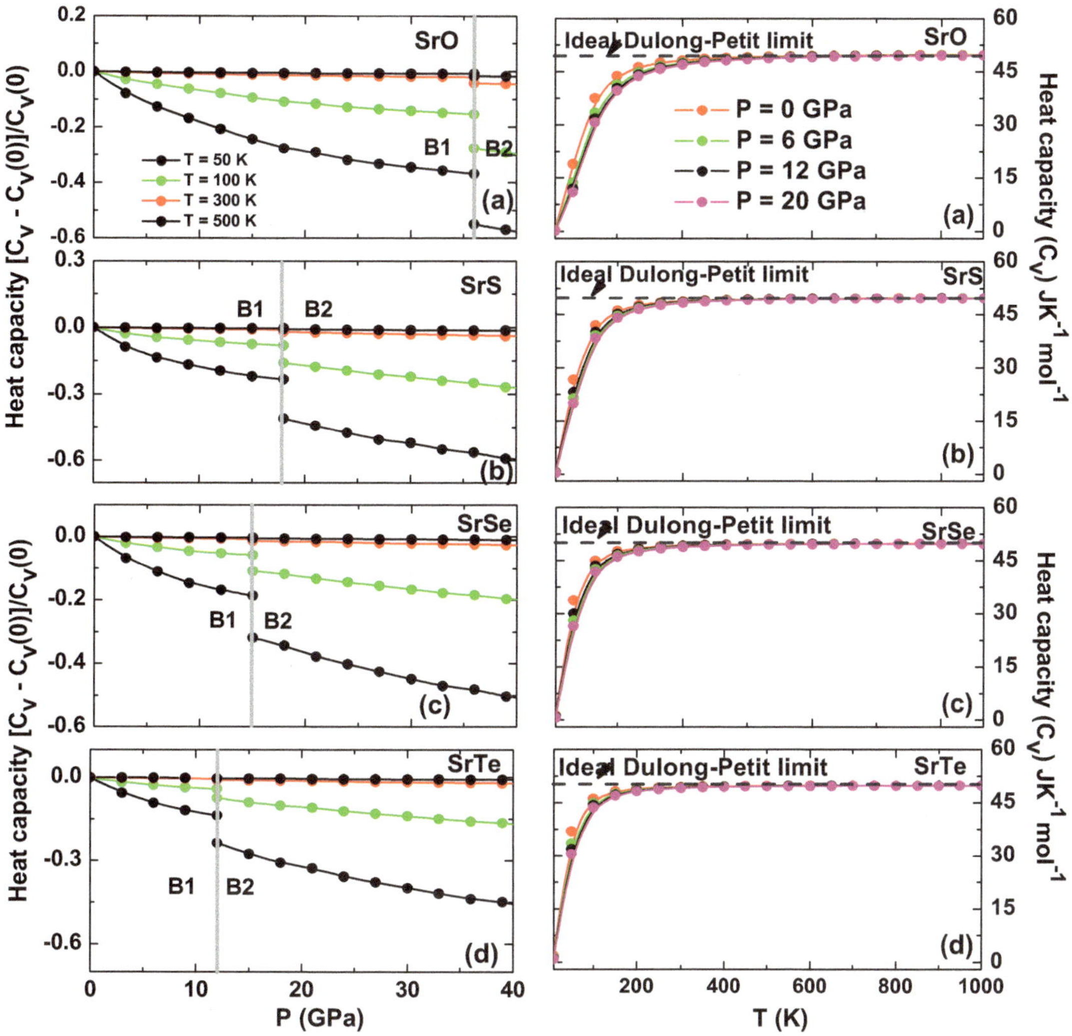

Fig. 15 Variation of heat capacity (C_v) with pressure at different temperatures and with temperature at different pressures

The Debye temperature calculated from elastic constants must have a close resemblance as to measure from specific heat measurements only at low temperatures. It is instructive to study the temperature-dependent behavior of heat capacity. Figure 15b illustrates the heat capacity at constant volume, C_v, behavior as functions of temperature for various pressures (0, 6, 12, 20 GPa) within the framework of quasi-harmonic model for SrX (X = O, S, Se, Te). It can be seen that below room temperature (300 K), C_v increases very rapidly with the temperature at all pressures. Above room temperature, C_v increases slowly with the temperature. It almost approaches a constant ideal gas limit—the Dulong–Petit limit, $C_v(T) = 3\ R$, at higher temperatures as well at all pressures for SrX (X = O, S, Se, and Te).

We further determine thermal expansion coefficient $\alpha_{\text{th.exp}}$ using Eq. (47) as

$$\alpha_{\text{th.exp}} = \frac{\gamma C_v}{B_T V} \tag{48}$$

Apart from the pressure dependence of heat capacity at constant volume C_v, the Grüneisen parameter and Bulk modulus is also needed to elucidate the thermal expansion coefficient ($\alpha_{\text{th.exp}}$). It is a measure of any alteration in phonon frequency depending on the lattice's expansion or contraction in volume as a result of temperature variation. Figure 16a illustrates the pressure dependence of $\alpha_{\text{th.exp}}$. For SrX (X = O, S, Se, and Te) in both rock salt and CsCl, i.e., $B1$ and $B2$ phase. A nonlinear decrease in $\alpha_{\text{th.exp}}$ is thus witnessed with pressure in both NaCl and CsCl phase. The decrease is more rapid in rock salt structure while a slow decrease is documented in CsCl structure. We note that at P_T, the thermal expansion coefficient $\alpha_{\text{th.exp}}$ is suppressed by 10–20 % in SrX (X = O, S, Se, Te) at temperatures 50, 100, 300, and 500 K, respectively.

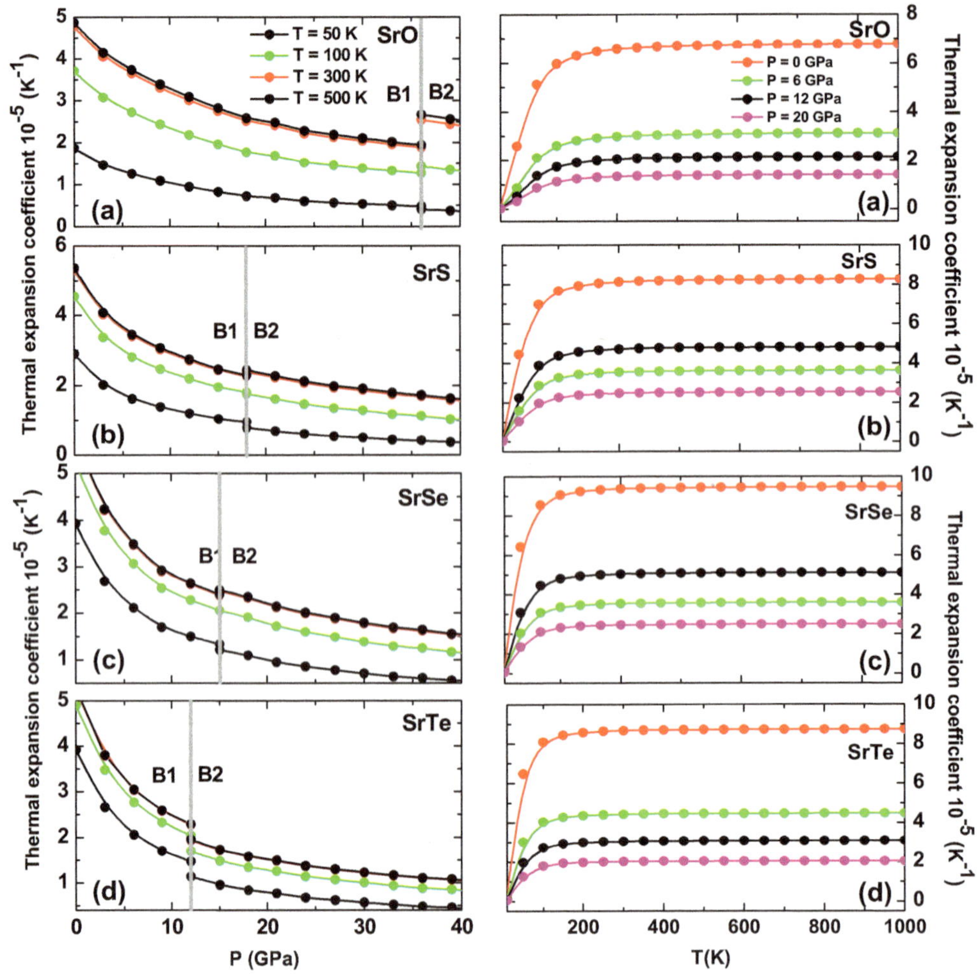

Fig. 16 Variation of the thermal expansion coefficient ($\alpha_{th.exp}$) with pressure at different temperatures and with temperature at different pressures

The variations of $\alpha_{th.exp}$ as functions of temperature at various pressures have been plotted in Fig. 16b for Sr*X* (*X* = O, S, Se, and Te). This figure shows that at low temperatures ($T < 200$ K), $\alpha_{th.exp}$ enhances rapidly with temperature at $P = 0$ and 20 GPa. At high temperatures (T > 200 K), a sharp increase of $\alpha_{th.exp}$ is witnessed at all pressures. This figure also suggests that with enhanced pressure, the increase of $\alpha_{th.exp}$ with temperature becomes smaller. In other words, the slope of $\alpha_{th.exp}$ gradually decreases at higher temperatures at all pressures except at $P = 0$ GPa. Returning back to pressure dependence of thermal expansion coefficient ($\alpha_{th.exp}$), we note a rapid decrease in rock salt structure with the increase of pressure. A comparison of $\alpha_{th.exp}$ (*P*) and $\alpha_{th.exp}$ (*T*) reveals that $\alpha_{th.exp}$ (*P*) values are smaller than that $\alpha_{th.exp}$ (*T*) for Sr*X* (*X* = O, S, Se, and Te) under same pressure and temperature. Thus $\alpha_{th.exp}$ (*P*) and $\alpha_{th.exp}$ (*T*) are differently sensitive to the Sr and *X* ions in Sr*X* (*X* = O, S, Se, and Te) lattice is mechanical hard due to bond strengthening and thermal soft due to bond weakening.

We may add that the elastic properties of CaN, SrN, and BaN compounds within the generalized gradient approximation are performed. The bulk and shear moduli, Young's moduli, Poisson's ratio, and brittle/ductile behavior for CaN, SrN, and BaN. The estimated anisotropy parameter, *A*, shows that SrN has higher degree of elastic isotropy in comparison to CaN and BaN [65]. Furthermore, the density functional theory has been used to compute the structural, electronic, and magnetic properties of Co2CrGe. The total energy calculations lead to stable magnetic configuration and the optimized lattice constant for Heusler alloys [66].

Conclusion

The present study thus addresses, the high-pressure and high-temperature-dependent structural, elastic, and thermodynamical studies in rock salt and cesium chloride structure of Sr*X* (*X* = O, S, Se, and Te). We have formulated an interatomic pair wise potential that incorporates the long-range Coulomb with charge transfer interactions, covalent nature of bonds, zero point energy effects and the short-range interactions as charge dipole–dipole and charge dipole-quadruple (van der Waals), as well overlap repulsive interaction up to second-neighbor ions.

From the knowledge of Gibbs's free energies in rock salt and CsCl structure, we determine the pressure induced first-order structural phase transition about 35.5 (SrO), 17.6 (SrS), 14.9 (SrSe), 11.9 (SrTe) GPa. As a next step we determine the cohesive energy as −33.93 (SrO), −60.70 (SrS), −61.43 (SrSe), and −67.26 (SrTe) eV and the volume collapse is about 13.0 (SrO), 11.4 (SrS), 10.7 (SrSe), and 11.1 (SrTe) %. Compressions in Sr*X* (*X* = O, S Se, Te) at higher pressure indicates the mechanical stiffening of lattice. The phase-transition pressure and volume collapse are consistent with earlier observations. We emphasize that agreement with experimental and theoretical data is not fortuitous, but it is attributed to proper parameterization and formulation of potential with non-central many-body forces in terms of the screening of the effective Coulomb potential through modified ionic charge.

Furthermore, we make effort to determine the second-order aggregate elastic constants C_{ij} under applied pressure and temperature with respect to finite strain. Once pressure and temperature dependence of C_{ij} is known, Cauchy discrepancy and elastic anisotropy in second-order elastic constants, melting temperature, third-order elastic constants C_{ijk}, Cauchy discrepancy and anisotropy in third-order elastic constants, isotropic shear modulii as G_H, G_V, and G_R, Young's modulus *E*, Poisson's ratio ν, Pugh's ratio ϕ and Vicker's hardness H_V to discuss ductile/brittle nature and mechanical stiffening/thermal softening of SrX lattice. In continuity, Lamé's constant (λ, μ), longitudinal (shear) wave velocity to enumerate the compressibility and shear stiffness of the material, Grüneisen constant, Debye temperature, isothermal compressibility, heat capacity and thermal expansion coefficient to shed light on anharmonicity of Sr*X* (*X* = O, S, Se, Te). We comment that incorporation of charge transfer interactions, covalent contribution and quantum effects leads to nonzero value of Cauchy discrepancy ($C_{12}-C_{44} \neq 0$) and is a corner stone of the proposed interatomic potential.

From the present investigations on Sr*X* (*X* = O, S, Se, and Te) we draw the following conclusions:

(a) The volume collapse (V_P/V_0) in terms of compressions in Sr*X* (*X* = O, S, Se, and Te) at higher pressure indicates the mechanical stiffenening of lattice. The expansion of Sr*X* (*X* = O, S, Se, and Te) lattice is inferred from steep increase in V_T/V_0 and is attributed to thermal softening of Sr*X* (*X* = O, S, Se, and Te) lattice.

(b) Larger deviation in Cauchy discrepancy Δ_1^2 emphasizes the importance of the many-body non-central (charge transfer and covalency) interaction and substantial anharmonic effects at high pressures.

(c) Melting temperature (T_M), Vicker's hardness (H_V), shear modulus (G_H), Young's modulus (E) and bulk modulus (B_T) increases with enhanced pressure infers the hardening or stiffening of the lattice and suppressed T_M, G_H, E and B_T variations in temperature suggests the weakening of the lattice as a result of thermal softening of SrX (X = O, S, Se, and Te) lattice.

(d) From the Pugh's ratio (ϕ) we classify SrO as ductile but SrS, SrSe, SrTe are brittle material while to that Poisson's ratio suggest Sr*X* (*X* = O, S, Se, and Te) its brittle nature. Hence, we conclude that SrS, SrSe, SrTe are brittle in nature at zero pressure the two empirical rules only differ on the exact border between the two types of behavior and SrO is a borderline case between the classes of ductile and brittle materials.

(e) Grüneisen parameter (γ_G) Debye temperature (θ_D), isothermal compressibility (β), heat capacity at constant volume (C_v) and thermal expansion coefficient ($\alpha_{th.\ exp}$) probes the importance of anharmonicity in Sr*X* (*X* = O, S, Se, Te) at high pressures and temperatures.

(f) Elastic properties showed $T_M(P) > T_M\ (T)$, $H_V(\text{-}P) > H_V\ (T)$, $E\ (P) > E\ (T)$, $\theta_D(P) > \theta_D(T)$ and $\alpha_{th.exp}\ (P) < \alpha_{th.exp}\ (T)$ Infers that these are differently sensitive to the Sr and *X* ions as Sr*X* lattice is mechanical hard due to bond strengthening and thermal soft due to bond weakening.

To an end, an interatomic pair wise potential for cubic SrX incorporating the non-central many-body forces as long-range Coulomb with charge transfer interactions, covalent nature of bonds (due to Sr–Sr, Sr–*X*, and *X*–*X* interacting electric fields), quantum effects, charge dipole–dipole and charge dipole-quadruple (van der Waals), and the short-range interactions as overlap repulsion up to second-neighbor ions successfully explains the pressure and temperature induced structural, elastic and thermodynamical properties of Sr*X* (*X* = O, S, Se, Te) consistent with available results.

Appendix

The expressions for the second order elastic constants for rock salt structure crystals follows:

$$C_{11} = \frac{e^2}{4a^4}\left\{-5.112Z_m^2 + A_1 + \frac{A_2+B_2}{2} + 9.3204Z(af'(r))\right\} \tag{49}$$

$$C_{12} = \frac{e^2}{4a^4}\left\{0.226Z_m^2 - B_1 + \frac{A_2-5B_2}{4} + 9.3204Z(af'(r))\right\} \tag{50}$$

$$C_{44} = \frac{e^2}{4a^4}\left\{2.556Z_m^2 + B_1 + \frac{A_2+3B_2}{4}\right\} \tag{51}$$

and

$$B_1 + B_2 = -1.165Z_m^2. \tag{52}$$

Thus second order elastic constants difference lead to finite value at Cauchy pressure:

$$C_{12} - C_{44} \neq 0 \tag{53}$$

Henceforth, polarizability of the ions has effect on the elastic constants. It should be noted that if charge transfer mechanism is not taken in to account, Cauchy relation: $C_{12} - C_{44} = 0$, The Cauchy violations ($C_{12} \neq C_{44}$) is seen by several crystals due to anisotropy in the electron distribution or angle bending. Here, C_{11} represents a measure of resistance to deformation by applied stress and C_{44} represents the measure of resistance to deformation with respect to applied shearing stress. The elastic constants C_{12} and C_{44} are related to the elasticity in shape, which is a shear constant.

The pressure derivatives of second order elastic constants under hydrostatic pressure P are obtained in the form

$$3\Omega\frac{dB_T}{dp} = -\{13.980Z_m^2 + C_1 - 3A_1 + C_2 - 3A_2 - 167.7648Z(af'(r)) + 41.9420Z(a^2f''(r))\} \tag{54}$$

$$2\Omega\frac{d\sigma}{dp} = -\left\{23.682Z_m^2 + C_1 + \frac{C_2+6A_2-6B_2}{4} - 50.0752Z(af'(r)) + 13.9808Z(a^2f''(r))\right\} \tag{55}$$

$$\Omega\frac{dC_{44}}{dp} = -\left\{-11.389Z_m^2 + A_1 - 3B_1 + \frac{C_2+2A_2-10B_2}{4} + 44.6528Z(af'(r))\right\}. \tag{56}$$

The notation Ω is

$$\Omega = -2.330Z_m^2 + A_1 + A_2 + 27.9612Z(af'(r)). \tag{57}$$

Finally, the third order elastic constants for rock salt structure crystal follows:

$$C_{111} = \frac{e^2}{4a^4}\left\{37.556Z_m^2 + C_1 - 3A_1 - \frac{3A_2+9B_2-C_2}{4} - 89.305Z(af'(r)) + 13.980Z(a^2f''(r))\right\} \tag{58}$$

$$C_{112} = \frac{e^2}{4a^4}\left\{-4.836Z_m^2 - \frac{3A_2+3B_2-C_2}{8} - 18.640Z(af'(r)) + 4.66Z(a^2f''(r))\right\} \tag{59}$$

$$C_{123} = \frac{e^2}{4a^4}\{2.717Z_m^2 + 16.692Z(af'(r))\} \tag{60}$$

$$C_{144} = \frac{e^2}{4a^4}\{2.717Z_m^2 + 5.564Z(af'(r))\} \tag{61}$$

$$C_{166} = \frac{e^2}{4a^4}\left\{-4.836Z_m^2 - 2(B_1+B_2) - \frac{3A_2-3B_2-C_2}{8} + 5.564Z(af'(r))\right\} \tag{62}$$

$$C_{456} = \frac{e^2}{4a^4}\{2.717Z_m^2 - (B_1+B_2)\}. \tag{63}$$

In view of equilibrium condition, the third order elastic constants difference also lead to finite value at Cauchy pressure inferring that polarizability of the ions has effect on the elastic constants:

$$C_{112} - C_{166} \neq 0 \tag{64}$$

$$C_{123} - C_{456} \neq 0 \tag{65}$$

$$C_{144} - C_{456} \neq 0 \tag{66}$$

$$C_{123} - C_{144} \neq 0. \tag{67}$$

Furthermore, the third order elastic constants satisfy the identity:

$$C_{123} + 2C_{456} - 3C_{144} = 0. \tag{68}$$

Various symbols appear in the above expressions are associated with the crystal energy and have the following form in rock salt structure

$$A_1 = \frac{8a^3}{e^2}\left[\frac{d^2}{dr^2}V_{ij}(r)\right]_{r=a}, \tag{69}$$

$$A_2 = \frac{16a^3}{e^2}\left[\frac{d^2}{dr^2}V_{ii}(r) + \frac{d^2}{dr^2}V_{jj}(r)\right]_{r=\sqrt{2}a}, \tag{70}$$

$$B_1 = \frac{8a^3}{e^2}\left[\frac{1}{r}\frac{d}{dr}V_{ij}(r)\right]_{r=a}, \tag{71}$$

$$B_2 = \frac{16a^3}{e^2}\left[\frac{1}{r}\frac{d}{dr}V_{ii}(r) + \frac{1}{r}\frac{d}{dr}V_{jj}(r)\right]_{r=\sqrt{2}a}, \tag{72}$$

$$C_1 = \frac{8a^3}{e^2}\left[r\frac{d^3}{dr^3}V_{ij}(r)\right]_{r=a}, \tag{73}$$

$$C_2 = \frac{16a^3}{e^2}\left[r\frac{d^3}{dr^3}V_{ii}(r) + \frac{d^3}{dr^3}V_{jj}(r)\right]_{r=\sqrt{2}a}. \tag{74}$$

Similarly for CsCl structure crystal, one determines the expressions for the elastic constants. It follows

$$C_{11} = \frac{e^2}{4a^4}\left\{0.7010Z_m^2 + \frac{A_1 + 2B_1}{6} + \frac{A_2}{2} + 3.1336Z(af'(r))\right\} \tag{75}$$

$$C_{12} = \frac{e^2}{4a^4}\left\{-0.5201Z_m^2 + \frac{A_1 - B_1}{6} + 3.1336Z(af'(r))\right\} \tag{76}$$

$$C_{44} = \frac{e^2}{4a^4}\left\{-0.3505Z_m^2 + \frac{A_1 + 2B_1}{6} + \frac{B_2}{2}\right\} \tag{77}$$

and

$$B_1 + B_2 = -0.3392Z_m^2. \tag{78}$$

The pressure derivatives of second order elastic constants under hydrostatic pressure P take the parametric form

$$3\Omega\frac{dB_T}{dp} = -\left\{2.0350Z_m^2 + \frac{C_1 - 3A_1 + C_2 - 3A_2}{18} + 24.4242Z(a^2f''(r)) - 56.4091Z(af'(r))\right\} \tag{79}$$

$$2\Omega\frac{d\sigma}{dp} = -\left\{-5.2229Z_m^2 + A_1 - B_1 + \frac{C_2}{2} + 47.9383Z(af'(r))\right\} \tag{80}$$

$$\Omega\frac{dC_{44}}{dp} = -\left\{1.2321Z_m^2 + \frac{A_1 - 7B_1 + C_1 + 3A_2 - 9B_2}{18} + 2.7138Z(a^2f''(r)) - 6.5791Z(af'(r))\right\} \tag{81}$$

and Ω is

$$\Omega = -0.3392Z_m^2 + \frac{A_1 + A_2}{2} + 9.4008Z(af'(r)) \tag{82}$$

Finally the expressions for the third order elastic constants of CsCl type crystals are

$$C_{111} = \frac{e^2}{4a^4}\left\{-4.5656Z_m^2 + \frac{C_1 - 3A_1 - 24B_1}{18} + \frac{C_2 - 3A_2}{2} + 13.1567Z(af'(r)) + 2.7138Z(a^2f''(r))\right\} \tag{83}$$

$$C_{112} = \frac{e^2}{4a^4}\left\{0.7001Z_m^2 + \frac{C_1 - 3A_1 + 3B_1}{18} + 6.2677Z(af'(r)) + 2.7138Z(a^2f''(r))\right\} \tag{84}$$

$$C_{123} = \frac{e^2}{4a^4}\left\{1.2000Z_m^2 + \frac{C_1 - 3A_1 + 3B_1}{18} - 15.9799Z(af'(r)) + 2.7138Z(a^2f''(r))\right\} \tag{85}$$

$$C_{144} = \frac{e^2}{4a^4}\left\{1.2000Z_m^2 + \frac{C_1 - 3A_1 + 3B_1}{18} - 5.3266Z(af'(r)) + 0.9046Z(a^2f''(r))\right\} \tag{86}$$

$$C_{166} = \frac{e^2}{4a^4}\left\{0.3609Z_m^2 + \frac{C_1 - 3A_1 + 3B_1}{18} - (B_1 + B_2) - 5.3266Z(af'(r)) + 0.9046Z(a^2f''(r))\right\} \tag{87}$$

$$C_{456} = \frac{e^2}{4a^4}\left\{1.2000Z_m^2 + \frac{C_1 - 3A_1 + 3B_1}{18}\right\} \tag{88}$$

Here, we define following form in CsCl structure

$$A_1 = \frac{16a^3}{e^2}\left[\frac{d^2}{dr^2}V_{ij}(r)\right]_{r=a\sqrt{3}}, \tag{89}$$

$$A_2 = \frac{8a^3}{e^2}\left[\frac{d^2}{dr^2}V_{ii}(r) + \frac{d^2}{dr^2}V_{jj}(r)\right]_{r=2a}, \tag{90}$$

$$B_1 = \frac{16a^3}{e^2}\left[\frac{1}{r}\frac{d}{dr}V_{ij}(r)\right]_{r=a\sqrt{3}}, \tag{91}$$

$$B_2 = \frac{8a^3}{e^2}\left[\frac{1}{r}\frac{d}{dr}V_{ii}(r) + \frac{1}{r}\frac{d}{dr}V_{jj}(r)\right]_{r=2a} \tag{92}$$

$$C_1 = \frac{16a^3}{e^2}\left[r\frac{\partial^3}{\partial r^3}V_{ij}(r)\right]_{r=a\sqrt{3}} \tag{93}$$

and

$$C_2 = \frac{8a^3}{e^2}\left[r\frac{d^3}{dr^3}V_{ii}(r) + r\frac{d^3}{dr^3}V_{jj}(r)\right]_{r=2a} \tag{94}$$

in terms of the short-range energy

$$V_{ij}(r) = \sum_{ij} b\beta_{ij}\exp\left(\frac{r_i + r_j - r_{ij}}{\rho}\right) - \sum_{ij}\frac{C}{r_{ij}^6} - \sum_{ij}\frac{D}{r_{ij}^8}. \tag{95}$$

The short-range interaction (SR) energy is expressed in terms of the overlap repulsion (first term) and the vdW $d{-}d$ and $d{-}q$ attractions (second and third terms), respectively.

References

1. Potzel, O., Taubmann, G.: The pressure induced B1–B2 phase transition of alkaline halides and alkaline earth chalcogenides. A first principles investigation. J. Solid State Chem. **184**, 1079 (2011)
2. Souadkia, M., Bennecer, B., Kalarasse, F., Mellouki, A.: Ab initio calculation of vibrational and thermodynamic properties of SrX (S, Se, Te) in the B1 (NaCl) and B2 (CsCl) structures. Comput. Mater. Sci. **50**, 1701 (2011)
3. Luo, H., Greene, R.G., Ruoff, A.L.: High-pressure phase transformation and the equation of state of SrSe. Phys. Rev. B **49**, 15341 (1994)
4. Sato, Y., Jeanloz, R.: Phase Transition in SrO. J. Geophys. Res. Solid Earth **86**, 11773 (1981)
5. Zimmer, H.G., Winzen, H., Sayassen, K.: High-pressure phase transitions in CaTe and SrTe. Phys. Rev. B **32**, 4066 (1985)
6. Son, P.R., Bartels, R.A.: CaO and SrO single crystal elastic constants and their pressure derivatives. J. Phys. Chem. Solids **33**, 819 (1972)
7. Syassen, K.: Pressure induced structural transition in SrS. Phys. Status Solidi A **91**, 11 (1985)
8. Khenata, R., Baltache, H., Rerat, M., Driz, M., Sahnoun, M., Bouhafs, B., Abbar, B.: First-principle study of structural, electronic and elastic properties of SrS, SrSe and SrTe under pressure. Phys. B **339**, 208 (2003)
9. Habas, M.P., Dovesi, R., Lichanot, A.: The phase transition in alkaline-earth oxides: a comparison of ab initio Hartree-Fock and density functional calculations. J. Phys. Condens. Matter **10**, 6897 (1998)
10. Yan, C., Lai-Yu, L., Ou-He, J., Xiang-Rong, C.: Phase transition and thermodynamic properties of SrS via first-principles calculations. Chin. Phys. B **17**, 1355 (2008)
11. Kholiya, K., Gupta, B.R.K.: Theoretical investigation of phase transition properties in strontium chalcogenides using potential model. Phase Transitions **81**, 403 (2008)
12. Tsuchiya, T., Kawamura, K.: Systematics of elasticity: ab initio study in B1-type alkaline earth oxides. J. Chem. Phys. **114**, 10086 (2001)
13. Born, M., Huang, K.: Dynamical theory of crystal lattices. Clarendon, Oxford (1956)
14. Vashishta, P., Kalia, R.K., Nakano, A.: Interaction potential for silicon carbide: a molecular dynamics study of elastic constants and vibrational density of states for crystalline and amorphous silicon carbide. J. Appl. Phys. **101**, 103515 (2007)
15. Stillinger, F.H., Weber, T.A.: Computer simulation of local order in condensed phases of silicon. Phys. Rev. B **31**, 5262 (1985)
16. Motida, K.: Szigeti charge and its correlation with hyperfine coupling constant of doped Mn^{2+} ion in divalent metal compounds. J. Phys. Soc. Jpn. **49**, 213 (1980)
17. Motida, K.: Effect of covalency on phonon dispersion relations in NaCl type alkali halide crystal. J. Phys. Soc. Jpn. **55**, 1636 (1986)
18. Slater, J.C., Kirkwood, J.G.: The vander Waals forces in gases. Phys. Rev. **37**, 682 (1931)
19. Tosi, M.P.: Cohesion of ionic solids in the born model. Solid State Phys. **16**, 1 (1964)
20. Hafemeister, D.W., Flygare, W.H.: Outer-shell overlap integral as a function of distance for halogen–halogen, halogen–alkali, and alkali–alkali ions in the alkali halide lattices. J. Chem. Phys. Soc. **43**, 795 (1965)
21. Lowdin, P.O.: Quantum theory of cohesive properties of solids. Adv. Phys. **5**, 1 (1956)
22. Lundqvist, S.O.: Three body potential for alkali halides. Ark. Fys. **12**, 263 (1957)
23. Varshney, D., Shriya, S.: Elastic, mechanical and thermodynamic properties at high pressures and temperatures of transition metal monocarbides. Int. J. Refract. Metals Hard Mater **41**, 375 (2013)
24. Singh, R.K.: Many body interactions in binary ionic solids. Phys. Rep. **85**, 259 (1982)
25. Varshney, D., Sharma, P., Kaurav, N., Shah, S., Singh, R.K.: Study of elastic properties and their pressure dependence of semi magnetic semiconductors. J. Phys. Soc. **74**, 382 (2005)
26. Varshney, D., Joshi, G., Kaurav, N., Singh, R.K.: Structural phase transition (zincblende–rocksalt) and elastic properties in AlY (Y = N, P and As) compounds: pressure-induced effects. J. Phys. Chem. Solids **70**, 451 (2009)
27. Varshney, D., Joshi, G., Varshney, M., Shriya, S.: Pressure induced structural phase transition and elastic properties in BSb, AlSb, GaSb and InSb compounds. Phys. B Condens. Matter **405**, 1663 (2010)
28. Varshney, D., Joshi, G., Varshney, M., Shriya, S.: Pressure dependent elastic and structural (B3–B1) properties of Ga based monopnictides. J. Alloys Compd. **495**, 23 (2010)
29. Varshney, D., Joshi, G.: High-pressure structural phase transition and elastic properties of Ga1-x InxAs semiconducting compounds. Eur. Phys. J. B **70**, 523 (2009)
30. Varshney, D., Kaurav, N., Kinge, R., Singh, R.K.: High-pressure structural (B1–B2) phase transition and elastic properties of II–VI semiconducting Sr chalcogens. Comput. Mater. Sci. **41**, 529 (2008)
31. Varshney, D., Shriya, S., Varshney, M., Singh, N., Khenata, R.: Elastic and thermodynamical properties of cubic (3C) silicon carbide under high pressure and high temperature. J. Theor. Appl. Phys. **9**, 221–249 (2015)
32. Varshney, D., Joshi, G., Varshney, M., Shriya, S.: Pressure induced mechanical properties of boron-based pnictides. Solid State Sci. **12**, 864 (2010)
33. Varshney, D., Kaurav, N., Kinge, R., Singh, R.K.: Pressure induced phase transition (B1–B2) and elastic properties in alkaline earth BaX (X = S, Se and Te) chalcogenides. Phase Transitions **81**, 1 (2008)
34. Varshney, D., Kaurav, N., Kinge, R., Singh, R.K.: High-pressure induced structural phase transition in alkaline earth CaX (X = S, Se and Te) semiconductors: NaCl-type (B1) to CsCl-type (B2). J. Alloys Compd. **484**, 239 (2009)
35. Varshney, D., Kaurav, N., Kinge, R., Singh, R.K.: Pressure induced phase transition and elastic properties of Y and Sc antimonides. J. Alloys Compd. **448**, 250 (2008)
36. Varshney, D., Singh, N.: Thermoelectric power of metallic Rb3C60: phonon-drag and carrier diffusion contributions. J. Theor. Appl. Phys. **6**(37), 1–9 (2012)
37. Varshney, D., Dodiya, N.: Metallic and semi-conducting resistivity behavior of $La_{0.7}Ca_{0.3\text{-}x}$ K_xMnO_3 ($x = 0.05$, 0.1) manganites. J. Theor. Appl. Phys. **9**, 45–48 (2015)
38. Tessman, J.R., Kahn, A.H., Shockley, W.: Electronic polarizabilities of ion in crystals. Phys. Rev. **92**, 890 (1953)
39. Shannon, R.D.: Dielectric polarizabilities of ions in oxides and fluorides. J. Appl. Phys. **73**, 348 (1993)
40. Weast, R.C. (ed.): Handbook of chemistry and physics, 63rd edn. CRC, Boca Raton (1982)
41. Murnaghan, F.D.: The compressibility of media under extreme pressures. Proc. Natl. Acad. Sci. **3**, 244 (1944)
42. Goel, P., Choudhury, N., Chaplot, S.L.: Superionic behavior of lithium oxide Li2O: a lattice dynamics and molecular dynamics study. Phys. Rev. B **70**, 174307 (2004)

43. Li, X.-F., Chen, X.-R., Meng, C.-M., Ji, G.-F.: Ab initio calculations of elastic constants and thermodynamic properties of Li_2O for high temperatures and pressures. Solid State Commun. **139**, 197 (2006)
44. Yun-Dong, G., Ze-Jin, Y., Qing-He, G., Zi-Jiang, L., Wei, D.J.: The phase transition, and elastic and thermodynamic properties of CaS derived from first-principles calculations. Phys. Condens. Matter **20**, 115203 (2008)
45. Vukcevich, M.R.: The elastic constants of cubic crystals with covalent and partially covalent bonds. Phys. Stat. Solidi (b) **54**, 435 (1972)
46. Barsch, G.R.: Relation between third-order elastic constants of single crystals and polycrystals. J. Appl. Phys. **39**, 3780 (1968)
47. Hill, R.: The elastic behaviour of a crystalline aggregate. Proc. Phys. Soc. (London) **65A**, 349 (1952)
48. Voigt, W.: Lehrbuch der Kristallphysik. Teubner, Leipzig (1928)
49. Reuss, A.: Calculation of the flow limits of mixed crystals on the basis of the plasticity of monocrystals. Angew Z. Math. Mech. **9**, 49 (1929)
50. Pugh, S.F.: Relations between the elastic moduli and the plastic properties of polycrystalline pure metals. Phil. Mag. **45**, 823 (1954)
51. Frantsevich, N., Voronov, F.F., Bokuta, S.A.: In: Frantsevich, I.N. (eds.) Elastic constants and elastic moduli of metals and insulators handbook. Naukova Dumka, Kiev, p. 60 (1983)
52. Schreiber, E., Anderson, O.L., Soga, N.: Elastic constants and their measurements. McGraw-Hill, New York (1973)
53. Harrison, W.A.: Electronic structure and properties of solids. Dover, New York (1989)
54. Bouhemadou, A., Khenata, R., Kharoubi, M., Seddik, T., Reshak, A.H., Al-Douri, Y.: FP-APW+ lo calculations of the elastic properties in zinc-blende III-P compounds under pressure effects. Comput. Mater. Sci. **45**, 474 (2009)
55. Maachou, A., Aboura, H., Amrani, B., Khenata, R., Bin Omran, S., Varshney, D.: Structural stabilities, elastic and thermodynamic properties of Scandium Chalcogenides via first-principles calculations. Comput. Mater. Sci. **50**, 3123 (2011)
56. Kleinman, L.: Deformation potentials in silicon. Phys. Rev. **128**, 2614 (1962)
57. Blackman, M.: The specific heat of solids. Proc. R. Soc. Lond. A **181**, 58 (1942)
58. Blackman, M.: The specific heat of solids. Proc. R. Soc. Lond. A **159**, 416 (1937)
59. Blackman, M.: The specific heat of solids. Proc. R. Soc. Lond. A **149**, 126 (1935)
60. Blackman, M.: The specific heat of solids. Proc. R. Soc. Lond. A **148**, 384 (1935)
61. Blackman, M.: The specific heat of solids. Proc. R. Soc. Lond. A **148**, 365 (1935)
62. Gopal, E.S.R.: Specific heats at low temperatures. Plenum Press, New York (1966)
63. Tari, A.: The specific heat of matter at low temperatures. Imperial College Press, London (2003)
64. Souadkia, M., Bennecer, B., Kalarasse, F.: Ab initio lattice dynamics and thermodynamic properties of SrO under pressure. J. Phys. Chem. Solids **73**, 129 (2012)
65. Sharifzadeh, H.S., Sharifzadeh, S.S., Kanjouri, F., Esmailian, A.: Mechanical properties of CaN, SrN, and BaN compounds by density functional theory. J. Theor. Appl. Phys. **7**(16), 1–5 (2013)
66. Rai, D.P., Shankar, A., Sandeep, Ghimire, M.P., Thapa, R.K.: Electronic and magnetic properties of a full-Heusler alloy Co_2-CrGe: a first-principles study. J. Theor. Appl. Phys. **7**(3), 1–6 (2013)

Exactly solvable reaction diffusion models on a Bethe Lattice through the empty-interval method

Laleh Farhang Matin

Abstract The most general reaction–diffusion model on a Bethe Lattice with nearest-neighbor interactions is introduced, which can be solved exactly through the empty-interval method. The stationary solutions of such models are discussed. For some special choice of reaction rates the dynamics of the system is also studied.

Keywords Reaction–diffusion · Bethe Lattice · Empty interval method

Introduction

Reaction–diffusion systems have been studied using various methods including analytical techniques, approximation methods, and simulation. Approximation methods are generally different in different dimensions, as for example the mean filed techniques, working good for high dimensions, generally do not give correct results for low-dimensional systems. A large fraction of analytical studies belong to low-dimensional (especially one-dimensional) systems, as solving low-dimensional systems should in principle be easier [1–11].

The Cayley tree or Bethe lattice is a tree (a lattice having no loops) where every site is connected to ξ nearest neighbor sites. This no—loops property may allow exact solvability for some models, for general coordination number ξ Reaction diffusion models on the Cayley tree are studied in, for example [12–17]. In [12, 13, 16] diffusion-limited aggregations, and in [14] two-particle annihilation reactions for immobile reactants have been studied. There are also some exact results for deposition processes on the Bethe lattice [17].

The empty interval method (EIM) has been used to analyze the one dimensional of diffusion-limited coalescence [18–21]. Using this method, the probability that n consecutive sites are empty has been calculated. This method has been used to study a reaction–diffusion process with three-site interactions [22]. EIM has been also generalized to study the Kinetics of the q-state one-dimensional Potts model in the zero-temperature limit [23]. In [18–21] one-dimensional diffusion-limited processes have been studied using EIM. There, some of the reaction rates have been taken infinite, and the models have been worked out on continuum. For the cases of finite reaction—rates, some approximate solutions have been obtained.

In [24, 25], all the one-dimensional reaction–diffusion models with nearest neighbor interactions which can be exactly solved by EIM have been found and studied. Conditions have been obtained for the systems with finite reaction rates to be solvable via EIM, and then the equations of EIM have been solved. In [25], general conditions were obtained for a single-species reaction–diffusion system with nearest neighbor interactions, to be solvable through EIM. Here solvability means that evolution for E_n (the probability that n consecutive sites be empty) is closed. It turned out there, that certain relations between the reaction rates are needed, so that system is solvable via EIM. The evolution equation of E_n is a recursive equation in terms of n, and is linear. It was shown that if certain reactions are absent, namely reactions that produce particles in two adjacent empty sites, the coefficients of the empty intervals in the evolution equation of the intervals are n-independent, so that the evolution equation can be easily solved. The criteria for solvability, and the solution

L. F. Matin (✉)
Department of Physics, Islamic Azad University, North Tehran Branch, Tehran, Iran
e-mail: laleh.matin@gmail.com

of the empty-interval equation were generalized to cases of multi-species systems and multi-site interactions in [26–28].

In this article, the most general single-species reaction–diffusion model with nearest-neighbor interactions on a Cayley tree is investigated, which can be solved exactly through the empty interval method. The scheme of the paper is as follows. In Sect. 2, the most general reaction–diffusion model with nearest neighbor interactions on a Cayley tree is studied, which can be solved exactly through EIM. The evolution equation of E_n, the probability that a connected collection of n sites be empty is also obtained. In Sect. 3 the stationary solutions of such models are discussed.

The dynamics of the system for special choice of reaction rates is studied in Sect. 4. Finally, Sect. 5 is devoted to concluding remarks.

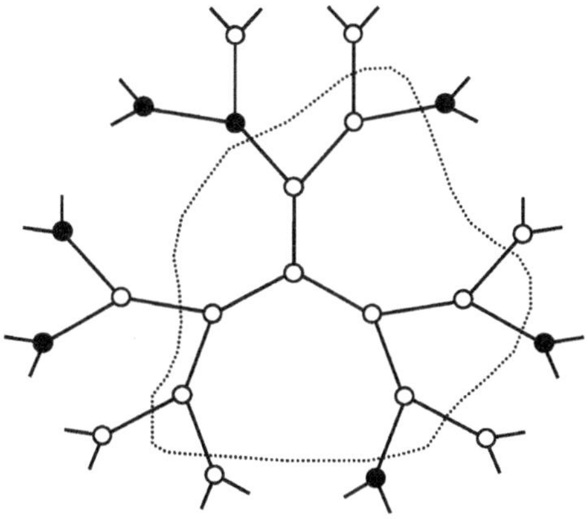

Fig. 2 An empty cluster with the links at the boundary, on a Cayley tree with $\xi = 3$

Models solvable through the empty interval method on a Cayley tree

The Cayley tree (a lattice without loops) where every site is connected to ξ sites (Figs. 1, 2). Two sites are called neighbors if they are connected through a link. Consider a system of particles on a Cayley tree. Each site is either empty or occupied by one particle. The interaction (of particles and vacancies) is nearest neighbor. The probability that a connected collection of n sites be empty is denoted by E_n. It is assumed that this quantity does not depend on the choice of the collection. An example is a tree where the probability that a site is occupied is p and is independent of the states of other sites. Then

$$E_n = (1-p)^n. \tag{1}$$

The following graphical representations help express various relations in a more compact form. An empty (occupied) site is denoted by o ($\bullet$). A connected collection of n empty sites is denoted by O_n.

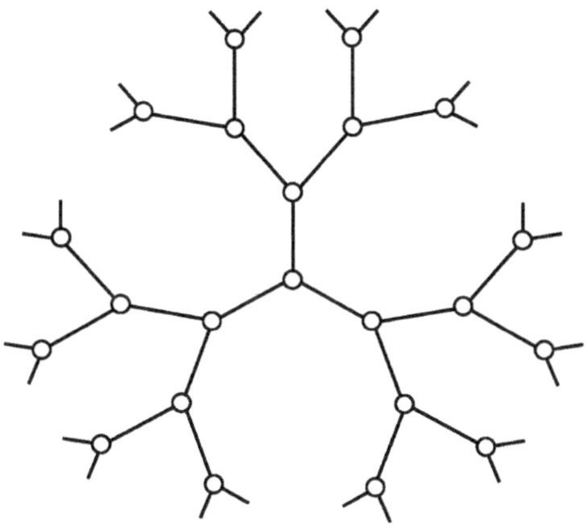

Fig. 1 The Cayley tree (Bethe lattice) with $\xi = 3$

There is no loop in a Cayley tree, so each site can only be connected to a single existing cluster site, by a single link. An empty cluster expands or shrinks through interaction on its boundary. It can also be destroyed through particle generations in the bulk. Assume that the interactions are nearest neighbor (two-site) of the following kinds, with the rates indicated.

$$\bullet\circ \to \left\{\begin{matrix} \bullet\bullet, & r_1 \\ \circ\bullet, & r_2 \end{matrix}\right\}, \circ\bullet \to \left\{\begin{matrix} \bullet\circ, & r_2 \\ \bullet\bullet, & r_1 \end{matrix}\right\}, \bullet\bullet \to \left\{\begin{matrix} \bullet\circ, & r_3 \\ \circ\bullet, & r_3 \end{matrix}\right\}, \tag{2}$$

And

$$\circ\circ \to \left\{\begin{matrix} \bullet\bullet, & r \\ \circ\bullet, & r_0 \\ \bullet\circ, & r_0 \end{matrix}\right\}. \tag{3}$$

The closeness of the evolution equation for E_n requires that the rate of change of anything to ($\circ\circ$) becomes zero. The probability of finding an empty cluster connected to an occupied site $P\,(\bullet - O_n)$ can be written in terms of E_n and E_{n+1}:

$$P(\bullet - O_n) + P(\circ - O_n) = P(O_n) \Rightarrow P(\bullet - O_n) = E_n - E_{n+1}. \tag{4}$$

To study the evolution of E_n, one considers states which can transform to O_n (the sources) and states O_n can transform to them (the sinks). These are the followings.

$$\begin{aligned} &\textit{Sources}: \circ - \bullet - O_{n-1}, \bullet - \bullet - O_{n-1}, \\ &\textit{Sinks}: \bullet - O_n, \circ - O_n, O_n. \end{aligned} \tag{5}$$

Using these, the time evolution equation for E_1 ($n > 1$) is seen to be

$$\frac{dE_n}{dt} = R_n[r_2 P(\circ - \bullet - O_{n-1}) + r_3 P(\bullet - \bullet - O_{n-1})] - R_n(r_1 + r_2)P(\bullet - O_n) - R_n(r + r_0)P(\circ - O_n) - (n-1)(r + 2r_0)P(O_n), \quad (6)$$

where R_n is the number of sites adjacent to a collection of n connected sites. A simple induction in [29] shows that

$$R_n = n(\xi - 2) + 2. \quad (7)$$

Using [4] and

$$P(\circ - \bullet - O_{n-1}) + P(\bullet - \bullet - O_{n-1}) = P(\bullet - O_{n-1}), \quad (8)$$

equation [6] recasts to

$$\frac{dE_n}{dt} = R_n[r_2(E_{n-1} - E_n) + (r_3 - r_2)P(\bullet - \bullet - O_{n-1}) - (r_1 + r_2)(E_n - E_{n+1}) - (r + r_0)E_{n+1}] - (n-1)(r + 2r_0)E_n, \quad n > 1. \quad (9)$$

In order that the evolution for E_n be closed, it is needed that

$$r_3 = r_2. \quad (10)$$

It is also seen that the evolution for E_1 takes a form similar to that of E_n $(n > 1)$, provided one defines

$$E_0 = 1. \quad (11)$$

Then we have

$$\frac{dE_n}{dt} = R_n[r_2(E_{n-1} - E_n) - (r_1 + r_2)(E_n - E_{n+1}) - (r + r_0)E_{n+1}] - (n-1)(r + 2r_0)E_n \qquad n \geq 1. \quad (12)$$

along with the boundary condition [11].

The stationary solution

The stationary solution of the system (E^s, for which the time derivative vanishes), satisfies

$$0 = R_n r_2 E^s_{n-1} - [R_n(r_1 + 2r_2) + (n-1)(r + 2r_0)]E^s_n + R_n(r_1 + r_2 - r - r_0)E^s_{n+1}, \quad n \geq 1, \quad (13)$$

with the boundary condition

$$E^s_0 = 1. \quad (14)$$

Consider two cases separately.

***r* = *r*₀ = 0**

One has

$$r_2 E^s_{n-1} - (r_1 + 2r_2)E^s_n + (r_1 + r_2)E^s_{n+1} = 0. \quad (15)$$

This is a second order linear difference equation with constant coefficients. To solve this, one puts the ansatz

$$E^s_n = C_1 z_1^n + C_2 z_2^n, \quad (16)$$

In [15], and obtains

$$E^s_n = C_1 + C_2\left(\frac{r_2}{r_1 + r_2}\right)^n. \quad (17)$$

The constants C_1 and C_2 depend on the initial conditions. So the stationary solution is not unique. Note that the result is independent of coordination number ξ. So the stationary behavior of the system, reaction diffusion model on a Cayley tree with arbitrary coordination number ξ, is similar to that of a reaction diffusion model on a one-dimensional lattice ($\xi = 2$), provided of course that there is no process which creates particles from two neighboring vacant sites. Using [17] for $n = 0$, and $n = 1$, together with the boundary condition [14], the constants C_1 and C_2 can be expressed in term of E^s_1

$r \neq 0$ or $r_0 \neq 0$

It is a difficult task to obtain a closed form for E^s_n. Things become, however, simpler for large n's. One puts the ansatz

$$E^s_n = C_n z^n, \quad (18)$$

In [13], assuming that C_n is slowly varying with respect to n. Keeping only the largest powers in n, one arrives at

$$f(z) = 0. \quad (19)$$

where f is defined through

$$f(z) := (\xi - 2)(r_1 + r_2 - r - r_0)z^2 - [(\xi - 2)(r_1 + 2r_2) + (r + 2r_0)]z + (\xi - 2)r_2. \quad (20)$$

There are two solutions for z, each giving rise to a stationary solution. The general stationary solution would be linear combination of these. A stationary solution however, is acceptable only if it is nonnegative and nonincreasing for all n's. Such a solution should not blow up at $n \to \infty$, which shows that for acceptable solutions

$$|z| \leq 1. \quad (21)$$

Using

$$f(0) = (\xi - 2)r_2 > 0,$$
$$f(1) = -(\xi - 2)(r + r_0) - (r + 2r_0) < 0, \quad (22)$$

It is seen that one of the roots of [20] (let's call it z_1) satisfies

$$0 < z_1 < 1. \quad (23)$$

For the other root, two cases occur:

(i) $r_1 + r_2 > r + r_0$.
Here $f(\infty) > 0$, and $z_2 > 1$. Hence this root gives rise to r_0 an unphysical solution. So the stationary solution is unique.

(ii) $r_1 + r_2 < r + r_0$.
Here $f(\infty) < 0$, and $z_2 < 0$. Using

$$z_1 + z_2 = \frac{(\xi - 2)(r_1 + 2r_2) + r + 2r_0}{r_1 + r_2 - r - r_0} < 0, \qquad (24)$$

One arrives at

$$|z_1| < |z_2|. \qquad (25)$$

This shows that in any combination $C_{1n}z_1^n + C_{2n}z_2^n$, with C_{1n} and slowly varying, for large n's the second term is the dominant term unless it is identical to zero. But the sign of this term changes with n for large n. So the sign of the whole combination changes with n for large n's unless the second term vanishes identically. Hence any combination of stationary solutions with a nonzero contribution corresponding to z_2 is unphysical. The stationary solution is again unique. Note that if one considers the evolution equation [12] on its own, without taking into account the fact that $E_n s$ are not totally independent of each other and cannot be negative, then there would be two stationary solutions. What eliminates one of these is that only special initial values are possible for E_n's, so that in the large time limit only one of the stationary solutions survives.

To summarize, if r or r' are nonvanishing, the stationary solution is unique, so the solution for large times is independent of the initial conditions.

There are two special subcases where [13] is simplified. In these cases [13], it is reduced to a first order difference equation.

$r_2 = 0$

Here

$$E_n^s = \delta_{n0} \qquad (26)$$

is a solution to [13] and [14], and as the stationary solution is unique, this is the stationary solution.

$r_1 + r_2 = r + r_0$

In this case [13] becomes a first order recursive equation for $E_{n'}^s$ from which on obtains

$$E_n^s = \prod_{j=1}^{n} \frac{(\xi - 2)r_2 j + 2r_2}{[(\xi - 2)(r_1 + 2r_2) + (r + 2r_0)]j + 2r_2 + r} \qquad (27)$$

Dynamical solutions

It is a difficult task to obtain the dynamical solution in general. One may seek solutions with exponential time dependence:

$$E_n^\varepsilon(t) = E_n^\varepsilon \exp(\varepsilon t) \qquad (28)$$

Putting this in [12], one arrives at

$$0 = R_n r_2 E_{n-1}^\varepsilon - [R_n(r_1 + 2r_2) + (n-1)(r + 2r_0) + \varepsilon]E_n^\varepsilon + R_n(r_1 + r_2 - r - r_0)E_{n+1}^\varepsilon, \quad n \geq 1. \qquad (29)$$

The boundary condition corresponding to [11] is

$$E_0^\varepsilon = 0. \qquad (30)$$

In general, it is difficult to solve these for large n, however, one can use arguments similar to those used in the previous section. Using an ansatz similar to [19], it is seen that ε does not change the equation governing z. Again the root z_2 (which is either greater than one or negative) give rise to unphysical solutions. The reason is that although the dynamical solution [29] need not be nonnegative and decreasing in n by itself, the sum of such a solution and the stationary solution can be a complete physical solution to [11] and [12] and hence should satisfy these conditions. For sufficiently large n, a contribution coming from z_2 would be dominant and either blows up or becomes negative. So the whole solution either blows up or becomes negative, unless there is no contribution corresponding to z_2 in the dynamical solutions. This leaves only one physical solution for (30). Such a solution does not necessarily satisfy (31). The condition (30) may be satisfied only for special values of ε. Hence the spectrum of the evolution operator is discrete, in other words the system has a finite relaxation time.

For two special choices of parameters, however, one can go further. These are the cases discussed in the previous section when the evolution equation becomes first order in n.

$r_2 = 0$

Here the reactants are immobile, and the coagulation rate is also zero. But there are nonzero rates for birth and decoagulation. So, one expects at large times all the sites to be occupied, and in the stationary states all the $E_n' s$ are zero as it was shown to be the case in the previous section. The evolution equation can be solved exactly for a special choice of initial condition, Eq. (12) recasts to

$$\frac{dE_n}{dt} = -(R_n r_1 + (n-1)(r + 2r_0)]E_n + R_n(r_1 - r - r_0)E_n + 1, \quad n \geq 1. \qquad (31)$$

Putting the ansatz

$$E_n(t) = E_1(t)[b(t)]^{n-1}, \tag{32}$$

In (32), one arrives at

$$\frac{db}{dt} = -[(\xi - 2) + r + 2r_0]b + (\xi - 2)(r_1 - r - r_0)b^2,$$
$$\frac{dE_1}{dt} = \xi[-r_1 + (r_1 - r - r_0)b]E_1. \tag{33}$$

These are readily solved and one obtains.

$$b(t) = \frac{b(0)\exp(-\beta t)}{1 + \alpha b(0)[1 - \exp(-\beta t)]},$$
$$E_1(t) = E_1(0)\exp(-\xi r_1 t)\left\{\frac{1}{1 + \alpha b(0)[1 - \exp(-\beta t)]}\right\}^{\frac{\xi}{\xi-2}}, \tag{34}$$

where

$$\alpha := \frac{(\xi - 2)(r + r_0 - r_1)}{(\xi - 2)r_1 + r + 2r_0}$$
$$\beta := (\xi - 2)r_1 + r + 2r'. \tag{35}$$

Using these, one obtains

$$E_n(t) = E_n(0)\exp[-\xi r_1 t - B(n-1)t] \times \left\{\frac{1}{1 + \alpha b(0)[1 - \exp(-\beta t)}\right\}^{\frac{\xi}{\xi-2}+n-1} \tag{36}$$

It is seen that for times, all $E_n s$ tend to zero. In fact they decay like

$$E_n(t) \sim \exp[-\xi r_1 t - \beta(n-1)t]. \tag{37}$$

A special case where the ansatz (33) works is the case of initially uncorrelated sites, so that each site is occupied with probability ρ regardless of other sites.

One has then

$$E_n(0) = (1 - \rho)^\wedge n, \tag{38}$$

So that

$$E_1(0) = 1 - \rho,$$
$$b(0) = 1 - \rho. \tag{39}$$

$\boldsymbol{r}_1 + \boldsymbol{r}_2 = \boldsymbol{r} + \boldsymbol{r}_0$

In this case (12) becomes

$$\frac{dE_n}{dt} = R_n r_2 E_{n-1} - [R_n(r_1 + 2r_2) - (n-1)(r + 2r_0)]E_n, \quad n \geq 1 \tag{40}.$$

This set of equations, together with the boundary condition [11], can be solved iteratively. The equation for $E_1(t)$ becomes

$$\frac{dE_1}{dt} = \xi r_2 - \xi(r_1 + 2r_2)E_1, \tag{41}$$

the solution to which is

$$E_1(t) = \frac{r_2}{r_1 + 2r_2} + \left[E_1(0) - \frac{r_2}{r_1 + 2r_2}\right]\exp[-\xi(r_1 + 2r_2)t]. \tag{42}$$

So, at large times each site is occupied whit the probability

$$1 - E_n(\infty) = \frac{r_1 + r_2}{r_1 + 2r_2} \tag{43}$$

and the average density relaxes to this value with the relaxation time

$$r = \frac{1}{\xi(r_1 + 2r_2)} \tag{44}$$

Concluding remarks

The most general one-dimensional single-species exclusion model on a Cayley tree was considered, for which the evolution of the empty—intervals is closed. The stationary solutions of such models were discussed. It was shown that except for special values of rates, the stationary solutions are unique, hence independent of initial conditions. The dynamics of such systems were also studied and it was shown that generally the spectrum is discrete and depends on n. For special choices of reaction rates, the dynamics was studied in more detail and closed form for the empty interval probabilities were obtained. Among the questions remaining one can mention the statics and dynamics of systems with arbitrary rates, as well as the possible existence of phase transitions. One can also investigate Cayley trees with boundaries, with injection and extraction at the boundaries.

References

1. Schütz, G.M.: Exactly solvable models for many-body systems far from equilibrium. In: Domb, C., Lebowitz, J. (eds.) Phase Transitions and Critical Phenomena, vol. 19. Academic press, London (2000)
2. Alcaraz, F.C., Droz, M., Henkel, M., Rittenberg, V.: Reaction-diffusion processes, critical dynamics, and quantum chains. Ann. Phys. (N.Y.) **230**, 250 (1994)
3. Krebs, K., Pfannmuller, M.P., Wehefritz, B., Hinrichsen, H.: Finite-size scaling studies of one-dimensional reaction-diffusion systems. J. Stat. Phys. **78**(FS), 1429 (1995)
4. Simon, H.: Concentration for one and two species one-dimensional reaction-diffusion. J. Phys. **A28**, 6585 (1995)
5. Privaman, V., Cadilhe, A.M.R., Glasser, M.L.: Exact solution of anisotropic diffusion limited reactions with coagulation and annihilation. J. Stat. Phys. **81**, 881 (1995)

6. Henkel, M., Orlandini, E., Schütz, G.M.: Equivalences between stochastic systems. J. Phys. **A28**, 6335 (1995)
7. Henkel, M., Orlandini, E., Santos, J.: Reaction-diffusion processes from equivalent integrable quantum chains. Ann. Phys. **259**, 163 (1997)
8. Lushnikov, A.A.: Binary reaction 1+1→0 in one dimension. Sov. Phys. JETP **64**, 811 (1986). [Zh. Eksp.Teor.Fiz.**91**(1986)1376]
9. Alimohammadi, M., Karimipour, V., Khorrami, M.: Exact solution of a one-parameter family of asymmetric exclusion processes. Phys. Rev. **E57**, 6370 (1998)
10. Alimohammadi, M., Karimipour, V., Khorrami, M.: A two-parametric family of asymmetric exclusion processes and its exact solution. J. Stat. Phys. **97**, 373 (1999)
11. Aghamohammadi, A., Khorrami, M.: Similarity transformation in one-dimensional reaction-diffusion systems: the voting model as an example. Phys. **A33**, 7843 (2000)
12. Vannimenus, J., Nickel, B., Hakim, V.: Models of cluster growth on the Cayley tree. Phys. Rev. **B30**, 391 (1984)
13. Krug, J.: Surface structure of random aggregates on the Cayley tree. J. Phys. **A21**, 4637 (1988)
14. Majumdar, S.N., Privman, V.: Annihilation of immobile reactions in the Bethe lattice. J. Phys. **A26**, L743 (1993)
15. Kelbert, M.Y., Suhov, Y.M.: The Markov branching random walk and systems of reaction-diffusion (Kolmogorov–Petrovskiĭ–Piskunov) equations. Commun. Math. Phys. **167**, 607 (1995)
16. Majumdar, S.N.: Traveling front solutions to directed diffusion limited aggregation, digital search trees and the Lempel–Ziv data compression algorithm. Phys. Rev. **E68**, 026103 (2003)
17. Cadilhe, A., Privman, V.: Random sequential adsorption of mixtures of dimers and monomers on a pre-treated Bethe lattice. Modern Phys. Lett. **B18**, 207 (2004)
18. Burschka, M.A., Doering, C.R., Ben-Avraham, D.: Transition in the relaxation dynamics of a reversible diffusion-limited reaction. Phys. Rev. Lett. **63**, 700 (1989)
19. Ben-Avraham, D.: The method of interparticle distribution functions for diffusion-reaction systems in one dimension. Phys. Lett. **B9**, 895 (1995)
20. Ben-Avraham, D.: Nonequilibrium statistical mechanics in one dimension. In: Privman, V. (ed), pp. 29–50. Cambridge University Press (1997)
21. Ben-Avraham, D.: Complete exact solution of diffusion-limited coalescence, A + A → A. Phys. Rev. Lett. **81**, 4756 (1998)
22. Henkel, M., Hinrichsen, H.: Exact solution of a reaction-diffusion process with three-site interactions. J. Phys. **A34**, 1561–1568 (2001)
23. Mobilia, M., Bares, P.A.: Generalized empty-interval method applied to a class of one-dimensional stochastic models. Phys. Rev. **E64**, 066123 (2001)
24. Aghamohammadi, A., Khorrami, M.: Models solvable through the empty-interval method. Eur. Phys. J. **B47**, 583586 (2005)
25. Alimohammadi, M., Khorrami, M., Aghamohammadi: Exactly solvable models through the empty-interval method. Phys. Rev. **E64**, 056116 (2001)
26. Khorrami, M., Aghamohammadi, A., Alimohammadi, M.: Exactly solvable models through the empty interval method, for more-than-two-site interactions. J. Phys. **A36**, 345 (2003)
27. Aghamohammadi, A., Alimohammadi, M., Khorrami, M.: Exactly solvable models through the generalized empty interval method, for multi-species interactions. Eur. Phys. J. **B31**, 371 (2003)
28. Aghamohammadi, A., Khorrami, M.: Exactly solvable modeles through the generalized empty interval method: multi species and more-than-two-site interactions. Int. J. Modern Phys. **B18**, 2047 (2004)
29. Rozikov, U.A.: A contour method on Cayley trees. J. Stat. Phys. **130**, 801–813 (2008)

24

Antibacterial effect, structural characterization, and some applications of silver chiral nano-flower sculptured thin films

Hadi Savaloni[1] · Fatemh Haydari-Nasab[1] · Abbas Abbas-Rohollahi[2]

Abstract Silver chiral nano-flowers with 3-, 4-, and 5-fold symmetry were produced using oblique angle deposition method in conjunction with the rotation of sample holder with different speeds at different sectors of each revolution. X-ray diffraction, atomic force microscopy, and scanning electron microscopy were employed to obtain the nanostructure and morphology of the films. Their antibacterial, electrical, and hydrophobic properties were investigated. Antibacterial properties were investigated against a range of microorganisms including *Escherichia coli* ATCC 8739, *Staphylococcus aureus* ATCC 25923, and *Candida albicans* PTCC 5027. Electrical conductivity of these films relative to that of bulk sample is reduced by a factor of about 10^3 due to porosity, surface roughness, and anisotropic structure of these films. Hydrophobicity results show dependence on the symmetry of these chiral nano-flowers.

Keywords Silver chiral nano-flower · Thin film · Antibacterial · Electrical conductivity · Hydrophobicity

Introduction

The metallic nanostructures with specific properties dependent on their size and geometrical shape have created an interesting research environment for scientist in different fields of application. In particular, noble metals (gold, silver, and copper) have found applications in optics [1, 2], electronic [3, 4], biomedicine and antibacterial activity [5–7], and quantum size domain [8].

This group of metallic nanostructures because of their ability in exhibiting localized surface Plasmon resonances (LSPR) has found many applications in biodetection [9]. LSPR excitation can considerably enhance the local electric field which is the main reported mechanism of surface-enhanced Raman spectroscopy (SERS) [10, 11].

Silver nanostructures, such as silver nanoparticles, provide a large surface-area-to-volume ratio and a high fraction of surface atoms [12], which enhance the antimicrobial activity of silver, even at low concentrations [13–18]. The antibacterial property of silver depends on the amount of silver and its reduction rate in the test environment. In recent years, silver coatings have shown to decrease bacterial adherence and growth in vitro [19–21]. The antibacterial activity of silver ions has been well established [22, 23], as lying in interactions with S, O, and N containing groups. However, the role of silver nanomaterials has not been as clearly determined and many contradictory findings are reported [24].

The cluster size, clusters cross-section, and its reaction rate with substrate have an important role in silver antibacterial activity [25]. Recent investigations have shown that enhanced antibacterial property can be achieved when silver nanocolumns/rods are used instead of normal silver thin films or silver nanoparticles. This is due to higher roughness of these structures and surface area ratio [26].

In addition, it is also well established that superhydrophobic surface can be achieved through combination of high surface roughness and low surface free energy [27–29]. A large number of physical techniques such as

✉ Hadi Savaloni
savaloni@khayam.ut.ac.ir

1 Department of Physics, University of Tehran, North-Kargar Street, Tehran, Iran

2 Department of Biology, University of Tehran, Enghelab Street, Tehran, Iran

lithographic patterning and etching, molding, and imprinting have been used to obtain rough surfaces in order to produce superhydrophobic surfaces [30–32]. On the other hand, direct deposition of micro/nanostructures of different materials such as metals [33], polymers [34], oxides [35], or carbon nanotubes [36] on substrates through chemical reactions and or assembly processes also have proved to show superhydrophobicity. Fan et al. [37] have found that as the height of the nanorod increased, the water contact angle of as grown Si nanorods decreased.

Layer-by-Layer (LbL) assembly was first introduced by G. Decher [38] in order to achieve superhydrophbic surface, as a versatile method to assemble layered nanostructures with tailored composition and architecture [39]. Savaloni and Esfandiar [40] achieved high degree of hydrophobicity by deposition of chiral-graded zigzag silver sculptured thin films. In recent years, oblique angle deposition (OAD) of thin films as a physical vapor deposition method has provided facilities for the production of variety of nanostructures with structural anisotropy which can be controlled by pre-design of the structure [41, 42].

OAD (incident angle <85°) and glancing angle deposition (GLAD) (incident angle >85°) methods together with the rotation of substrate about its surface normal can be used to produce differently shaped nanostructures such as chirals [43], zigzags [44], S-shaped, and other shapes [45]. Sculptured nanostructures (range in size between 1 and 100 nm) fabricated by these methods have many applications such as optical filters [46], photonic crystals [47], catalysts [48], magnetic storages [49], micro-batteries [50], bioscaffolds [51], and microchannels [52].

Oblique silver nanocolumns grown using OAD method without rotation of substrate have shown surface-enhanced Raman spectra (SERS) [53] and surface-enhanced fluorescence (SEF) [54] which can give useful and desired results in detection of minute amount of chemicals and viruses. Silver chiral nano-flowers with different symmetries have also shown a high degree of enhancement in obtaining surface Raman Spectra [11]. The structure of sculptured thin films (STFs) consists of rough surface with a much higher void fraction than the films produced using conventional deposition methods.

The use of OAD technique together with rotation of substrate and change of rotation rate during each revolution of the substrate by 2 N ($N = 1$–5) provides facilities to produce nanostructures with N-fold symmetry called chiral nano-flowers [55].

In this work, we have used this technique and produced silver chiral nano-flowers with different symmetries and investigated their antibacterial, hydrophobicity, and electrical properties as well as structural characteristics. This work is to the best of our knowledge at the time of submission, the first report on the above-mentioned properties of sculptured silver chiral nano-flowers with different orders of symmetry.

Experimental details

Silver (99.99 % purity) chiral nano-flower-shaped sculptured thin films with different symmetries were deposited on glass (15×15 mm^2 microscope slide) substrates by resistive evaporation from tungsten boats with a circular outlet of 6 mm in diameter at room temperature. An Edwards (Edwards E19 A3) coating plant with a base pressure of 2×10^{-7} mbar was used. The deposition angle was fixed at (80°) and a deposition rate of 2.5 Ås^{-1} was chosen. In order to achieve a uniform deposition on the substrates, the distance between the evaporation source and the substrate was set at 30 cm. The deposition process was repeated a few times and the reproducibility of the results was confirmed. The movement of the stepper motor and its speed of revolution as well as facility for dividing each revolution to different sectors are controlled through interface to a computer in which the related software is written and installed [56].

Nano-flowers with 3-, 4-, and 5-fold symmetries were produced, considering that for an N-fold symmetry each revolution of the substrate holder should be divided to 2 N sectors. In this work, in order to produce 3- and 4-fold symmetry nano-flower, the smaller sector was chosen as ($\Theta_L = 24°$) and for production of 5-fold symmetry, it was chosen as ($\Theta_L = 18°$) which rotates with a speed of $R_L = 0.0151$ (rev/s), while the other sectors size depends on the order of chosen symmetry and its speed (R_H) was chosen to be 32 times of R_L.

Prior to deposition, all substrates were ultrasonically cleaned in heated acetone then ethanol. The surface roughness of the substrates was measured by a Talysurf profilometer and AFM, and the rms substrate surface roughness R_q obtained using these methods was 0.3 and 0.9 nm, respectively.

The deposition rate was controlled using a quartz crystal monitor (Sigma Instruments, SQM-160, USA) positioned close to the substrate and at almost the same azimuthal angle as that of the substrate. This was corrected after obtaining the film thickness using field emission electron microscope (FESEM; Hitachi S-4100 SEM, Japan). FESEM samples were coated with a very thin layer of gold to prevent the charging effect. The surface physical morphology and roughness was obtained by means of atomic force microscope (AFM) (NT-MDT SOLVER, with a Si tip of 10 nm radius in contact mode) analysis.

Antibacterial activity of silver nano-flower thin films with different symmetries of 3-, 4-, and 5-folds against the *Escherichia coli* ATCC 8739, *staphylococcus aureus* ATCC 25923, and *Candida albicans* PTCC 5027 bacteria was studied using the so-called diffusion assay method. The bacteria were chosen because gram-negative bacteria are responsible for more than 80 % of all infections, with *E. coli* being responsible for more infections than all other genera combined. *S. aureus* and *C. albicans* were also utilized as representative members of gram-positive and fungal genera [57]. Among the bacteria studied, *E. coli* is gram-negative, whereas *S. aur* is gram-positive.

Prior to performing the microbiological experiment, all glass wares and samples were sterilized by autoclaving at 120 °C for 15 min. All films were cleaned by ethanol and acetone. For sterilization, they were placed into a sterilized Petri dish and irradiated by UVC lamp for 30 min. The microorganisms were cultured on a nutrient agar plate and incubated at 37 °C for 24 h. The cultured organisms were added in 10 ml saline solution to reach the concentration of bacteria to 10^8 colony forming units per milliliter (CFU/ml) corresponding to MacFarland scale. 100 μl of the saline solution containing the microorganisms was added to nutrient agar plate and spread with a sterile cotton swab. Then for antibacterial test, each thin film was placed onto a cultured nutrient agar plate and incubated at 37 °C for 24 h. After 24 h area of inhibition zone of growth was measured.

Results and discussions

FESEM and AFM (nanostructure)

In our earlier work [58], we reported on the structural characteristics of similar thin films obtained using FESEM and AFM analysis. Since the results of this work are similar to those reported earlier, we do not see any point in repeating such results. However, it should be mentioned that in this work, the deposition angle was 80° and for the production of 5-fold symmetry nano-flower in this work we used $\Theta_L = 18°$, while in the earlier work $\Theta_L = 24°$ was applied. In order to get the structural information one may consult Figs. 2 and 3 in [58]. In Table 1 we have included the results obtained for grain size and surface void fraction for the three types of silver chiral nano-flowers produced in this work.

Table 1 Results of grain/petal size, average surface roughness, and surface void fraction in the Ag chiral nano-flowers of different symmetries on glass substrates

Ag/glass nano-flower	D_{XRD} (nm)	D_{AFM} (nm)	R_{ave} (nm)	F_v (%)
3-fold	15.0	82.0	12.0	41.5
4-fold	17.0	70.0	15.9	32.4
5-fold	24.0	66.0	10.9	30.1

F_v is the surface void fraction obtained from 2D AFM images

Crystal structure

The XRD patterns of the samples produced in this work for three different symmetries are given in Fig. 1. Crystalline structure of the samples is consistent with the JCPDS card No: 04-0783. Zhou et al. [59] reported crystalline structure for their silver nanorod samples even when they reduced the substrate temperature to −40 °C.

Polycrystalline films deposited on substrates generally show preferred orientation, with a strength which depends on the deposition method, film material, and deposition conditions, including residual gas pressure. However, one should consider the fact that in case of oblique angle deposition or slanted films, the texture axis can be tilted out of the surface normal, and the evaluation of intensity distribution of various reflections for determining the texture can be completely misleading. In that case, the texture had to be analyzed by taking pole figures at fixed diffraction angle [31, 60]. Hence the procedure usually used for films deposited at normal deposition angle [61] cannot be applied to oblique angle or glancing angle deposited films. A program of work on this subject is currently under investigation in our group.

Crystallite size (coherently diffracting domains) D was obtained by applying the Scherrer formula [62] to measure the full width at half maximum (FWHM) of the dominant peak of silver nano-flower films [i.e., Ag(111)]. The results with 10 % accuracy are given in Table 1. The results show that the crystallite size (coherently diffracting domains) increases with increasing the order of symmetry of the chiral nano-flower.

Antibacterial property of Ag chiral nano-flowers

Nano-silver structures such as silver nanoparticles that have a large surface-to-volume ratio have shown a high degree of antibacterial property [63]. The antimicrobial activity of silver-based materials depends on the nanostructure that should provide a high surface-area-to-volume ratio and hence a high fraction of surface atoms [12, 26], which enhances the antibacterial property of silver. Pal et al. [64] investigated the antibacterial properties of silver nanoparticles with different shapes. They concluded that the differences in the observed trends in *E. coli* ATCC 8739 inhibition can be explained in terms of the percentage of active facets present in nanoparticles of different shapes. They also speculate that the action of silver nanoparticles is broadly similar to that of silver ion. It may be anticipated

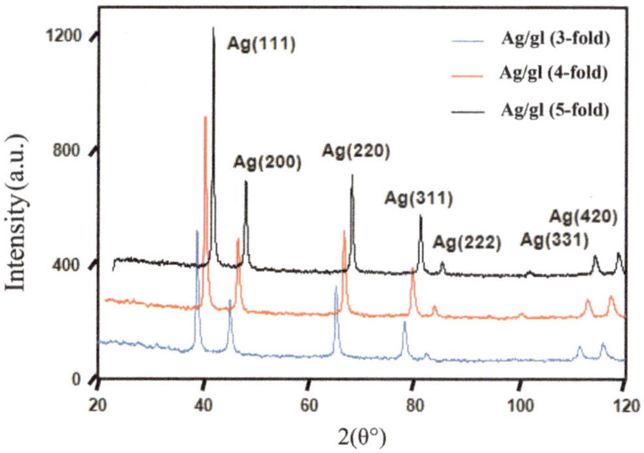

Fig. 1 XRD patterns of silver chiral nano-flower sculptured thin films with different symmetries

that a bacterial cell in contact with silver nanoparticles takes in silver ions, which inhibit a respiratory enzyme(s), facilitating the generation of reactive oxygen species and consequently damaging the cell [62]. It is also demonstrated that the reactivity of silver is favored by high-atom-density facets, such as {111} [15, 65, 66]. Hence it may reiterate that by increasing the surface-to-volume ratio and surface roughness of thin films, the probability of releasing silver atoms/ions in the medium increases, and since the Ag(111) facet consists of the maximum number of atoms in the silver structure, it should provide the highest antibacterial effect.

It is generally believed that heavy metals such as silver interact with the thiol groups of proteins that are important for the bacterial respiration and the transport of significant substances through the cell [67, 68]. In addition, silver ions can bind to the bacterial cell wall (slightly negative) and also alter the function of the bacterial cell membrane [69, 70]. Experimental evidence also proposes that silver ion interaction with bacteria causes DNA molecules to lose their replication capability [70]. Therefore, considering the structural and morphological results obtained for our silver chiral nano-flower thin films, presented in the preceding sections, we expect that our nano-sculptured thin films should show an enhanced antibacterial property compared to thin films deposited at normal incident angle.

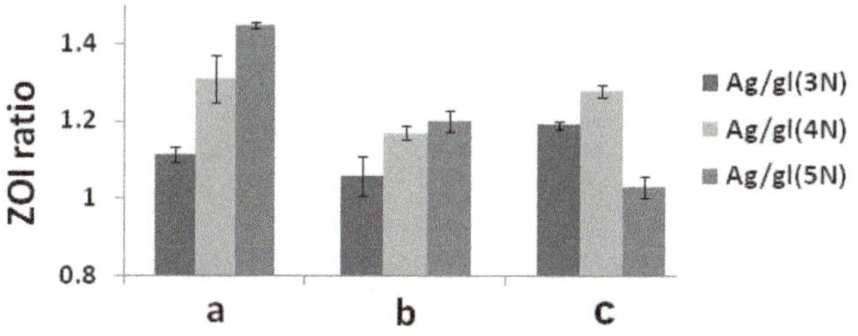

Fig. 3 The zone of inhibition for silver chiral nano-flower films of different symmetries against different test strains is compared after 24 h incubation; *a Escherichia coli* ATCC 8739, *b Staphylococcus aureus* ATCC 25923, *c Candida albicans* PTCC 5027

Figure 2 shows typical zone of inhibition (ZOI) for 3-fold symmetry silver chiral nano-flower on Petri dish for two different bacteria and a fungus, namely *E. coli* ATCC 8739, *S. aur* ATCC 25923, and *C. albicans* PTCC 5027 after 24 h incubation. ZOI was recorded as the ratio of the area of antibacterial activity to the area of the sample.

In Fig. 3, the ZOI obtained for silver chiral nano-flower films of different symmetries against different test strains are compared. In Fig. 3, columns indicated by (1) are the results of *E. coli* ATCC 8739, columns indicated by (2) are for *S. aur* ATCC 25923, and columns indicated by (3) are for *C. albicans* PTCC 5027, respectively. It can be seen that in case of the first two bacteria, the 5-fold symmetry nano-flower antibacterial effect dominates the other two structures. This may be due to the high intensity of (111) orientation of this sample, as discussed in "Crystal structure" and higher surface-to-volume ratio due to increased number of petals in this film. However, for *C. albicans* PTCC 5027, the 4-fold symmetry nano-flower is more active than the other two structures. In addition it should also be mentioned that all three nanostructures used in this work have more antibacterial effect on *E. coli* ATCC 8739, relative to the *S. aur* ATCC 2592. This is because the *E. coli* ATCC 8739 is a gram-negative bacteria.

Electrical property of Ag nano-flowers

In this section, we discuss the results obtained from the measurements of the dc resistance. In order to investigate the influence of possible (low frequency) charging effects

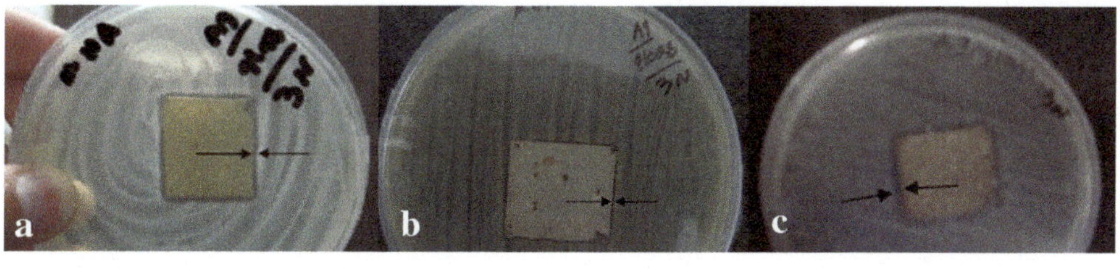

Fig. 2 Typical zone of inhibition (ZOI) for 3-fold symmetry silver chiral nano-flower for two different bacteria (*Escherichia coli* ATCC 8739 and *Staphylococcus aureus* ATCC 25923) and a fungus (*Candida albicans* PTCC 5027) after 24 h incubation

Table 2 Results of conductivity measurement for Ag/glass nano-flower sculptured thin films with different symmetries in S/cm unit

	Ag/glass (3-fold)	Ag/glass (4-fold)	Ag/glass (5-fold)
0°	7.83×10^3	5.81×10^3	6.11×10^3
45°	7.03×10^3	5.85×10^3	6.34×10^3
90°	7.54×10^3	5.51×10^3	6.09×10^3
135°	6.92×10^3	6.12×10^3	6.64×10^3
Anisotropy between 0° and 90°	0.0188	0.0265	0.0016

at the electrical contacts and leads, current–voltage (I–V) curves were recorded, scanning the voltage both in increasing and decreasing increments. In order not to heat up the samples, during the measurements, the current was kept low enough. A linear I–V curve was obtained, independent of the scan direction, and there was no indication of hysteresis effect. In order to investigate the anisotropy effect in our samples, the I-V curves measurements were also carried out in four different directions on the samples in 45° increments. The average values of conductivity for three silver chiral nano-flower samples with different symmetries are given in Table 2. Results in Table 2 show that conductivity of each sample changes with the direction of measurement, hence showing the effect of anisotropy in the structure of chiral nano-flower silver thin films as expected. The anisotropy in the data was obtained using.

$$\text{Anisotropy} = \frac{A_x - A_y}{A_x + A_y}$$

where x and y are related to two measurements at x and y directions, namely 0° and 90° measurements. It can be seen that the sample with 4-fold symmetry shows the highest anisotropy, this is expected since the petals of nano-flower with 4-fold symmetry are positioned with exactly 90 degree spacing between them and may have introduced the bundling-like effect which usually occurs in the oblique deposition technique into the structure of the grown thin film.

In addition, we can also observe that the sample with 3-fold symmetry shows the highest conductivity and the sample with 4-fold symmetry shows the lowest conductivity. The reasons for these observations may be explained as follows; it is well known that the conductivity of a sample depends on the number of grain boundaries, surface roughness, impurities, film cross-section, and number and size of voids in the electron path that cause scattering of electrons resulting in the loss of energy. As expected from the design of the chiral nano-flower structures, the 3-fold symmetry consists of largest grains/petals and the 5-fold symmetry contains smallest grains/petals (Table 1, column 3), hence considering the element of size, it can be concluded that the 3-fold symmetry sample should have the highest conductivity. However, the difference between grain/petal sizes of 4-fold and 5-fold symmetry samples may not be significant enough for causing the different conductivities obtained for these two samples. The sample with 4-fold symmetry has the highest surface roughness (Table 1, column 4). Therefore, it may cause a high degree of electron scattering from the surface which in turn decreases the film conductivity. In addition as our electrical measurements at different directions mentioned above showed this sample has the highest anisotropy which we related this phenomenon to the bundling-like effect in this sample, hence this may also be another parameter in the decreased conductivity of this sample relative to 3-fold and 5-fold symmetry samples. The size and number of voids (grain boundaries) are other parameters that we should consider with regard to the changes in the conductivity of the samples. As mentioned in the experimental section, the sections with high rotation rate (i.e., voids) for 3- and 4-fold symmetry nano-flowers were produced with 24° angles, while for 5-fold symmetry, this was 18° (a reduction by 25 %). Hence we expect that the size of voids in the 5-fold symmetry sample be smaller than the other two samples, while the number of voids (grain boundaries) between petals of the chiral nano-flower in 5-fold symmetry is increased. Therefore, in general, as the symmetry order increases, the θ_h section of the growing film becomes smaller and the growing sculpture thin film seems to resemble a circle. In particular, in our 5-fold symmetry chiral nano-flower case, we also reduced the section (θ_h) relative to 4-fold symmetry sample. This can be the reason for higher conductivity of this film relative to the 4-fold symmetry sample which has larger voids between petals.

Hydrophobicity effect of silver chiral nano-flower sculptured thin films

It is well established that superhydrophobic surface can be achieved through combination of high surface roughness and low surface free energy [27–29]. Since the surfaces of chiral nano-flowers produced in this work consist of pointed petals and recesses between petals, it is expected to be rougher than thin films produced using normal deposition procedure, hence they may show higher hydrophobicity effect. The results of surface roughness shown in Table 1 and results of hydrophobicity test on our silver chiral nano-flower sculptured thin films with 3-, 4-, and

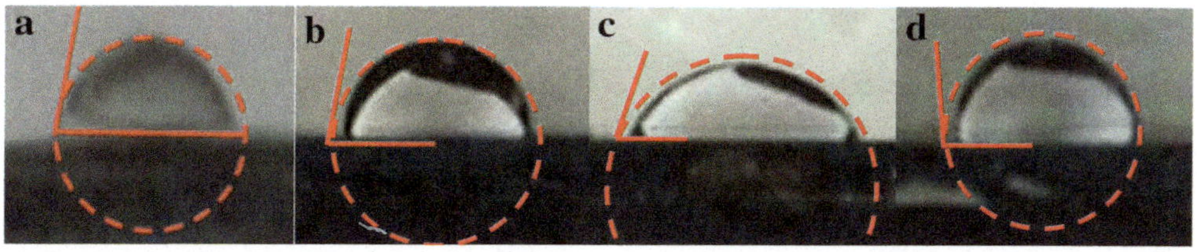

Fig. 4 Contact angles for water droplets on, **a** 300 nm silver thin film deposited at normal angle to the substrate without rotation of substrate holder; **b** silver chiral nano-flower with 3-fold symmetry; **c** silver chiral nano-flower with 4-fold symmetry; **d** silver chiral nano-flower with 5-fold symmetry

Table 3 Contact angle for Ag/glass nano-flower sculptured thin films with different symmetries

Structure	Ag/gl (normal)	Ag/gl (3-fold)	Ag/gl (4-fold)	Ag/gl (5-fold)
Contact angle	69°	85°	72°	106°

5-fold symmetries and a silver film of 300 nm thickness deposited under conventional condition of normal incidence angle shown in Fig. 4 are compared. It can be observed that the contact angle for all chiral nano-flower samples is greater than that obtained for the film deposited at normal incidence angle (Table 3). Obviously, these results cannot be accepted as superhydrophobic achievement, but they indicate that their surface hydrophobicity is increased relative to the conventionally deposited silver film. 4-fold symmetry silver chiral nano-flower shows smaller contact angle than those obtained for 3- and 5-fold symmetry silver chiral nano-flowers. This can be due to increased symmetry axes in this sample, though the film surface roughness for this film is greater than 3- and 5-fold symmetry nano-flowers. As discussed throughout this paper, the morphology (length and shape) of nanocolumns as well as interspaces and air diffusion between nanocolumns change. These variations of structural parameters are the main factors for the increase of contact angle [71]. Hence in this work, we have achieved enhanced hydrophobic surface (i.e., for 3- and 5-fold symmetry chiral nano-flowers) by engineering the sample surface via sculptured thin film deposition to produce chiral nano-flowers.

Conclusions

We produced and obtained the antibacterial effect, electrical, and hydrophobicity properties of silver chiral nano-flowers with three different symmetries by applying oblique angle deposition method in conjunction with the rotation of sample holder with different speeds at different sectors of each revolution. Nanostructure and morphology of the produced samples were obtained using XRD, AFM, and FESEM. Antibacterial properties were investigated against a range of microorganisms including *E. coli* ATCC 8739, *S. aureus* ATCC 25923, and *C. albicans* PTCC 5027. It was observed that the 5-fold symmetry chiral nano-flower antibacterial effect is more pronounced than 3- and 4- fold symmetry chiral nano-flowers in acting on *E. coli* ATCC 8739 and *S. aur* ATCC 25923, which can be related to its crystallographic structure [high intensity of (111) orientation] and higher surface-to-volume ratio. Electrical conductivity of these films relative to that of bulk sample is reduced by a factor of about 10^3 due to porosity, surface roughness, and anisotropic structure of these films. Hydrophobicity results showed dependence on the symmetry of chiral nano-flower.

Acknowledgments This work was carried out with the support of the University of Tehran and the Iran National Science Foundation (INSF). HS is grateful to the Centre of Excellence for Physics of Structure and Microscopic Properties of Matter, Department of Physics, University of Tehran for partial support of this work.

References

1. Feng, J., Zhao, X., Liu, B., Zhou, X.: Microstructural characterization and optical polarization of glass with needle-like micro–nano silver oriented arrangement. Opt. Commun. **281**, 5041–5044 (2008)
2. Chu, T.C., Liu, W.C., Tsai, D.P.: Enhanced resolution induced by random silver nanoparticles in near-field optical disks. Opt. Commun. **246**, 561–567 (2005)
3. Park, K., Seo, D., Lee, J.: Conductivity of silver paste prepared from nanoparticles. Coll. Surf. A. **313**, 351–354 (2008)
4. Alqudami, A., Annapoorni, S., Sen, P., Rawat, R.S.: The incorporation of silver nanoparticles into polypyrrole: conductivity changes. Synth. Met. **157**, 53–59 (2007)
5. Chen, C.Y., Chiang, C.L.: Preparation of cotton fibers with antibacterial silver nanoparticles. Mater. Lett. **62**, 3607–3609 (2008)

6. Nakamura, H., Tanaka, M., Shinohara, S., Gotoh, M., Karube, I.: Development of a self-sterilizing lancet coated with a titanium dioxide photocatalytic nano-layer for self-monitoring of blood glucose. Biosens. Bioelectron. **22**, 1920–1925 (2007)
7. Beyth, N., Yudovin-Farber, I., Bahir, R., Domb, A.J., Weiss, E.I.: Antibacterial activity of dental composites containing quaternary ammonium polyethylenimine nanoparticles against Streptococcus mutans. Biomaterials **27**, 3995–4002 (2006)
8. Jiang, L., Wang, W., Wu, D., Zhan, J., Wang, Q., Wu, Z., Jin, R.: Preparation of silver quantum dots embedded water-soluble silica/PAAc hybrid nanoparticles and their bactericidal activity. Mater. Chem. Phys. **104**, 230–234 (2007)
9. Alivisatos, P.: The use of nanocrystals in biological detection. Nat. Biotechnol. **22**, 47–52 (2004)
10. Willets, K.A., Van Duyne, R.P.: Localized surface plasmon resonance spectroscopy and sensing. Annu. Rev. Phys. Chem. **58**, 267–297 (2007)
11. H Savaloni and R Babaei, Surface-enhanced Raman spectroscopy and structural characterization of Ag/Cu Chiral Nano-flower Sculptured Thin films. Appl. Surf. Sci. (2013) in print
12. Ichinose, N., Ozaki, Y., Kashu, S.: Superfine particle technology. Springer, New York (1992)
13. Lee, H.J., Yeo, S.Y., Jeong, S.H.: Antibacterial effect of nanosized silver colloidal solution on textile fabrics. J. Mater. Sci. **38**, 2199–2204 (2003)
14. Lok, C.N., Ho, C.M., Chen, R., He, Q.Y., Yu, W.Y., Sun, H., Tam, P.K.H., Chiu, J.F., Che, C.M.: Proteomic analysis of the mode of antibacterial action of silver nanoparticles. J. Proteome Res. **5**, 916–924 (2006)
15. Morones, J.R., Elechiguerra, J.L., Camacho, A., Holt, K., Kouri, J.B., Ramirez, J.T., Yacaman, M.J.: The bactericidal effect of silver nanoparticles. Nanotechnology **16**, 2346–2353 (2005)
16. Maneerung, T., Tokura, S., Rujiravanit, R.: Impregnation of silver nanoparticles into bacterial cellulose for antimicrobial wound dressing. Carbohydr. Polym. **72**, 43–51 (2008)
17. Mahapatra, S.S., Karak, N.: Silver nanoparticle in hyperbranched polyamine: synthesis, characterization and antibacterial activity. Mater. Chem. Phys. **112**, 1114–1119 (2008)
18. Xu, K., Wang, J.X., Kang, X.L., Chen, J.F.: Fabrication of antibacterial monodispersed Ag–SiO_2 core–shell nanoparticles with high concentration. Mater. Lett. **63**, 31–33 (2009)
19. Leung, J.W.C., Lau, G.T.C., Sung, J.J.Y., Costerton, J.W.: Decreased bacterial adherence to silver-coated stent material: an in vitro study. Gastrointest. Endosc. **38**, 338–340 (1992)
20. Liedberg, H., Lundeberg, T.: Silver alloy coated catheters reduce cafheter-associated bacteriuria. Urol. Res. **25**, 937–952 (1991)
21. Boswald, M., Girisch, M., Greil, J., Spies, T., Stehr, K., Krall, T., Guggenbichler, J.P.: Antimicrobial activity and biocompatibility of polyurethane and silicone catheters containing low concentrations of silver: a new perspective in prevention of polymer-associated-foreign-body-infections. Zentralbl. Bakteriol. **283**, 187–200 (1995)
22. Slawson, R.M., Van Dyke, M.I., Lee, H., Trevors, J.T.: Germanium and silver resistance, accumulation, and toxicity in microorganisms. Plasmid **27**, 72–79 (1992)
23. Russell, A.D., Chopra, I.: Understanding antibacterial action and resistance. Hemel Hempstead, chapter 3. Hertfordshire: Ellis Horwood, UK (1996)
24. Marambio-Jones, C., Hoek, E.M.V.: A review of the antibacterial effects of silver nanomaterials and potential implications for human health and the environment. J. Nanopart. Res. **12**, 1531–1551 (2010)
25. Silvestry-Rodriguez, N., Sicairos-Ruelas, E.E., Gerba, C.P., Bright, K.R.: Silver as a disinfectant. Rev. Environ. Contam. Toxicol. **191**, 23–45 (2007)
26. Akhavan, O., Ghaderi, E.: Capping antibacterial Ag nanorods aligned on Ti interlayer by mesoporous TiO_2 layer. Surf. Coat. Technol. **203**, 3123–3128 (2009)
27. Wenzel, R.N.: Resistances of solid surfaces to wetting by water. J. Ind. Eng. Chem. **28**, 988–994 (1936)
28. Cassie, A.B.D., Baxter, S.: Wettability of porous surfaces. Trans. Faraday Soc. **40**, 546–551 (1944)
29. Zhao, N., Shi, F., Wang, Z., Zhang, X.: Combining layer-by-layer assembly with electrodeposition of silver aggregates for fabricating superhydrophobic surfaces. Langmuir **21**, 4713–4716 (2005)
30. Nosonovsky, M., Bhushan, B.: Biomimetic Superhydrophobic Surfaces: multiscale Approach. Nano Lett. **7**, 2633–2637 (2007)
31. Cho, W.K., Choi, I.S.: Fabrication of hairy polymeric films inspired by geckos: wetting and high adhesion properties. Adv. Funct. Mater. **18**, 1089–1096 (2008)
32. Winkleman, A., Gotesman, G., Yoffe, A., Naaman, R.: Immobilizing a drop of water: fabricating highly hydrophobic surfaces that pin water droplets. NanoLett. **8**, 1241–1245 (2008)
33. Larmour, I.A., Bell, S.E.J., Saunders, G.C.: Sheets of large superhydrophobic metal particles self assembled on water by the cheerios effect. Angew. Chem. Int. Ed. **47**, 5043–5045 (2007)
34. Balu, B., Breedveld, V., Hess, D.W.: Fabrication of "roll-off" and "sticky" superhydrophobic cellulose surfaces via plasma processing. Langmuir **24**, 4785–4790 (2008)
35. Namavar, F., Cheung, C.L., Sabirianov, R., Mei, W.N., Zeng, X.C., Wang, G., Haidar, H., Garvin, K.L.: Lotus effect in engineered zirconia. Nano Lett. **8**, 988–996 (2008)
36. Pastine, S.J., Pkawa, D., Kessler, B., Rolandi, M., Llorente, M., Zettl, A., Frechet, J.M.: A facile and patternable method for the surface modification of carbon nanotube forests using perfluoroarylazides. J. Am. Chem. Soc. **130**, 4238 (2008)
37. Fan, J.G., Dyer, D., Zhang, G., Zhao, Y.P.: Nanocarpet effect: pattern formation during the wetting of vertically aligned nanorod arrays. Nano Lett. **4**, 2133–2138 (2004)
38. Decher, G., Hong, J.D.: Buildup of ultrathin multiplayer films by a self—assembly process: consecutive adsorption of anionic and cationic bipolar amphiphiles on charged surfaces. Makromol. Chem. Macromol. Symp. **46**, 321–327 (1991)
39. Decher, G., Schlenoff, J.B.: Multilayer thin films: sequential assembly of nanocomposite materials. Wiley, Weinheim (2002)
40. Savaloni, H., Esfandiar, A.: Fabrication, characterization and some applications of graded chiral zigzag shaped nano-sculptured silver thin films. Appl. Surf. Sci. **257**, 9425–9434 (2011)
41. Robbie, K., Brett, M.J.: Chiral sculptured thin films. Nature (London) **384**, 616 (1996)
42. Robbie, K., Brett, M.J.: Sculptured thin films and glancing angle deposition: growth mechanics and applications. J. Vac. Sci. Technol. A **15**, 1460–1465 (1997)
43. Dick, B., Brett, M.J., Smy, T.: Controlled growth of periodic pillars by glancing angle deposition. J. Vac. Sci. Technol. B **21**, 23–28 (2003)
44. Kesapragada, S.V., Victor, P., Nalamasu, O., Gall, D.: Nanospring pressure sensors grown by glancing angle deposition. Nano Lett. **6**, 854–857 (2006)
45. Messier, R., Venugopal, V.C., Sunal, P.D.: Origin and evolution of sculptured thin films. J. Vac. Sci. Technol. A **18**, 1538–1545 (2000)
46. Wu, Q., Hodgkinson, I.J., Lakhtakia, A.: Circular polarization filters made of chiral sculptured thin films: experimental and simulation results. Opt. Eng. **39**, 1863–1868 (2000)
47. Kennedy, S., Brett, M.J., Miguez, H., Toader, O., John, S.: Optical properties of a three-dimensional silicon square spiral photonic crystal. Photon. Nanostruct. Fundam. Appl. **1**, 37–42 (2003)

48. Harris, K.D., McBride, J.R., Nietering, K.E., Brett, M.J.: Fabrication of porous platinum thin films for hydro- carbon sensor applications. Sens. Mater. **13**, 225–234 (2001)
49. Dick, B., Brett, M.J., Smy, T.J., Freeman, M.R., Malac, M., Egerton, R.F.: Periodic magnetic microstructures by glancing angle deposition. J. Vac. Sci. Technol. A **18**, 1838–1844 (2000)
50. Fleischauer, M.D., Li, J., Brett, M.J.: Columnar thin films for three-dimensional microbatteries. J. Electrochem. Soc. **156**, A33–A36 (2009)
51. Demirel, M.C., So, E., Ritty, T.M., Naidu, S.H., Lakhtakia, A.: Fibroblast cell attachment and growth on nanoengineered sculptured thin films. J. Biomed. Mater. Res. B Appl. Biomater. **81**, 219–223 (2007)
52. Harris, K.D., Brett, M.J., Smy, T.J., Backhouse, C.: Microchannel surface area enhancement using porous thin films. J. Electrochem. Soc. **147**, 2002–2006 (2000)
53. Suzuki, M., Maekita, W., Wada, Y., Nagai, K., Nakajima, K., Kimura, K., Fukuoka, T., Mori, Y.: Ag nanorod arrays tailored for surface-enhanced Raman imaging in the near-infrared region. Nanotechnology **19**, 265304 (2008)
54. Abdulhalim, I., Karabchevsky, A., Patzig, C., Rauschenbach, B., Fuhrmann, B., Eltzov, E., Marks, R., Xu, J., Zhang, F., Lakhtakia, A.: Surface-enhanced fluorescence from metal sculptured thin films with application to biosensing in water. Appl. Phys. Lett. **94**, 063106 (2009)
55. Zhao, Y.P., Ye, D.X., Wang, G.C., Lu, T.M.: Novel nano-column and nano-flower arrays by glancing angle deposition. Nano Lett. **2**, 351–354 (2002)
56. Savaloni, H., Babaei, F., Song, S., Placido, F.: Characteristics of sculptured Cu thin films and their optical properties as a function of deposition rate. Appl. Surf. Sci. **255**, 8041–8047 (2009)
57. Davis, C.P., Wagle, N., Anderson, M.D., Warren, M.M.: Bacterial and fungal killing by iontophoresis with long-lived electrodes. Antimicrob. Agents Chemother. **35**, 2131–2134 (1991)
58. Savaloni, H., Haydari-Nasab, F., Malmir, M.: Nano-structural characteristics and optical properties of silver chiral nano-flower sculptured thin films. Appl. Surf. Sci. **257**, 9044–9055 (2011)
59. Zhou, Q., Li, Z., Yang, Y., Zhang, Z.: Arrays of aligned, single crystalline silver nanorods for trace amount detection. J. Phys. D Appl. Phys. **41**, 152007 (2008)
60. Savaloni, H., Moradi, G.R., Player, M.A.: Texture development and residual stress in UHV evaporated silver films on glass substrates as a function of substrate temperature. Vacuum **77**, 245–257 (2005)
61. Savaloni, H., Gholipour-Shahraki, M., Player, M.A.: A comparison of different methods for X-ray diffraction line broadening analysis of Ti and Ag UHV deposited thin films: nanostructural dependence on substrate temperature and film thickness. J. Phys. D Appl. Phys. **39**, 2231–2247 (2006)
62. Savaloni, H., Modiri, F.: Surface nano-structural modifications and characteristics in nitrogen ion implanted W as a function of temperature and N+ energy. Appl. Surf. Sci. **253**, 1135–1142 (2006)
63. Jeon, H.J., Yi, S.C., Oh, S.G.: Preparation and antibacterial effects of Ag–SiO_2 thin films by sol–gel method. Biomaterials **24**, 4921–4928 (2003)
64. Pal, S., Tak, Y.K., Song, J.M.: Does the antibacterial activity of silver nanoparticles depend on the shape of the nanoparticle? a study of the gram-negative bacterium escherichia coli. Appl. Env. Mic. **73**, 1712–1720 (2007)
65. Ajayan, P.M., Marks, L.D.: Quasimelting and phases of small particles. Phys. Rev. Lett. **60**, 585–587 (1988)
66. Hatchett, D.W., White, H.S.: Electrochemistry of sulfur adlayers on the low-index faces of silver. J. Phys. Chem. **100**, 9854–9859 (1996)
67. Cho, K.H., Park, J.E., Osaka, T., Park, S.G.: The study of antimicrobial activity and preservative effects of nanosilver ingredient. Electrochim. Acta **51**, 956–960 (2005)
68. Ivan, S., Branka, S.S.: Silver nanoparticles as antimicrobial agent: a case study on *E. coli* as a model for Gram-negative bacteria. J. Coll. Interface Sci. **275**, 177–182 (2004)
69. Percival, S.L., Bowler, P.G., Russell, D.: Bacterial resistance to silver in wound care. J. Hosp. Infect. **60**, 1–7 (2005)
70. Feng, Q.L., Wu, J., Chen, G.Q., Cui, F.Z., Kim, T.N., Kim, J.O.: A mechanistic study of the antibacterial effect of silver ions on *Escherichia coli* and *Staphylococcus aureus*. J. Biomed. Mater. Res. **52**, 662–668 (2000)
71. Cheng, Y.H., Chou, C.K., Chen, C., Cheng, S.Y.: Critical length of nanowires for hydrophobic behavior. Chem. Phys. Lett. **397**, 17–20 (2004)

25

Charged particles in curved space-time

K. Mehdizadeh[1] · O. Jalili[2]

Abstract Considering the dynamics of geometry and the matter fields, dynamical equations of geometry and the matter fields are re-derived. The solutions of these equations are studied. We focus on a charged particle and explain the axiomatic approach to drive the electromagnetic self-force on its motion, then the energy conservation is considered. A new mathematical concept, which is introduced in axiomatic approach in general, is discussed.

Keywords Charged particles motion · Curved space · Radiation

Introduction

In considering the plasma around a neutron star, we encounter a charged gas in the curved space. In statistical formalism of such problem usually we just need the first velocity moment of Vlasov or Boltzmann equation. In the equation of first velocity moment we study the average motion of particles in all elements. It is clear that every particle's trajectory is not in full adjustment to the element's trajectory. Some physical phenomena come back to this mismatching. For example, the first velocity moment cannot explain the Landau damping. Apart from this, the very charged particle motion is an important problem that contains some mathematical complexity. Radiation field of a charged particle motion in curved space, in general, can be divided into two parts: electromagnetic field and gravitational field. In this paper we only consider the first one. The solution of Maxwell's equation depends on several parameters, e.g., on the presence of other charged particles and the spacetime geometry. We focus on a particle that, even small, changes the spacetime geometry. One of the known ways to solve the differential equations is the use of Green functions. The Green functions that is appropriate for our problem are bitensors. Bitensor is a mathematical object that is defined in two distinct points of a manifold. The method we specially focus on it using another mathematical object: difference of two tensors belonging to two distinct tangent spaces. In this manner we come to a new mathematical concept in the formulation of the problem. These new mathematical concepts need further research and studies. We discuss some aspects of these new concepts in this paper. Our paper contains following aspects. First we describe the charge particle motion in the curved spacetime and discuss some of its complexities then in Sect. 3, the electromagnetic self-force is discussed by axiomatic approach. In Sect. 4, the Conservation of energy is considered that is the extension of energy conservation theorem to the curved space. Finally, in Sect. 5 we have discussed the new mathematical concepts that are introduced in the problem.

Particles motion in the curved space

To avoid the complexity of extended charged matters, we turned to the point particles. We can principally obtain the extended matter's fields from point's field (even it is not

✉ O. Jalili
omid_jalili@yahoo.com

K. Mehdizadeh
kazem_mehdizadeh93@yahoo.com

1 Department of physics, Ayatollah Amoli Branch, Islamic Azad University, Amol, Mazandaran, Iran

2 Department of physics, North Tehran Branch, Islamic Azad University, Tehran, Iran

the case for gravitational waves, because the Einstein equations are not linear). Choice of the point assumption causes a problem: the point matters are not compatible with general relativity, because it creates black hole (a spherical matter distribution with the fixed mass and charge when its radius vanishes becomes a black hole). On the other hand, we have point particle in the world (leptons). Thus the general relativity is not a complete theory by itself, it needs to join with the quantum theory, i.e., we need quantum gravity. So far we have two candidates for quantum gravity: string theory and loop quantum gravity. In the first one, we exclude the point concept and introduce the extended object (string). In the second one, we exclude the continuum structure of spacetime and introduce the spacetime network instead of manifold structure. In the present paper, we study some small distribution of matter to simplify our considerations.

In the absence of the interaction between matter field, particle and radiation, the action is given by:

$$\begin{aligned} S &= S_{\text{Gravity}} + S_{\text{Matter}} + S_{\text{Radiation}} + S_{\text{Single Particle}} \\ &= -\frac{1}{16\pi G}\int \sqrt{-g}R\mathrm{d}x + \int \mathscr{L}\sqrt{-g}\mathrm{d}x \\ &\quad + \int \frac{-1}{4}F_{\alpha\beta}F^{\alpha\beta}\sqrt{-g}\mathrm{d}x + \int \mathrm{d}s. \end{aligned} \tag{1}$$

$\mathscr{L}$ is the Lagrangian density of the matter field. The dynamical variable in the S_{Gravity} is the metric $g_{\mu\nu}$, in the S_{Matter} is the $\phi(x)$ (matter field), in the $S_{\text{Radiation}}$ is $A_\mu(x)$ (Maxwell field) and in the last term is $x(\tau)$ (particle's path). Minimizing the action (1) with respect to $g_{\mu\nu}$ gives the Einstein's equations:

$$G_{\mu\nu} = 8\pi T_{\mu\nu}|_{\text{Matter}} + 8\pi T_{\mu\nu}|_{\text{Radiation}} + 8\pi T_{\mu\nu}|_{\text{Single Particle}} \tag{2}$$

$G_{\mu\nu}$ is Einstein's tensor. It is obvious that both matter and the radiation, and of course the presence of the particle, contribute in the forming of the metric. Minimizing the action (1) with respect to $\phi(x)$ give the Euler–Lagrange Eq. for ϕ;

$$\delta S|_{\text{with respect to } \phi} = 0 \Rightarrow \nabla_\alpha \frac{\partial \mathscr{L}_m}{\partial \phi_{,\alpha}} - \frac{\partial \mathscr{L}_m}{\partial \phi_\alpha} = 0. \tag{3}$$

Minimizing the action (1) relative to A_μ give the Euler equation for the Maxwell field:

$$\begin{aligned} \delta S|_{\text{with respect to A}} = 0 &\Rightarrow \nabla_\alpha \frac{\partial \mathscr{L}_r}{\partial A_{,\alpha}} - \frac{\partial \mathscr{L}_r}{\partial A_\alpha} = 0 \\ &\Rightarrow \Box A^\alpha - R^\alpha_\beta A^\beta = 0, \end{aligned} \tag{4}$$

where $\Box = g^{\alpha\beta}\nabla_\alpha\nabla_\beta$ is the covariant wave operator and R^α_β is the Ricci tensor. Note the presence of metric in both of Eqs. (3) and (4). Note also that the matter contributes in the formation of the metric (Eq. 2). Then even the non-charged matter creates electromagnetic fields, this was the point that motivated Einstein to join electromagnetism with the gravity because one of them creates the other. Minimizing the action (1) with respect to $x(\tau)$ gives the geodesic of the $g_{\mu\nu}$ manifold:

$$\nabla_U U = 0 \quad \Rightarrow \quad \ddot{x}^\alpha + \Gamma^\alpha_{\beta\gamma}\dot{x}^\beta\dot{x}^\gamma = 0 \quad (\text{U is the four-velocity}). \tag{5}$$

It is seen the path of particle is the geodesic of the manifold, namely the manifold that the particle itself contributes in its metric formation (Eq. 2)! The action (1) is a non-interacting action, namely the fields are free but, as we have seen, the nature of gravity is such that even the non-interacting action has the interacting behaviors: non-gravitational fields contribute in the creation of metric and this metric contribute in the other fields and this is the interaction between the fields and the particle with the metric! However, it is not the only possible interaction. Actually it is the possible minimum interaction. If the interaction terms are added, we have:

$$\begin{aligned} S &= S_{\text{Gravity}} + S_{\text{Matter}} + S_{\text{Radiation}} + S_{\text{Single Particle}} \\ &\quad + S_{\text{A and x interaction}} + S_{\text{A and } \phi \text{ interaction}}, \end{aligned} \tag{6}$$

where $S_{\text{A and x interaction}}$ is the action term of interaction of electromagnetic and one charged particle and $S_{\text{A and } \phi \text{ interaction}}$ is the action term of interaction of electromagnetic and matter field. What about the gravity–gravity interaction term namely $S_{\text{G and G interaction}}$? Let us first continue without this term. With the interaction (6), Eqs. (3), (4) and (5) are modified. Note that Eq. (2) is not changed but Eqs. (3) and (4) will have the right term, for example Eq. (4) will get the new form

$$\Box A^\alpha - R^\alpha_\beta A^\beta = -4\pi J^\alpha|_{\phi \text{ field}} - 4\pi J^\alpha|_{\text{Single particle}}, \tag{7}$$

namely the charged matter enters as currents to create the electromagnetic fields. Finally, Eq. (5) became

$$\ddot{x}^\alpha + \Gamma^\alpha_{\beta\gamma}\dot{x}^\beta\dot{x}^\gamma = \frac{e}{m}F^\alpha_\beta U^\beta. \tag{8}$$

This means that the particle's path is no longer a geodesic. It is seen that dynamical structure is more fundamental than the geometrical concepts! Here the path of particle is not a geodesic. The path of particle is obtained by the minimizing the appropriate action. Then the action approach is more general than the geometrical approach. We can design a suitable Lagrangian for an electromagnetic field and represent the whole theory in a mechanical manner.

Let us continue. General relativity has introduced the notions of geodesic and metric instead of acceleration and force, but we can continue the same classical notions. For this, we rewrite the geodesic equation in the following manner:

$$U^{\mu}\nabla_{\mu}U = 0 \Longrightarrow \frac{dx^{\mu}}{d\tau}\nabla_{\mu}U = 0 \Longrightarrow \frac{dx^{\mu}}{d\tau}\left(\frac{dU^{\alpha}}{dx^{\mu}} + \Gamma^{\alpha}_{\mu\nu}U^{\nu}\right) = 0$$
$$\Longrightarrow \frac{d^2x^{\alpha}}{d\tau^2} + \frac{dx^{\mu}}{d\tau}\Gamma^{\alpha}_{\mu\nu}U^{\nu} = 0 \Longrightarrow m_0\frac{d^2x^{\alpha}}{d\tau^2} = -m_0\frac{dx^{\mu}}{d\tau}\Gamma^{\alpha}_{\mu\nu}U^{\nu}. \tag{9}$$

Thus, we can interpret the right hand side as gravitational force. Note that the radiation, matter and the single particle, all of these contribute in formation of this force because all of them contribute in making the $\Gamma^{\alpha}_{\mu\nu}$ tensor. In the presence of the interaction, from Eq. (8), the particle's path can be written as:

$$\nabla_U P^{\alpha} = eF^{\alpha}_{\beta}P^{\beta} \Longrightarrow m_0\ddot{x}^{\alpha} = -m_0\Gamma^{\alpha}_{\beta\gamma}\dot{x}^{\beta}\dot{x}^{\gamma} + eF^{\alpha}_{\beta}U^{\beta} \quad (\mathrm{Pisthefour-momentum}). \tag{10}$$

It is seen that electromagnetic force has appeared twice: once, implicitly in determining the Γ (the first term in the right) and the second times as the explicit term (second term in the right). However, in the literature usually when speaking about the force, it means only the explicit force, namely:

$$f = eF^{\alpha}_{\beta}U^{\beta} = \nabla_U P^{\alpha}. \tag{11}$$

Now we come back to our question about the $S_{\mathrm{G\ and\ G\ interaction}}$ term. It is known that the gravity has energy [1], then it can create gravity. It is the one of the nonlinear features of gravity (other nonlinearity aspect of gravity is the very nonlinearity of Einstein's tensor). But there is no such thing as energy tensor for gravity. This is because according to equivalence principle, we can choose a locally flat spacetime such that the gravitational energy vanishes. But if a tensor is zero in one frame it is so in all frames. Thus we must take into account the gravitational energy in some different sort. Such consideration is important in some cosmological large-scale structure [2]. Here, we work in some smaller scales such that this nonlinearity aspect is absent.

Electromagnetic self-force in axiomatic approach

To obtain first the radiation field of a point charged particle we must solve Eq. (7) for one particle source, namely:

$$\Box A^{\alpha} - R^{\alpha}_{\beta}A^{\beta} = -4\pi J^{\alpha}|_{\mathrm{Single\ particle}}, \tag{12}$$

where the source is singular at the particle position and vanishes elsewhere. The singular behavior of the source term enters into the solution and the field will be singular at the position of particle. We can regularize the infinity at the particle's mass and obtain normal value for the self-force. Such procedure is equivalent to using the regular potential rather than the usual retardation potential [3]:

$$AR = \frac{1}{2}(A_{\mathrm{ret}} - A_{\mathrm{adv}}), \tag{13}$$

where A_{ret} and A_{adv} are the retarded and advanced potential, respectively. Both of these fields are singular at the particle's position, but the A_{R} field is not (because it is the difference of two singular fields). Actually the regular potential satisfies the homogenous equation, hence its regular behavior is not surprising:

$$\Box A^{\alpha}_{R} - R^{\alpha}_{\beta}A^{\beta}_{R} = 0. \tag{14}$$

Historically Dewitt and Brehme [5] extended the Dirac' flat space result for the radiation of charged point particle [4]. Their result in normal coordinate is (corrected by Hobbs several years after that [6]):

$$\begin{aligned}
F^{\pm}_{\alpha'\beta'} = 2e[&r^{-2}u_{[\alpha'}\Omega_{\beta']} - \frac{1}{2}r^{-1}(a^{\alpha}\Omega_{\alpha})u_{[\alpha'}\Omega_{\beta']} \\
&+\frac{1}{2}r^{-1}a_{[\alpha'}u_{\beta']} + \frac{3}{8}(a^{\alpha}\Omega_{\alpha})^2u_{[\alpha'}\Omega_{\beta']} \\
&-\frac{1}{6}u^{\alpha}u^{\beta}\Omega^{\gamma}\Omega^{\delta}R_{\alpha\gamma\beta\delta}u_{[\alpha'}\Omega_{\beta']} - \frac{3}{4}(a^{\alpha}\Omega_{\alpha})a_{[\alpha'}u_{\beta']} \\
&+\frac{1}{6}\Omega_{[\beta'}R_{\alpha']\sigma\alpha\tau}u^{\alpha}\Omega^{\sigma}\Omega^{\tau} + \frac{1}{8}u_{[\alpha'}\Omega_{\beta']}a^2 - \frac{1}{2}\dot{a}_{[\alpha'}\Omega_{\beta']} \\
&\pm\frac{2}{3}\dot{a}_{[\alpha'}u_{\beta']} + \frac{1}{12}u_{[\alpha'}\Omega_{\beta']}R - \frac{1}{12}u_{[\alpha'}R_{\beta']\gamma}\Omega^{\gamma} \\
&+\frac{1}{2}\Omega_{[\alpha'}R_{\beta']\gamma}u^{\gamma} + \frac{1}{12}u_{[\alpha'}\Omega_{\beta']}R_{\gamma\delta}\Omega^{\gamma}\Omega^{\delta} \\
&+\frac{1}{2}R_{[\alpha'|\gamma|\beta']\delta}u^{\gamma}\Omega^{\delta} - \frac{1}{12}u_{[\alpha'}\Omega_{\beta']}R_{\gamma\delta}u^{\gamma}u^{\delta} \\
&+\frac{1}{6}u_{[\alpha'}R_{\beta']\gamma\delta\epsilon}u^{\gamma}u^{\delta}\Omega^{\epsilon} \mp \frac{1}{3}u_{[\alpha'}R_{\beta']\gamma}u^{\gamma}] \\
&\pm e\int_{\tau^{\pm}}^{\pm\infty}\nabla_{[\beta'}G^{\pm}_{\alpha']\alpha''}u^{\alpha''}(\tau'')d\tau'' + O(r)
\end{aligned} \tag{15}$$

The last term, tail term, represents the failure of Huygens principle. In the this equation e is the particle's charge, r the distance from particle position (affine parameter of space-like curve from the particle position to observation point), a the particle's acceleration, $\dot{a}$ the time derivative of acceleration and the others are some geometrical quantities. At the particle's position the first three terms of above equation are singular and the rest are regular. It is seen that these singular terms depend only on four-velocity and four-acceleration and nothing else such as curvature or derivative of acceleration. Thus with the equal velocity and acceleration, in the difference of two such expression the singular terms can be eliminated. Actually the force exerted on these two particles have the same infinity structure such that in the difference they will be canceled. Thus in our particle size limit we can postulate the following axioms [7]:

Electromagnetic Axiom 1 (Comparison axiom) Consider two points, P and $\widetilde{P}$, each lying on time-like world lines in possibly different spacetimes which contain Maxwell fields F^{ab} and $\widetilde{F}^{ab}$ sourced by particles of charge e on the world lines. If the four-accelerations of the world lines at P and $\widetilde{P}$ have the same magnitude, and if we identify the neighborhoods of P and $\widetilde{P}$ via Riemann normal coordinates such that the four-velocities and four-accelerations are identified, then the difference in the electromagnetic forces f^a_{EM} and $\widetilde{f}^a_{\mathrm{EM}}$ is given by the limit as $r \to 0$ of the Lorentz force associated with the difference of the two fields averaged over a sphere at geodesic distance r from the world line at P.

$$f^a_{\mathrm{EM}} - \widetilde{f}^a_{\mathrm{EM}} = \lim_{r\to 0}(e\langle F^{ab} - \widetilde{F}^{ab}\rangle)u_b \tag{16}$$

As an application of above axiom, we can compare the force applied on a charged particle in an arbitrary spacetime to Minkowski spacetime. It is a good choice to consider field which its force approaches to zero. The symmetry consideration shows that the force of the symmetric field, $F_{ab} = \frac{1}{2}(F^-_{ab} + F^+_{ab})$, may vanish. Thus we accept the following axiom [7]:

Electromagnetic Axiom 2 (flat spacetime axiom) If (M, g_{ab}) is Minkowski spacetime, the world line is uniformly accelerating, and F_{ab} is the half-advanced, half-retarded solution, $F_{ab} = \frac{1}{2}(F^-_{ab} + F^+_{ab})$ then $f^a = 0$ at every point on the world line.

To obtain the F_{ab}, we can simply use Eq. (15) with the tail term and delete all terms that contain the curvature. Then we obtain the following expression for a charged particle self-force in an arbitrary spacetime:

$$\begin{aligned} f^a_{\mathrm{EM}} &= e(F^{\mathrm{in}})^{ab}u_b + \frac{2}{3}e^2(\dot{a}^a - a^2u^a) \\ &+ \frac{1}{3}e^2(R^a_b u^b + u^a R_{bc}u^b u^c) \\ &+ e^2 u_b \int_{-\infty}^{\tau^-} \nabla^{[b}(G^-)^{a]c'} u_{c'}(\tau')\mathrm{d}\tau' \end{aligned} \tag{17}$$

where $(F^{\mathrm{in}})^{ab} = F^{ab} - (F^-)^{ab}$. Equation (17) is the same as the result that Dewitt and Brehme [5] gave before, but here we derive it with the less and easy manipulation.

With the given force we can simply drive an expression for acceleration:

$$a = \frac{f^a_{\mathrm{EM}}}{m} \tag{18}$$

But the right hand side depends on acceleration too! There is a known recipe to solve it. In the same manner as Born series in scattering, we can substitute the acceleration in the right side from the left side and iterate this work. Then we obtain the following expression for the first several terms in that series:

$$\begin{aligned} a^a &= \frac{e}{m}(F^{\mathrm{in}})^{ab}u_b + \frac{2}{3}\frac{e^2}{m}\Big(\frac{e}{m}u^c\nabla_c(F^{\mathrm{in}})^{ab}u_b \\ &+ \frac{e^2}{m^2}(F^{\mathrm{in}})^{ab}F^{\mathrm{in}}_{bc}u^c - \frac{e^2}{m^2}u^a(F^{\mathrm{in}})^{bc}u_c F^{\mathrm{in}}_{bd}u^d\Big) \\ &+ \frac{1}{3}\frac{e^2}{m}(R^a_b u^b + u^a R_{bc}u^b u^c) \\ &+ \frac{e^2}{m}u_b \int_{-\infty}^{\tau^-} \nabla^{[b}(G^-)^{a]c}u_c(\tau')d\tau' \end{aligned} \tag{19}$$

Conservation of energy

In our problem the total energy of the system is:

$$\begin{aligned} E_{\mathrm{Total}} &= E_{\mathrm{Particle}} + E_{\mathrm{Gravity}} + E_{\mathrm{Electromagnetism}} \\ &+ E_{\text{Interaction of one particle to EM field}}. \end{aligned} \tag{20}$$

For the energy concept to be a reasonable concept, the spacetime must have the time-like Killing vector [8]. Even for such spacetime it is practically hard to write explicitly an energy expression for every term in the right had side of Eq. (20). For example, for E_{Particle} we need to exactly know spacetime metric, one metric that the particle has some role in creation of it! However, in some asymptotical case, there exists a simple relation. For example, for asymptotically flat spacetime in the comparison of the initial and final energy we have:

$$\Delta E_{\mathrm{Particle}} + \Delta E_{\mathrm{Electromagnetism}} + \Delta E_{\text{Particle--Electromagnetism Interaction}} = 0. \tag{21}$$

It have been proved that [9]:

$$\int_{\mathcal{I}^+} t^a T_{ab}\varepsilon^b_{cde} - \int_{\mathcal{I}^-} t^a T_{ab}\varepsilon^b_{cde} = \int_{\mathcal{M}} t^a f_a \mathrm{d}\tau, \tag{22}$$

where t^a is time-like Killing vector, $\mathcal{I}^+(\mathcal{I}^-)$is space-like hyper-surface at $t \to +\infty(t \to -\infty)$, T_{ab} the particle's energy momentum tensor and f_a the electromagnetic self-force, i.e., Eq. (17). The left hand side of this equation is the change in particle's energy and the right hand side is the work done on it. It simply means:

$$\Delta E_{\mathrm{Particle}} + \Delta E_{\text{Particle--Electromagnetism Interaction}} = 0. \tag{23}$$

There is no electromagnetic energy change and no gravitational energy change: $\triangle E_G = 0, \triangle E_{\text{EM}} = 0$.

Results and discussion

The radiation of charged particles in curved space is an important subject in general relativistic kinetic theory. In formulating it, we see the nonlinear nature of general relativity that comes through two ways. We have seen that even the non-interacting particles have some kind of interaction: the particle changes the geometry and the new geometry acts on other particles motion. The general relativity replaces geodesic with old Newtonian concept, i.e., the force. However, we can use the force concept as before. This issue is based on a fundamental fact: the algebraic methods (the mechanical approach) is more generic than geometric method. Where the Einstein's geodesic method fails, the dynamical approach works. The radiation of charged particle in curved space like any plasma problem is the simultaneous solving of several equations: the set of Maxwell equations, the particles motion equations and the geometry equations. With some conditions, like the fixed background assumption, we need to solve the two sets of these three sets. However, we can solve these sets with the usual methods such as the Green function method and so on; we confront a new mathematical concept in the Green function method, the bitensor. In general, the physical quantities are the local concepts that belong to the tangent space of every spacetime points. Even the definition of derivative that is the comparison of quantities at two different points must be changed to become a local quantity. But the bitensors are not a local quantity. A bitensor is not built simply by juxtaposing of two tangent space quantities but these quantities multiplied as the numbers, thus this juxtaposing is not the direct product or tensor product! These new manipulations of geometrical quantities enter again in the axiomatic approach. In this paper, we focused on axiomatic approach to obtain the self-electromagnetic force. We have seen in this approach we need to subtract the geometrical quantities belong to two different manifolds. In the bitensor approach, the product quantities belong to different points of one manifold, but in the axiomatic approach these two points belong to different manifolds. Thus the operations are not simply juxtaposing the quantities or anything else but the difference of these quantities! In physical point of view, when we can subtract two quantities which are the same kinds also must have equal dimension. Thus here that we are subtracting two things belong to different manifolds, we come to a new concept that needs more investigations. This is a new geometrical concept. We may ask some questions: can we make these subtracted quantities a new tensorial quantity by suitable assumptions about each quantity? If such extension is present, what is the derivative of it?... We see that there are some technical questions about these new quantities that need some further researches.

Conclusions

To obtain electromagnetic radiation of a charged particle in curved space, we must solve the differential equation of electromagnetic potential $A_\mu(x)$ with the single particle source term. The traditional ways to do it are the method of Green function and the conformal mapping method [10]. Since the self-field and the self-force become infinity, we must then regularize them by renormalized mass. After that we obtain finite values for the field and force. The energy conservation can be obtained with some subtlety. In curved space we can reobtain the energy work theorem. In a parallel and equivalent manner with some assumption about space-time manifold, we can obtain aforementioned result but with very less calculation. This parallel approach is the axiomatic approach. In this approach, we compare tensorial object belonging to two different manifolds. The comparison of tensorial object belonging to a specific manifold are well known. But comparison of quantities that belong to two different manifolds is new. This new phenomenon is very interesting and valuable for further researches.

Acknowledgments The authors would like to convey their gratitude to express their thanks to professor M.V. Takook and professor M. Saravi for their comments and useful discussions.

References

1. Arnowitt, R., Deser, S., Misner, C.W.: Dynamical structure and definition of energy in general relativity. Phys. Rev. **116**, 1322 (1959)
2. de Felice, F., Clarke, C.J.S.: Relativity on Curved Manifolds. Cambridge University Press, Cambridge (1992)
3. Poisson, E., Pound, A., Vega, I.: The motion of point particles in curved spacetime. Living Rev. Relativ. **14**. doi:10.12942/lrr-2011-7
4. Dirac, P.A.M.: Classical theory of radiating electrons. Proc. Roy. Soc. A **167**, 148 (1938)
5. DeWitt, B.S., Brehme, R.W.: Radiation damping in a gravitational field. Ann. Phys. **9**, 220–259 (1960)
6. Hobbs, J.M.: Radiation damping in conformally flat universes. Ann. Phys. **47**, 141 (1968)
7. Quinn, T.C., Wald, R.M.: Axiomatic approach to electromagnetic and gravitational radiation reaction of particles in curved spacetime. Phys. Rev. D **56**, 3381 (1997)
8. Parker, L., Toms, D.: Quantum Field Theory in Curved Spacetime. Cambridge University Press, Cambridge (2009)
9. Quinn, T.C., Wald, R.M.: Energy conservation for point particles undergoing radiation reaction. Phys. Rev. D **60**, 064009 (1999)
10. Bicák, J., Krtouš, P.: Fields of accelerated sources: born in de Sitter. J. Math. Phys. **46**, 102504 (2005)

Non-equilibrium phase transition in a two-species driven-diffusive model of classical particles

Mohammad Ghadermazi[1] · Farhad H. Jafarpour[2]

Abstract A two-species driven-diffusive model of classical particles is introduced on a lattice with periodic boundary condition. The model consists of a finite number of first class particles in the presence of a second class particle. While the first class particles can only hop forward, the second class particle is able to hop both forward and backward with specific rates. We have shown that the partition function of this model can be calculated exactly. The model undergoes a non-equilibrium phase transition when a condensation of the first class particles occurs behind the second class particle. The phase transition point and the spatial correlations between the first class particles are calculated exactly. On the other hand, we have shown that this model can be mapped onto a two-dimensional walk model. The random walker can only move on the first quarter of a two-dimensional plane and that it takes the paths which can start at any height and end at any height upper than the height of the starting point. The initial vertex (starting point) and the final vertex (end point) of each lattice path are weighted. The weight of the outset point depends on the height of that point while the weight of the end point depends on the height of both the outset point and the end point of each path. The partition function of this walk model is calculated using a transfer matrix method.

✉ Mohammad Ghadermazi
m.ghadermazi@iausdj.ac.ir

1 Plasma Physics Research Center, Science and Research Branch, Islamic Azad University, Tehran, Iran

2 Physics Department, Bu-Ali Sina University, 65174-4161 Hamedan, Iran

Keywords Driven-diffusive model · Walk model · Correlation function · Zero-range processes

Introduction

One of the most studied models which shows non-equilibrium phase transitions is asymmetric simple exclusion process (ASEP). In many literatures, the ASEP in the presence of an impurity on a ring have been studied. The role of the impurity is to investigate the motion of the shock fronts in the ASEP. The single impurity in [1, 2] hops in the opposite direction relative to the ordinary particle of the ASEP while in [3] the impurity moves in the same direction as the ordinary particles of the ASEP. In both cases the phase structure of the models have been studied extensively.

In this paper, we study the effects of the presence of a single impurity on the ASEP on a ring where the second class particle (impurity) is allowed to hop in the both directions, relative to the ordinary particle of the ASEP, with the rates q and p. It has been shown that the steady-state distribution of all one-dimensional exclusion models whose steady-states have a simple factorized form, can be written in a matrix product form. The matrices which are necessary for this purpose satisfy a generalized quadratic algebra [4]. In [5], the authors have introduced a mathematical tool for studying of correlations in the models whose steady-states have a simple factorized form. They use the matrices which satisfy a generalized quadratic algebra. In this paper, we introduce an infinite-dimensional matrix representation which satisfies the quadratic algebra of the model. The canonical partition function of the model is calculated exactly. Using a canonical ensemble, the phase structure of the model is studied in thermodynamic limit. The steady-state distribution of our model has a factorized form hence two matrix representations are presented which satisfy the generalized quadratic algebra of the model. An infinite-dimensional matrix representation and a 2-dimensional matrix representation. Using this matrix representations the calculations seem to be very straightforward. It has been shown that using the grand canonical partition function one can analyze the phase structure of the model. The transition point can be calculated numerically or analytically [6, 7]. By assigning the fugacity z to the first class particles in a grand canonical ensemble, we shall find the exact phase structure and calculate the density profile and the correlations of the first class particles precisely. Some of the critical exponents of the model in phase transition is also obtained.

In recent years, many studies have been done on connections between the one-dimensional driven-diffusive systems and the two-dimensional walk models [8–10]. It has been shown that the partition function of some of the one-dimensional driven-diffusive models with open boundaries obtained using a matrix product method is equal to the partition function of a two-dimensional walk model obtained using a transfer matrix method [8, 11, 12]. In [11], the author has introduced a lattice path with specific dynamical rules where walker can start from origin and end at any height upper than origin and it has been shown that the partition function of this two-dimensional walk model is exactly equal to that of a driven-diffusive system defined on a discrete lattice with periodic boundary conditions that can be mapped to a zero-range process [13, 14]. In this paper, a two-dimensional walk model is introduced in which the walker can only move on the first quarter of a two-dimensional plane. Random walker can start moving at any height upper than the origin and end at any height upper than the starting point. All the paths made by the random walker are weighted. The weight of a given path will be equal to the product of the weights of the consecutive steps in that path and the weight of the starting and end points. The partition function of this walk model is the sum of the unnormalized weights of different paths consisting of $t-1$ steps and is calculated using a transfer matrix method.

The paper is organized as follows: In "The driven-diffusive model" a two-species driven-diffusive model is introduced. In "The canonical partition function" the canonical partition function of the model is calculated in the thermodynamic limit and the phase structure of the model is investigated. In "The spatial correlations" the correlation functions and the critical exponent of the model are calculated. In "The walk model" we introduce a two-dimensional walk model related to the two-species driven-diffusive model.

The driven-diffusive model

In [11], the author has introduced a one-dimensional driven-diffusive model of classical particles with hardcore interactions. The model consists of a single particle of type A (called the second-class particle) and $M-1$ particles of type B (called the first-class particles). The particles move on a one-dimensional lattice of length t with periodic boundary condition. The particle of type A hops from the lattice site i to $i+1$ with the rate p provided that the target site is empty. A particle of type B hops from the lattice site i to $i+1$ with the rate 1 provided that the target site is empty.

In this paper, we assume that the second-class particle is also allowed to hop backward with the rate q. If an empty lattice site is denoted by $\emptyset$, we can summarize the reaction rules at a pair of lattice sites i and $i+1$ as follows

$$\begin{aligned} A\emptyset &\rightarrow \emptyset A \quad \text{with rate } p \\ \emptyset A &\rightarrow A\emptyset \quad \text{with rate } q \\ B\emptyset &\rightarrow \emptyset B \quad \text{with rate } 1. \end{aligned}$$

In the long-time limit the system attains a non-equilibrium steady-state. It can be shown that the probability distribution can be obtained using a matrix product method. For this purpose, we label the particle of type A with 1 and label the particles of type B with $2, 3, \ldots, M$. If the number of empty lattice sites in front of the i'th particle is denoted by n_i, a general configuration of the model can be written as $\{n\} = \{n_1, n_2, \ldots, n_M\}$. If the lattice site is occupied by the particle of type A, the matrix D_1 is attributed to it. If a lattice site is occupied by a particle of type B, the matrix D_2 is attributed to it. In the steady-state the probability of finding the system in a general configuration $\{n\} = \{n_1, n_2, \ldots, n_M\}$ is given by

$$P(\{n\}) = \frac{1}{Z_{t,M}(p,q)} Tr(D_1 E^{n_1} D_2 E^{n_2} \ldots D_M E^{n_M}) \tag{1}$$

in which $Z_{t,M}(p,q)$ is the normalization factor which is also called the canonical partition function of the model and that it should be calculated by considering the conservation of the number of empty sites i.e., $\sum_{v=1}^{M} n_v = t - M$. A sufficient condition for (1) to be the steady-state probability distribution of the model is

$$\begin{aligned} pD_1E - qED_1 &= D_1 \\ D_2E &= D_2. \end{aligned} \tag{2}$$

Details of the proof is given in [15]. It can easily be verified that the above algebra has the following infinite-dimensional matrix representation

$$D_1 = \sum_{i=0}^{\infty}\sum_{i'=i}^{\infty}\binom{i'}{i}\frac{q^i}{p^{i'}}|i\rangle\langle i'|,$$
$$D_2 = \sum_{i=0}^{\infty}|0\rangle\langle i|,$$
$$E = \sum_{i=0}^{\infty}|i+1\rangle\langle i|$$

in which $|i\rangle_j = \delta_{i,j}$ for $i,j = 0,1,\ldots,\infty$.

The canonical partition function

The number of empty lattice sites is a conserved quantity and does not change by the dynamical rules; therefore, using (1) one can calculate the canonical partition function of the model as follow

$$Z_{t,M}(p,q) = \sum_{n_1,\ldots,n_M=0}^{t-M} Tr[D_1E^{n_1}D_2E^{n_2}\ldots D_2E^{n_M}]\delta_{\sum_{v=1}^{M} n_v,(t-M)}. \tag{3}$$

Using the matrix representations of D_2 and E, one can write

$$D_2E^n = D_2.$$

Using the above equation we can rewrite the canonical partition function of the model as follows

$$Z_{t,M}(p,q) = \sum_{n_1,\ldots,n_M=0}^{t-M} Tr[D_1E^{n_1}D_2]\delta_{\sum_{v=1}^{M} n_v,(t-M)}. \tag{4}$$

Using the matrix representation of the matrices D_1, D_2 and E and the definition of trace of a matrix one can obtain

$$Tr[D_1E^{n_1}D_2] = \sum_{i=0}^{\infty}\langle i|D_1E^{n_1}D_2|i\rangle = \sum_{i=0}^{n_1}\binom{n_1}{i}\frac{q^i}{p^{n_1}}.$$

Inserting the above equation into (4), the canonical partition function of the model can be written as

$$Z_{t,M}(p,q) = \sum_{n_1=0}^{t-M}\sum_{i=0}^{n_1}\binom{n_1}{i}\binom{t-n_1-2}{M-2}\frac{q^i}{p^{n_1}}. \tag{5}$$

It can be seen that the partition function of the model in thermodynamic limit $M,t \to \infty$ behaves as

$$Z_{t,M}(p,q) \simeq \begin{cases} p^{t(\rho-1)}(1+q)^{t-1}(1+q-p)^{2-M} & \text{for } \rho < \dfrac{1-p+q}{1+q} \\ \dbinom{t}{M}\dfrac{p\rho^2}{p-(1-\rho)(1+q)} & \text{for } \rho > \dfrac{1-p+q}{1+q} \end{cases} \tag{6}$$

in which ρ is the density of the first-class particle and is given by $\rho = (M-1)/t \simeq M/t$. It can be seen that a phase transition occurs at $\rho = (1-p+q)/(1+q)$. To investigate the phase behavior of the model we calculate the mean number of the empty lattice sites in front of the second-class particle. Given that the total number of empty lattice sites on the lattice is $t-M$, the probability that the number of empty lattice sites in front of the second-class particle is n_1, is given by

$$P_{t,M}(n_1) = \frac{1}{Z_{t,M}(p,q)}\sum_{i=0}^{n_1}\frac{q^i}{p^{n_1}}\binom{n_1}{i}\binom{t-n_1-2}{M-2} \tag{7}$$

Hence the average number of empty lattice sites in front of the particle of type A is

$$\bar{n}_1 = \sum_{n_1=0}^{t-M} n_1P_{t,M}(n_1) = -p\frac{\partial \ln Z_{t,M}(p,q)}{\partial p}. \tag{8}$$

Using (6), the average number of empty lattice sites in front of the particle of second-class particle in the thermodynamic limit is as follows

$$\bar{n}_1 = \begin{cases} t\left(1-\rho-\dfrac{p\rho}{1+q-p}\right) & \text{for } \rho < \dfrac{1-p+q}{1+q} \\ \dfrac{p}{p-(1-\rho)(1+q)} - 1 & \text{for } \rho > \dfrac{1-p+q}{1+q} \end{cases}. \tag{9}$$

As can be seen there is a phase transition from a phase in which the mean number of empty lattice site in front of the second-class particle is of order t to another phase where it is a constant.

The spatial correlations

In [4] the author has shown that the steady state of a disordered driven-diffusive system consisting of M different type of particles, that can be mapped onto the Zero Range Process, can be obtained using the matrix method in which the matrices should satisfy the following generalized quadratic algebra

$$D_\mu E^{n_\mu}D_{\mu'} = f_\mu(n_\mu)D_{\mu'} \qquad \text{for } \mu,\mu' = 1,2,\ldots,M \tag{10}$$

in which $f_\mu(n_\mu)$ is a function of transition rates and can be constructed using pairwise balance condition [16]. Our model is a two-species driven-diffusive model of classical particles on a lattice with periodic boundary condition with the following dynamic

$$\begin{aligned} &\mu'\,\underbrace{0\cdots0}_{n_{\mu'}}\,\mu\,\underbrace{0\cdots0}_{n_\mu}\,\mu'' \to \mu'\,\underbrace{0\cdots0}_{n_{\mu'}+1}\,\mu\,\underbrace{0\cdots0}_{n_\mu-1}\,\mu'' \quad \text{with the rate } u_\mu(n_\mu) \\ &\mu'\,\underbrace{0\cdots0}_{n_{\mu'}}\,\mu\,\underbrace{0\cdots0}_{n_\mu}\,\mu' \to \mu'\,\underbrace{0\cdots0}_{n_{\mu'}-1}\,\mu\,\underbrace{0\cdots0}_{n_\mu+1}\,\mu'' \quad \text{with the rate } v_\mu(n_\mu) \end{aligned} \tag{11}$$

where $u_\mu(n_\mu)$ is the hopping rate of the particle μ to its right neighboring lattice site and $v_\mu(n_\mu)$ is the hopping rate of the particle μ to its left neighboring lattice site i.e.,

$$\begin{aligned} u_1(n_1) &= p, \quad v_1(n_1) = q, \\ u_2(n_2) &= 1, \quad v_2(n_2) = 0. \end{aligned} \tag{12}$$

It can be checked that by defining $f_1(n_1) = (\frac{1+q}{p})^{n_1}$ and $f_2(n_2) = 1$ and requiring $f_{1,2}(0) = 1$, the following infinite-dimensional matrix representation satisfies the quadratic algebra (10)

$$\begin{aligned} D_1 &= \sum_{i=0}^{\infty} f_1(i)|0\rangle\langle i|, \\ D_2 &= \sum_{i=0}^{\infty} |0\rangle\langle i|, \\ E &= \sum_{i=0}^{\infty} |i+1\rangle\langle i|. \end{aligned} \tag{13}$$

Using (13) the grand-canonical partition function of the model can be written as

$$Z_t(p,q,z) = Tr[D_1 C^{t-1}] = \sum_{i=0}^{\infty} f(i)\langle i|C^{t-1}|0\rangle \tag{14}$$

where the matrix $C = E + zD_2$ and that z is the fugacity of the first-class particles. According to the matrix representations (13) it can be verified that

$$C^{t-1}|j\rangle = \sum_{i=0}^{t-2} z(z+1)^{t-i-2}|i\rangle + |j+t-1\rangle. \tag{15}$$

Now the grand-canonical partition function $Z_t(p,q,z)$ can be calculated using (15)

$$Z_t(p,q,z) = \frac{pz\left(\frac{1+q}{p}\right)^{t-1}}{1+q-p(1+z)} - \frac{pz(1+z)^{t-1}}{1+q-p(1+z)} + \left(\frac{1+q}{p}\right)^{t-1}. \tag{16}$$

The fugacity z has to be fixed by density of the first-class particles which is given by the following equation

$$\rho(z) = \frac{z}{t}\frac{\partial}{\partial z}\ln Z_t(p,q,z). \tag{17}$$

It is known that the real positive values of the fugacity are of physical interest hence it is necessary that $p < 1+q$. Using (16) it can be shown that in the thermodynamic limit the density of first-class particles can be written as follows

$$\rho(z) = \begin{cases} \dfrac{z}{1+z} & \text{for } z > \dfrac{1+q-p}{p} \\ 0 & \text{for } z < \dfrac{1+q-p}{p} \end{cases}. \tag{18}$$

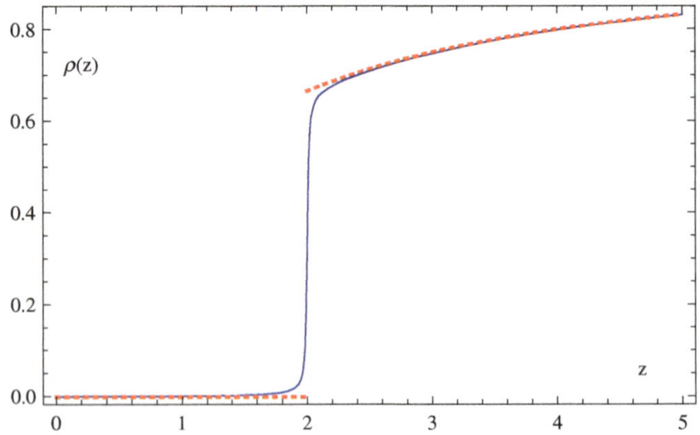

Fig. 1 The density of the first-class particles as a function of fugacity z obtained from the exact solution (the *blue line*) and in the thermodynamic limit (the *red dotted line*) for $t = 200$, $p = 1$ and $q = 2$

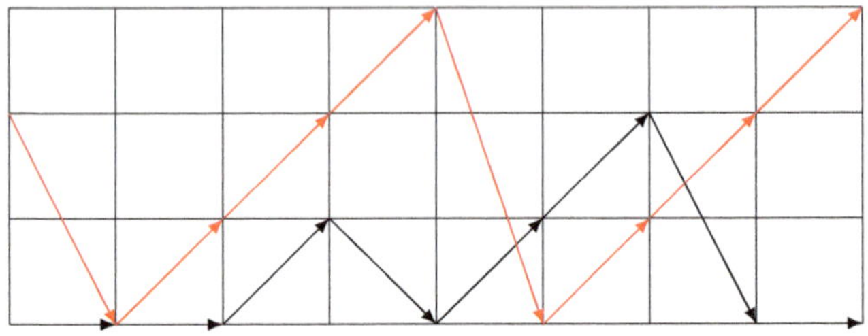

Fig. 2 Two lattice path that start from (0, 0) and (0, 2)

According to (18) it turns out that there is a critical fugacity $z_c = \frac{1+q-p}{p}$ at which the density of the first-class particles shows a finite discontinuity. The behavior of $\rho(z)$ for $z < z_c$ and $z > z_c$ are different and the system undergoes a first-order phase transition provided that $p < 1+q$. In Figs. 1 and 2 exact expression of the density of the first-class particles and its thermodynamic limit are plotted as a function of the fugacity z. As can be seen, in the thermodynamic limit both plots overlap. At z_c there is a finite discontinuity while for $z > z_c$, $\rho(z)$ grows with z until it saturates.

In [5] the authors have studied the spatial correlations in exclusion models corresponding to the Zero Range Processes. They have shown that the spatial correlations of the exclusion models that can be mapped onto the Zero Range Processes can be expressed in terms of 1-point and 2-point correlation functions $G_i^{(1)}$ and $G_{i,j}^{(2)}$. Given that the only impurity is at site 1, the density of the first-class particles at the lattice site i can be written as

$$G_i^{(1)} = \langle \rho_i \rangle = \frac{1}{Z_t(p,q,z)} Tr\left[D_1 C^{i-2}(zD_2)C^{t-i}\right]. \tag{19}$$

Calculating (19) using (15) is straightforward and the result is

$$\langle \rho_i \rangle = A_1 + A_2 \exp\left(\frac{-i}{\xi}\right) \tag{20}$$

in which ξ is a correlation length which is given by

$$\xi^{-1} = \ln \frac{p(1+z)}{1+q}. \tag{21}$$

The coefficients A_1 and A_2 in (20) are functions of the transition rates p, q and also the system size t

$$A_1 = \frac{pz^2(1+z)^{t-2}}{pz\left[(1+z)^{t-1} - \left(\frac{1+q}{p}\right)^{t-1}\right] + (1+q-p-pz)\left(\frac{1+q}{p}\right)^{t-1}},$$
$$A_2 = \frac{z\left(\frac{1+q}{p}\right)^{-2}(1+q-p)(1+z)^t}{p\left(\frac{1+q}{p}\right)^2 z\left[(1+z)^{t-1} - \left(\frac{1+q}{p}\right)^{t-1}\right] + (1+q-p-pz)\left(\frac{1+q}{p}\right)^{t-1}}. \tag{22}$$

The density of the first class particles increases exponentially from the vicinity of the second-class particle. In the thermodynamic limit the density of the first class particles $\langle \rho_i \rangle$ behaves as (18) far from the second class particle.

It should be noted that in addition to infinite dimensional matrix representation (13) the quadratic algebra (10) has a 2-dimensional matrix representation. In [17] the authors have shown that the quadratic algebra (10) has a finite-dimensional representation which depends on the number of types of particles. The dimension of the matrix is M if the number of types of the particles is equal to M. Hence for our model with two species of particles the quadratic algebra (10) has a 2-dimensional matrix representation given by

$$D_1 = \begin{pmatrix} 1 & 0 \\ 1 & 0 \end{pmatrix}, \quad D_2 = \begin{pmatrix} 0 & 1 \\ 0 & 1 \end{pmatrix}, \quad E = \begin{pmatrix} \frac{1+q}{p} & 0 \\ 0 & 1 \end{pmatrix}. \tag{23}$$

According to (23) the matrix $C = E + zD_2$ in (14) can be written as

$$C = \begin{pmatrix} \frac{1+q}{p} & z \\ 0 & 1+z \end{pmatrix}. \tag{24}$$

It can be seen that the correlation length (21) can be written as a function of the eigenvalues of the 2-dimensional matrix C as

$$\xi^{-1} = \ln\left(\frac{\lambda_1}{\lambda_2}\right) \tag{25}$$

where $\lambda_1 = 1 + z$ and $\lambda_2 = \frac{1+q}{p}$, in agreement with the known results obtained in [18]. The 2-point correlation function $G_{i,j}^{(2)}(z) = \langle \rho_i \rho_j \rangle$ can be written as

$$G_{i,j}^{(2)}(z) = \frac{z^2}{Z_t(p,q,z)} Tr\left[D_1 C^{i-2} D_2 C^{j-i-1} D_2 C^{t-j}\right]. \tag{26}$$

Using (15) and (20) and after some straightforward calculations one can obtain $G_{i,j}^{(2)}(z)$ explicitly

$$G_{i,j}^{(2)}(z) = \frac{z}{1+z} \langle \rho_i \rangle. \tag{27}$$

The $(n+1)$-point correlations are written as $G_{j_1 \ldots j_n}^{(n+1)} = \langle \rho_i \rho_{i+j_1} \cdots \rho_{i+j_1+\cdots+j_n} \rangle$

$$G_{j_1 \ldots j_n}^{(n+1)} = \frac{1}{Z_t(p,q,z)} Tr[D_1 C^{i-2}(zD_2) C^{j_1-i-1}(zD_2) C^{j_2-j_1-1}(zD_2) \ldots C^{t-n}]. \tag{28}$$

Using (19) and (26) we can express the above equation in terms of $G_i^{(1)}$ as follows

$$G_{j_1 \ldots j_n}^{(n+1)} = \left(\frac{z}{1+z}\right)^n \langle \rho_i \rangle. \tag{29}$$

We can calculate the critical exponents of model at the phase transition point. To find the critical exponent defined by $\rho \propto (z - z_c)^\beta$, we only need to consider the behavior of the density of the first-class particles as a function of fugacity z at the critical point $z_c = \frac{1+q-p}{p}$ in the thermodynamic limit. According to (17), it can be seen that the density of the first-class particles in the vicinity of z_c can be expressed as $\rho \propto (z - z_c)^{-1}$. Hence the critical exponent $\beta = -1$. Near the critical fugacity, $(1+z) \to \frac{1+q}{p}$. According to (20), it can be seen that in the thermodynamic limit the density profile $\langle \rho_i \rangle \propto (z - z_c)^{-1}$. Hence, the critical exponent α defined by $\langle \rho_i \rangle \propto (z - z_c)^\alpha$ is $\alpha = -1$. With the correlation function given asymptotically by $G_{i,j}^{(2)} \sim (j-i)^{-D+2-\eta} \exp(\frac{-(j-i)}{\xi})$ in which D is the dimension of the system, we find $\eta = 1$.

The walk model

In this section, we show that there exists a walk model which is equivalent to the driven-diffusive model explained in the previous sections. We consider a two-dimensional walk model in which a random walker can start from any height upper than the origin $(0, j)$ in which j is an integer $j \geq 0$. We assume that the random walker can take a finite number of steps on $Z_+^2 = \{(i,j) : i,j \geq 0 \quad \text{are integers}\}$ according to the rules which will be explained later. For the reasons that will become clear shortly we assume that the length of the lattice path is equal to $t-1$. After taking a finite number $t-1$ of consecutive steps, the random walker can get to the lattice site $(t-1, j')$ where $j' = j, j+1, \ldots, j+t-1$. The initial vertex (starting point) and the final vertex (end point) of the lattice path are weighted. This type of lattice path is introduced in [19]. For any path the weight of the start and end points depend on the height of these points. There are different ways that after taking the finite number of steps $t-1$, the random walker can get to the lattice site $(t-1, j')$. The weight of a given path will be equal to the product of the weights of the start and end points and the consecutive steps. The random walker moves according to the following rules:

1. The random walker can start from any height upper than the origin as $(0, j)$ where $j = 0, 1, 2, \ldots, \infty$.
2. A path that starts from the height $(0, j)$, after $t-1$ steps might terminate at any height such as $(t-1, j')$ where $j' = j, j+1, j+2, \ldots, j+t-1$.
3. The weight of the initial vertex (starting point) for the path that starts from the height $(0, j)$ is q^j.
4. The weight of the final vertex (end point) for the path that starts from the height $(0, j)$ and terminates to the lattice site $(t-1, j')$, is $\binom{j'}{j}\frac{1}{p^{j'}}$
5. For $i \geq j$ and from the lattice site (i, j) to $(i+1, j+1)$ the steps have the weight 1(*upward* steps).
6. For $i \geq j$ and from the lattice site (i, j) the random walker can drop to the surface $(i+1, 0)$. These steps have the weight 1 (*jump* steps for $j > 0$ and *horizontal* steps for $j = 0$).

In Fig. 2 we have plotted two different paths of length 8 according to the above mentioned rules. We will be interested in those paths of fixed length which contain a certain number of *jumps* and *horizontal* steps (equivalently *upward* steps); therefore, for our later convenience we introduce an *ad hoc* fugacity z and change the last rule as follows: for $i \geq j$ from the lattice site (i, j) random walker can drop to the surface $(i+1, 0)$ with the weight z.

The position of the random walker in lattice path will be denoted by the vector $|j\rangle$ in which j is the height relative to the horizontal plane which is a number between 0 and ∞. These vectors have the following properties

$$\begin{aligned} |j\rangle_k &= \delta_{j,k} \qquad \text{for } j, k = 0, 1, \ldots, \infty, \\ \langle j|j'\rangle &= \delta_{j,j'} \qquad \text{for } j, j' = 0, 1, \ldots, \infty, \\ \sum_{j=0}^{\infty} |j\rangle\langle j| &= \mathcal{I} \end{aligned} \tag{30}$$

in which $\mathcal{I}$ is an infinite-dimensional identity matrix. We assume that the random walker starts from the height $|j\rangle$ in which $j = 0, 1, \ldots, \infty$. After taking $t-1$ steps the random walker can get to the lattice site $(t-1, j')$ in which $j' = j, j+1, \ldots, j+t-1$. There are different paths to get to the lattice site $(t-1, j')$. Each of these paths has its own weight. We now calculate the weight of a given path $\mathbf{p}$ as follow

$$W_{\mathbf{p}} = w^i \left[\prod_{i=1}^{t-1} w^{\text{step}}(e_i)\right] w^f \tag{31}$$

in which w^i and w^f are the weights of the start and end points and $w^{\text{step}}(e_i)$ is the weight of the i'th step in the path. We know that the transfer matrix updates the state of the random walker hence according to the rules of the steps in the lattice path and their weights, the transfer matrix corresponding to this lattice path can be written as follow

$$C|j\rangle = z|0\rangle + |j+1\rangle. \tag{32}$$

The matrix representation of the transfer matrix C is

$$C = \begin{pmatrix} z & z & z & z & & \cdots \\ 1 & 0 & 0 & 0 & 0 & \cdots \\ 0 & 1 & 0 & 0 & 0 & \cdots \\ 0 & 0 & 1 & 0 & 0 & \cdots \\ \vdots & \vdots & \vdots & \vdots & & \end{pmatrix}. \tag{33}$$

The partition function of the lattice path

As we mentioned the random walker can start from the any height upper than the origin $|j\rangle$ in which $j = 0, 1, \ldots, \infty$. We have also assumed that the total number of steps is $t-1$. After taking these steps the random walker can get to the lattice site $(t-1, j')$ where $j' = j, j+1, \ldots, j+t-1$ through different paths. Each of these paths has its own weight. The partition function of the walk model is the sum of the unnormalized weights of different paths consisting of $t-1$ steps that start from different heights $|j\rangle$ and get to the different heights $|j'\rangle$ where $j = 0, 1, \ldots, \infty$ and $j' = j, j+1, \ldots, j+t-1$. To obtain the partition function of the lattice path, we calculate the sum of the weights of all paths that start from the height $|j\rangle$ and, according to the mentioned rules, after $t-1$ successive steps get to the height $|j'\rangle$. This sum is given by the following equation

$$Z_{j,j'} = w^i \langle j'|C^{t-1}|j\rangle w^f \tag{34}$$

in which $Z_{j,j'}$ is the sum of unnormalized weights of different paths that start from the lattice site $(0, j)$ and after taking $t-1$ steps get to the lattice site $(t-1, j')$. According to the mentioned rules the weight of the start and end points for each path that starts from the height $|j\rangle$ and ends at the height $|j'\rangle$ are given by

$$w^i = q^j, \quad w^f = \binom{j'}{j}\frac{1}{p^{j'}}. \tag{35}$$

Using the above equations the $Z_{j,j'}$ can be written as

$$Z_{j,j'} = \binom{j'}{j}\frac{q^j}{p^{j'}}\langle j'|C^{t-1}|j\rangle. \tag{36}$$

Considering that the lattice path can start from different height $|j\rangle$ in which $j = 0, 1, \ldots, \infty$ and end at different height $|j'\rangle$ where $j' = j, j+1, \ldots, j+t-1$ then the partition function of the lattice path can be written as

$$Z = \sum_{j=0}^{\infty}\sum_{j'=j}^{j+t-1}\binom{j'}{j}\frac{q^j}{p^{j'}}\langle j'|C^{t-1}|j\rangle. \tag{37}$$

Using (15) the partition function of the lattice path is given by the following relation

$$Z_t(p,q,z) = \sum_{j=0}^{\infty}\sum_{j'=j}^{j+t-1}\binom{j'}{j}\frac{q^j}{p^{j'}}\left(\sum_{k=0}^{t-2} z(z+1)^{t-k-2}\delta_{j',k} + \delta_{j',j+t-1}\right). \tag{38}$$

Hence, the partition function of the lattice path can be rewritten as

$$Z_t(p,q,z) = \sum_{j=0}^{\infty}\sum_{j'=j}^{j+t-1}\binom{j'}{j}\frac{q^j}{p^{j'}}z(z+1)^{t-j'-2} + \sum_{j=0}^{\infty}\binom{j+t-1}{j}\frac{q^j}{p^{j+t-1}}. \tag{39}$$

Using Newton's binomial expansion the above equation can be rewritten as follows

$$Z_t(p,q,z) = \sum_{j=0}^{\infty}\sum_{j'=j}^{j+t-1}\binom{j'}{j}\frac{q^j}{p^{j'}}z(z+1)^{t-j'-2} + \frac{1}{p^{t-1}}\left(1-\frac{q}{p}\right)^{-t}. \tag{40}$$

Note that all parameters in the summand are non-negative thus using Tonelli's theorem we can interchange the summations, as $\left(\sum_{j=0}^{\infty}\sum_{j'=j}^{t-1}\right) \equiv \left(\sum_{j'=0}^{\infty}\sum_{j=0}^{j'}\right)$. Hence, the partition function of the lattice path can be written as

$$Z_t(p,q,z) = \sum_{j'=0}^{\infty}\sum_{j=0}^{j'}\binom{j'}{j}\frac{q^j}{p^{j'}}z(z+1)^{t-j'-2} + \frac{1}{p^{t-1}}\left(1-\frac{q}{p}\right)^{-t}. \tag{41}$$

We are interested in the partition function of the original walk model in the special case, that after taking $t-1$ successive steps, the random walker has taken a certain number of upward steps. We study the case in which after $t-1$ successive steps, the random walker can be at the heights between 0 and $t-M$ where $M \le t$. To find the partition function of the model in this case, let us have a closer look at the role of the fugacity z. The weight associated with a horizontal or downward movement is proportional to z; therefore, the coefficient of z^{M-1} in (41) is equal to the partition function of the walk model which consists of at most $t-M$ upward steps. The result is

$$(z+1)^{t-j'-2} = \sum_{i=0}^{t-j'-2}\binom{t-j'-2}{i}z^i.$$

Using the above equation, the coefficient of the z^{M-1} can be easily calculated as follows

$$Z_{t,M}(p,q) = \sum_{j'=0}^{t-M}\sum_{j=0}^{j'}\binom{j'}{j}\binom{t-j'-2}{M-2}\frac{q^j}{p^{j'}}. \tag{42}$$

One can interpret this partition function as the sum of the weights of all paths that have the length $t-1$ which contain $t-M$ upward steps (or equivalently $M-1$ horizontal and downward steps).

The phase behavior of the lattice path in the thermodynamic limit

As a relevant quantity, one can investigate the mean height of the random walker. The probability that the paths who starts from the height $|j\rangle$, and after $t-1$ successive steps according to the rules of the lattice path end at the height $|j'\rangle$, is given by

$$P_{t,M}(j,j') = \frac{1}{Z_{t,M}(p)}\frac{q^j}{p^{j'}}\binom{j'}{j}\binom{t-j'-2}{M-2}. \tag{43}$$

Hence the average height of all possible paths in the lattice path is

$$\langle h\rangle = \sum_{j'=0}^{t-M}\sum_{j=0}^{j'} j' P_{t,M}(j,j'). \tag{44}$$

It should be noted that in the above equation the arrangement of the index has been changed with respect to the Tonelli theorem. According to (42), the average of the height in lattice path is given by the following relation

$$\langle h\rangle = -p\frac{\partial \ln Z_{t,M}(p)}{\partial p}. \tag{45}$$

In the thermodynamic limit due to the behavior of the partition function of the lattice path, it turns out that the mean height of the random walker is given by

$$\langle h\rangle = \begin{cases} t\left(1-\rho-\dfrac{p}{1+q-p}\rho\right) & \text{for } p-q<1-\rho(1+q) \\ \dfrac{p}{p-(1-\rho)(1+q)}-1 & \text{for } p-q>1-\rho(1+q) \end{cases}. \tag{46}$$

As can be seen in the thermodynamic $t \to \infty$, there is a phase transition from a phase in which the mean height of the random walker is of order t to another phase where it is

$(1 - \rho)(1 + q)/[p - (1 - \rho)(1 + q)]$. If $q = 0$ the results are exactly those obtained in [11].

Concluding remarks

In this paper, we have introduced a two-species driven-diffusive model of classical particles defined on a one-dimensional lattice with periodic boundary condition which can be mapped onto a zero-range process. The canonical partition function of the model is calculated and phase behavior of this model is investigated. After calculating the grand canonical partition function, the critical fugacity is obtained at which the model undergoes a first-order phase transition. The density profile of the model is calculated exactly and the spatial correlations of the model are obtained in terms of 1-point correlation function. We have introduced a two-dimensional walk model in which the random walker, in contrast with the lattice path introduced in [11], can start from any height upper than the origin and that the end point of the lattice path can be at any height upper than the start point. This type of lattice path is introduced in [19]. The partition function of the lattice path is calculated using the transfer matrix method. Comparing this partition function with that of the driven-diffusive model we have shown that these two model are equivalent. It should be noted that the walk model introduced in [11] and the one introduced in present work can be mapped onto zero-range process.

References

1. Lee, H.-W., Popkov, V., Kim, D.: Two-way traffic flow: exactly solvable model of traffic jam. J. Phys. A Math. Gen. **30**, 8497 (1997)
2. Fouladvand, M.E., Lee, H.-W.: An exactly solvable two-way traffic model with ordered sequential update. Phys. Rev. E. **60**, 6465 (1999)
3. Mallick, K.: Shocks in the asymmetry exclusion model with an impurity. J. Phys. A Math. Gen. **29**, 5375 (1996)
4. Masharian, S.R., Jafarpour, F.H.: A heterogeneous zero-range process related to a two-dimensional walk model. Int. J. Mod. Phys. B **26**(09) (2011)
5. Basu, U., Mohanty, P.K.: Spatial correlations in exclusion models corresponding to the zero-range process. J. Stat. Mech. L03006 (2010)
6. Jafarpor, F.H.: The application of the Yang–Lee theory to study a phase transition in a non-equilibrium system. J. Stat. Phys. **113**, 112 (2003)
7. Arndt, P.F.: Yang–Lee theory for a nonequilibrium phase transition. Phys. Rev. Lett **84**, 814 (2000)
8. Brak, R., de Gier, J., Rittenberg, V.: Nonequilibrium stationary states and equilibrium models with long range interactions. J. Phys. A Math. Gen. **37**, 4303 (2004)
9. Brak, R., Essam, J.W.: Asymmetric exclusion model and weighted lattice paths. J. Phys. A Math. Gen. **37**, 4183 (2004)
10. Blythe, R.A., Janke, W., Johnston, D.A., Kenna, R.: Continued fractions and the partially asymmetric exclusion process. J. Phys. A Math. Theor. **42**, 325002 (2009)
11. Jafarpour, F.H.: Bose–Einstein condensation and a two-dimensional walk model. Phys. Rev. E **83**, 041112 (2011)
12. Zeraati, S., Jafarpour, F.H.: Multiple-transit paths and density correlation functions in partially asymmetric simple exclusion process. Phys. Rev. E **83**, 041133 (2010)
13. Evans, M.R., Hanney, T.: Nonequilibrium statistical mechanics of the zero-range process and related models. J. Phys. A Math. Gen. **38**, R195 (2005)
14. Evans, M.R.: Phase transitions in one-dimensional nonequilibrium systems. Braz. J. Phys. **30**, 42 (2000)
15. Evans, M.R.: Bose–Einstein condensation in disordered exclusion models and relation to traffic flow. Europhys. Lett. **36**, 13 (1996)
16. Schtz, G.M., Ramaswamy, R., Barma, M.: Pairwise balance and invariant measures for generalized exclusion processes. J. Phys. A Math. Gen. **29**, 837 (1995)
17. Mohammad Ghadermazi., Farhad H Jafarpour.: A new family of exactly solvable disordered reaction–diffusion systems. J. Stat. Mech. P09023 (2013)
18. Hinrichsen, H., Sandow, S., Peschel, I.: On matrix product ground states for reaction-diffusion models. J. Phys. A Math. Gen. **29**, 2643–2649 (1996)
19. Brak, R., Essam, J.W.: Asymmetric exclusion model and weighted lattice paths. J. Phys. A Math. Gen **37**, 14 (2004)

Permissions

All chapters in this book were first published in JTAP, by Springer International Publishing AG.; hereby published with permission under the Creative Commons Attribution License or equivalent. Every chapter published in this book has been scrutinized by our experts. Their significance has been extensively debated. The topics covered herein carry significant findings which will fuel the growth of the discipline. They may even be implemented as practical applications or may be referred to as a beginning point for another development.

The contributors of this book come from diverse backgrounds, making this book a truly international effort. This book will bring forth new frontiers with its revolutionizing research information and detailed analysis of the nascent developments around the world.

We would like to thank all the contributing authors for lending their expertise to make the book truly unique. They have played a crucial role in the development of this book. Without their invaluable contributions this book wouldn't have been possible. They have made vital efforts to compile up to date information on the varied aspects of this subject to make this book a valuable addition to the collection of many professionals and students.

This book was conceptualized with the vision of imparting up-to-date information and advanced data in this field. To ensure the same, a matchless editorial board was set up. Every individual on the board went through rigorous rounds of assessment to prove their worth. After which they invested a large part of their time researching and compiling the most relevant data for our readers.

The editorial board has been involved in producing this book since its inception. They have spent rigorous hours researching and exploring the diverse topics which have resulted in the successful publishing of this book. They have passed on their knowledge of decades through this book. To expedite this challenging task, the publisher supported the team at every step. A small team of assistant editors was also appointed to further simplify the editing procedure and attain best results for the readers.

Apart from the editorial board, the designing team has also invested a significant amount of their time in understanding the subject and creating the most relevant covers. They scrutinized every image to scout for the most suitable representation of the subject and create an appropriate cover for the book.

The publishing team has been an ardent support to the editorial, designing and production team. Their endless efforts to recruit the best for this project, has resulted in the accomplishment of this book. They are a veteran in the field of academics and their pool of knowledge is as vast as their experience in printing. Their expertise and guidance has proved useful at every step. Their uncompromising quality standards have made this book an exceptional effort. Their encouragement from time to time has been an inspiration for everyone.

The publisher and the editorial board hope that this book will prove to be a valuable piece of knowledge for researchers, students, practitioners and scholars across the globe.

List of Contributors

M. R. Kiyani and Y. S. Jalili
Department of Physics, Science and Research Branch, IAU. P. O. Box: 14665-678, Tehran, Iran

J. Pakravan and M. V. Takook
Department of Physics, Razi University, Kermanshah, Iran

Alireza A. Ganjovi
Photonics Research Institute, Institute of Science, High Technology & Environmental Sciences, Graduate University of Advanced Technology, Kerman, Iran

Christian Illg, Michael Haag and Manfred Fähnle
Max Planck Institute for Intelligent Systems, Heisenbergstr. 3, 70569 Stuttgart, Germany

Nicolas Teeny
Max Planck Institute for Nuclear Physics, Saupfercheckweg 1, 69117 Heidelberg, Germany

Jens Wirth
Institut für Analysis, Dynamik und Modellierung, Universität Stuttgart, Pfaffenwaldring 57, 70569 Stuttgart, Germany

A. Niknam and A. A. Rajabi
Physics Faculty, Shahrood University, P.O. Box 3619995161-316, Shahrood, Iran

M. Solaimani
Department of Physics, Faculty of Science, Qom University of Technology, Qom, Iran

Mohsen Emami-Razavi
Plasma Physics Research Center, Science and Research Branch, Islamic Azad University, P.O. Box 14665-678, Tehran, Iran

Won Sang Chung
Department of Physics and Research Institute of Natural Science, College of Natural Science, Gyeongsang National University, Jinju 660-701, Korea

N. Fatahi
Department of Physics, Science and Research Branch, Islamic Azad University, Tehran, Iran

Samaneh S. Hoseini and Edris Faizabadi
School of Physics, Iran University of Science and Technology, Tehran 16846, Iran

C. A. Onate
Theoretical Physics Section, Department of Physics, University of Benin, Benin, Nigeria

J. O. Ojonubah
Mathematics Department, Federal College of Education Okene, Okene, Nigeria

Sudipta Mukherjee, Dipan Bandyopadhyay, Pranab Kishore Dutta and Subir Kumar Sarkar
Department of E.T.C.E, Jadavpur University, 188, Raja S. C. Mallick Road, Kolkata 700032, West Bengal, India

Ali Shokuhi Rad
Department of Chemical Engineering, Qaemshahr Branch, Islamic Azad University, Qaemshahr, Iran

Bahniman Ghosh
Microelectronics Research Center, University of Texas at Austin, 10100 Burnet Road, Bldg. 160, Austin, TX 78758, USA
Department of Electrical Engineering, Indian Institute of Technology Kanpur, Kanpur 208016, India

Akash Gramin
Department of Electrical Engineering, Indian Institute of Technology Kanpur, Kanpur 208016, India

Bharat Borah and M. Ansari
Department of Mathematics, North-Eastern Hill University, Shillong 793022, India

Bahniman Ghosh
Microelectronics Research Center, University of Texas at Austin, 10100, Burnet Road, Bldg. 160, Austin, TX 78758, USA
Department of Electrical Engineering, Indian Institute of Technology Kanpur, Kanpur 208016, India

Kshitij Dwivedi
Department of Electrical Engineering, Indian Institute of Technology Kanpur, Kanpur 208016, India

Dinesh Varshney and Neha Dodiya
Materials Science Laboratory, School of Physics, Vigyan Bhawan, Devi Ahilya University, Khandwa Road Campus, Indore 452001, India

P. Valipour
Department of Textile and Apparel, Qaemshahr Branch, Islamic Azad University, Qaemshahr, Iran

S. E. Ghasemi
Young Reseachers and Elite Club, Sari Branch, Islamic Azad University, Sari, Iran

Mohammad Reza Khosravani
Chair of Solid Mechanics, University of Siegen, Paul-Bonatz- Str. 9-11, 57068 Siegen, Germany

D. D. Ganji
Department of Mechanical Engineering, Babol University of Technology, Babol, Iran

W. K. Le, Z. Z. Yuan and X. Y. Zhang
State Key Laboratory of Advanced Processing and Recycling of Non-ferrous Metals, Lanzhou University of Technology, Lanzhou 730050, Gansu, China

Babatunde J. Falaye
Applied Theoretical Physics Division, Department of Physics, Federal University Lafia, P. M. B. 146, Lafia, Nigeria

Sameer M. Ikhdair
Department of Physics, Faculty of Science, An-Najah National University, New campus, P. O. Box 7, Nablus, West Bank, Palestine
Department of Electrical Engineering, Near East University, Nicosia, Northern Cyprus, Mersin 10, Turkey

Majid Hamzavi
Department of Physics, University of Zanjan, Zanjan, Iran

Nahomi Kan
National Institute of Technology, Gifu College, Motosu-shi, Gifu 501-0495, Japan

Kiyoshi Shiraishi
Graduate School of Science and Engineering, Yamaguchi University, Yamaguchi-shi, Yamaguchi 753-8512, Japan

Rahul Vishwakarma
Department of Physics and Electronics, Dr. R M L Avadh University, Faizabad, UP, India

Dinesh Varshney and S. Shriya
Materials Science Laboratory, School of Physics, Vigyan Bhavan, Devi Ahilya University, Khandwa Road Campus, Indore 452001, India

S. Jain
Materials Science Laboratory, School of Physics, Vigyan Bhavan, Devi Ahilya University, Khandwa Road Campus, Indore 452001, India
Department of Physics, S. D. Bansal College of Engineering, Indore 453331, India

R. Khenata
Laboratoire de Physique Quantique et de Mode´lisation Mathématique (LPQ3 M), Départementde Technologie, Universitéde Mascara, 29000 Mascara, Algeria

Laleh Farhang Matin
Department of Physics, Islamic Azad University, North Tehran Branch, Tehran, Iran

Hadi Savaloni and Fatemh Haydari-Nasab
Department of Physics, University of Tehran, North-Kargar Street, Tehran, Iran

Abbas Abbas-Rohollahi
Department of Biology, University of Tehran, Enghelab Street, Tehran, Iran

K. Mehdizadeh
Department of physics, Ayatollah Amoli Branch, Islamic Azad University, Amol, Mazandaran, Iran

O. Jalili
Department of physics, North Tehran Branch, Islamic Azad University, Tehran, Iran

Mohammad Ghadermazi
Plasma Physics Research Center, Science and Research Branch, Islamic Azad University, Tehran, Iran

Farhad H. Jafarpour
Physics Department, Bu-Ali Sina University, 65174-4161 Hamedan, Iran

Index